Neuroscience and the Person

Scientific Perspectives on Divine Action

A Series on "Scientific Perspectives on Divine Action"
Robert John Russell, General Editor

First Volume
Quantum Cosmology and the Laws of Nature:
Scientific Perspectives on Divine Action
Edited by Robert John Russell, Nancey Murphy,
and C.J. Isham

Second Volume
Chaos and Complexity:
Scientific Perspectives on Divine Action
Edited by Robert John Russell, Nancey Murphy,
and Arthur Peacocke

Third Volume
Evolutionary and Molecular Biology:
Scientific Perspectives on Divine Action
Edited by Robert John Russell, William R. Stoeger, S.J.,
and Francisco J. Ayala

Fourth Volume
Neuroscience and the Person:
Scientific Perspectives on Divine Action
Edited by Robert John Russell, Nancey Murphy,
Theo C. Meyering, and Michael A. Arbib

Future Scientific Topics
Quantum Physics and Quantum Field Theory

Jointly published by the Vatican Observatory and
the Center for Theology and the Natural Sciences

Supported in part by a grant from
the Wayne and Gladys Valley Foundation

Neuroscience and the Person
Scientific Perspectives on Divine Action

Robert John Russell
Nancey Murphy
Theo C. Meyering
Michael A. Arbib

Editors

Vatican Observatory
Publications,
Vatican City State

Center for Theology and
the Natural Sciences,
Berkeley, California

2002

Robert John Russell (General Editor) is Professor of Theology and Science in Residence at the Graduate Theological Union, and Founder and Director of the Center for Theology and the Natural Sciences, in Berkeley, California, USA.

Nancey Murphy is Professor of Christian Philosophy at Fuller Theological Seminary, Pasadena, California, USA.

Theo C. Meyering is Associate Professor in the Department of Philosophy at Leiden University, The Netherlands.

Michael A. Arbib is Director of the USC Brain Project, Professor and Chairman of Computer Science, and Professor of Neuroscience, Biomedical Engineering, Electrical Engineering, and Psychology at the University of Southern California, Los Angeles, California, USA.

First edition 1999
First impression 2002

Copyright © Vatican Observatory Foundation 2002

Jointly published by the Vatican Observatory and the Center for Theology and the Natural Sciences

Distributed (except in Italy and the Vatican City State) by
    The University of Notre Dame Press
    Notre Dame, Indiana 46556
    U.S.A.

Distributed in Italy and the Vatican City State by
    Libreria Editrice Vaticana
    V-00120 Citta del Vaticano
    Vatican City State

ISBN 0-268-01490-6 (pbk.)

ACKNOWLEDGMENTS

The editors wish to express their gratitude to the Vatican Observatory and the Center for Theology and the Natural Sciences for co-sponsoring this research. Particular appreciation goes to George Coyne and William Stoeger, whose leadership and vision made this series of conferences possible.

Editing for this volume began with an initial circulation of papers for critical responses and pre-conferences in Rome and Berkeley (1997) before the main conference in Pasierbiec, Poland (June, 1998), and continued with interactions between editors and authors after the main conference. The editors express their gratitude to all the participants for written responses to pre-conference drafts and for their enthusiastic discussions during the conferences.

Special thanks goes to Kirk Wegter-McNelly, CTNS Editing Coordinator, who devoted meticulous attention and long hours to every stage in the post-conference phases of the production of this volume while pursuing doctoral studies at the GTU. In consultation with the editors, Kirk was responsible for copy-editing the text, preparation of figures, layout of the volume, cover design, and the indices. Thanks go to George Coyne for overseeing printing and distribution.

On behalf of all the participants, the editors also thank staff members at CTNS, the Vatican Observatory, the retreat center at Pasierbiec, student assistants from the Pontifical University in Cracow, and especially Michael Heller for their work in organizing the main conference and pre-conferences.

The editors thank Fergus Kerr and the publisher for permission to reprint chapter one of Fergus Kerr, *Theology After Wittgenstein*, 2nd ed. (London: SPCK, 1997).

CONTENTS

CONTENTS

INTRODUCTION

Nancey Murphy

1 Background to the Volume

In June of 1998 the fourth conference in a series cosponsored by the Vatican Observatory and the Center for Theology and the Natural Sciences was held in Pasierbiec, Poland. Its purpose was to explore relations between the cognitive neurosciences and Christian theology. Most of the essays in this volume were discussed and debated there, and then revised for publication.

This series of conferences represents a response to the call from Pope John Paul II in 1979 for an interdisciplinary collaboration of scholars to seek a fruitful concord between science and religion. Responding to this call, the Vatican Observatory, a century-old astronomical research institution now located in the papal villa at Castel Gandolfo, Italy, sponsored a number of events, culminating in a major international conference in September, 1987.[1] George Coyne, Director of the Vatican Observatory, then proposed a series of five conferences for the decade of the 1990s and invited CTNS to become a cosponsor. Coyne convened a planning meeting at the Observatory, at which a wide variety of topics was proposed for the series of conferences. It became apparent, however, that one theological problem was integrally related to many of the more specific topics; this is the problem of how to understand the role of divine action in the natural world. Thus, the problem of divine action became the organizing theological theme for the series, to be considered from the point of view of a variety of recent scientific advances.

It is appropriate that scientific developments should be brought to bear in reflecting on the topic of divine action since it was the rise of modern science that made the issue problematic in a way that it had not been in earlier periods. According to traditional Christian (and Jewish) views, God acts in all of the ordinary events in the natural and human worlds—for example, making the "sun to rise on the evil and the good, and [sending] rain on the righteous and on the unrighteous" (Matthew 5:45 NRSV). In addition, God performs special acts such as leading Israel out of Egypt and, most dramatically, raising Jesus from death. Before the rise of modern science it was not difficult to reconcile divine action, whether of the general or special sort, with natural causation: in the vocabulary of medieval theologians such as Thomas Aquinas, God is the primary cause and natural entities are secondary causes, much like a carpenter and the carpenter's tools.

With the rise of modern science in the sixteenth and seventeenth centuries, the law-like character of natural processes was seen as a testament to God's sovereignty: just as humans obey (imperfectly) God's moral law, so too nature obeys (perfectly) God's "natural laws." But as the concept of the laws of nature became a pillar of modern understanding, problems arose for traditional theological accounts. While it was easy to maintain a role for God as creator and sustainer of the whole natural order, it became difficult to understand special divine acts, that is, those that could not be seen as mediated by the *regular* workings of the laws of nature. Many theists

---

[1] The proceedings of this conference are published in Robert John Russell, William R. Stoeger, and George V. Coyne, eds., *Physics, Philosophy and Theology: A Common Quest for Understanding* (Vatican City State: Vatican Observatory Publications, 1988).

rejected the concept of miracles as irrational intrusions into God's orderly creation. Much of the development of modern theology of a more liberal sort can be seen as an attempt to maintain Christian teaching about God's involvement in "salvation history" without resorting to accounts of supernatural intervention in the natural order. A common approach was to make a sharp distinction between the human and natural worlds—between history and culture on one hand, and nature on the other.

Many of the participants in the present series of conferences are convinced of the necessity of holding on to some concept of "special divine acts," and this is for a variety of reasons: to give meaning to the idea of God's intentional, personal involvement in people's lives; to account for revelation as more than mere human discovery of religious truths; to avoid the unhappy situation of attributing all events equally, both good and bad, to divine intent. Yet these same scholars have been reluctant to achieve this by taking the route of the more conservative theologians who insist on God's miraculous contravention of the laws of nature, a route that often results in inattention to the results of science or, worse, to their outright rejection.[2]

However, if the rise of modern science was the source of the problem of divine action, happily, twentieth-century science may contribute to its solution. The mechanistic conception of nature sponsored by Newtonian physics has given way to a much richer and more complex view of the world, the implications of which we are only beginning to grasp. What does science have to say on philosophical issues such as the nature of causation and of time? on the limits of human knowledge? on the place of human life and consciousness in nature? Even less has contemporary theology taken account of all of these developments. Thus, a crucial concern for this series of conferences was the issue of the nature and origin of the laws of nature themselves. The specific scientific topics chosen for the meetings were: (1) quantum cosmology (combined with the topic of the "fine-tuning" of the laws of nature); (2) chaos theory; (3) molecular and evolutionary biology; (4) neuroscience and the mind-body problem; and (5) quantum physics and quantum field theory. The first and third conferences were held at Castel Gandolfo in 1991 and 1996; the second was held at the Center for Theology and the Natural Sciences in Berkeley.[3] The present conference was held in Pasierbiec, Poland, at the invitation of Michael Heller, professor of philosophy at the Pontifical Academy of Theology in Cracow. The final conference in the series is to be held in Castel Gandolfo in June, 2000.

The success of a venture such as this depends almost as much on the structure and organization of the events as it does on the scholarly abilities of the participants. Scientists and theologians inhabit vastly different worlds of discourse. How is it

---

[2] However, the question is sometimes raised whether "noninterventionist special divine action" is a coherent idea. See, for instance, Michael A. Arbib, "Towards a Neuroscience of the Person" (in this volume), sec. 5.

[3] Papers from these first three conferences can be found in: Robert John Russell, Nancey Murphy, and C.J. Isham, eds., *Quantum Cosmology and the Laws of Nature: Scientific Perspectives on Divine Action* (Vatican City State: Vatican Observatory; Berkeley, Calif. Center for Theology and the Natural Sciences, 1993); Robert John Russell, Nancey Murphy, and Arthur Peacocke, eds., *Chaos and Complexity: Scientific Perspectives on Divine Action* (Vatican City State: Vatican Observatory; Berkeley, Calif.: Center for Theology and the Natural Sciences, 1995); and Robert John Russell, William R. Stoeger, S.J., and Francisco J. Ayala, eds., *Evolutionary and Molecular Biology: Scientific Perspectives on Divine Action* (Vatican City State: Vatican Observatory; Berkeley, Calif.: Center for Theology and the Natural Sciences, 1998). For a more complete history of this project and a more detailed account of the problem of divine action, see Russell's introductions to each of these volumes.

possible to avoid talking past one another? For this reason the long-range planning committee[4] has attempted to maintain continuity from one conference to the next. Thus, a group of scholars with expertise in theology, philosophy, or the relations between theology and science have participated in the entire process, while specialists in the various scientific disciplines as well as in theology, philosophy, history, and science have been brought in for each conference. An additional strategy to enhance interaction is to begin work on each topic with smaller "pre-conferences" in which specialists present relevant information from each discipline—before conference papers are drafted. These drafts are circulated for comments from all participants and then revised both before and after the main conference—so the participants are to be thanked for undertaking a vast amount of work! Each conference is planned by a committee composed of long-term participants and one or more experts in the relevant scientific field. The organizers and editors for the present conference were Robert Russell, William Stoeger, Michael Arbib, Theo Meyering, and Nancey Murphy.[5]

2 Why Neuroscience and Divine Action?

One might think that a touch of prescience was involved in planning in 1990 for a conference on neuroscience and theology—anticipating the dramatic developments to come in "the decade of the brain," and the theological implications of the consequent revisions in human self-understanding. Actually, the interest in neuroscience came by a less direct route: it was thought that the investigation of divine *action* would profit from a study of the philosophy of action, which, in turn, draws upon the philosophy of mind, and thence from the neurosciences.[6] In the intervening years, however, it has become apparent that theories of human nature provide a natural meeting place for dialogue between theology and philosophy of religion, on the one hand, and the cognitive neurosciences on the other.

The link between neuroscience and the problem of divine action can be stated as follows: Most Christian theologians in the modern era followed René Descartes (as earlier theologians followed Plato) in assuming a dualistic account of human nature—humans composed of body *and* soul or mind. The problem of divine action in the natural world could then be minimized by concentrating on God's action in the human sphere—the sphere of *Geist,* spirit. Here God could interact freely and directly with human souls. But as the neurosciences now increasingly throw their considerable weight behind arguments for the unity of the human being—a purely physical organism—they have called theology's comity agreement into question. To put it baldly, if God has to do with humans at all, it must be by means of interaction with their bodily substance and, more particularly, with their brains. The problem of divine action in the human world thus depends upon an answer to the problem of divine action in the natural world, because humans, considered in terms of their composition, are thoroughly a part of the natural order. Thus, locating the relevance of the neurosciences for theology is very much a matter of considering the impact of

---

[4] William Stoeger, Robert Russell, and Nancey Murphy, in consultation with George Coyne.

[5] See p. 475 for a list of contributors and conference participants.

[6] For an excellent overview of these related issues, see Thomas F. Tracy, *God, Action, and Embodiment* (Grand Rapids: Eerdmans, 1984).

these sciences on an ongoing debate throughout Western intellectual history regarding the nature of the human person.

3 Human Nature in Western History

This section offers a brief historical sketch of discussions of the nature of the human person, to which the present volume aims to contribute.[7] There are a variety of disciplinary perspectives from which this story can be told, but the focus of this historical account will be on three: theology, the natural sciences, and philosophy.

3.1 Theological Perspectives

It would be reasonable to begin this historical section with an account of scriptural teaching on the nature of the person. However, this is one of the most vexing issues in the discussion, in part because our current questions were not the questions of the writers of Scripture, but also because of a long history of projecting onto the texts interpretations and even translations that reflect later writers' concerns and assumptions. Joel Green's essay in this volume gives an account of scriptural views on the nature of the person, as well as a discussion of the methodological pitfalls involved in interpreting the relevant texts.[8] Green argues that the New Testament texts do not present as dualistic an account of human nature as has been supposed for most of the intervening centuries.

The question of the metaphysical makeup of the human person has not been perceived, throughout the course of Christian history, as a matter central to Christian teaching; the issue has arisen at several major turning points in church history but at other times it has been relegated to the status of background assumption. One point at which it became prominent was in the adaptation of early Jewish-Christian teaching to the gentile world. Here a variety of Greek and Roman philosophical positions were adopted and adapted for theological purposes. For example, Tertullian (160–220) followed the Stoics in teaching that the human soul is corporeal and is generated with the body. Origen (185–254) followed Plato in teaching that the soul is incorporeal and eternal, pre-existing the body. After the time of Jerome (c. 347–420) the soul was generally thought to be created at the time of conception.

Augustine (354–430) has probably been the most influential teacher on these matters because of his legacy in both Protestant and Catholic theology and also because of his importance in the development of Christian spirituality. Augustine's conception of the person is a modified Platonic view: a human being is an immortal (not eternal) soul using (not imprisoned in) a mortal body. Foreshadowed in Augustine's thought is the modern problem of causal interaction. If the soul is to the body as an agent to a tool, then it is inconsistent to say that the body affects the soul. Consequently Augustine was never able to give a satisfactory account of sensory knowledge. Augustine's notion that one knows one's own soul directly would come to have striking repercussions in modern epistemology.

---

[7] This section is an adaptation of material from my chapter, "Human Nature: Historical, Scientific, and Religious Issues," in *Whatever Happened to the Soul? Scientific and Theological Portraits of Human Nature*, Warren S. Brown, Nancey Murphy, and H. Newton Malony, eds. (Minneapolis: Fortress, 1998), 1–29.

[8] Green's essay was solicited after the conference at Pasierbiec to provide a more detailed account of these issues than can be addressed in this introduction.

Augustine was much influenced by the Neoplatonists, who had incorporated Platonic philosophy into religious systems emphasizing the care and development of the soul as the means of salvation. Augustine and other early theologians similarly influenced bequeathed this emphasis on the soul to subsequent spiritual writers. It is by cultivating the higher faculties of the soul (and often by repressing the lower faculties and the body) that one develops the capacity for knowledge of and relation to God.

A second point at which questions of human nature became prominent was during the Aristotelian revival of the high Middle Ages, occasioned by the introduction of Muslim scholarship into European intellectual life. In the writings of Thomas Aquinas (1225–74) we have the most systematic development of an Aristotelian alternative to the largely Platonic accounts influenced by Augustine. Thomas took up both Aristotelian hylomorphic metaphysics and Aristotle's thesis that the soul is the form of the body. On this account, all physical entities are composed of prime matter and *form*. Form is an immanent principle that makes the entity to be what it is, giving it its characteristic properties and powers of operation. The form of a living being is its soul. Thomas had an elaborate account of the hierarchically ordered faculties or powers of the soul, which served not only theology but psychology for centuries.

For all of its repercussions elsewhere in theology, the Reformation seems not to have brought this issue to the fore. Controversies regarding purgatory led to disputes about the "intermediate state" between death and the final resurrection. Martin Luther (1483–1546) and some of the radical reformers argued that the soul either dies with the body or "sleeps" until the general resurrection; John Calvin (1509–64) wrote a treatise titled *Psychopannychia* (1542) to contest such views.[9] Yet even those who argued for the death of the soul were obviously presupposing a dualistic conception of the person.

The development of biblical criticism and critical church history in the modern period has again required examination of presuppositions concerning the nature of the person. Critical church history provided moderns with a sense of the historical development of thought, which allowed questions to arise in a new way about the consistency of present church teachings with those of the Bible.

Historical criticism of the Bible itself has had a major impact on modern conceptions of the person, but there have been contradictory tendencies. In the eighteenth and especially the nineteenth centuries many New Testament scholars cast doubt on the historicity of Jesus' resurrection.[10] Skepticism about resurrection led to increased emphasis among theologians on the immortality of the soul as the basis for Christian hope in an afterlife. Immanuel Kant's transcendental proof of the soul's immortality surely reinforced this move.

Meanwhile biblical scholars began to question whether body-soul dualism was in fact the position to be found in Scripture. One important contribution here was the work of H. Wheeler Robinson, an Old Testament scholar whose book, *The Christian Doctrine of Man* went through three editions and eight printings between 1911 and 1952. Robinson argued that the Hebrew idea of personality is that of an animated body, in contrast to the Greek idea of an incarnated soul. Surveying the synoptic

---

[9] "Psychopannychia" translates as a watchful or sentient "wake" of the soul.

[10] Among the most prominent was D.F. Strauss, whose *The Life of Jesus Critically Examined* was published in 1835.

gospels and the Pauline and Johannine writings, he argued that the psychological terminology and ideas of the New Testament are largely continuous with the Old Testament in conceiving of the person as a unity rather than dualistically. However, he also said that the most important advance in the New Testament is the belief that the essential personality (whether called the *psychē* or the *pneuma*) survives bodily death. This soul or spirit may be temporarily disembodied, but it is not complete without the body, and its continued existence after bodily death is dependent upon God rather than a natural endowment of the soul. So despite his claim that understanding of the history of anthropological dogma requires attention to the points of contrast and conflict between Hebrew and Greek conceptions of human nature, the contrasts in Robinson's own account between New Testament and Platonic conceptions turn out not to be so great as they might first appear.

Theological thinking on these issues around the time Robinson wrote can only be described as confused. This can be seen by comparing related entries in reference works from early in the twentieth century. In *The New Schaff-Herzog Encyclopedia of Religious Knowledge* (1910) there is a clear consensus on a dualist account of Scriptural teaching[11]: C.A. Beckwith in "Biblical Conceptions of Soul and Spirit" claims that "the human soul is indeed bound to corporeality, yet it survives death because it possesses the Spirit of God as its immanent principle of life," and states that *sōma* (body) is opposed to *psychē* (mind or soul) as *sarx* (flesh) is to *pneuma* (spirit) (11:12–14). In the same reference work, E. Schaeder describes resurrection in terms of God's giving new bodies to risen souls (9:496–97). G. Runze in "Immortality" claims that soul and body were sharply distinguished in the New Testament and that at death the soul rests in God until it receives a new glorified body (5:459, 462).

Yet in a slightly earlier work, *A Dictionary of the Bible* (1902),[12] two sharply opposed views appear: J. Laidlaw in his article on "Soul" says that throughout most of the Bible, "soul" is simply equivalent to the life embodied in living creatures (4:608). E.R. Bernard's article on "resurrection," however, describes resurrection as "the clothing of the soul with a body which has to be reconstituted," and points out that "the only self which we know is a self constituted of a body as well as a soul" (4:236).

This tendency to juxtapose incompatible accounts of biblical teaching continued through the middle of the century, when several new factors gave the issue greater prominence. One was the rise of neo-orthodox theology after World War I. Karl Barth and others made a sharp distinction between Hebraic and Hellenistic conceptions, and strongly favored the former. Barth also contributed to recognition of the centrality of the resurrection in Christian teaching. The biblical theology movement in the mid-twentieth century continued to press for the restoration of earlier, Hebraic understandings of Christianity. A decisive contribution was Rudolf Bultmann's claim in his *Theology of the New Testament* that Paul uses *sōma* (body) to characterize the human person as a whole.[13] The following year John A.T. Robinson published *The Body: A Study in Pauline Theology.*[14] Although he was

---

[11] Samuel Macauley Jackson, ed. (New York and London: Funk and Wagnalls Company, 1910).

[12] James Hastings, ed. (Edinburgh: T. & T. Clark, 1902).

[13] Rudolf Bultmann, *Theology of the New Testament*, vol. 1 (New York: Scribner, 1951).

[14] John A.T. Robinson, *The Body: A Study in Pauline Theology* (London: SCM, 1952).

much influenced by H. Wheeler Robinson, we can see that the meaning of "holistic" anthropology has shifted decisively: whereas the earlier Robinson found a qualified dualism in Paul's writings—what some authors call "holistic dualism"—in Bultmann's and the later Robinson's view, Paul's teaching (as well as the Old Testament) is seen as thoroughly physicalist.

A casual survey of the literature of theology and biblical studies throughout this century, then, shows a gradual replacement of a dualistic account of the person, along with a view of the afterlife based on the doctrine of the immortality of the soul, first, by a recognition of the holistic character of biblical conceptions of the person, often while still presupposing temporarily separable "parts," and then by a holistic but also physicalist account of the person. One way of highlighting this shift is to note that in *The Encyclopedia of Religion and Ethics* (1909–21) there is a lengthy article on "soul" and no entry for "resurrection."[15] In *The Anchor Bible Dictionary* (1992) there is no entry for "soul" but a very long set of articles on "resurrection."[16]

The foregoing picture of twentieth-century thought is an oversimplification, however, for several reasons. First, as will appear in Fergus Kerr's essay (in this volume), the replacement of substance dualism by a *metaphysically* holist or physicalist anthropology has not generally been accompanied by critical reflection on the philosophical baggage that dualism brought into theology, such as radical individualism and antipathy toward the body.

Second, this century has seen the development in Protestantism of two fairly distinct theological traditions. The account given above traces developments in what we may loosely call the liberal tradition. Meanwhile, the tendency among conservatives has been to maintain a dualist (or trichotomist—body, soul, and spirit) account of the person, but also to stress resurrection, which is described as the restoration of a body to the soul. This tendency among conservative biblical scholars and theologians is not necessarily due to overlooking the textual studies mentioned above. Robert Gundry, for instance, has offered a point-by-point rebuttal to Bultmann's arguments in his book, *Soma in Biblical Theology.*[17]

A third complication is Catholic thought. Here there appears little difference between Catholic and Protestant biblical scholarship, but considerable difference between (official) Catholic theology and that of Protestant thinkers. In his *Dictionary of the Bible* (1965) John McKenzie discusses both *nepheš* and *psychē* under the heading of "soul," but emphasizes that in neither case is meant either a Platonic conception of soul or what it means "in common speech." The closest we can come to an adequate English translation of *nepheš* in the Old Testament is "life" or perhaps "self." In the New Testament, he says, the *psychē* is the totality of the self as a living and conscious subject, and "it is the totality of the self which is saved for eternal life."[18] In a *Theological Dictionary* published in the same year, however, the authors distinguish the Christian conception of the soul from a Platonic view—the Platonic soul is a being, whereas the Christian soul is a "real principle of being." This "personal spirituality cannot be derived from matter." Because it "stands over

---

[15] James Hastings, ed. (New York: Charles Scribner's Sons, 1909–21).

[16] David Noel Freedman, ed. (New York: Doubleday, 1992).

[17] Robert H. Gundry, *Soma in Biblical Theology: With Emphasis on Pauline Anthropology* (Grand Rapids: Zondervan, 1987).

[18] John L. McKenzie, S.J., *Dictionary of the Bible* (New York: Macmillan, 1965), 839.

against and independent of mere natural being," it does not cease to exist at death.[19] This theological account is clearly indebted to a Thomistic metaphysic.[20]

Pope John Paul II issued a statement recently that bears on these issues. The purpose of the address to the plenary session of the Pontifical Academy of Sciences in October of 1996 was to reaffirm the teaching of Pius XII regarding the compatibility of evolutionary biology with Catholic teaching, so long as certain qualifications are borne in mind. One of these qualifications concerns human origins. John Paul quoted Pius XII's statement that "if the human body takes its origin from pre-existent living matter, the spiritual soul is immediately created by God."[21]

Pius's teaching, then, affirms (indirectly) a dualist account of the human person and uses this account to delimit the scope of science's investigation of human origins. Many readers take John Paul II to affirm the same dualist anthropology and the same limits to the scientific study of humankind. However, as George Coyne points out, after the quotation from Pius XII, the word "soul" does not reappear in the document.[22] Rather than speaking of the moment when the soul is created, John Paul II speaks of "the moment of transition to the spiritual."[23] This transition is not scientifically observable; science can discover valuable signs indicating what is specific to the human being, but only theology can explain the ultimate meaning of these distinctively human features.

So John Paul II recognizes limits to the scientific study of human nature, yet without a compartmentalization based on dualistic anthropology. His chief concern is one that was hotly debated at the Pasierbiec conference: reductionism. He says that "theories of evolution which, in accordance with the philosophies inspiring them, consider the mind as emerging from the forces of living matter, or as a mere epiphenomenon of this matter, are incompatible with the truth about man."[24] (More on reductionism below, in section 3.3.)

Perhaps it should not be a surprise that there is lack of theological consensus on the nature of the person when we consider the wide number of theological issues that need to be addressed if a physicalist account of the person is substituted for body-soul dualism. The preceding historical survey shows the close tie to eschatological issues. Resurrection of the body has been a mere adjunct to a doctrine of the immortality of the soul for centuries.[25] If there is no substantial soul to survive bodily death then what is to be made of doctrines, formalized at the time of the Reformation, specifying that the dead enjoy conscious relation to God prior to the general

---

[19] Karl Rahner and Herbert Vorgrimler, *Theological Dictionary* (New York: Herder and Herder, 1965), 442–43.

[20] See Stephen Happel's essay in this volume, sec. 3.

[21] Pope John Paul II, "Message to the Pontifical Academy of Sciences," 22 October, 1996; reprinted in Russell et al., eds., *Evolutionary and Molecular Biology*, 2–9; 6.

[22] George V. Coyne, S.J., "Evolution and the Human Person: The Pope in Dialogue," in Russell et al., *Evolutionary and Molecular Biology*, 11–17.

[23] "Message," 6.

[24] Ibid. Had the Pope referred to theories that consider the mind as emerging *solely* from the forces of living matter there would be complete agreement between his statement and the position of many of the authors here. Such a qualification is not an impossible reading of the Pope's intent.

[25] See the essay by Ted Peters in this volume.

resurrection?[26] There are at least two options here. Some consider the biblical evidence for an intermediate state to be both scanty and ambiguous, and claim that the entire person simply disintegrates at death to be recreated by God at the general resurrection. Another approach is to question the meaningfulness of a time-line in discussing eschatological issues. That is, we presume that God is, in some sense, "outside" of time. If those who have died are "with God" we cannot meaningfully relate their experience to our creaturely history.

The metaphysical makeup of the person is but one aspect of a much broader topic of theological concern, now designated "Christian anthropology." One component of the larger task is to trace the consequences of a physicalist account of the person for a variety of issues such as the place of humankind in the rest of nature, the source and nature of human sinfulness, and the claim that humans are made in the image of God.[27]

Recognition of both the centrality of resurrection to Christian teaching and the continuity of humans with the whole of nature calls for reconsideration of the scope of God's final transformative act. There is increased motive to agree with theologians such as Wolfhart Pannenberg who argue that the resurrection of Jesus is a foretaste of the transformation awaiting the entire cosmos.[28] Paul hints at this in Romans: "For the creation waits with eager longing for the revealing of the children of God; for the creation was subjected to futility, not of its own will but by the will of the one who subjected it, in hope that the creation itself will be set free from its bondage to decay and will obtain the freedom of the glory of the children of God" (Romans 8:19–21 NRSV).

Finally, as Nicholas Lash has argued, a doctrine of God is always correlative to anthropology. When the human person is identified with a solitary mind, God tends to be conceived as a *disembodied* mind, as in the case of classical theism. Much of Lash's own writing argues for the recovery of an embodied and social anthropology in order to recapture a more authentic account of religious experience, but also of a thoroughly trinitarian concept of God.[29]

There are equally important issues to be re-examined in related areas of Christian thought. The concept of the soul has played a major role in the history of Christian ethics for centuries, for example, as justification for the prohibition of abortion and for the differential treatment of animals and humans. Where do these arguments stand with a revised concept of the nature of the person?

The soul has also long been the focus of spiritual direction and pastoral counseling. What becomes of traditional concepts of religious experience if the person is understood to be purely physical?[30] How is God's revelation to humans to

---

[26] This teaching was made official for Catholicism by the Fifth Lateran Council in 1513. John Calvin supported the doctrine of the intermediate state in his *Psychopannychia* (1545), and this has settled the issue for many Protestants.

[27] See the essay by Joel Green in this volume.

[28] Wolfhart Pannenberg, *Jesus—God and Man* (Philadelphia: Westminster, 1968). See also the essay by Peters in this volume.

[29] See, for example, Nicholas Lash, *Easter in Ordinary: Reflections on Human Experience and the Knowledge of God* (Charlottesville, Virginia: University Press of Virginia, 1988). See also the essay by Kerr in this volume.

[30] See the essays in this volume by Fraser Watts and by Wesley Wildman and Leslie Brothers.

be understood if humans are body rather than "spirit"?[31] There have been reactions in recent years against the asceticism fostered by Platonic dualism as well as against the tendency to distinguish between saving souls and caring for people's physical needs. Feminist writers have been critical of accounts of gender relationships in which a superior rational soul has been associated with the masculine, and a subordinate material body with the feminine.[32] There is much room for development of more holistic approaches to all of these issues.

3.2 Scientific Perspectives

Science, throughout its history, has been a potent factor in shaping human self-conception. Our ability to relate current theories of human nature to ancient accounts is limited in particular by the fact that our very concept of *matter* has changed so considerably through the intervening centuries. Early modern physics contributed to a more radical form of dualism than had been seen in the Middle Ages, while more recent developments in biology have contributed to arguments for physicalism. Some brief background in physics and biology is developed here; the more dramatic contributions of current neuroscience will be seen throughout the volume.

3.2.1 Modern Physics

Galileo (1564–1642) is famous for his role in the Copernican revolution. However, he played a comparable role in a development that has had equally revolutionary consequences: the substitution of a corpuscular or atomist conception of matter for ancient and medieval hylomorphism. The revolution in astronomy called for adjustments in physics and chemistry, and even in theology, ethics, and political theory. Atomism in metaphysics and physics eventually affected chemistry, psychology, the social sciences, and, indirectly, theology.[33] The animosity toward any sort of teleology that one still encounters in contemporary philosophy of biology seems to be residual heat from the explosive collision of these two worldviews.

No less were there consequences in philosophy. René Descartes (1596–1650) is considered the originator of modern philosophy. We can see why he would turn away from the Catholic account of the soul as the substantial *form* of the body and propose a radical substance dualism more akin to Platonism when we recall that Descartes was also involved in the atomist revolution in physics.[34] Descartes distinguished two basic kinds of realities, extended substance (*res extensa*) and thinking substance (*res cogitans*); the latter included angels and human minds.

Notice the shift from "souls" to "minds." Latin has two words, *anima* and *animus*, each translated "soul," but *animus* is also translated "mind" while *anima* is also the principle of life. Recent English-language philosophy has used the term "mind" rather than "soul."

---

[31] See Arthur Peacocke's essay, sec. 4, and George Ellis's essay, sec. 5.4.

[32] See Ian Barbour's essay, sec. 2.3.

[33] See Nancey Murphy, *Anglo-American Postmodernity: Philosophical Perspectives on Science, Religion, and Ethics* (Boulder, Colo.: Westview Press, 1997), chap. 1.

[34] Joel Green notes that Descartes's radical substance dualism has so influenced current thinking on these issues that we tend to read Plato himself with Cartesian presuppositions. See Green's essay in this volume, n. 12.

The shift from hylomorphism to atomism and substance dualism created what is now seen by many to be an insoluble problem: the interaction between mind and body. Whereas for Aristotle and his followers the mind/soul was but one instance of form, in modern thought the mind becomes something of an anomaly in an otherwise purely material world of nature. Furthermore, the very conception of matter has changed. Before, matter and form had been correlative concepts—matter was that which had the potential to be activated by form. Matter (at least as unformed, prime matter) was entirely passive. For early moderns, matter is also passive, inert. But now, instead of being moved by immanent forms, it is moved by external forces—physical forces. Now there is a dilemma: hold on to the immateriality of mind, and there is then no way to account for its supposed ability to move the body; interpret it as a quasi-physical force and its effects ought to be measurable and quantifiable as is the case with any other force in nature. But nothing of the latter enters into modern or twentieth-century physics.

As Owen Flanagan points out, the problem, which emerged within modern physics, has to do with the law of conservation of energy:

> If Descartes is right that a nonphysical mind can cause the body to move, for example, we decide to go to a concert and go, then [there must be a transfer of energy from our mind to] our body, since we get up and go to the concert. In order, however, [for energy to be transferred to any physical] system, it has to have been transferred from some other physical system. But the mind, according to Descartes, is not a physical system and therefore it does not have any energy to transfer. The mind cannot account for the fact that our body ends up at the concert.
>
> If we accept the principle of the conservation of energy we seem committed either to denying that the nonphysical mind exists, or to denying that it could cause anything to happen, or to making some very implausible ad hoc adjustments in our physics. For example, we could maintain that the principle of conservation of energy holds, but that every time a mind introduces new energy into the world—thanks to some mysterious capacity it has—an equal amount of energy departs from the physical universe—thanks to some perfectly orchestrated mysterious capacity the universe has.[35]

Lest one conclude that the problems of mind-body interaction are merely the result of too crude a view of physical interactions in modern physics, it is important to note that the law of conservation of mass-energy remains part of the framework of twentieth-century physics.

It is worth mentioning briefly the epistemological problems created by this metaphysical shift. For Aristotelians, sensory knowledge resulted from the transference of the form of the thing perceived into the intellect of the perceiver, whose mind was, literally, in-*formed* by exactly that which makes the object to be what it is. Thus, exact knowledge of the essences of things was possible on the basis of very little observation. Perceptual *error* is what needed explanation.

In a world composed of atoms, sensation must result from the impinging of atoms on the sensory membranes, and then from coded information conveyed to the brain (and for dualists thence to the mind). Ideas in the mind are no longer identical to the forms inherent in things, but mere representations produced by a complicated process of transmission, encoding, and decoding. Thus arises modern skepticism with regard to sense perception.[36]

---

[35] Owen Flanagan, *The Science of the Mind*, 2nd ed. (Cambridge: MIT Press, 1991), 21.

[36] See Theo C. Meyering, *Historical Roots of Cognitive Science: The Rise of a Cognitive Theory of Perception from Antiquity to the Nineteenth Century* (Dordrecht: Kluwer Academic Publishers, 1989).

Descartes's solution was to begin with the Augustinian notion that we know our own souls/minds directly.[37] But for early modern philosophers that is *all* we know directly. Descartes reassured himself of the possibility of (indirect) knowledge of the external world by arguing that a benevolent creator would not have constructed us so as to be constantly deceived.

3.2.2 Biology

Darwinian evolution has had an impact on contemporary culture comparable to that of the revolution in physics and astronomy that heralded the beginning of modern science. One of the many issues it raised was the continuity between humans and (other) animals. One effect was to provide additional reason to question any form of body-soul or body-mind dualism.

The rejection of hylomorphism represented by the development of modern physics meant the rejection of animal (and plant) souls, understood as the substantial forms of their bodies. Descartes described animals in purely material terms. Against this background, the recognition of human kinship with and development from lower animals warranted the conclusion, in the eyes of many, that humans, too, are purely material.

Many theologians evaded this materialist conclusion by means of the same strategy employed by Pope Pius XII, granting that the human body is a product of biological evolution, but maintaining that God creates a soul for each individual at conception. This intellectual maneuver runs into difficulties, however, when we ask when the *human* species appeared. Contemporary biologists now offer a very complex account of human origins in which there is no clear distinction between animals and humans. Were our first hominid ancestors human, or are only modern humans truly human, or did the change take place somewhere in between? What about hominid species that are not in the direct line of descent to modern humans? To claim that humans alone have the gift of a soul seems to force an arbitrary distinction where there is much evidence for continuity.

If human distinctiveness cannot be attributed to the unique possession of a soul or immaterial mind, in what *does* it consist? This has become an intriguing philosophical and theological issue—one likely to benefit from continuing scientific investigation of actual similarities and differences between ourselves and the other higher primates. Development of the science of genetics has contributed to the discussion of human nature initiated by the theory of evolution. Not only did the biochemical explanation of heredity solidify the evolutionary account of human origins, but it also contributed new evidence for human continuity with the other species, since all other life forms possess DNA, and there may not be *any* genes unique to the human species.[38]

Against this background, the significance of recent progress in neuroscience is clear. Francisco J. Ayala writes that Darwin completed the Copernican revolution by bringing the origin and diversity of living things (including humans) into the realm of science.[39] Yet Darwin's revolution is complete only when the whole human,

---

[37] Flanagan uses results from current neuroscience to call into question this philosophical assumption concerning the priority of self-knowledge. See *The Science of the Mind*, 194–200.

[38] See V. Elving Anderson, "A Genetic View of Human Nature," in Brown, et al., *Whatever Happened to the Soul?*, 49–72; 50.

[39] Francisco J. Ayala, "Darwin's Devolution: Design without Designer," in Russell, et al.,

both the physical and the mental aspects, become the object of scientific investigation. This is exactly what the cognitive neurosciences have attempted.

3.3 Philosophical Issues

Ancient philosophical accounts of human nature have already been mentioned, as well as the influence of modern physics on Descartes's philosophy of mind. This section, then, will focus on subsequent philosophical developments.

There have been psycho-physical monists of various sorts throughout the modern period. Descartes's contemporary, Thomas Hobbes (1588–1679), was a materialist, who described thinking as "motions about the head" and emotions as "motions about the heart." George Berkeley (1685–1753) took the opposite tack and resolved all of reality into ideas (perceptions). Nonetheless, the most common position throughout the earlier years of the modern period was dualism. Yet, ever since Descartes proposed the pineal gland as the locus of mind-brain interaction, the problem of the relation of the mind and body has occupied philosophers.

A variety of solutions to the problem of causal interaction have been tried. Psychophysical parallelism is the view that physical events cause physical events, mental events cause mental events, and that the *appearance* of causal interaction between the mental and the physical is an illusion created by the fact that there is a pre-established harmony between these two independent causal chains. This harmony is either established at the beginning by God or is the result of constant divine interventions. Gottfried Wilhelm Leibniz (1646–1716) was one proponent of this theory. The inspiration for his monadology is said to have come from observation that clocks on various towers throughout the city kept the same time yet had no causal influence on one another. This position was never widely accepted, and lost its appeal with the growing atheism of modern philosophy. As Flanagan says, "to have to introduce God to explain the workings of the mind, however, is to introduce a big Spirit in order to get rid of the perplexities of a world of little spirits, and to magnify the complications one presumably set out to reduce."[40]

Another attempt to solve the problem of mental causation is epiphenomenalism. This is the theory that conscious mental life is a causally inconsequential by-product of physical processes in the brain. This position has two drawbacks. First, why should causation from the physical to the mental be thought any less problematic than from the mental to the physical? Second, there seems to be overwhelming evidence for interaction. These objections are not thought fatal, however, and some say that there is now solid scientific evidence for the thesis.[41]

As a consequence both of the problems with mind-body interaction and of the scientific developments to be surveyed below, the balance has shifted in the philosophy of mind from dualism to a variety of forms of materialism or physicalism. Logical behaviorism, widely discussed from the 1930s to the early 1960s, claimed that talk of mental phenomena is a shorthand (and misleading) way of talking about actual and potential behavior. Gilbert Ryle (1900–76) ridiculed the Cartesian mind

---

Evolutionary and Molecular Biology, 101–16; 101.

[40] Flanagan, *The Science of the Mind*, 64.

[41] This is one interpretation of research by Benjamin Libet on "readiness potential" in motor activity. See discussions in, for example, Ned Block, Owen Flanagan, and Güven Güzeldere, eds., *The Nature of Consciousness: Philosophical Debates* (Cambridge: MIT Press, 1997), passim. For a contrary view, see Flanagan, *The Science of the Mind*, 344–48.

as "the ghost in the machine," claiming that the view of the mind as a substance or object rests on a "category mistake" of assuming that because "mind" is a noun there must be an object that it names. While Ryle's critique of dualism is widely accepted as definitive, it has not proven possible to translate language about the mind into language about behavior and dispositions.

A still-current option is the mind-brain identity thesis. There are various versions: the *mind* is identical with the brain; mental *properties* (such as the property of being in pain, or believing the proposition that *p*) are identical with physical properties; or mental *events* are identical with brain events. The first of these is an infelicitous expression of the identity thesis since it makes the very mistake for which Ryle criticized the dualists.

An important distinction in the philosophy of mind is that between "type identity" and "token identity." Token identity is the thesis that every particular mental event or property is identical with *some* physical event or other; type identity is a stronger thesis to the effect that every type of mental event is identical with a type of physical event. So, for instance, a type of sensation, such as pain, is identical with a particular type of neuron firing. Type identity entails the reducibility of the mental descriptions to physical descriptions. Flanagan says:

> The implication that follows from the latter assumption is this: if type-type identity theory is true then reduction of psychology to neuroscience will eventually be possible. It is easy to see why reduction requires that all the concepts of the science to be reduced be translatable into the concepts of the reducing science. These translations are called "bridge laws" and once they are in place reduction merely involves replacing, synonym for synonym. Type-type identity statements, of course, are precisely the necessary bridge laws.[42]

This strong identity thesis may be philosophically unobjectionable in cases such as pain sensations, but it runs into problems with higher-order mental states such as believing some proposition. When conjoined with the thesis that brain events obey causal laws, type identity implies not only that beliefs could be redescribed in purely neurological language, but also that our beliefs are governed by neurophysiological laws. Some philosophers are perfectly sanguine about this outcome, and in fact look forward to the day when neuroscience replaces our "folk psychology."[43] Others, however, object that such an account makes nonsense of our views that beliefs are (ordinarily) held for reasons rather than out of causal necessity.

The distinction between type and token identity theories makes it possible to state the difference between reductive and nonreductive materialism or physicalism. A variety of philosophers have held that the functioning brain is indeed the seat of mental capacities, and so every mental event is identical with *some* physical event. Yet, because there are no type identities between the mental and the physical (no psychophysical laws) the mental cannot be reduced to the physical. Donald Davidson's "anomalous monism" is the best known current version of nonreductive physicalism. He claims that there are no strict laws at the mental level; beliefs are related instead by principles of rationality. Because there *are* causal laws at the physical level, beliefs must be only token identical with brain states. Whether or not his anomalous monism does in fact avoid reductionism, Davidson is responsible for

---

[42] Flanagan, *The Science of the Mind*, 218.

[43] For example, see Patricia Smith Churchland, "A Perspective on Mind-Brain Research," *Journal of Philosophy* 77 (1980): 185–207; and Paul M. Churchland, "Eliminative Materialism and Propositional Attitudes," *Journal of Philosophy* 78 (1981): 67–90.

a valuable conceptual innovation in the philosophy of mind—introduction of the concept of *supervenience*. Supervenient properties (for example, mental properties) are taken to be dependent on subvenient properties (for example, brain properties) but without being strictly reducible to them. Essays in this volume by William Stoeger, Nancey Murphy, Theo Meyering, and Philip Clayton all exploit this conceptual development. It is interesting, though, that this notion found little favor with the neuroscientists at the conference. Thus, much interesting work remains to be done to create a philosophical framework that is adequate to or provides helpful clarification of the actual scientific practices, whether more or less reductionist, of cognitive neuroscientists.

4 Overview of the Volume

4.1 Resources

Previous volumes in this series have included background essays on the relevant scientific and philosophical issues. In this volume we provide both scientific and theological background. The latter (described in section 4.1.1) is provided by essays commissioned after the Pasierbiec conference, whereas the scientific snapshots (section 4.1.2) are based on presentations made at the conference. If this book can be said to have any single thesis or argument, it is that the theology-science dialogue finds valuable points of convergence in the construction of a contemporary view of the human person. From disciplines as different as theology and biblical studies, on the one hand, and basic research in the neurosciences on the other, comes the recognition that the dualist, individualist, and cognitivist view of the human person, bequeathed by Western philosophy and theology, needs revision. Thus, essays in this section will establish the poles between which the dialogue in this volume moves.

4.1.1 Biblical and Theological Background

New Testament scholar **Joel B. Green** was asked to write an essay reflecting current scholarship on biblical views of human nature. In "Restoring the Human Person: New Testament Voices for a Wholistic and Social Anthropology," Green laments the fact that recent investigations of "biblical anthropology" have focused either on the question of body-soul dualism or on a series of topics oriented around human sin and its remedies. This is unfortunate because Scripture is largely unconcerned with speculative questions about human nature; the attempt to find answers to current philosophical questions can obscure the biblical writers' own central concerns.

Green focuses his investigation on the writings of Paul and Luke—Paul because he is regarded as the most important theologian of the apostolic age; Luke because he is arguably the only Gentile author represented in the New Testament. This suggests that if any New Testament material were to reflect the dualism alleged to characterize Hellenistic thought, it would be that of Luke.

Luke's concern with human nature arises within the context of his understanding of salvation. Luke raises questions about what needs to be saved, and what "saved existence" would look like. Answers to these two questions point to his understanding of authentic human existence. Green examines as paradigmatic Luke's narrative of Jesus' healing of the woman with the hemorrhage (Luke 8:42b–48). Green finds here a *holistic* and *social* anthropology, evidenced by the fact that healing involves not merely reversal of her physical malady, but also restoration of her place in both the social world and the family of God. The importance of this (and other relevant texts) is to call into question two closely related tendencies in the twentieth-century West: to think of salvation fundamentally in "spiritual terms," and, with respect to issues of healing and health, to think primarily in terms of bodies. Green supports his claim for the holistic and social anthropology of the Bible by examining other Lukan texts, as well as many of the Pauline and Genesis texts that have been used in the past to warrant body-soul dualism. In addition, he criticizes the popular "word study" method of biblical interpretation that has allowed body-soul dualism to achieve a prominence in Christian thought far out of proportion to the scriptural evidence. Such an approach too easily lends itself to reading contemporary meanings into biblical terms.

Green concludes that Christians today who embrace a monistic account of humanity place themselves centrally within a biblical understanding. At the same time, he says, biblical faith resists any suggestion that our humanity can be *reduced* to our physicality. Furthermore, an account of the human person that takes seriously the biblical record will deny that human nature can be understood "one person at a time," and will focus on the human capacity and vocation for community with God, with the human family, and in relation to the cosmos.

"The Modern Philosophy of Self in Recent Theology" by **Fergus Kerr** is reprinted here from Kerr's book, *Theology after Wittgenstein*,[44] because it ably demonstrates the extent to which Christian theology carries a "metaphysical load"—an account of the human person derived from Cartesian philosophy. It is ironic that the modern philosophical conception of the self sprang, as Kerr notes, from explicitly theological concerns. In the process of demonstrating the existence of God and the immortality of the soul, Descartes articulated a conception of human nature according to which the self is *essentially* a thinking thing, thus redefining what it is to be human in terms of consciousness. Descartes, together with Immanuel Kant, bequeathed a picture of the self-conscious and self-reliant, self-transparent and all-responsible individual, which continues to permeate contemporary thought even where Descartes's substance dualism has been repudiated.

Kerr examines a number of authors to show how this picture of the self shows up in recent theology, and this despite the fact that some eminent theologians, such as Karl Barth and Eberhard Jüngel, have argued that the Cartesian turn to the subject has nearly ruined theology. Kerr considers the role of the Cartesian ego in the works of Karl Rahner, Hans Küng, Don Cupitt, Schubert Ogden, Timothy O'Connell, and Gordon Kaufman.

It is always as the cognitive subject that people first appear in Rahner's theology. "Students alerted to the bias of the Cartesian legacy would suggest that language or action, conversation or collaboration, are more likely starting points." Rahner's theology depends heavily on the notion of self-transcendence: when self-conscious subjects recognize their own finitude, they have already transcended that finitude. This process of self-reflection produces a dynamic movement of "ceaseless self-transcendence towards the steadily receding horizon which is the absolute: in effect, anonymously, the deity." While Kerr recognizes the theological payoff of this move, making arguments for the existence of God redundant, it is at the expense of an account of humans as "deficient angels"—that is, as attempting to occupy a standpoint beyond immersion in the bodily, the historical, and the institutional.

From his survey of Rahner and other examples, Kerr concludes that "in every case, though variously, and sometimes very significantly so, the model of the self is

---

[44] Fergus Kerr, *Theology after Wittgenstein*, 2nd ed. (London: SPCK, 1997), chap. 1.

central to some important, sometimes radical and revisionary, theological proposal or program. A certain philosophical psychology is put to work to sustain a theological construction. Time and again, however, the paradigm of the self turns out to have remarkably divine attributes." The philosophy of the self that possesses so many modern theologians is a view which philosophers today are working hard to destroy. Kerr's essay ends with a brief survey of the post-Wittgensteinian philosophers who pursue this task—most notably, Bernard Williams and Charles Taylor.

4.1.2 Snapshots from the Neurosciences

The cognitive neurosciences are expanding and developing so quickly that it is not practical to attempt an overview of the entire field in one essay. However, in planning the conference we invited scientists with a broad array of interests in order to convey some flavor of the burgeoning field. In addition to their own contributions to the theme of the conference, these participants were asked to provide "snapshots" of their research at Pasierbiec for publication in this volume.

Joseph E. LeDoux specializes in the use of animal models for studying emotion. In "Emotions: How I've Looked for Them in the Brain," LeDoux describes his work on fear conditioning in rats. The rats are conditioned to associate a sound with a noxious stimulus. This sound then elicits the behavioral responses accompanying the emotional experience of fear: muscle tension, release of stress hormones, and so forth. Note that LeDoux distinguishes between the behavioral system and subjective feelings. It is the former, he argues, that should be seen as essential to understanding the function of emotions.

LeDoux uses a variety of techniques to relate fear behavior to specific circuits in the brain. First, lesion studies (selective damage to parts of the brain) and brain imaging techniques make it possible to locate the general regions involved. Next the circuits activated in fear responses can be followed by injecting tracer substances into those areas and recording the "firing patterns" of neurons in relation to various emotional states under a variety of learning paradigms. In this way, LeDoux has confirmed the crucial role of the amygdala, a distinctive cluster of neurons found deep in the anterior temporal lobe of each hemisphere. Inputs to the amygdala from sensory processes in the thalamus and cortex are key to processing fear stimuli, while projections from the amygdala to brainstem areas are involved in control of the behavioral, autonomic, and hormonal responses that constitute fear behavior. LeDoux notes that a variety of other brain systems are also involved in the various feeling states we term "emotions" in humans; only empirical research will show whether his work on fear generalizes to other emotions.

Peter Hagoort specializes in the study of the neural underpinnings of language. In "The Uniquely Human Capacity for Language Communication: From POPE to [po:p] in Half a Second," he points out that the sophisticated capacity for language unique to humans and performed in various forms such as speaking, listening, writing, reading, and sign language, rests on a tripartite architecture: coding for meaning, for syntax, and for sound structures. A central component of language skills is the mental lexicon, a part of declarative memory that stores the meaning, syntactic properties, and sounds of roughly 40,000 words.

Hagoort has studied the order in which information is retrieved from the mental lexicon—for example, when one recognizes the image of a well-known person. Words are not discrete units, each to be found localized in some small circuit in the brain; the various components of the ability to use words are all stored differently.

First there is a conceptual selection and specification process, followed by retrieval of syntactic information, and then by retrieval of a sound pattern—all of this resulting in the utterance "pope." The different retrieval processes occur with high speed, and are temporally orchestrated with millisecond precision.

One of the ways in which the sequence of events involved in word retrieval has been studied is by recording electrical brain activity, using a series of electrodes attached to the scalp. The brain regions involved (mainly in the left hemisphere) have been localized by means of neurological data and brain imaging techniques.

Hagoort notes that the understanding of the neural substrate of language is an essential ingredient in an understanding of the human person, not only because sophisticated linguistic ability is unique to humans, but also because language itself mediates our sense of self.

In "The Cognitive Way to Action," **Marc Jeannerod** describes research on the generation of voluntary action. He begins with an historical overview of theories in the field. Already in the 1930s researchers noted that even the simplest movements produced by the nervous system of a frog appeared to be organized purposefully. So the question was, How are these actions represented in the brain? An important advance was the recognition that behavior is guided by internal models of the external world, with predictions built in as to how the external world will be modified by the organism's behavior and how the organism itself will be affected by the action. The existence of such models is supported by ethological studies showing that certain behavioral sequences unfold blindly and eventually reach their goal after they have been triggered by external cues. Localized brain stimulation can also trigger similarly complex actions.

Early accounts hypothesized serial steps in the neural generation of actions. However, current brain studies suggest simultaneous activation in cortical and subcortical levels of the motor system. This distributed model of action-generation raises the issue of a central coordinator to determine the temporal structure of the motor output. The behavior of patients with damage to the frontal lobes suggests to Jeannerod that the supervisor system is associated with this region.

New light is now being shed on the problem of the neural substrates of action-generation by the study of mentally-simulation action. Jeannerod and his colleagues instructed subjects to imagine themselves grasping objects. Using Positron Emission Tomography (PET) and functional Magnetic Resonance Imaging (fMRI), they identified the cortical and subcortical areas involved. They were then able to show that forming the mental image of an action produces a pattern of cortical activation that resembles that involved in intentionally executing the action.

Jeannerod expects research such as this to shed light on the neural underpinnings of central aspects of the self such as intentionality and self-consciousness. He notes that there is now neuropsychological evidence for the moral dictum that "to intend is to act."

In "A Neuroscientific Perspective on Human Sociality," **Leslie A. Brothers** describes recent findings on the neural substrate of social behavior. A wide variety of evidence points to the role of the amygdala in processing information crucial for social interactions: (1) Monkeys with experimental lesions of the anterior temporal lobes (where the amygdala is located) have particular difficulty responding to the social signals of other monkeys. (2) Human patients with lesions of the amygdala have difficulty interpreting facial expressions, direction of eye gaze, and tone of voice. (3) Tests during neurosurgery show that a person's ability to identify facial expressions can be disrupted by electrical stimulation to temporal-lobe regions. (4)

Researchers studying vision in monkeys found temporal-lobe neurons that seemed to be responsive only to social visual stimuli such as faces. Brothers' own research involved recording the activity of individual neurons in the region of the amygdala while monkeys watched video clips of other monkeys engaged in a number of activities. Her results showed that some nerve cells are particularly attuned to respond to movements that bear social significance, such as the specific yawn that males use to signal dominance.

The picture that is emerging from human and monkey studies, says Brothers, is that representations of features of the outside social world are first assembled in the temporal lobe cortices of the primate brain. Meaningful social events are registered when a host of signals and relevant contextual information are integrated. Our brains need to tell us the difference between someone approaching with friendly intent and someone whose aims are hostile, for example. The visual features of a face have to be put together to yield an image of a particular individual so that past interactions with this individual can be recalled. Next, movements of the eyes and mouth indicate the person's disposition. Information from head position and body movement tell where this person is looking or going, providing raw material for the representation of a mental state such as his or her goal or desire. As these processes are taking place, the neural representation of others' social intentions must be linked to an appropriate responsive behavior in the perceiver. Response dispositions should be set into play "downstream" from the temporal cortices, where face-responsive neurons have been found, in structures such as the amygdala. The amygdala, together with several other interconnected structures, receives sensory information and in turn projects directly to somatic effector structures such as the hypothalamus, brainstem, and primitive motor centers, making it a candidate for the link between social perception and response.

Brothers notes that human social interaction depends on the ability to employ the concept of *person*—a mind-body unit. What the research summarized here suggests is that the evolution of our brains has made it possible for us to construct and participate in the language-game of personhood; we have brains specially equipped for social participation.

It may seem strange for a single section of a book to incorporate essays on both neuroscience and biblical studies, yet there are striking points of convergence. From entirely different directions, the background material from biblical studies, theology, and neuroscience points toward a view of the human person that might best be identified as "anti-Cartesian." The obvious sense in which this is the case, of course, is the rejection of mind-body or body-soul dualism. But there are subtler issues as well. Descartes's identification of the person with the (inward, individual) mind sponsored an individualism that always threatened to become solipsism. Kerr wryly criticizes the Cartesian predicament in his characterization of Gordon Kaufman's early view of interpersonal knowledge: "Kaufman, until he thought better, supposed that it was only when the other opened his mouth and spoke that one realized that a person lay hidden within the middle-sized, lightly sweating and gently palpitating object on the other side of the dinner table."[45] But here we find the biblical claim that we are *essentially* social beings supported by the research described by Brothers— we are equipped neurobiologically to recognize persons as persons rather than having to infer the presence of consciousness from their speech or behavior.

---

[45] In this volume, sec. 3.6.

Kerr criticizes Descartes's replacement of the ancient and medieval concept of humans as rational *animals* with the notion that we are instead essentially thinking *things*, thus making a sharp distinction between animals and humans. Yet Green points out that in the creation narratives humans are distinguished from animals only in that they participate in a unique relationship with God. The viability of animal models for neuropsychological research (Jeannerod, Brothers, LeDoux) provides empirical testimony to our kinship with other animals.

Descartes's metaphor of "viewing ideas in the mind" has led to a concept of human cognition that overemphasized contemplation as opposed to action. Jeannerod's research (as well that of Michael Arbib, described below) rightly underscores the essentially action-oriented character of much of human cognition.[46] Descartes has been accused of sponsoring a negative view of the emotional aspects of human life; LeDoux's research rightly emphasizes the survival value of emotion and its centrality in constituting the human person. Kerr appreciatively cites Karl Barth's argument for the importance of imagination as well as intellect. These emphases would be supported, as well, by a survey of biblical texts.

Hagoort's research highlights the capacity for sophisticated language use as a distinctive human feature. Both the Jewish and Christian traditions emphasize that "in the beginning was the word." Whether or not one believes in a role for God's word, it is undeniable that many of our social institutions, including religions, are intimately rooted in the human capacity for language. God's giving Adam the power to name the animals and the legend of the tower of Babel are further pointers to the centrality of language in humanity's relation with God. Thus, current attempts to determine possible pathways for the evolution of the "language-readiness" of the human brain are very much relevant to attempts to link neuroscience to our understanding of divine action.[47]

4.2 From Neuroscience to Philosophy

The structure of the rest of this volume reflects the editors' perception that insights from the cognitive neurosciences on the one hand, and religion on the other, *converge* in a theory of human nature. However, insights from science are generally *mediated* by a philosophical account of the human person. Thus, we envision a "flow of argument" something like the following:

cognitive   → philosophy → theory of   ← Christian   ← theology
neurosciences   of mind   the person   anthropology

So in this section we include essays that bridge the fields of cognitive neuroscience and philosophy of mind. These essays are of two types: The first begins with findings or theories in science and raise philosophical questions from those points of view. The second addresses current philosophical accounts of the person intended to be consistent with the general physicalist outlook of the neurosciences.[48] Essays in the

---

[46] Jeannerod's research, reported in his essay in the following section, even calls into question the Cartesian dogma that one knows one's own mind immediately and indubitably.

[47] I thank Michael Arbib for this insight.

[48] While current philosophy of mind is generally consistent with the physicalism of neuroscience, many philosophers of mind overlook the "anti-Cartesian" emphases noted above that derive from the empirical study of cognitive functions. That is, they attempt to create physicalist accounts of the self that maintain all but Descartes's dualist metaphysic.

following section (*From Science and Philosophy to Christian Anthropology*) then bring the results of science and of philosophical reflection together with considerations from Christian theology. The final section (*Contrasting Reflections on the Theological Context*) pairs two essays with conflicting answers to the question whether or not a theological account (a theistic worldview) is needed in order to make sense of the various dimensions of human nature canvassed in this volume.

4.2.1 Scientific Starting Points

Michael A. Arbib, in his essay "Towards a Neuroscience of the Person," provides an excellent framework for relating the neurosciences to the concerns of the human sciences and theology. The organizing idea of his essay is the following: a discussion of what neuroscience has to contribute to an emerging science of the person will provide a bridge between the narrow foci of individual researchers' efforts in the cognitive neurosciences, on the one hand, and the far broader but less scientifically grounded considerations of humanists, including theologians, as they seek to explicate the nature of the person.

Arbib begins with a survey of topics on which neuroscience has offered insights into mental phenomena such as memory, emotion and motivation, social behavior, and language. This sampling of scientific developments raises the question whether the cognitive neurosciences will eventually provide a framework for understanding all of the phenomena that define human nature. In particular there is the question whether the study of the brain can explain the religious dimension of human life, or whether the subject-matter of theology will always elude neuroscientific investigation. Arbib maintains that a complete science of the person must take account of theology, but argues that theology ought to be understood not as the science of God but as the study of human belief in God. This latter understanding would open the discussion for nonbelievers (such as Arbib himself) but would incorporate the former understanding of theology if God in fact exists. Neuroscience cannot address the concept of God directly but can make progress toward theological questions, especially if theology is defined in the broad sense.

Another important issue is the relation of neuroscience to questions of morality. Arbib notes that both religion (even on a nontheistic account) and neuroscience can provide insight. Neuroscience cannot answer questions of right and wrong, but it can elucidate aspects of morality such as decision-making, empathy, and social behavior.

Arbib then sketches the possible role of computational neuroscience in bridging levels between neuron and person. Schema theory provides a link between "cognition-level" and "neuron-level" descriptions of the person. Basic schema theory operates at the level of cognitive science, and explains mental operations and behavior in terms of functional units. There are schemas for recognition of objects, planning and control of actions, and more abstract operations as well. Mental life and behavior result from the dynamic interaction, cooperation, and competition of many schema instances. The individual can be understood as a self-organized "schema encyclopedia." Schema theory provides a bridge between neuroscience and the humanities: it can be extended "downward" by studying the neural realizations of simple schemas; it can be extended "upward" by recognizing that schemas have an

---

Daniel Dennett is a strong critic of this tendency, but the resources for a post-Cartesian account of the self are already available in the later writings of Ludwig Wittgenstein, as Kerr argues in *Theology after Wittgenstein*.

external social reality in collective patterns of thought and behavior. Arbib claims that while schema theory can contribute to many open questions regarding the dependence of aspects of the person on the brain, Christian teaching parts company with science on the issue of the resurrection.[49]

In his second essay for this volume, "Emotions: A View through the Brain," **Joseph E. LeDoux** provides an argument for his claim that the scientific investigation of emotion requires a distinction between emotional behavior or associated physiological responses and the subjective feelings experienced by humans. Emotional behavior can be understood by the evolutionist in terms of the function it serves in human and animal life. Emotional feelings must be seen as secondary since emotional behavior is present in organisms that do not have the capacity for conscious awareness. LeDoux defines emotional feelings as a result of sophisticated brains being aware of their own activities—in this case, being aware that an emotion system, such as the fear system, is activated. The problem of explaining emotional feelings is a part of the *single* problem of the explanation of consciousness. However, different emotional behavior or response systems may involve different brain mechanisms. Here LeDoux is critical of the "limbic-system" theory, a theory that sought to identify a single set of brain structures involved in all emotional responses.

LeDoux summarizes in more detail here the results of research on fear conditioning in rats (see above), and notes that studies of the effects of damage to the amygdala in humans, as well as fMRI studies, show that the amygdala is the key to the fear-conditioning system in humans, as well. However, the association of fear not with the original stimulus but with the environmental context in which the stimulus was encountered appears to depend on the hippocampus.

The persistence of learned fear responses is obviously valuable for survival. However, the inability to inhibit unwarranted fear responses can have devastating consequences, as in phobias and post-traumatic stress disorder. Thus, research on the probable role of neocortical areas in extinction of fear responses may be of great value in treating these disorders.

Less is known about other basic emotion such as anger or joy; it remains to be seen whether the amygdala is involved in these as well. Far less is known about "higher-order" emotions such as jealousy. And, as mentioned above, an account of emotional feelings awaits an adequate account of consciousness in general. However, LeDoux notes that working memory receives a greater number and variety of inputs in the presence of an emotional stimulus than otherwise, due to the variety of neural pathways involved; he speculates that this excess stimulation is what adds the affective charge to representations in working memory that we associate with felt emotions.

In his second essay, "Are There Limits to the Naturalization of Mental States?" **Marc Jeannerod** brings further neuropsychological research to bear on the topic of intentional action and its role in constituting self-awareness. He notes that humans are social beings, and that communicating with others is a basic feature of human behavior. A long-standing philosophical question is how it is possible for one person to recognize the mental states of others. A key insight here comes from neuroscience:

---

[49] But see Peters's essay, where he argues that the notion of bodily resurrection is less directly in conflict with neuroscience than the notion of a disembodied soul, since science and theology agree that (apart from an act of God) the self cannot survive the dissolution of the brain.

the neural system one uses for detecting intentions of other agents is part of the neural system that generates one's own intentions. Evidence for this comes from studies with monkeys showing the existence of neuronal populations in several brain areas that selectively encode postures and movements performed by conspecifics. Much of this population of neurons overlaps with those involved in the generation of the monkey's own movements. This same sort of overlapping of function is suggested by PET-scan studies in humans. When subjects were told to watch an action with the purpose of imitating it, parts of the motor cortex were activated, whereas this was not the case if subjects were told to watch only for the purpose of later recognition.

The research summarized here sheds light on the problem of other minds, but in so doing raises a new philosophical problem: if the intention of another's action is represented in *my* neural system by means of the same neural activity as my own intention to act, how does this intention get attributed to the right agent? Jeannerod shows that having a neural representation of an intention and attributing it to myself are two different processes, which are not automatically linked. Jeannerod reports further research that highlights this problem. Experimental situations have been devised in which it is not obvious to the subjects whether they are seeing an image of their own hand or that of the experimenter, moving in response to instructions. When the experimenter's hand movement departed from the instructions, subjects had no difficulty recognizing it was not their own. But in thirty percent of cases when the experimenter's hand followed the instructions, normal subjects mistook it for their own. Schizophrenic patients misattributed the experimenter's movements to themselves eighty percent of the time. This is consistent with clinical reports that schizophrenics suffer from a tendency to incorporate external events into their own experience.

Jeannerod ends his essay with a reflection on the limits of human abilities to know other minds. A person's individuality resides in the fact that no two individuals ever share all of the same experiences. Thus, no two people's global neural states will ever be the same. If neuroscientific understanding is based on similar or identical neural representations, then some aspects of personal identity are beyond the realm of scientific inquiry.

These three essays together provide a tantalizing glimpse of the possibilities for new approaches to philosophical questions from the perspectives of the neurosciences. As did the neuroscience "snapshots" in the previous section, they contribute to an understanding of the human self that is anti- or post-Cartesian. Their bearing on the issue of dualism versus physicalism is obvious. In addition, Jeannerod's essay supports (as did those in the previous section) a concept of human nature as essentially social and, as noted above, reverses the modern problem of other minds: we are apparently endowed with neural equipment ("mirror neurons") for recognizing others' intentions, but when associated neural processes fail, we may mistake others' actions for our own—thus falsifying the Cartesian dictum that one knows one's own mind (intentions) directly and indubitably. LeDoux's emphasis on the survival value of emotional responses calls into question the long-standing suspicion in Western philosophy of the emotions as anti-rational. Arbib's emphasis on the action-oriented character of schemas supports a pragmatic as opposed to a contemplative approach to epistemology.[50] On this account, there is room for

---

[50] See Richard Rorty, *Philosophy and the Mirror of Nature* (Princeton: Princeton University Press, 1979).

significant dialogue between neuroscientists and those theologians who are working toward a theological view of human nature (and of God) freed from the Cartesian picture of the self as an autonomous rational consciousness, indifferent toward community and antipathetic toward the body.[51]

The central question raised by all three essays, but most explicitly by Arbib's, is the question of reductionism: given the fact that so many human attributes, including language, emotion, morality, and perhaps even religious awareness, fall within the purview of neuroscientific research, is it the case that the neurosciences of the future (or perhaps an ideal but never actual neuroscience) can give an exhaustive explanation of all human attributes? This is the central issue addressed by the following three essays. Yet this juxtaposition with the foregoing essays should not be allowed to conceal a discrepancy between a point of view shared by many of the neuroscientists at the conference and one shared by the philosophers (and theologians). It is tempting to express this disagreement by saying that the scientists advocate reductionism while the philosophers argue against it. Yet the disagreement is not so neat; it may better be expressed as an inability to see the *point* of the other's argument. So Arbib might say: "Why would you *want* to argue against reductionism? It is clear that (much of) psychology can be recast as cognitive science; and that much of cognitive science can be embedded within cognitive neuroscience and thus reduced (more or less) to the neural level, and so on down to physics. This does not imply that the languages of psychology and cognitive science are being replaced by the language of neuroscience, but rather that science combines multiple levels of explanation in different mixes depending on the data to be explained."[52]

Theo Meyering might respond: "Yes, we agree that *much of* psychology (your own qualification) can be reduced to neuroscience, and that science will always need to employ a combination of multiple levels of explanation. Yet you speak in your second essay of social interaction having top-down influence on the microstructure of the brain. So why would you want to argue against my *anti*-reductionist stance?"[53]

Some of the issues at stake can be clarified by distinguishing among a variety of reductionist theses—ontological and epistemological sorts, in a variety of strengths—as Meyering and others will do in the following subsection. Yet these distinctions still do not account for the conference participants' ability to talk past one another on this issue. Perhaps a more helpful distinction is that between reductionist *theses* (of all sorts) and methodological reductionism—a research *strategy* to which cognitive neuroscientists must be committed in principle.[54]

4.2.2 Philosophical Theories of Human Nature

In "The Mind-Brain Problem, the Laws of Nature, and Constitutive Relationships," **William R. Stoeger, S.J.** argues that a correct understanding of the meaning of

---

[51] See especially Kerr, *Theology after Wittgenstein*; and Nicholas Lash, *Theology on the Way to Emmaus* (London: SCM Press, 1986); and idem, *Easter in Ordinary*.

[52] This is a paraphrase of a comment by Arbib on Theo Meyering's essay.

[53] This reply is my own creation, but I think it adequately represents Meyering's position.

[54] Francisco Ayala employs this distinction to interpret what appears to have been a similarly perplexing set of disagreements at a conference on philosophy of biology in 1972. See his "Introduction" in Ayala and Theodosius Dobzhansky, eds., *Studies in the Philosophy of Biology: Reduction and Related Issues* (Berkeley and Los Angeles: University of California Press, 1974), vii–xvi.

"laws of nature" is essential for clarifying issues associated with the mind-brain problem. He distinguishes between "the laws of nature" as the regularities, relationships, and processes that obtain in nature, and "our laws of nature" as our provisional, incomplete, and imperfect models of these regularities. In some areas of science our models give fairly adequate accounts of the actual regularities and relationships; in others adequate models are still lacking. Modeling mental processes and their relations to brain processes seems especially problematic due to the subjective and holistic character of mental phenomena; in fact, it is not yet clear what would *count* as an adequate model for explaining the mental in terms of brain processes.

The sense of "laws of nature" that one intends has a bearing on the meaning of essential terms in the philosophy of mind, such as "emergence" and "supervenience," and on an even deeper issue underlying the mind-brain problem: the very meaning of "physical" or "material," versus "nonphysical" or "immaterial." "Matter" is not a scientific term and the meaning of "material" is historically contingent. In common usage, Stoeger takes it to refer to that which we can model, describe, and understand using the resources of the natural sciences. Correlatively, the immaterial is that which transcends the regularities known by science. Thus, the identification of the mental with the immaterial does not mean that the mental could not be a property of neurologically highly organized matter.

Stoeger draws attention to the "constitutive relationships" that account for the hierarchical structure of reality, such that higher levels are composed by complex ordering of lower-level entities. The constitutive relationships of a complex whole are all of those connections, relationships, and interactions that either incorporate its lower-level components into that more complex whole, relate that whole to higher-level unities in such a way as to contribute essentially to its character, or maintain its connection to the Ground of its existence. Stoeger's insight is that insofar as there are constitutive relationships of the sort that relate an entity to higher-level systems, those entities are not reducible either causally or mereologically (that is, as mere aggregates are reducible to their parts). Thus, Stoeger concludes that mental states cannot be reduced to brain-states: there are constitutive relationships not just among the brain-states that realize them, but also relating the mental states they determine with one another and with historical and environmental conditions. These external constitutive relations play a role in determining the sequences and clustering of mental states.

Stoeger ends by reflecting on the Aristotelian and Thomist accounts of *form* and *soul* as that which makes an entity to be what it is. He notes that a scientifically accessible correlate of these notions is his own account of constitutive relationships.

In "Supervenience and the Downward Efficacy of the Mental: A Nonreductive Physicalist Account of Human Action," **Nancey Murphy** sets out to answer the question: If mental events are intrinsically related to (supervene on) neural events, how can it *not* be the case that the contents of mental events are ultimately governed by the laws of neurobiology? The main goal of her essay, then, is to explain why, in certain sorts of cases, complete causal reduction of the mental to the neurobiological fails. To do so, she first considers the concept of *supervenience,* offering a definition that runs counter to the "standard account." The concept of supervenience was introduced in ethics to describe the relation between moral and nonmoral (descriptive) properties; the former are not identical with the latter, but one is a "good" person *in virtue of* possessing certain nonmoral properties such as generosity. Supervenient properties are multiply realizable; that is, (in the moral case) there are a variety of lifestyles each of which constitutes one a good person. Murphy criticizes typical attempts at formal definitions of "supervenience" for presuming that subvenient properties alone are sufficient to determine supervenient properties. She argues that many supervenient properties are codetermined by context—this is a move similar to Stoeger's in recognizing constitutive relationships not only at the subvenient level but also at the supervenient level itself or between the level in question and even higher levels of organization.

Murphy argues that it is this participation of entities in higher causal orders by virtue of their supervenient properties that accounts for the fact of downward causation. In Donald Campbell's original example, it is the functional properties of the termites' jaw structure—their relation to a higher-level causal order—that allows for environmental feedback, resulting in modifications at the (subvenient) genetic level. These modifications are a result of selection among lower-level causal processes.[55]

Murphy then turns to the issue of mental causation: How do *reasons* get their grip on the causal transitions among neural states? The key to answering this question is the fact that neural networks are formed and reshaped (in part, at least) by feedback loops linking them with the environment; the environment selectively reinforces some neural connections but not others. Murphy points out that it is not only the physical environment that plays a downward causal role in configuring neural nets, but also the *intellectual* environment. It is the fact that mental states supervene, in Murphy's sense of the term, on brain-states—that is, that they are co-constituted by both brain-states and their intellectual context—that makes the occurrence of the brain-states themselves subject to selective pressures from the intellectual environment.

Theo C. Meyering, in "Mind Matters: Physicalism and the Autonomy of the Person," takes yet a third approach to the issue of reduction. He states that "if (true, downward) mental causation implies nonreducibility [as Stoeger and Murphy argue] and physicalism implies the converse, it is hard to see how these two views could be compatible." Meyering distinguishes three versions of reductionism: radical (industrial strength) physicalism; ideal (regular strength) physicalism, and mild or token physicalism. Radical physicalism asserts that all special sciences are reducible to physics in the sense that their laws can be deduced via bridge laws from those of physics. Ideal physicalism asserts that while it is *practically* impossible to reduce the special sciences, such reduction would be possible were there an ideally complete physics.[56] Token physicalism is ontologically reductionist: there are no events that are not "token-identical" with *some* physical event or other.[57] However, there are no identities between higher-level and lower-level *types* of events; consequently some events described by the special sciences have no physical *explanation* at all.

All of these reductionist positions are to be contrasted with compositional (milder than mild) physicalism, which asserts that some higher-level events are not even token-identical with physical events because the higher-level event (say, a crash

---

[55] While Murphy takes feedback and selection among lower-level causal processes to be the essential ingredient in downward causation, Arthur Peacocke, in his essay in this volume, assimilates it to "whole-part influence."

[56] This distinction parallels Stoeger's recognition that epistemological reducibility is relative to the meaning of "laws of nature."

[57] See above, sec. 3.3.

in the stock market) is constituted by innumerable physical particulars in all sorts of states and interactions.

Meyering then surveys some of the existing arguments for the nonreducibility of the special sciences. One of the most important is the argument from multiple realizability. The claim is that economics, for example, is not reducible to physics because economic concepts (for example, *monetary exchange*) are "wildly multiply realizable" (for example, using coins, strings of wampum, signing a check). Thus, there can be no bridge laws and no reduction. Such an argument, however, only cuts against radical physicalism, not the weaker (and a priori more plausible) ideal physicalism.

A stronger argument for the indispensability of special-science explanations is based on the role of functional explanations. For example, the functional description of aspirin as an analgesic is in some instances a more useful explanation of its causal role (relieving a headache) than is its description as the chemical level.

Meyering's own contribution focuses neither on multiple realizability of supervenient properties nor on multiple "fillers" of functional roles, but on "multiple supervenience." In particular, a *single* subvenient state of affairs (for example, a cloud of free electrons permeating the metal of which a ladder is constructed) may realize a variety of supervenient *dispositional* properties (in this case, electrical conductivity, thermal conductivity, opacity). An explanation (say, of the cause of a deadly accident) requires reference to the dispositional property (electrical conductivity), not merely to the subvenient property. Meyering argues that it is this possibility of multiple supervenience, not multiple realizability, that gives arguments against reduction based on functional properties their real force. Downward causation, then, can be understood in terms of selective activation of one of several dispositional properties of a lower-level state, and thus can be assigned a stable place in our picture of how the world is organized without upsetting our conception of physics as constituting a closed and complete system of physical events.

So these three essays, in different but closely related ways, argue that "nonreductive physicalism" is *not* an oxymoron. It is possible to conceive of mental states as supervenient on brain-states, to maintain the integrity of causal explanation at the neurobiological level, and yet to maintain a distinctive causal role for the mental *qua* mental. Several essays in the next section (by Philip Clayton, Arthur Peacocke, and Ian Barbour) advocate similar positions.

4.3 From Science and Philosophy to Christian Anthropology

The essays in this section proceed from neuroscientific or philosophical accounts of human nature to theological interpretations. Again, we have two subsections. The first deals with general theological issues; the second looks at the specific issue of religious experience.

4.3.1 Anthropological and Theological Issues

Philip Clayton, in "Neuroscience, the Person, and God: An Emergentist Account," provides a fine overview of issues spanning the range from neuroscience, through philosophy of mind, to theology. Beginning, as does Arbib, with a list of some of neuroscience's achievements in understanding human phenomena, Clayton points out that either of two extreme positions, if true, would block any significant dialogue between the neurosciences and theology. On the one hand, strong forms of dualism that make mind into a separate substance remove mental phenomena forever from the realm of scientific study. On the other hand, eliminative materialism—the view that "folk" psychological entities such as beliefs and desires do not exist—resolves the debate with theology by removing theology altogether from the realm of possible theories. For this reason, Clayton's essay challenges the "Sufficiency Thesis," according to which neuroscientific explanations will finally be sufficient to fully explain human behavior.

Once these two extreme views, ontological dualism and radical reductionism, have been dismissed, a wide range of interesting possibilities remains for integrating neuroscientific results and theological interpretations into a theory of the person. One set of issues hinges on how one understands the epistemological status of theology. Clayton advocates the view that while religious beliefs are not subject to proof or confirmation from science, they need to be answerable to scientific advances in the weaker sense of not being counter-indicated by the empirical sciences.

The more difficult issues have to do with interpretation of the results of neuroscience. It is clear that neural states are major determinants of subjective experience and thought, yet Clayton takes the "structural couplings" between the conscious organism and its environment, the phenomena of reference and meaning, and the experience of "qualia" (the subjective side of conscious experience) to suggest that mental events or properties are not thoroughly reducible to neural states. Clayton, along with others in this volume, understands mental events as supervenient on their physical substrates; however, along with Murphy, he challenges the standard accounts of supervenience that seem inevitably to result in causal reduction of the mental to the physical. Clayton's version of "soft" or "emergentist" supervenience defines a property F as emergent if, and only if, there is a law to the effect that all systems with this microstructure have F, but F cannot, even in theory, be deduced from the most complete knowledge of the basic properties of the components of the system. If mental properties supervene on physical properties in this manner, Clayton concludes, there is room for genuine mental causation—not all causes of human behavior are purely neuronal causes.

Clayton's account of supervenience leads to an *emergentist-monist* account of the person: "monist" because, while there are many types of properties encountered in the world, there is only one natural system that bears all those properties; "emergentist" because, while mental phenomena result from an (incredibly complex) physical system, the brain, they represent a genuinely new causal and explanatory level in the world. He notes that emergentist monism is open to theological applications and interpretations, although it does not require a theistic outlook—an issue that will be joined by Arbib and George Ellis in the final section of this volume.

In "The Sound of Sheer Silence: How Does God Communicate with Humanity?" **Arthur Peacocke** advocates an emergentist-monist account of the natural world: its unity is seen in the fact of its hierarchical ordering such that each successive level is a whole constituted of parts from the level below. This world exhibits emergence in that the properties, concepts, and explanations relevant to higher levels are not logically reducible to those of lower levels. An emergentist-monist account of the human person fits consistently within this worldview. It is important to note that, unlike many philosophers, Peacocke does not identify mental properties with brain properties. Rather, he recognizes the mental or personal as an emergent level above the (purely) biological, and attributes mental properties to the unified whole that is the "human-brain-in-the-body-in-social-relations."[58]

---

[58] Brothers and other authors in this volume would agree in emphasizing both the embodied

More important than the logical irreducibility of levels in the hierarchy of complex systems is causal irreducibility. Peacocke discusses the concept of downward causation and a variety of related concepts of causal processes in complex systems, one of which is the distinction between structuring and triggering causes. A structuring cause is an ongoing state of a system (for example, the hardware conditions in a computer) that makes it possible for an event (the triggering cause; for example, striking a key) to have the effect that it does. Peacocke concludes that the term "whole-part influence" best captures what is common to all of these insights.

This essay elaborates on Peacocke's earlier work on divine action, which regards the entire created universe as an interconnected system-of-systems, and adopts a panentheistic account of God's relationship to the world such that God is understood as immanent within the whole of creation, yet the world is seen as "contained" within the divine. Thus, God's action is to be understood on the analogy of whole-part influence.

The foregoing account of divine action lays necessary groundwork for an account of revelation: until we can postulate ways in which God can effect "instrumentally" particular events and patterns of events in the world, we cannot hope to understand how God's intentions and purposes might be known "symbolically." There are a variety of ways God is taken to be made known: general revelation through the order of nature; through the resources of religious traditions; through the "special revelations" that serve as the foundation of religious traditions; and in the religious experience of ordinary believers. While dualist anthropologies allowed for direct contact between God and the soul or spirit, Peacocke concludes that when the person is understood in an emergentist-monist way it is more consistent with what we know of God's relation to the rest of creation to suppose that God's communication is always mediated, even if only by affecting the neural networks that subserve human memories and other sorts of experience, including the feeling of God's presence. Thus, all of these forms of revelation can be understood as the result of God acting through the mediation of the human and natural worlds.

So the first essays in this section advocate an "emergentist-monist" account of the human person as being compatible with both neuroscience and theology, and distinguish their views from physicalist views. A word about terminology: "physicalism" is used in two ways in philosophy. One designates a materialist metaphysical theory or worldview—all entities are physical entities. This ontological claim is often combined with a claim about the explanatory supremacy of physics. The second use of "physicalism" is in philosophy of mind, and here it designates a variety of positions, all opposed to substance dualism, but differing in the status they ascribe to mental properties or events—all the way from eliminativism to a nonreductive physicalism that countenances top-down mental causation. The physicalism that Clayton rejects is primarily the physicalism associated with the explanatory primacy of physics. However, both Clayton and Peacocke choose "emergentist monism" over the "nonreductive physicalism" currently used in the philosophy of mind due to their recognition that many so-called "nonreductive" physicalists are in fact quite thoroughly reductionist.

However, **Ian Barbour,** in "Neuroscience, Artificial Intelligence, and Human Nature: Theological and Philosophical Reflections," develops a somewhat different position that is ontologically monist but "dipolar" rather than emergentist. In his

---

and social character of mind and personhood.

three-stranded argument, Barbour sets out to show that it is consistent with neuroscience, computer science, and a theological view of human nature to understand a person as a multilevel psychosomatic unity who is both a biological organism and a responsible self. He considers the themes of embodiment, emotions, the social self, and consciousness.

Barbour surveys biblical and theological accounts of the person that emphasize the integration of body and mind, reason and emotion, individual and social groups. He then cites work by neuroscientists that highlights these same features, including Arbib's action-oriented schema theory, LeDoux's work on emotions, and Brothers' work on the neural bases of social interaction. The ways in which computers fall short of human capacities provides additional insight into human nature: to approach the level of human functioning, computers require analogues to embodiment, learning and socialization, and emotion. The question of the possibility of consciousness in a computer is particularly problematic. Barbour shows that the concepts of *information, dynamic systems, hierarchical levels,* and *emergence* are valuable for integrating insights from neuroscience and AI research with that of theology in a theory of human nature.

Barbour argues that process philosophy provides a supportive metaphysical framework for understanding the concept of human nature that he has developed in this essay. Alfred North Whitehead's philosophy emphasizes processes or events rather than substances. These events are all of one kind (thus, monism) but are all dipolar—they have both an objective and a subjective phase. Thus, in attenuated form, experience can be attributed not only to humans and animals, but also to lower forms of life, and even to atoms. In its own way, process philosophy emphasizes the same themes that Barbour traced through theology, neuroscience, and AI research. So Barbour concludes that a dipolar monism based on process philosophy is supportive of a biblical view of the human as a multilevel unity, an embodied social self, and a responsible agent with capacities for reason and emotion.

In "The Soul and Neuroscience: Possibilities for Divine Action," **Stephen Happel** puts three notions into conversation with one another: Edmund Husserl's philosophical interpretation of inner time-consciousness; Thomas Aquinas's theological language of the soul; and contemporary neuroscientific analyses of human agency, memory, and bodily knowing.

Happel argues that medieval soul-language is not simply a devotional leftover from a discredited dualist substance philosophy. The concept of *soul* was a medieval attempt to explain the living experience of the cognitive, embodied subject. In his analysis of the role of the soul in human knowledge, Aquinas makes a variety of philosophical claims that are relevant to current research and discussion: First, human knowing is an active as well as a receptive process, dependent on the empirical world, yet critical in relationship to the world and to its own operations. Second, this knowing only takes place with the intimate cooperation of the individual's body. Third, intelligence is open-ended; it wonders and inquires about everything within its horizon. Fourth, this intelligence can reflect upon itself. Fifth, open-ended human intelligence can go beyond the senses, intending and estimating, even understanding the reality of God. Sixth, human intelligence rightly apprehends reality through its senses and makes correct judgments on the basis of the evidence provided.

Time-consciousness is central to Husserl's phenomenological description of human subjectivity. The *agency* of human consciousness is found in the retention, present awareness, and expectation that allow humans to be aware of temporally-extended objects of consciousness. There is a flow of interactions among memories, present consciousness, and future expectations that gives consciousness its unity. Happel shows that Husserl's notion of subjective time consciousness coheres with Aquinas's metaphysical vocabulary regarding intellectual powers: the world of interiority that Husserl examines turns from the consciousness of the subject to a self-reflexive knowledge of that subject; the unified body and soul, for Aquinas, becomes a self-conscious subject, examining itself introspectively.

Contemporary neuroscience examines time-memory, embodiment, and human initiative in the empirical subject, the knower who examines both self and the world through models and experiments. Reflecting on current theories of long-term and working memory, schema theory, somatic markers, and the hermeneutics of sense perception, Happel raises questions about human agency. He sees Husserl's analysis of time-consciousness as a possible hypothesis for experiment and verification in the neurosciences, and he challenges the neurosciences to think about mind and consciousness not only as initiators but as radically open to their constitution as a social reality.

The examination of human consciousness in three major disciplines—philosophy, theology, and neuroscience—has as its goal the criticism of modern individualistic (solipsistic, autonomous) concepts of the human subject. Happel reasons that if the subject can be conceived as open to finite transcendence (that is, to the reality of the other in and to the subject) this should shed light on how God operates through the interaction of finite subjects in our world to bring about divine ends.

In "Resurrection of the Very Embodied Soul?" **Ted Peters** argues that the Christian understanding of eternal salvation is not threatened by the rejection of substance dualism. In fact, the rejection of dualism by both the cognitive neurosciences and the Christian tradition represents an important area of consonance between theology and science—namely, that human reality is embodied selfhood. Peters notes that this issue deserves attention because some theorists, in both cognitive science and philosophy, claim two things: first, the findings of the neurosciences regarding the brain's influence on the mind demonstrate that the human soul cannot be thought to exist apart from a physical body and, second, that this physicalist interpretation so undermines the doctrine of the immortal soul that the Christian view of eternal salvation becomes counter-scientific.

Peters points out that until recently theologians have not been forced to clarify the distinction between two overlapping ways of conceiving personal salvation: One, rooted primarily in the ancient Hebrew understanding, pictures the human person as entirely physical, as dying completely, and then undergoing a divinely effected resurrection. The other, a later view influenced by Greek metaphysics, pictures the human person as a composite of body and soul; when the body dies the soul survives independently until reunited with a body at the final resurrection. In both pictures, however, the resurrection of the body is decisive for salvation. Now, however, to the extent that the dualistic vocabulary and conceptuality inherited by Christian theology from the Platonic tradition begins to look too much like Cartesian substance dualism, theology is in error.

In approaching the constructive question of how best to relate cognitive theory and theology, Peters first examines and rejects two "blind alleys": the notion of the "humanizing brain" developed by James Ashbrook and Carol Albright, and the artificial intelligence model of the human soul as disembodied information processing developed by Frank Tipler. In contrast to Tipler's view, Peters notes that belief in the resurrection, for Christian theology, does not depend on any natural process identifiable by science or philosophy, but on the witnessed resurrection of Jesus Christ at the first Easter. The Christian promise points toward an eschatological transformation—a new creation—to be wrought by God. Peters follows Wolfhart Pannenberg in connecting the resurrection to God's eschatological act wherein time is taken up into eternity, and wherein God provides for continuing personal identity even when our bodies disintegrate.

4.3.2 Neuroscience and Religious Experience

In "Cognitive Neuroscience and Religious Consciousness," **Fraser Watts** notes that when divine action is considered in relation to the physical sciences the rationality of Christian faith may be at stake, but when God's action is considered in relation to the cognitive neurosciences the credibility of daily religious life and practice may be at stake as well: How do humans relate to God as persons who are not mere minds, souls, spirits? Two major issues raised are the validity of revelation and the nature and possibility of religious experience.

There are both scientific and theological reasons for attending to the brain when attempting to understand religious experience. However, Watts resists the question of whether religious experience is caused by the brain or by God. Theological and neurological explanations are complementary; one is free to privilege the level of explanation that is most relevant in a particular context.

Watts considers two developments in attempts to understand the involvement of neural processes in religious experience. The first is based on claims that temporal lobe epilepsy (TLE) patients have more religious preoccupations than others; this has given rise to the further claim that religious experience should be linked with the neural basis of TLE. However, Watts disputes both the data and this interpretation. A second attempt to link religious experience and the cognitive neurosciences is that of Eugene d'Aquili and colleagues. Watts finds this research of more interest in that it involves a somewhat more sophisticated theory of religious experience and ties it to a theory of more general cognitive functioning—d'Aquili's theory of "cognitive operators."

Watts's own thesis is that a truly adequate cognitive theory of religious experience would benefit from attention to analogies between religious and emotional experience. The most valuable cognitive theories of emotion are multi-level, for example, distinguishing the sensory-motor aspects from the interpretation of the experience, and further distinguishing between intuitive perceptions of meaning and the ability to describe the experience propositionally. Watts speculates that this latter distinction, in particular, will shed light on the phenomena of religious life.

An attempt to understand the role of God in religious experience will be hampered, according to Watts, by too narrow a focus on "divine action." Any analogy with human "action" needs to be balanced with other metaphors that keep before our mind the fact that God's action is constant rather than episodic, interactive rather than controlling. He suggests the concept of "resonance" or "tuning" as an image for understanding the divine-human interaction. Conscience might then be understood in terms of resonance with the will of God.

In "A Neuropsychological-Semiotic Model of Religious Experiences," **Wesley J. Wildman** and **Leslie A. Brothers** observe that the neurosciences have largely succeeded, through their analyses of brain structure and function, in portraying that which is distinctively human as continuous with the laws and forms of complexity observed throughout the natural world. This generally accepted conclusion about human beings reconfigures the whole theory of religious experience by proposing explanations for them that are independent of the assumption that they are experiences of anything properly called a religious object. This reductionistic challenge is not different in philosophical terms from earlier challenges, but it does invite theories of religious experience that attend to the neurosciences.

As Fraser Watts points out in his essay, religious experience is notoriously difficult to define and delimit. Wildman and Brothers choose the term "experiences of ultimacy" both to focus their study on a subset of the broader category of religious experience, and also to avoid prejudicing their treatment in favor of theistic religions that focus on (putative) experiences of God.

The goal of this essay, then, is to present a richly textured interpretation of experiences of ultimacy. The authors develop this interpretation in two phases. First, they describe these experiences as objectively as possible, combining the descriptive precision of phenomenology, informed by the neurosciences, with a number of more obviously perspectival insights from psychology, sociology, theology, and ethics. Their hope is that the resulting taxonomy will be compelling enough to support constructive efforts in theology and philosophy that depend on an interpretation of religious experience—including those in this volume that attempt to speak of divine action in relation to human consciousness.[59]

The authors make two constructive ventures on the basis of this description. In the first, inspired by existing social processes used to identify authentic religious experiences, they describe a procedure whereby genuine experiences of ultimacy can be distinguished from mere claims to such experiences. They recognize a variety of "markers" that together point toward authenticity: subjects' descriptions (considered within their socio-linguistic contexts), phenomenological characteristics, judgments by experts in discernment or psychology, conformity with theological criteria, and ethical transformation. Judgments of this sort bring such experiences into the domain of public, scientific discussion as much as they can be, and the authors speculate that this will encourage more mainstream discussion of such experiences by scientists and others.

The second constructive venture is the authors' attempt to evaluate claims made concerning the cause and value of experiences of ultimacy. The modeling procedure they adopt makes use of semiotic theory to plot the "traces" of causal interactions in the form of sign transformations, though not the causal interactions themselves. In the language of semiotic theory, these causal traces take the form of "richly intense sign transformations." This proposal keeps ontological presuppositions to a minimum by focusing on causal traces rather than the causes themselves. Nevertheless, the authors contend, it does offer a religiously or spiritually positive way of interpreting authentic ultimacy experiences. At the end of the essay the authors offer a suggestion about the nature of the ultimate reality that might leave such causal traces.

4.4 Contrasting Reflections on the Theological Context

So far this introduction has largely emphasized points of convergence between neuroscience and theology in the views of human nature that they sponsor (the issue

---

[59] See especially the essays by Watts, Peacocke, and Ellis.

of reductionism being one important exception). This section, however, highlights that fact that however much agreement there may be, for example, on the embodied and social aspects of the human being, this does not entail agreement on the central issues of theology itself: God and God's relation to humankind. The two essays in this section take contrasting positions on the extent to which human experience requires a theistic explanation. This section thus extends Wildman and Brothers' inquiry concerning the cause of religious *experience* to a question about the cause of religious *belief* itself and to a consideration of the theological implications of human experience more broadly.

In the first part of "Crusoe's Brain: Of Solitude and Society," **Michael A. Arbib** develops a thesis regarding social influences on brain function and hence on brain structure. Social schema theory attempts to understand how social schemas, constituted by collective patterns of behavior in a society, provide an external reality from which a person acquires schemas "in the head." There is thus a top-down influence of social interaction on the microstructure of the brain through evolutionary processes, with brain action effectuated through perceptual and motor schemas. Conversely, it is the collective effect of behaviors that express schemas held by many individuals that constitutes and changes this social reality.

The learning of language provides an example of how individuals interiorize social schemas. Current research on mirror neurons (neurons that are active not only when an action is performed but also when the action is being perceived) provides a hypothesis that language specialization in humans derives from an ancient mechanism related to the observation and execution of motor acts. Arbib rejects Noam Chomsky's hypothesis that language learning depends on innate universal grammar. Instead, based on work with Jane Hill, he argues that language in children begins with repetition of words and phrases, shaped by the use of very rudimentary grammatical schemas that develop by means of ("neo-Piagetian") assimilation and adaptation. The richness of the metaphorical character of language can be interpreted in terms of schema theory: a word or phrase is an impoverished representation of some schema assemblage. Thus, extraction of meaning is a virtually endless dynamic process.

In the second part of Arbib's essay he applies social schema theory to a discussion of ideology and religion. Social schemas include those that we take to be representations of the world, but others that we do not, such as ideals of human life that are never realized and models that are false but useful. While schema theory has no implications for the question of the existence of God, it does offer new and useful vocabulary for discussing the projection theory of religion, found already in the writings of Ludwig Feuerbach and Sigmund Freud. An ideology can be viewed as a very large social schema. It is, like language, something that the child comes to as an external reality and internalizes to become a member of society. While it is central to schema theory to analyze the mechanisms whereby social construction and reality depiction are dynamically interlinked, it is important to note that many "realities" are socially defined rather than "physical." Thus, social schema theory provides a way of asking whether the "reality" of God is both external reality and social construction, or whether "God" is merely a social construct. Arbib suggests that the wide variation among religious beliefs argues for the latter conclusion. He offers this argument as an antidote to the "unabashedly Christian worldview of many other contributors to this volume, for whom the reality of divine action is taken as a given."

In "Intimations of Transcendence: Relations of the Mind to God," **George F.R. Ellis** explores a strongly theistic interpretation of religious experience. He aims to show the logical coherence of a particular "kenotic" theological position, as well as its consistency with current views within both physics and neuroscience. After outlining the position he takes on fundamental issues such as the role of models in science, the hierarchical structuring of science, and the relations between causal explanations from different levels, Ellis presents a summary of the theological-ethical position that he developed with Nancey Murphy.[60] "Kenosis" is a term from Christology that referred originally to Christ's "emptying" himself of divine attributes. Ellis and Murphy extend the meaning of the term, using it to describe God's loving self-sacrifice as revealed in the life and death of Jesus. On this view, kenosis is an overall key to the nature of creation because it is the nature of the Creator. A kenotic ethic of self-giving love reflects the ultimate nature and power of God, manifest most clearly in the resurrection of Jesus.

Evidence for the theological vision proposed here comes from a variety of human experiences, which Peter Berger has termed "intimations of transcendence." Ellis argues that while there may be evolutionary or functional explanations of moral behavior, human creativity, aesthetic appreciation, love and joy, in all of these cases there seems to be an excess. Humans, for example, sacrifice themselves not only for kin but for strangers, and human love goes beyond the bounds of the practical. However, none of these intimations is sufficient to yield a detailed account of the nature of the transcendent. Thus, Ellis asserts the need for a channel of revelation. A major goal of this essay is to argue that a view of divine action (revelation) through the mediation of the human brain is consistent with contemporary neuroscience. He speculates that the causal gap revealed by quantum theory allows for a "causal joint" whereby information may be made available to human consciousness without violation of energy conservation. However, Ellis's argument does not depend critically on the role of quantum phenomena in consciousness, but rather on the coherence and explanatory scope of the theological vision he proposes.

In summary, then, all the essays described here contribute to the view, set forth earlier in this introduction, that insights from the cognitive neurosciences, on the one hand, and Christian theology on the other, converge in a theory of human nature. Despite differing views on the scope of reduction in the human sciences, and despite differences in religious belief, the participants came to agreement on the vital importance for neuroscience, philosophy, and Christian theology of an account of the human person in which the embodiment, sociality, and rich emotionality of mind all play a crucial role.

---

[60] Nancey Murphy and George F.R. Ellis, *On the Moral Nature of the Universe* (Minneapolis: Fortress Press, 1996).

I. RESOURCES

Religious:

Joel B. Green
Fergus Kerr

Scientific:

Joseph E. LeDoux
Peter Hagoort
Marc Jeannerod
Leslie A. Brothers

RESTORING THE HUMAN PERSON:
NEW TESTAMENT VOICES FOR A
WHOLISTIC AND SOCIAL ANTHROPOLOGY

Joel B. Green

"Biblical anthropology," as an area of theological exploration, has since the mid-twentieth century focused either on the possibility of a body-soul dualism, especially in Pauline thought, or on a series of topics oriented around the problem of human sin and its remedies.[1] This is unfortunate for several reasons, the most important being that many of us have thus been blinded to one of the pivotal areas where Scripture speaks most clearly and prophetically about human nature. This is the biblical concern with the fundamentally social or relationally oriented nature of humanity, a concern that has been too easily eclipsed by modernist impulses toward an individualistic account of the human person.

It is also true that, on the whole, Scripture is unconcerned with speculative questions about the nature of humanity, and tends toward a narrative presentation of its anthropology which does not allow for easy division of its concerns into topics. Moreover, as recent work in pragmatics and sociolinguistics has demonstrated, much of what is communicated in any discourse situation is not actually verbalized, since acts of communication depend on shared pools of presuppositions; hence, what is textualized in Scripture is only the tip of the iceberg, as it were, of what is being communicated. It is important to understand how an utterance draws on and/or mitigates conventional understandings within its own discourse situation. These considerations indicate both why more work remains to be done with regard to biblical anthropology and the direction such work needs to take.

In this essay, I will argue that the anthropology of two New Testament writers, Paul and Luke, is fundamentally social and wholistic in its presentation. Since all New Testament writers (albeit to varying degrees) self-consciously fashion their theology in terms of continuity with and interpretation of Israel's Scriptures, this agenda will of necessity require an accounting of some aspects of the anthropology of Israel's Scriptures. In an attempt to take seriously the terms of the debate regarding body-soul dualism as this relates to New Testament texts and especially to Paul, I will also have recourse to occasional forays into methodological questions regarding how issues of anthropology have been framed and how the evidence has been

---

[1] On the question of body-soul dualism, see Rudolf Bultmann, *Theology of the New Testament*, 2 vols. (New York: Scribner, 1951/55), 1:192–203; J.K. Chamblin, "Psychology," in *Dictionary of Paul and His Letters*, G.F. Hawthorne, R.P. Martin, and D.G. Reid, eds. (Downers Grove, Ill,: InterVarsity, 1993), 765–75; John W. Cooper, *Body, Soul, and Life Everlasting: Biblical Anthropology and the Monism-Dualism Debate* (Grand Rapids, Mich.: Wm. B. Eerdmans, 1989); Robert H. Gundry, *Sōma in Biblical Theology with Emphasis on Pauline Anthropology*, Society of New Testament Studies Monograph 29 (Cambridge: Cambridge University Press, 1976); Robert Jewett, *Paul's Anthropological Terms: A Study of Their Use in Conflict Settings*, Arbeiten zur Geschichte des antiken Judentums und das Urchristentums 10 (Leiden: E.J. Brill, 1971); Peter Müller, *Der Soma-Begriff bei Paulus: Studien zum paulinischen Menschenbild und seine Bedeutung für unsere Zeit* (Stuttgart: Urachhaus, 1988); John A.T. Robinson, *The Body: A Study in Pauline Theology*, 2d ed., Studies in Biblical Theology (London: SCM, 1977). For a topical approach to biblical anthropology, see, Udo Schnelle, *Neutestamentliche Anthropologie: Jesus—Paulus—Johannes*, Biblisch theologische Studien 18 (Neukirchen-Vluyn: Neukirchener, 1991).

sifted. However, my primary concern is with Paul and Luke. I am interested in Paul because he is generally regarded as the most important theologian of the apostolic age. I focus on Luke, author of the Gospel of Luke and the Acts of the Apostles, because he is arguably the only Gentile author represented in the New Testament canon; this fact allows a *prima facie* case that, if any New Testament materials were to reflect the dualism alleged to characterize the Hellenistic world, it would be his.

1 Finding a Starting Point

1.1 The Potential Contribution of Linguistics

Until recently, one of the mainstays in the conversation about biblical anthropology has been the contribution of Hebrew and Greek lexicography.[2] Certain words, vested with particular meanings, have been said to point to certain conclusions regarding the make-up of the human person. As in statistics, however, so in linguistics, the same evidence base, in different hands, can sometimes lead to opposing results. This is the case in the discussion of the human person, in which Hebrew terms (such as *nepheš, bāśār, lēb,* and *rūah*) and Greek terms (such as *sōma, psychē, pneuma,* and *sarx*) are investigated for their meaning. Unfortunately for this debate, these words are each polysemous, and are capable of a range of translations into English. Thus, depending on context, *nepheš,* though often identified with the idea of a "soul," might be translated into English as "life," "person," "breath," "inner person," "self," "desire," or even "throat." *Bāśār* might be translated with the English terms "flesh," "body," "meat," "skin," "humankind," or "(the) animal (kingdom)." Translations of *lēb* might include "heart," "mind," "conscience," and "inner life." Finally, *rūah* might be taken as a reference to "wind," "breath," "seat of cognition and/or volition," "disposition," "spirit," or "point on a compass."

In Israel's Scriptures, the Hebrew term *nepheš* is used with reference to the whole person as the seat of desires and emotions, not to the "inner soul" as though this were something separate from one's being. *Nepheš* can be translated in many places as "person," or even by the personal pronoun (Leviticus 2:1; 4:2; 7:20). It denotes the entire human being, but can also be used with reference to animals (Genesis 1:12, 24; 2:7; 9:10). From time to time, the Hebrew term *bāśār* stands in parallel with but not in contrast to *nepheš*—the one referring to the external being of the person, the other to the internal (Isaiah 10:18). Indeed, although *bāśār* frequently refers to the fleshly aspect of a person (Psalm 119:73; Isaiah 45:11–12), this term is also prominent as an expression of the spiritual. *Bāśār* and *nepheš* "... are to be understood as different aspects of man's [*sic*] existence as a twofold unity."[3] The related term, *gewiyya,* refers to the human being in her wholeness, though usually in a weakened condition; typically, it is used to denote the body of a human only in its state as a corpse or cadaver.[4] The Scriptures of Israel employ other terms, too, when speaking of humans from the perspective of their varying

---

[2] For the recent appearance of such data, see, Cooper, *Body, Soul, and Life Everlasting,* 42–49; Paul K. Jewett with Marguerite Shuster, *Who We Are: Our Dignity as Humans—A Neo-Evangelical Theology* (Grand Rapids, Mich.: Eerdmans, 1996), 35–46.

[3] N.P. Bratsiotis, "בָּשָׂר," in *Theological Dictionary of the Old Testament,* vol. 2, G.J. Botterweck and H. Ringgren, eds. (Grand Rapids, Mich.: Eerdmans, 1975), 313–32 (326).

[4] See Heinz-Josef Fabry, "גְּוִיָּה," in *Theological Dictionary of the Old Testament,* 2:433–38 (esp. 435–36).

functions—for example, *lēb*, with reference to human existence, sometimes in its totality (Genesis 18:5; Ezekiel 13:22), sometimes with reference to the center of human affect (Proverbs 14:30) or perception (Proverbs 16:9);[5] and *rūah*, used with reference to the human from the perspective of his being imbued with life (Genesis 2:7; Job 12:10; Isaiah 42:5).

Similar polysemy is found among the relevant Greek terms: *sōma* is capable of translation into English as "body," "physical being," "church," "slave," and even "reality"; *psychē* as "inner self," "life," and "person"; *pneuma* as "spirit," "ghost," "inner self," "way of thinking," "wind," and "breath"; and *sarx* as "flesh," "body," "people," "human," "nation," "human nature," and, simply, "life."

Given this polysemy, we would be mistaken to assume that the word *psychē*, which someone might wish to translate as "soul," actually *means* "soul" or requires an identification with the concept of "soul" defined as the spiritual part of a human distinct from the physical or as an ontologically separate entity constitutive of the human person. Nor should we imagine that in any give utterance *psychē* refers to "inner life," "life," *and* "person"—or to even one of these possible referents. In the same way, we would not expect native speakers of English to confuse a "light blue" with a "blue mood" or a "light switch." In the end, studies of the human person oriented toward the semantics of biblical Hebrew or Greek are capable of only limited and primarily negative results. We can show, for example, that words like *nepheš* or *psychē* do not necessarily refer to ontologically separate (or separable) parts of the human person. On the other hand, neither can such study show that, in individual texts, the opposite is necessarily the case.[6]

Words take their meaning fundamentally not from etymology nor even from the dictionary, but from usage; what matters most, then, is not how a term originated or is defined in the lexicon, but how it is used in a given utterance, within a given discourse. The idea that one could simply pile up all of the references in Scripture to "body" or "soul," and from this deduce "the biblical understanding of the human person" is misguided on linguistic grounds. We must face the reality that neither the Old nor the New Testament writers developed a specialized or technical, denotative vocabulary for theoretical discussion of the human person. And if this is so, then contemporary interpreters ought to exercise care when reading the biblical materials in light of specialized language that has developed subsequently.

1.2 Made in the Image of God

The nature of humanity is not often a topic of speculative concern in Scripture, and it is at least partially for this reason that readers of the Bible continue to debate the question. If we take as our beginning point "the beginning," the creation material in Genesis 1–2, the debate is not thereby circumvented, but two affirmations are

---

[5] See H.-J. Fabry, "לֵב," in *Theological Dictionary of the Old Testament*, vol. 7, Botterweck, Ringgren, and Fabry, eds. (Grand Rapids, Mich.: Eerdmans, 1995), 399–437.

[6] This précis is enough to suggest the imprecision with which the anthropological vocabulary of the Scriptures of Israel is utilized. Some scholars further suggest that, although the Scriptures of Israel provide no particularly "scriptural" vocabulary for anthropological analysis, they do draw on the common terminology of the ancient near east in depicting the human person as an integrated whole. See, e.g., Fabry, "לֵב," 412–13; Brevard S. Childs, *Biblical Theology of the Old and New Testaments: Theological Reflection on the Christian Bible* (Minneapolis: Fortress, 1992), 566, 571–72; Eduard Schweizer, "Body," in *The Anchor Bible Dictionary*, 6 vols., D.N. Freedman, ed. (New York: Doubleday, 1992), 1:767–72 (esp. 768).

indisputable. First, what makes the human being uniquely human is *not* the possession of an entity separate from the body, such as a "soul" or "spirit." Second, according to the Genesis account, the singular vocation of our embodied existence is to reflect the image of God.

Some may take issue with this first affirmation. After all, does it not say in Genesis 2:7 that "the Lord God formed the human being of the dust of the ground, breathed into his nostrils the breath of life, and the human being became a living soul (*nepheš*)"?[7] Might not one argue reasonably from this text that the lifeless human being is nothing but a mere "body" in need of a "soul," and thus that it is this "soul" that is necessary for human existence? Discussion of this point might turn for some on the meaning of the Hebrew term employed in the narrative, *nepheš*—a possibility we have already rejected. We may go on to observe that, however one translates *nepheš* in this text, the same term is used only a few verses earlier with reference to "every beast of the earth," "every bird of the air," and "everything that creeps on the earth"—that is, to everything "in which there is life (*nepheš*)" (1:30). Clearly, then, the possession of *nepheš* is not a unique characteristic of the human person. What is more, unless one is ready to grant that animals have "souls" in the same way that humans are alleged to have, then we might better conclude that the Genesis account is referring to the divine gift of *life*: "the human being became a living person."

If the possession of a "soul"—that which separates human beings from the rest of creation—is not the distinguishing mark of the human person, what is? On this issue, the Genesis account is transparent. Unlike other members of creation, animate and inanimate, humanity is created by God "in his own image":

> Then God said, "Let us make humanity in our image, after our likeness. Let them have dominion over the fish of the sea, over the birds of the air, over the cattle, over the whole earth, and over every creeping thing that creeps upon the earth." So God created humanity in his own image, in the image of God he created them; male and female he created them. And God blessed them, and said to them, "Be fruitful and multiply, and fill the earth and subdue it. Have dominion over the fish of the sea, the birds of the air, and over every living thing that moves upon the earth." (Genesis 1:26–27)

Of all the creatures, only humanity is created after God's own likeness, in God's own image (*imago Dei*). Only to humanity does God speak directly. Humanity alone receives from God this divine vocation.

The relative succinctness of this text from Genesis cannot be equated with clarity, however, with the result that the *imago Dei* tradition has been the focus of diverse interpretations among Jews and Christians—ranging from some physical characteristic of humans (such as standing upright) to a way of knowing (especially the human capacity to know God), and so on. Taken within its immediate setting in Genesis 1, "the image of God" in which humanity is made transparently relates to the exercise of dominion over the earth on God's behalf. But this observation only begs the question, for we must then ascertain what it means to exercise dominion in this way—that is, in a way that reflects God's own style of interaction with his creatures. What is more, this way of putting the issue does not grapple with the profound word spoken over humanity and about humanity, that human beings in themselves (and not only in what they do) reflect the divine image.

What is this quality that distinguishes humanity? God's words affirm the creation of the human family in its relation to himself, as his counterpart, so that the nature of humanity derives from the human family's relatedness to God. The concept of the

---

[7] Unless otherwise noted, biblical translations are my own.

imago Dei, then, is fundamentally relational, and takes as its ground and focus the graciousness of God's own covenantal relations with humanity and the rest of creation. The distinguishing mark of *human* existence when compared with other creatures is thus the whole of human existence and not some "part" of the individual. Humanity is created uniquely in relationship to God and finds itself as a result of creation in covenant with God. Humanity is given the divine mandate to reflect God's own covenant love in relation with God, within the covenant community of all humanity, and with all that God has created.

Outside of Genesis, creation "in the image of God" otherwise plays little role in the Old Testament, though it is mentioned in Deuterocanonical materials (for example, Wisdom of Solomon 2:23–24; Sirach 17:1–13) and later texts from Hellenistic Judaism. In the New Testament, Paul's thought is closest to the interpretation of the *imago Dei* expressed in the Wisdom of Solomon, which uses the phrase with reference to the actual expression of the "image of God" in a human life rather than to human capacity or potential (as in Sirach). Paul develops the motif of Christ as the "image of God" (2 Corinthians 4:4; Colossians 1:15; cf. Philippians 2:6) and, as its corollary, the conformation of human beings into the "image of Christ" (Romans 8:29; 1 Corinthians 15:49; 2 Corinthians 3:18). Accordingly, "in Christ" believers have access to the ultimate purpose of God for humanity. Through his creative and reconciling activity and in his ethical comportment, Christ both reveals the nature of God and manifests truly the human vocation (cf. Luke 6:35–36).

Subsequent Christian theology has emphasized the triune nature of God—that is, the fundamental relatedness of God within the Godhead.[8] Father, Son, and Holy Spirit—these constitute the community of the Godhead, from which all revelation flows and on the basis of which all creation exists and has its meaning. It would be easy to imagine that the community of the Godhead is replicated in a tripartite understanding of the human person: body, soul, and spirit. This would be a mistake in category. The human person, understood as an individual, is not a reflection of the Godhead, as though the human person were complete in her- or himself. God's words at the creation of humankind are spoken not over a human person, but over the human family. In fact, according to the Genesis text, God (singular) created humanity (singular) as men and women (plural), as "them" (plural). We reflect the community of the Triune God not as individuals but as the human community, whose life is differentiated from and yet bound up with nature, and whose common life springs from and finds its end in God's embrace.

The affirmation of human beings as bearers of the divine image in Genesis, together with the interpretation of the *imago Dei* tradition at the hands of Paul, points unquestionably to the uniqueness of humanity in comparison to all other creatures. This tradition does not locate this singularity in the human possession of a "soul," however, but in the human capacity to relate to God as his partner in covenant, and to join in companionship within the human family and in relation to the whole cosmos in ways that reflect the covenant love of God. "Humanness," in this sense, is realized in and modeled by Jesus Christ.

---

[8] On the "social" nature of the Trinity and its implications, see John D. Zizioulas, *Being as Communion: Studies in Personhood and the Church* (Crestwood, New York: St. Vladimir's Seminary Press, 1993); Jürgen Moltmann, *The Trinity and the Kingdom: The Doctrine of God* (San Francisco: Harper & Row, 1981); Miroslav Volf, *After Our Likeness: The Church as the Image of the Trinity* (Grand Rapids, Mich.: Eerdmans, 1998).

2 Body-Soul Dualism in the First-Century Roman World

Someone reviewing the history of discussion regarding the contribution of the New Testament to our understanding of the human person would with justification imagine that the chief issue is whether the New Testament writers were influenced more by Greek or Hebrew thought. This presumed dichotomy, at least, has provided the terms of the debate for most concerned. Such a focus is wrongheaded on several accounts. First, an affirmation of the monism of Israel's Scriptures is not itself proof that "Hebraic thought" disallowed duality. Second, "Greek thought" cannot be reduced to a single viewpoint on our question. Third, in those cultural circles in which the New Testament documents were generated and preserved, "Judaism" itself was not monochromatic and, indeed, had intermingled with Hellenism for some three centuries. Even in Palestine, "Judaism" exists on a continuum, more or less Hellenized.[9] Fourth, we must therefore consider that Israel's Scriptures would have been mediated to early Christianity via Hellenistic Jewish (and Hellenistic) readings.

By way of constructing further the framework of influence in relationship to which New Testament writers worked to portray the nature of the human person, I will, first, briefly recall the diversity of Hellenistic views of the human person. Following this, I will summarize some of the ways in which the human person was understood within Hellenistic Judaism. Insofar as the New Testament materials are themselves representative of the diversity of Hellenistic Judaism, I will mention selected New Testament texts as well.

2.1 Hellenistic Views

By and large, the Greeks never took the path René Descartes would take—namely, juxtaposing corporeal and incorporeal as if this were the same thing as juxtaposing material and immaterial (or physical and spiritual). Although belief in a form of body-soul duality was widespread in philosophical circles, most philosophers regarded the soul as composed of "stuff." Aristotle, for example, considered the soul, the basis of animate life, as part of nature, so that psychology and physics ("nature") could not be segregated. For him, "soul" was not immaterial. Even if "soul" is not the same thing as body, neither is it "nonmatter"; it can still occupy "space." Even Plato thought that the soul was constructed from elements of the world, though he argued for a radical distinction between body and soul. Within Epicureanism, mind and spirit were understood to be corporeal because they act on the body, and all entities that act or are acted upon are bodies. Borrowing in part from Aristotle, Stoicism taught that everything that exists is corporeal; accordingly, only non-existent "somethings" (like imagined things) could be incorporeal.

Following the demise of the Platonic academy as an institution in the early first century B.C.E., neo-Platonism took many forms, especially as influenced by Stoicism. As Martin notes, "When we analyze the Platonism—or perhaps we should say the Platonisms—that were around [in the first century C.E.], we encounter self-styled Platonists whose ideas of body and soul look to us remarkably like the monisms of Aristotle and the Stoics."[10] When one moves from the work of these

---

[9] See Martin Hengel, *Judaism and Hellenism: Studies in Their Encounter in Palestine during the Early Hellenistic Period*, 2 vols. in 1 (Philadelphia: Fortress, 1974); idem, *The "Hellenization" of Judaea in the First Century after Christ* (Philadelphia: Trinity, 1989).

[10] Dale B. Martin, *The Corinthian Body* (New Haven: Yale Univ. Press, 1995), 12.

philosophers to the views of ancient medical writers (who were, themselves, philosophers of a sort), one finds a keen emphasis on the inseparability of the internal processes of the body ("psychology," in modern parlance) and its external aspects ("physiology"). This is not because of tendencies to think in terms of "psychosomatic conditions" (to use concepts that are quite anachronistic), but because any differentiation between inner and outer was fluid and permeable. Finally, we inquire, What happens after we die? Cicero summarizes the two primary, competing views: either body and soul are annihilated at death or the soul separates from the body.[11]

In short, although some may find it useful to speak of a body-soul duality in the Greco-Roman world as a lowest common denominator in educated circles, this hardly relates the whole story. The Hellenism on the horizon of early Christians and the New Testament writers cannot be reduced so easily to a common denominator on questions of body and soul. This means that one cannot solve the problem of the relationship between body and soul in earliest Christianity merely by referring to parallels of thought or cultural settings. Such parallels and settings are themselves too complex for such decisions, and the ingredients available to those early Christian writers were more diverse than is usually thought.[12]

2.2 Hellenistic Judaism

Shaping the cultural map of the NT world in a far-reaching way were the innovations and upheavals in the Near East that followed in the wake of the military successes of Alexander the Great (356–23 B.C.E.) in the last half of the fourth century B.C.E.. When Alexander died, he had opened up the near east to the migration of Greek language and cultural expressions. Closely boundaried, more parochial societies were gradually joined by increasing cultural intercourse symbolized above all by the spread of Attic Greek as the *lingua franca*. Hellenism thus refers to the spread and embrace of the Greek language, but also to the way of life, business, education, and ethos mediated through the spread and use of that language. Within Judaism, this led to the development of significant diversity, a corollary of the range of responses to the challenges of Hellenism.

One of the areas in which Hellenistic influence is notable is in the developing belief among the Jewish people, first, that humans have "souls" and, second, that these "souls" had a prior existence before taking up residence in material bodies. This way of thinking may have roots in the Scriptures of Israel (for example, Jeremiah 1:5), but it was under the influence of some strands of Greek philosophy that Jewish literature would exhibit this doctrine. In Wisdom of Solomon 8:19–20, we read, "As a child I was naturally gifted, and a good soul fell to my lot; or rather, being good, I entered an undefiled body" (NRSV). Written later, some think as early as the late first century C.E., *2 Enoch* notes that "all souls are prepared for eternity, before the composition of the earth" (23:5).[13] Evidence of this doctrine apparently

---

[11] Cicero *Tusculan Disputations*, 1.11.23–24.

[12] The ease with which decisions of this sort have been made in this century derive in part from our failure to perceive the depth of Descartes's innovations. The Cartesian view of humanity was understood to have embraced ancient ways of thinking, with the result that few seemed to notice when Plato (for example) was conscripted to support Cartesian categories. Other Hellenistic writers, as well as the New Testament materials themselves, have similarly been read through Cartesian lenses.

[13] Translation from the Slavonic from Francis I. Andersen, "2 (Slavonic Apocalypse of)

surfaces in the Gospel of John, where the disciples ask Jesus concerning a man born blind, "Rabbi, who sinned, this man or his parents, that he was born blind?" (9:2).

Given the sociocultural context of Judaism in the Second Temple period, we should not be surprised to find in New Testament texts such as Matthew 10:28: "Do not fear those who put the body to death but are unable to kill the soul; fear rather the one who is able to destroy both soul and body in Gehenna." This saying echoes the martyr-theology of such Hellenistic Jewish texts as Wisdom of Solomon 16:13–14; 2 Maccabees 6:30, in which it is maintained that persecutors have access to the body, but only God has power over the whole person. Must this saying of Jesus be read in a way that points to anthropological dualism? It is worth exploring whether such texts make use of a metaphorical rather than an ontological dualism, in which the "inner" and "outer" aspects of a human being are separated for the sake of mitigating the power of those who would persecute the faithful.

Indeed, when examining the relevance of the New Testament materials to the monism-dualism controversy, the distinction between ontological and rhetorical dualisms must always be kept in mind.[14] With respect to that biblical evidence, it is important methodologically to cultivate awareness of the possibility that readers of Scripture have been led to various forms of anthropological dualism by, first, overlooking the possibility that biblical authors have employed conceptual and/or rhetorical distinctions as heuristic devices for speaking of what is in fact indivisible. "Inner" and "outer," or "body and "soul"—linguistic pairs such as these may have metonymic or synecdochic function in particular contexts, just as the pair in Scripture, "flesh and blood," does not designate two different parts of the human being but rather "human agency" (as opposed to divine). Similarly, one's "dispositions" and "practices" are inseparably conjoined,[15] even if, for purposes of discussion and analysis, one may occupy attention more than the other. Failing to account for the use of literary devices such as these in Scripture, some readers of Scripture go on, secondly, to compound their interpretive error by attributing technical language to the lexemes in question, with the result that what may well have been indivisible (though conceptually or rhetorically differentiated) becomes within the eye of the interpreter divisible (even if functionally inseparable).

In the present case, Matthew 10:28, the parallel in Luke 12:4 opens a further possibility for making sense of Jesus' saying—namely, that he is saying no more than that those who are persecuted should take comfort in knowing that martyrdom is only the end of one's existence in this world, not the end of one's life. According to this reading, the Greek term *psychē* would refer not to "soul" but to "vitality."

Enoch: A New Translation and Introduction," in *The Old Testament Pseudepigrapha*, 2 vols., J.H. Charlesworth, ed. (New York: Doubleday, 1983), 1:91–213 (140). The dating of *2 Enoch* in the late first century C.E. is an editorial gloss (91); in his introduction, Andersen notes that any attempt to date the document is speculative.

[14] At some level of theological abstraction concerning human existence, the questions with which we are struggling here become less pressing. This is because the two positions that are most easily supported within Christian theology—i.e., forms of nonreductive physicalism and functional wholism—are capable of sponsoring similar notions of soteriology (wholistic, refusing talk of "saving souls" in favor of the restoration of human existence within the cosmos) and ethical comportment. We may grant that support for either position may be found in Scripture, and that Christians throughout the last two millennia have gravitated to both.

[15] Cf. P. Bourdieu, *Language and Symbolic Power* (Cambridge: Harvard U. Press, 1991).

Even less likely is the notion that Paul's benediction in 1 Thessalonians 5:23 provides testimony to a tripartite understanding of human nature (spirit, soul, and body). The parallelism of the two phrases,

> May the God of peace himself sanctify you completely, and
> may your spirit and soul and body be preserved in entirety, free from blame...

signifies that Paul uses these terms to expand on the idea of "completely"—that is, in order to emphasize the completeness of the sanctification for which he prays. This is not a list of "parts," then, but a reference to "your whole being."[16]

The Alexandrian Jew Philo provides a witness of an altogether different sort, however. A Hellenistic Jewish philosopher, Philo is one of our best known examples of a first-century C.E. neo-Platonist. In his writings one finds the preservation of a body-soul duality, though the "soul" was not for him an "immaterial substance" in the Cartesian sense. The first-century Jewish historian Josephus also held to the independent existence of the soul, destined for immortality. Philo's writings are well-preserved because of their interest among early Christians, as were those of Josephus, but this is not to say that their views are representative. To the contrary, most Jews during this period would have rejected this form of anthropological duality in favor of a more "integrated" anthropology.[17]

In short, the popular notion that one must contrast a "Hebraic view" of the human person with a "Greek view," and that the Old Testament manifests the former while the New Testament reflects the latter, is mistaken on at least three grounds: (1) there never was a singular "Greek view"; (2) Hellenistic Judaism in the first century C.E. was located on a continuum representing more-or-less Hellenism, more-or-less Judaism; and (3) New Testament writers are as likely to have been influenced by the conceptual patterns of the Scriptures of Israel as by Hellenistic dualists. Two consequences flow naturally from these data. First, one cannot begin one's exploration of the New Testament data with clear presumptions about what one will find. Second, New Testament writers operated in a world characterized by diversity of opinion on the nature of the human person, and had themselves to make choices in the midst of that diversity. What choices did those writers make?

3 Soteriology in Luke-Acts

Usually regarded as the only Gentile author in the New Testament, and also one of the more Hellenized, Luke is responsible for having written more than one-fourth of what would become the New Testament. Although he never deals in a direct way with the nature of the human person, the major theme of his writing is "salvation," and it is through this emphasis that we are able to gain insight into how he understands human nature. That is, Luke's soteriology of necessity raises questions about "what needs to be saved" and "what saved existence would look like," and these point toward his understanding of authentic human existence. In order to gain

---

[16] This is argued by numerous commentators—e.g., F.F. Bruce, *1 and 2 Thessalonians*, Word Biblical Commentary 45 (Waco, Texas: Word, 1982), 129–30; Charles A. Wanamaker, *Commentary on 1 and 2 Thessalonians*, New International Greek Testament Commentary (Grand Rapids, Mich: Eerdmans; Exeter: Paternoster, 1990), 206–7.

[17] N.T. Wright, *Christian Origins and the Question of God*, vol. 1: *The New Testament and the People of God* (Minneapolis: Fortress, 1992), 254–55.

our bearings, we will examine in sequence two Lukan passages—one concerned
with healing as salvation, the other depicting the nature of eschatological existence.[18]

3.1 Healing as Salvation (Luke 8:42b–48)

The larger unit in which this text is set begins with Luke 8:40 and continues through
v. 56, thus including Luke's narration of the raising of a dead girl. The most obvious
and important structural feature of this larger unit is the intercalation of the two
episodes: the narrative of the healing of the woman suffering from a hemorrhage
(8:42b–48) has been embedded into the narrative of the raising of Jairus's daughter
(8:40–42a, 49–56). The relationship between these two episodes transcends
concerns of structure and includes numerous commonalities.[19]

Through the technique of intercalation, the Evangelist presents the simultaneous
unfolding of these two narrative events. Moreover, the interruption of the one story
of healing by the other heightens the drama of the first. The little girl is dying; does
she not need immediate attention? Taken together, these two episodes document the
sort of faith for which Jesus has been looking. Moreover, the completion of the one
incident prepares for the finale of the other. After the abundance of healing power
available in the case of the woman with a hemorrhage, might we not anticipate
Jesus' ability to raise a dead girl to life?

The text we will consider reads as follows:

> As he went, the crowds pressed in on him. (43) Now there was a woman who had been
> suffering from hemorrhages for twelve years; and though she had spent all she had on
> physicians, no one could cure her. (44) She came up behind him and touched the fringe
> of his clothes, and immediately her hemorrhage stopped. (45) Then Jesus asked, "Who
> touched me?" When all denied it, Peter said, "Master, the crowds surround you and press
> in on you." (46) But Jesus said, "Someone touched me; for I noticed that power had gone
> out from me." (47) When the woman saw that she could not remain hidden, she came
> trembling; and falling down before him, she declared in the presence of all the people
> why she had touched him, and how she had been immediately healed. (48) He said to her,
> "Daughter, your faith has made you well; go in peace." (NRSV)

As Jesus makes his way to the home of Jairus, a woman appears, whose behavior
redirects Jesus' attention away from the needs of Jairus and his daughter. The
woman whom Luke introduces provides the Evangelist with yet another opportunity
to define "the poor" to whom the good news is brought (4:18–19; 7:22; 8:1). The
simple fact that she is a woman in Palestinian society already marks her as one of
relatively low status. In addition to this, she is sick, and her sickness, while
apparently not physically debilitating,[20] was socially devastating. Her hemorrhaging
rendered her ritually unclean,[21] so that she lived in a perpetual state of impurity.
Although her physical condition was not contagious, her ritual condition was, forcing
her to live in isolation from her community these twelve years. Her prospects for

---

[18] See, additionally, Joel B. Green, *The Theology of the Gospel of Luke*, New Testament
Theology (Cambridge: Cambridge Univ. Press, 1995); idem, *The Gospel of Luke*, New
International Commentary on the New Testament (Grand Rapids, Mich.: Eerdmans, 1997).

[19] These include falling before Jesus (vv. 41, 47), daughter (vv. 42, 48, 49), twelve years
(vv. 42, 43), desperate circumstances (vv. 42, 43, 49), the fact and immediacy of healing (vv.
44, 47, 55), touching (vv. 44, 45, 46, 47, 53), impurity (flow of blood, v. 43; corpse, vv. 53,
54), fear (vv. 45, 47, 50), and the inseparability of faith and salvation (vv. 48, 50).

[20] After all, she has suffered, presumably from uterine bleeding, for twelve years.

[21] See Lev 15:19–31; 11Q Temple 48:14–17; Mishnaic tractate *Niddah*.

renewed social intercourse had dropped to nil with her lack of help from the physicians. Whether her doctors had been the celebrated physicians whose exorbitant fees made them accessible only to the elite or the quacks that exploited members of a naive and needy public,[22] the outcome is the same. To her otherwise sorry condition is now added a further factor: her material impoverishment.[23]

Her degraded status *vis-à-vis* the larger crowd could hardly be more pronounced; the same, of course, could be said of her need, which has been depicted as indeed grave. Just as the Geresene demoniac had dwelled among the dead (Luke 8:27), so this woman exists outside the boundaries of the socially alive in her community. The press of the crowds guarantees that she will infect others with her impurity, and her aim to touch Jesus is a premeditated act that will pass her uncleanness on to him. What is it that motivates her to risk the rebuff of the crowds, of the synagogue ruler (who, we may presume, is walking with Jesus back to his home), and of Jesus on account of her social impropriety? This is the story of her resolution to cross the borders of legitimate behavior to gain access to divine power.[24]

The effect of touching Jesus' garment is immediate. Her bleeding stops, and so she experiences a reversal of her malady. As we shall see, however, though her physical problem may be cured, she is not yet healed.

In fact, the importance of this Lukan text for our purposes resides in its capacity to call into question two closely related tendencies in the twentieth-century West— namely, to think of "salvation" fundamentally in "spiritual terms" and, with respect to issues of healing and health, to think primarily in terms of bodies. Far from speaking of deliverance in soulish or spiritual terms, the Gospel of Luke, together with the other Synoptic Gospels, presents Jesus' salvific ministry in ways that foster wholistic thinking about soteriology. To step outside of the Gospel of Luke momentarily, we can reset our bearings by observing how the Gospel of Matthew portrays the healing of the sick in chapters 8–9. Miraculous events seem to be lined up, one after the other, with little connection between them or progression among them. Even if read in this way, what is said of Jesus' ministry is significant. Repeatedly, Matthew depicts him as one who makes available the presence and power of God's dominion to those dwelling on the periphery of Jewish society in Galilee—a leper, the slave of a Gentile army officer, an old woman, the demon-possessed, a paralytic, a collector of tolls, a young girl, and the blind. Chronicles of restoration to physical health such as those Matthew narrates also recount the restoration to status within one's family and community, the faith-full reordering of life around God, and the driving back of demonic forces. Thus, for example: cleansing a leper allows him new access to God and to the community of God's people (Matthew 8:1–4; see Leviticus 13–14); healing a paralytic is tantamount to forgiving his sins (Matthew 9:2–8); extending the grace of God to toll collectors and sinners illustrates the work of a physician (Matthew 9:9–13); and, as throughout the biblical tradition, recovery of sight serves too as a metaphor for the insight of faith (Matthew 9:27–31). Similar accounts abound in the Gospels of Mark and Luke, where physical needs are correlated with

---

[22] See Howard C. Kee, *Medicine, Miracle and Magic in New Testament Times*, Society of New Testament Studies Monograph 55 (Cambridge: Cambridge Univ. Press, 1986), 64.

[23] In addition, given her condition it is difficult to imagine that this woman is married. In this case, it is possible that we are to imagine that she had inherited money on which to live, money that is now in the hands of her unsuccessful physicians.

[24] See Gerd Theissen, *The Miracle Stories of the Early Christian Tradition*, Studies of the New Testament and Its World (Philadelphia: Fortress, 1983), 43–45, 74–80.

spiritual needs, and spiritual needs are bundled with social needs—or, better, where spiritual, social, and physical needs are simply regarded as human needs.

Nevertheless, because of the lens through which we tend to read the accounts in the Gospels, we typically fixate on the physical (or biomedical) aspect of healing and neglect other ways in which healing is defined and practiced. Robert Hahn, an epidemiologist at the U.S. Centers for Disease Control, proposes a three-part framework that accounts for inter- and cross-cultural variety in experiencing, conceptualizing, assessing, and treating sicknesses: *Disease Accounts*—which focus on the patient's body as the source of sickness, with disease located within the body, at or beneath the skin, and generally "below" the mind; *Illness Accounts*—which focus on the patient's body and social environment as the source of sickness and the place of its occurrence, with relief requiring attention to persons and their social worlds; and *Disorder Accounts*—which focus on the patient's body, social environment, and the cosmic order as the source of sickness and the place of its occurrence, with relief requiring intervention at all three levels.[25] If we take seriously this insight from ethnomedicine, we are in a more competent position to view the Gospel healing accounts in general, and Luke's account of the restoration of the hemorrhaging woman in particular, in ways that make sense of the whole of what is being narrated.

To return to Luke 8, the significance of the woman's action is highlighted by the fourfold appearance of the verb "to touch" in vv. 44–47. Its first importance is its ambiguity. Why did this unclean, disgraceful woman presume to touch one to whom even a synagogue ruler had bowed (v. 41)? Even when interpreted in the most obvious (and negative) way available within the narrative, the damage she has done is not irreversible; the law contained a remedy: rites of purification for Jesus, a reprimand for the woman. But Jesus does not adopt this reading; instead, he recognizes that her touch instigated a transfer of "power."

At this juncture the real test of this woman begins, for Jesus calls upon her to acknowledge her actions to the whole crowd. In fact, at this point Luke's account is largely concerned with her movement from seclusion—the isolation first of ritual impurity and now of denial—to public proclamation. The crowds press in, ready to choke faith as it sprouts; will she give in to her fear or respond in faith? At this point, the narrative is emphatic: all (including this woman) denied having touched Jesus.

Why does she hide; and why, when she realizes that hiding is futile, does she let herself be known in "trembling"? Remember that her touching Jesus was irregular and thus open to interpretation. How would the stifling crowds respond? How would the synagogue ruler? How would Jesus? Crossing the boundaries from the nonhuman world of socioreligious quarantine into the human world, and extending beyond the human world so as to access divine power is, on the one hand, a violation of the biblical purity code. On the other, it is an act of faith—or so it is interpreted by Jesus. For that faith to express itself fully, however, it must transverse the perimeters of the holiness code and overcome the strangle-hold of social banishment. In actuality, given her social position, her hiding and trembling are expected behaviors.

What is unprecedented and unanticipated is her touching Jesus in the first place, and now, even more so, her public announcement. Luke spares her no potential

---

[25] Robert A. Hahn, *Sickness and Healing: An Anthropological Perspective* (New Haven: Yale Univ. Press, 1995), 27–38. Critiques of the tendency to regard biomedicine as the universal, "true" account of human health and healing are numerous—cf., e.g., Byron J. Good, *Medicine, Rationality, and Experience: An Anthropological Perspective*, The Lewis Henry Morgan Lectures 1990 (Cambridge: Cambridge Univ. Press, 1994).

embarrassment. Her proclamation is before *all* of the people. Note, too, the content of her declaration. She is concerned with *why* she touched him; that is, she is presented as a hermeneut and not simply as one who chronicles what had happened.

Only now, in response to her public testimony, does Jesus commend the woman and pronounce that she is whole. Her cure was realized in the privacy and anonymity afforded by the crowds, yet her real problem was a public one. Hence, he has her make a public declaration of her actions and her understanding of what she had done. Then, he confirms her story and verifies her healing, ruling out all possible interpretations of her unconventional behavior save one—namely, that it was an expression of her faith. Jesus' actions are calculated to signal, first, that her faith, tested by the boundaries of ritual purity legitimated by community sanctions, is genuine. Its authenticity is manifest in her willingness to cross the barriers of acceptable behavior in order to obtain salvation. Second, he signals his unwillingness to leave her cured according to biomedical definitions only. He embraces her in the family of God by referring to her as "daughter," thus extending kinship to her and restoring her to the larger community—not on the basis of her ancestry (cf. 3:7–9), but as a consequence of her active faith. Now she is not the only one who knows what God has done for her; so do the crowds gathered around Jesus. Because he has pronounced her whole, they are to receive her as one restored to her community and to the people of God.

3.2 Eschatological Ambiguity (Luke 16:19–31)

Those theologians and biblical scholars who continue to champion a dualist account of human nature tend to do so for eschatological reasons.[26] As I will demonstrate below, speculation about the afterlife does not provide us with a very fruitful point of departure for coming to terms with the nature of humanity in the New Testament materials. In the interim, it is useful to examine a well-known Lukan text that has sometimes been deployed in support of an "intermediate state," between death and final judgment.

(19) There was a rich man who was dressed in purple and fine linen and who feasted sumptuously every day. (20) And at his gate was tossed[27] a poor man named Lazarus, covered with sores, (21) who longed to satisfy his hunger with what fell from the rich man's table; even the dogs would come and lick his sores. (22) The poor man died and was carried away by the angels to be with Abraham. The rich man also died and was buried. (23) In Hades, where he was being tormented, he looked up and saw Abraham far away with Lazarus by his side. (24) He called out, "Father Abraham, have mercy on me, and send Lazarus to dip the tip of his finger in water and cool my tongue; for I am in agony in these flames." (25) But Abraham said, "Child, remember that during your lifetime you received your good things, and Lazarus in like manner evil things; but now he is comforted here, and you are in agony. (26) Besides all this, between you and us a great chasm has been fixed, so that those who might want to pass from here to you cannot do so, and no one can cross from there to us." (27) He said, "Then, father, I beg you to send him to my father's house—(28) for I have five brothers—that he may bear witness to them,[28] so that they will not also come into this place of torment." (29) Abraham replied, "They have Moses and the prophets; they should listen to them." (30) He said, "No, father Abraham; but if someone goes to them from the dead, they will repent." (31) He said to him, "If they do not listen to Moses and the prophets, neither will they be convinced even if someone rises from the dead." (NRSV)

---

[26] See esp. Cooper, *Body, Soul, and Life Everlasting*.

[27] NRSV: "lay."

[28] NRSV: "warn them."

Within its narrative context, the parable of Luke 16:19–31 is told to Pharisees who have called into question Jesus' fidelity to the law. In this setting, the parable serves as a counterchallenge, indicating both his faithfulness before the law and Pharisaic duplicity. In one sense, the parable is concerned with wealth and its manifestations: a wealthy man engages in conspicuous consumption without regard for a poor man, in spite of the fact that this beggar who resides at his gate is quite literally his "neighbor" (vv. 19–21; cf. 10:29–37); and the rich and poor experience the eschatological reversal forecast in 6:20–24 (v. 25). In another sense, it is focused on the law (and, more broadly, the Scriptures) which, the parable informs us, is very much concerned with the state of the poor. In this case, a wealthy man realizes too late that he has ignored the words of Moses and the prophets concerning the poor.

Our concerns here are more focused, however, on the eschatological picture painted by the parable—especially what it might say about an "intermediate state" and "individual eschatology."

The stage of Jesus' parable is set by the extravagant parallelism of the two main characters. The social distance between the two is continued through to the end, symbolized first by the gate and then by the "distance" (v. 23) and the "great chasm" fixed between them (v. 26). The rich man is depicted in excessive, even outrageous terms, while Lazarus is numbered among society's "expendables," a man who had fallen prey to the ease with which, even in an advanced agrarian society, persons without secure land holdings might experience devastating downward mobility.

Jesus' comparison of these two characters in life continues in death. Both Lazarus and the wealthy man are apparently in Hades, though segregated from each other. Thus, while Lazarus is in a blissful state, numbered with Abraham, the wealthy man experiences Hades as torment and agony. This portrait has many analogues in contemporary Jewish literature, where Hades is represented as the universal destiny of all humans, sometimes with the expected judgment and separation of persons as wicked or righteous already mapped out.[29]

This parable is often taken as instruction on "the intermediate state,"[30] often with reference to the state of a disembodied soul; or as a manifestation of Luke's "individual eschatology."[31] Although this text probably assumes an intermediate state, (1) it does so largely in order to make use of the common motif of the "messenger to the living from the dead," only to deny the sending of a messenger;[32] (2) "disembodied soul" must be read into the story since the characters in Hades act as human agents with a corporeal existence (note, for example, the rich man's request for water to drink); (3) *Testament of Abraham* 20:14—where Abraham's bosom

---

[29] See *1 Enoch* 22; 4 Ezra 7:74–101; cf. the helpful summary in Richard J. Bauckham, "Hades, Hell," in *Anchor Bible Dictionary*, 3:14–15.

[30] Cf. Cooper, *Body, Soul, and Life Everlasting*, 136–39. A.J. Mattill Jr., *Luke and the Last Things: A Perspective for the Understanding of Lukan Thought* (Dillsboro, N.C.: Western North Carolina Press, 1979), 26–32; he notes that many readers of Luke have found evidence here for "the platonizing of Luke-Acts," but argues that this is not the case.

[31] See Jacques Dupont, "Die individuelle Eschatologie im Lukasevangelium und in der Apostelgeschichte," in *Orientierung an Jesus: Zur Theologie der Synoptiker*, Paul Hoffman, ed. (Freiburg: Herder, 1973), 37–47. Cf., however, John T. Carroll, *Response to the End of History: Eschatology and Situation in Luke-Acts*, Society of Biblical Literature Dissertation Series 92 (Atlanta, Ga.: Scholars Press, 1988), 64–68.

[32] See Richard J. Bauckham, "The Rich Man and Lazarus: The Parable and the Parallels," *New Testament Studies* 37 (1991): 225–46 (esp. 236–44).

and his descendants are already in paradise, yet Abraham is to be taken to paradise—bears witness to imprecision in statements about the afterlife; and (4) neither Luke nor other Christian writers (like Paul) seem to think that discussion of an individual's fate negates a more thoroughgoing apocalyptic (corporate, future) eschatology.[33]

3.3 Portraits of Human Nature in Luke-Acts

Luke-Acts provides no direct testimony regarding the Lukan anthropology, but everywhere, not least in its soteriology, assumes an understanding of the human person. If we were able to pursue our interpretive agenda in detail throughout Luke-Acts, we would see that the indications that have already surfaced would be underscored again and again. Luke regularly depicts human beings with a need, often physical, that turns out to be set within a much more complex interrelationship with what it means to be human; repeatedly, then, Luke disallows, say, the physical to be distinguished from other aspects of human existence, or the human person to be reduced to his or her "body." As the preceding discussion portends, Luke has no concept of a disembodied soul, either in present or in eschatological existence, and yet apparently has no difficulty visualizing an "intermediate state." This undercuts the argument of some that belief in an intermediate state requires the divine provision of a "soul" that survives the decaying body. More central to Luke's thinking is the inseparability of humans in their embodied and communal existence, with the result that his soteriology is oriented radically around the restoration of old or provision of new relations in the community of God's people.[34]

4 Pauline Anthropology

The subject of Pauline anthropology is of course much too complex to be treated in the space available here. Given our current concerns, however, two issues need to be explored. The first has to do with Paul's understanding of the resurrection (since this has often been taken as evidence for an anthropological dualism in Paul). The second has to do with the use of the metaphor "body" in Paul, which urges from yet another perspective the essentially social nature of the human person.

4.1 Paul and the Resurrection

As we have observed, many of those theologians and biblical scholars who continue to champion a dualist account of human nature tend to do so for eschatological reasons. Given the finality of death for the physical body, without recourse to a separate entity or personal "essence"—that is, a soul—that survives death, how can Christians maintain a reasonable doctrine of the afterlife? Without denying the critical nature of this issue, we must also take into account the necessarily speculative and analogical nature of all language, scriptural and otherwise, available to us for dealing with life beyond the grave. We have no existential or phenomenological access to such realities, and thus no way to lexicalize with precision the nature or anthropology of that experience. It is also true that modern attempts to ground ancient anthropology in eschatology tend to oversimplify the diversity of

---

[33] This last point is made by Carroll, *End of History*, 64–68.

[34] See further, Green, *Theology of the Gospel of Luke*; idem, "'Salvation to the End of the Earth' (Acts 13:47): God as Saviour in the Acts of the Apostles," in *Witness to the Gospel: The Theology of Acts*, I.H.M. & D. Peterson, eds. (Grand Rapids: Eerdmans, 1998), 83–106.

ways in which Hellenistic Judaism grappled with notions of the afterlife, and to assume too easily that "what happens when we die" formed a significant existential question for persons in the Greco-Roman world. As Neil Gillman observes, the idea that "the human body is mortal, but that every human possesses a soul which separates from the body at death and enjoys a continued existence with God" is absent from the Israel's Scriptures, but is found in some Jewish texts of the Second Temple period.[35] In fact, Judaism in the Second Temple period is quite diverse on this issue. Some texts speak of the soul's immortality, some of resurrection, while still others fail to speak of an afterlife or simply reject the idea of one.[36] Speculation about the afterlife provides no solid ground on which to formulate anthropology.

At the same time, those who hold to a dualist position are right to urge that monists spell out the implications of their physicalism with reference to eschatology. John Polkinghorne has attempted such a narrative. He writes, "The Christian hope... is for me not the hope of *survival* of death, the persistence *post mortem* of a spiritual component which possesses, or has been granted, an intrinsic immortality. Rather the Christian hope is of death and *resurrection*." He continues, urging that it is intrinsic to true humanity "...that we should be embodied. We are not apprentice angels, awaiting to be disencumbered of our fleshly habitation. Our hope is of the resurrection of the *body*"—not the resuscitation of a corpse.

> We are such timebound creatures that our minds easily balk at thinking how time and 'time' [that is, eschatological time] might relate to each other, but here is one of the few points at which the mathematically minded may have a theological advantage, for they are routinely able to conceive how such relationships might be expressed. I do not want to labor the point, but the imagery of vector spaces orthogonal to each other (the old and new creations) but connected by mapping or projections (resurrection, redemptive transformation) or even intersections (resurrection appearances) affords a conceptual framework within which such notions could be contained.[37]

Similarly, noted New Testament scholar F.F. Bruce has observed, "The tension created by the postulated interval between death and resurrection might be relieved today if it were suggested that in the consciousness of the departed believer there is no interval between dissolution and investiture, however long an interval might be measured by the calendar of earth-bound human history."[38]

Where does Paul stand on such issues? The apostle writes in 2 Corinthians 5:1–3, "For we know that if our earthly tent is dismantled, we have a house from God—a dwelling not made with human hands, eternal in the heavens. In view of this we sigh, longing to put on our heavenly house, assuming, of course, that when we take it off we will not be found naked." These words have often been taken to refer to a state of disembodiment between death and the general resurrection. Can such a reading

---

[35] Neill Gillman, *The Death of Death: Resurrection and Immortality in Jewish Thought* (Woodstock, Vt.: Jewish Lights, 1997), 83–112 (83).

[36] See George W.E. Nicklesburg Jr., *Resurrection, Immortality, and Eternal Life in Intertestamental Judaism*, Harvard Theological Studies 26 (Cambridge: Harvard Univ. Press, 1972); E.P. Sanders, *Judaism: Practices and Beliefs (63 BCE—66CE)* (London: SCM, 1992), 298–303. On Greco-Roman views, see Ramsey MacMullen, *Paganism in the Roman Empire* (New Haven: Yale Univ. Press, 1981), 53–57; Martin, *Corinthian Body*, 108–17.

[37] John Polkinghorne, *The Faith of a Physicist: Reflections of a Bottom-Up Thinker* (Princeton, N.J.: Princeton Univ. Press, 1994), 163, 164, 173.

[38] F.F. Bruce, *Paul: Apostle of the Heart Set Free* (Grand Rapids, Mich.: Eerdmans, 1977), 312 n.40.

be sustained? It is worth noticing at the outset that even if some Greeks looked forward to death, for it was followed by the flight of the soul into the desirable goal of immortality, this is not Paul's understanding here. Even if we imagine that Paul is thinking in terms of a bodiless, interim period, then, he looks upon the possibility of this "nakedness" (*gymnos*) with abhorrence; within wider Judaism, after all, to be found "naked" was to suffer humiliation, to lose one's identity as a human.

To press further, Paul's opening comment in 2 Corinthians 5:1, "For we know that...," urges us to recognize that he is not charting new territory here, but rather calling to mind former instruction. Even if we might allow for the possibility that, in 2 Corinthians 5:1–10, Paul wants to clarify his earlier guidance in these matters, we should nonetheless anticipate that the message of these verses will not substantively depart from his message in 1 Corinthians 15.

In that earlier letter, Paul was concerned with divisions within the Corinthian community—divisions that were both social and philosophical. Those of relative wealth would have had access to itinerant philosophers; following customary practice in the Roman world, persons of high status in Corinth might have extended hospitality to such persons and thus have been exposed to more sophisticated notions about the afterlife. For them, Paul's talk of "the waking of the dead" would have been reminiscent of fables about the resuscitation of corpses, the stuff of popular myths. Taught to degrade the body, they would have found Paul's teaching about the resurrection incomprehensible, even ridiculous. Those of low status, however, would have been incapable of welcoming itinerant philosophers into their homes and, thus, would have lived apart from their influence. They would have had closer contact with superstitions and popular myths, including those relating the resuscitation of corpses and the endowment of those corpses with immortality. Since Paul's primary objective in 1 Corinthians is to restore unity (1:10), Paul's challenge is to represent the resurrection belief of early Christianity with enough sophistication to communicate effectively with those of high status while not alienating those of lower status.

It is in 1 Corinthians 15:38–58 that Paul discusses the nature of the resurrection, and in doing so he affirms the following: (1) There is a profound continuity between present life in this world and life everlasting with God. For human beings, this continuity has to do with bodily existence. That is, Paul cannot think in terms of a free-floating soul separate from a body. (2) Present human existence, however, is marked by frailty, deterioration, weakness, and is therefore unsuited for eternal life. Therefore, in order for Christian believers to share in eternal life, their bodies must be transformed. Paul does not here think of "immortality of the soul." Neither does he proclaim a resuscitation of dead bodies that might serve as receptacles for souls that had escaped the body in death. Instead, he sets before his audience the promise of the transformation of their bodies into glorified bodies (cf. Philippians 3:21). (3) Paul's ideas are, in part, rooted in images from the natural world and, in part, related to the resurrection of Jesus Christ. As it was with Christ's body, Paul insists, so it will be with ours: the same, yet not the same; transformed for the new conditions of life with God forever. (4) For Paul, this has important meaning for the nature of Christian life in the present. For example, this message underscores the significance of life in this world—a fact that many Christians at Corinth had not taken seriously. We should not imagine that our bodies are unimportant, then, or that what we do to our bodies or with our bodies is somehow unrelated to eternal life (cf. Colossians 1:24). The idea of eternal life is not "escapism." Rather, it provides the Christian both with hope and with a vision of what is important to God; as a result, we may look forward to the future while also allowing this vision of the future to help determine the nature of our lives in the present.

Returning to 2 Corinthians 5, we may go on to inquire into the meaning of Paul's language: "earthly tent...dismantled," "a house from God," and "longing to put on our heavenly house" (vv. 1–2). The metaphor of "tent" is deeply rooted in Israel's past, where the tabernacle does give way to a temple, which was itself destroyed and replaced; indeed, for Paul's audience, the temple had been replaced yet again, this time by the Christian community itself, a temple "not made with human hands." This language is then correlated with another metaphor—"putting on" or "being clothed," which refers to baptism (cf. Romans 13:11–14; Galatians 3:23–29; Colossions 3:9–10). Paul's language is dualistic, then, but not in an anthropological sense. He thinks of an eschatological dualism, contrasting the now and the not-yet: having put on Christ in baptism, we now yearn for a life that conforms to his, one in which the church is authentically and wholly the dwelling place of God. In this case, our hope that we not be found naked refers to the time of the final judgment (see v. 10), when we will experience the consummation of our new life in Christ rather than the "exposure" that comes in condemnation. As A.E. Harvey paraphrases, "But we are confident, because death cannot be any kind of disgrace or disqualification; indeed, death is the transition to the condition of solidarity with Christ in the heavenly church-dwelling which is what we are actually yearning and working for all the time..."[39]

Although often read against the backdrop of body-soul dualism, and thus taken as further support for body-soul dualism, 2 Corinthians 5:1–10 actually points in an altogether different direction. The dualism with which it is concerned is eschatological rather than anthropological. When read in relation to Paul's former teaching in 1 Corinthians 15 (to which the apostle himself draws attention), this text is seen to be fully consistent with a monistic account of the human person.

4.2 Paul and the "Body"

As I have already intimated, probably the most significant contribution the New Testament (and the Bible as a whole) has to make within the current discussion on the nature of the human person is grounded in its unrelenting witness to the necessary relatedness of humans with God on the one hand, and with both human and nonhuman creatures on the other. One of the problems in the current debate on the nature of the human person is its focus on the human being as an individual entity. As we have seen, however, human "health," according to the witness of the Gospels, is profoundly communal. Moreover, it must not be overlooked that Paul employs the term "body" (sōma) for the community of believers. Paul's own eschatological perspective, furthermore, is that the fate of human beings is intimately tied to that of the cosmos (see Romans 8).

Although Paul is capable of employing the term sōma in more personal ways, in reference to individual persons, far and away the majority of his uses of the term are analogical, used in reference to the Christian congregation, a "body" of believers. Even if this is not evidence of Pauline genius, his choice of this language is of anthropological interest as it invites reflection on significant areas of congregational concern. Margaret Mitchell has demonstrated the degree to which comparisons

---

[39] A.E. Harvey, *Renewal through Suffering: A Study of 2 Corinthians*, Studies of the New Testament and Its World (Edinburgh: T. & T. Clark, 1996), 66–69 (69).

between the human body and human societies were a commonplace in deliberative rhetoric,[40] especially as a form of social control. The analogy of the "body politic" was a useful means of explaining and sanctioning unity in the presence of diversity and discord, and for legitimating status inequality in a given community. In Paul's letters, especially in Romans and 1 Corinthians, this common metaphor surfaces to promote unity, as one would expect, but it does so in uncommon ways. First, with the metaphor of the body, Paul does not abandon diversity, but actually demonstrates how diversity can and ought to coexist with unity: "Indeed, the body does not consist of one member but of many" (1 Corinthians 12:14). Second, rather than appealing to the common-sense argument that, just as some parts of the body are more important than others, so some members of the body politic rightfully possess greater power and privilege, Paul uses the image of the body actually to demolish hierarchical patterns of behavior: "On the contrary, the members of the body that seem to be weaker are indispensable, and those members of the body that we consider less honorable we clothe with greater honor..." (1 Corinthians 12:22–23).

In 1 Corinthians 11:17–34 (cf. 10:16–17), an additional and foundational affirmation is made, for here Paul develops his notion that the community of believers is not a "body" on account of shared geography or a common political system. Building one analogy on the other—from the "body of Jesus" who died "for you," to the eucharistic "body" of Christ," to the community of believers as the "body of the Lord"—he constructs what is distinctive about this "body"—namely, that it is "of Christ." Its unity is grounded in the covenant sacrifice of Jesus and in the common experience of the indwelling Spirit, and expressed in the reflection of Christ's own character in the life of the congregation.

Hence, what happens to the one happens to the many; the fate of one person is tied to the fate of the whole. Body imagery thus pervades 1 Corinthians—from concern over sexual relations in and outside of marriage, to eating and drinking, to presentation of the body with regard to head coverings and hairstyles, to the nature of the resurrection body[41]—with Paul's basic hypothesis transparent. In Christ, human beings are "members one of another" (Romans 12:5). That Paul makes such a claim immediately following his discussion of Israel and the people of God in Romans 9–11 indicates his concern to root Christian identity in the metaphor of the body of Christ as an expression of the eschatological people of God.

4.3 Humanity in Paul

Paul's anthropology is far more textured than I have been able to suggest in these few comments. In an important sense, though, we have come full circle. Even if Paul does not here draw on the tradition of the "divine image," he is very much concerned with what we might call the "new peoplehood" of those who are in Christ, and he develops his understanding with reference to a renewed sense of essentially corporate identity. Enough has been presented to indicate the significant degree to which Paul's conception of the human being is of a fully embodied person whose redeemed existence is social.

---

[40] Margaret Mary Mitchell, *Paul and the Rhetoric of Reconciliation: An Exegetical Investigation of the Language and Composition of 1 Corinthians* (Louisville, Ky: Westminster/John Knox, 1992), esp. 157–64.

[41] See Martin, *Corinthian Body;* Jerome H. Neyrey, *Paul, in Other Words: A Cultural Reading of His Letters* (Louisville, Ky: Westminster/John Knox, 1990), 102–46.

5 Conclusion

How human beings construct "the self" is a longstanding concern. Given the strength of Cartesian categories and the experience of many in the West since the Enlightenment, it is perhaps not surprising to see the degree to which humanity has come to be understood "one person at a time," so to speak. This is true of current debates on whether dualistic or monistic accounts of human nature are most adequate, as well as of principled or functional affirmations of what Robert Bellah refers to as the "ontological individualism" characteristic of mainstream American society. This is the pattern of dispositions and practices that defines life's ultimate goals in terms of personal choice, freedom in terms of being left alone by others to believe and act as one wishes, and justice as a matter of equal opportunity for individuals to pursue happiness as each person has defined it for her- or himself.[42] It is also seen in our tendencies to narrate the successes and failures of our lives in predominately biographical or autobiographical terms: What decisions did she make that resulted in her current homeless status? What inner courage he must have shown to escape his desperate situation! We think in the first place that the problem (and the solution) is probably with the individual.[43] Many Protestants have learned to frame the evangelistic task as "winning souls." It is easy today to articulate the life of faith as a preeminently personal affair. However, those who have ears to hear may discern the confluence of numerous voices calling for a construction of the human self along radically different lines. One of those voices is clearly the voice of Scripture.

Though I have not sketched the history of Christian thought on the matter of the "soul," had I done so we would have seen the central place body-soul dualism has had in Christian thought from the patristic period on. Body-soul dualism has never been the *only* position, however, and a rehearsal of the biblical evidence suggests the degree to which body-soul dualism has achieved a prominence far out of proportion to the scriptural basis on which it was alleged to have been built. Christians who today embrace a monistic account of humanity may do so as persons assured that this position actually places them more centrally within the biblical material than has usually been granted over the past two millennia. At the same time, biblical faith would naturally resist any suggestion that our humanity can be reduced to our physicality. And biblical faith challenges those, past and present, who insist that the human person can ever be understood "one person at a time." If we would articulate an account of the human person that takes with utmost seriousness the biblical record, we would have far less conversation about the existence or importance of "souls" and far more about the human capacity and vocation for community with God, with the human family, and in relation to the cosmos.[44]

---

[42] Robert N. Bellah, *et al.*, *Habits of the Heart: Individualism and Commitment in American Life* (Berkeley, Calif.: Univ. of California Press, 1985).

[43] Robert N. Bellah *et al.*, *The Good Society* (New York: Alfred A. Knopf, 1991), 11.

[44] This essay is an adaption and expansion of earlier publications: Joel B. Green, "'Bodies—That Is, Human Lives': A Re-examination of Human Nature in the Bible," in *Whatever Happened to the Soul? Scientific and Theological Portraits of Human Nature*, W.S. Brown, N. Murphy, and H.N. Malony, eds. (Minneapolis: Fortress, 1998), 149–73; idem, "Scripture and the Human Person: Further Reflections," *Science and Christian Belief* 11 (January, 1999): 51–63; idem, "Monism and the Nature of Humans in Scripture," *Christian Scholar's Review* (forthcoming).

THE MODERN PHILOSOPHY OF SELF IN RECENT THEOLOGY

Fergus Kerr

It is very *remarkable* that we should be inclined to think of civilization—houses, streets, cars, etc.—as distancing man from his source, from what is sublime, infinite and so on. Our civilized environment, along with the trees and plants in it, then seems as though it were cheaply wrapped in cellophane and isolated from everything great, from God, as it were. That is a remarkable picture that forces itself on us.[1]

1 The Turn to the Subject

According to some theologians, the metaphysical load that Christian practice and discourse carry needs little exploration. They would say, for example, that confessing the doctrine of the Trinity or the resurrection of Christ is much more important than worrying about the consequences of the ancient controversy between realism and idealism. They would say, even after thinking about it, that the epistemological bias of the age need not interfere with biblical exegesis or systematic theology.

To disabuse them would be a difficult task. I would say only that, if theologians proceed in the belief that they need neither examine nor even acknowledge their inherited metaphysical commitments, they will simply remain prisoners of whatever philosophical school was in the ascendant thirty years earlier, when they were first-year students; or, more likely, 350 years earlier, which takes us neatly to René Descartes (1596–1650) and his famous turn to the first-person perspective: "I think, therefore I am."

The modern conception of the self sprang from explicitly theological concerns. The *Meditations,* which Descartes published in 1641, bore the subtitle: "in which are demonstrated the existence of God and the immortality of the soul." This work, dedicated to the Paris theology faculty, started the turn to the self. In the First Meditation, trying to get at what must be regarded as absolutely certain and indubitable, Descartes articulates the thought that one may be completely deceived about everything:

> I will suppose therefore that not God, who is supremely good and the source of truth, but rather some malicious demon of the utmost power and cunning has employed all his energies in order to deceive me. I shall think that the sky, the air, the earth, colors, shapes, sounds and all external things are merely the delusions of dreams which he has devised to ensnare my judgment. I shall consider myself as not having hands or eyes, or flesh, or blood or senses, but as falsely believing that I have all these things.[2]

In the Second Meditation the famous conclusion follows:

> In that case I too undoubtedly exist, if he is deceiving me; and let him deceive me as much as he can, he will never bring it about that I am nothing so long as I think that I am

---

[1] Ludwig Wittgenstein, *Culture and Value,* G.H. von Wright, ed., Peter Winch, trans. (Oxford: Blackwell, 1980), 50.

[2] René Descartes, *The Philosophical Writings*, vol. 2, J. Cottingham, R. Stoothoff, and D. Murdoch, trans. (Cambridge: Cambridge University Press, 1985), 15.

something. So after considering everything very thoroughly, I must finally conclude that this proposition, *I am, I exist*, is necessarily true whenever it is put forward by me or conceived in my mind.[3]

Descartes then seeks an understanding of what this "I" is, the existence of which he has shown. He finds that he has to break with his inherited ideas—for example, "that I was nourished, that I moved about, and that I engaged in sense-perception and thinking; and these actions I attributed to the soul." Instead of that account of the soul, which would have made good sense to Thomas Aquinas or Aristotle, and is very much what Wittgenstein retrieves, Descartes systematically thinks away every attribute of the soul until he reaches bedrock:

> At last I have discovered it—thought; this alone is inseparable from me... I am, then, in the strict sense only a thing that thinks; that is, I am a mind, or intelligence, or intellect, or reason... a thinking thing.[4]

I can make this thought-experiment, thinks Descartes: I can "peel off" everything, my previous beliefs, my senses, my body, my confidence even that the external world really exists, and I shall find, in the end, that I am *essentially* a thinking thing.

As he says, he no longer thinks of himself as a man or even as a rational animal; he has redefined what it is to be human in terms of *consciousness,* and his perspective is completely *egocentric.* Thus the Cartesian "I," as a thing that thinks, comes into the philosophical tradition.

Theology and spirituality, certainly in the Roman Catholic tradition, has been permeated with Cartesian assumptions about the self. The key figure, in transmitting the paradigm, is no doubt Malebranche (1638–1715). It is Kant (1724–1804), however, in his heroic efforts to reconcile the rationalism of the Enlightenment with his Lutheran inheritance, who has produced the most influential variation on the Cartesian paradigm—memorably described by Iris Murdoch:

> How recognizable, how familiar to us, is the man so beautifully portrayed in the *Grundlegung,* who confronted even with Christ turns away to consider the judgment of his own conscience and to hear the voice of his own reason. Stripped of the exiguous metaphysical background which Kant was prepared to allow him, this man is with us still, free, independent, lonely, powerful, rational, responsible, brave, the hero of so many novels and books of moral philosophy.[5]

As the remainder of this essay will show, the picture of the self-conscious and self-reliant, self-transparent and all-responsible individual which Descartes and Kant between them imposed upon modern philosophy may easily be identified, in various guises, in the work of many modern theologians. It is a picture of the self that many modern philosophers, Wittgenstein certainly among them, have striven to revise, incorporate into a larger design, or simply obliterate.

2 The Diaphanous Self

William James (1842–1910) was one of the philosophers to whom Wittgenstein often referred. His *Principles of Psychology,* published in 1890, is alluded to more frequently than any other text in the whole course of Wittgenstein's *Philosophical Investigations. To* see where the story of the Cartesian self had reached towards the

---

[3] Ibid., 17.

[4] Ibid., 18.

[5] Iris Murdoch, *The Sovereignty of the Good* (London: Routledge & Kegan Paul, 1970), 80.

end of the nineteenth century, it is instructive to note a comment that James offers on a sentence by G.E. Moore (1873–1958), who was to become one of Wittgenstein's lifelong friends.

James writes as follows:

> [W]e are supposed by almost every one to have an immediate consciousness of consciousness itself. When the world of outer fact ceases to be materially present, and we merely recall it in memory, or fancy it, the consciousness is believed to stand out and to be felt as a kind of palpable inner flowing, which, once known in this sort of experience, may equally be detected in presentations of the outer world.[6]

James then cites this remark by Moore:

> The moment we try to fix our attention upon consciousness and to see *what*, distinctly, it is, it seems to vanish. It seems as if we had before us a mere emptiness. When we try to introspect the sensation of blue, all we can see is the blue; the other element is as if it were diaphanous. Yet it *can* be distinguished, if we look attentively enough, and know that there is something to look for.[7]

The first person plural should not mislead us: Moore is attempting to perform a very Cartesian act of self-discovery. He is trying to discover, by introspection, what the sensation of blue is—but "that which makes the sensation of blue a mental fact seems to escape us: it seems, if I may use a metaphor, to be transparent—we look through it and see nothing but the blue."[8] But what the metaphor suggests, if I am not mistaken, is the possibility that one is somehow able to stand back from one's body, so to speak, to look through it at the world beyond. We look through our eyes as through a window, which, by being transparent, is all but invisible.

William James certainly detected the implications:

> I believe that "consciousness," when once it has evaporated to this estate of pure diaphaneity, is on the point of disappearing altogether.... Those who still cling to it are clinging to a mere echo, the faint turnout left behind by the disappearing "soul" upon the air of philosophy.[9]

It is a conclusion, as he recognizes, that many of his readers would resist. He formulates their objection in words that would make sense to many people today:

> "All very pretty as a piece of ingenuity," they will say, "but our consciousness itself intuitively contradicts you. We, for our part, *know* that we are conscious. We *feel* our thought, flowing as a life within us, in absolute contrast with the objects which it so unremittingly escorts. We cannot be faithless to this immediate intuition."

James goes on, in his characteristically bluff and boisterous fashion, to declare that the stream of consciousness, which he too is aware of having, consists chiefly of the stream of his breathing:

> The "I think" which Kant said must be able to accompany all my objects, is the "I breathe" which actually does accompany them.[10]

So much, in outline, for the state of the question at the beginning of the twentieth century. Clearly, if William James's account of the self as a fictitious entity that has been philosophically generated out of the sensation of one's breathing is the

---

[6] William James, *Essays in Radical Empiricism and A Pluralistic Universe* (New York: E.P. Dutton), 25.

[7] G.E. Moore, *Philosophical Studies* (London: Kegan Paul, Trench, Trubner, 1922), 25.

[8] Ibid., 20.

[9] James, *Radical Empiricism*, 4.

[10] Ibid., 22.

alternative on offer, people with theological interests are going to opt for something more like Moore's vision of consciousness as an elusive diaphaneity. The question is, however, whether such a sublime philosophy of consciousness has much to do with the men and women that we are.

3 Theologians in the Cartesian Era

It is not difficult to find theologians who acknowledge the great significance for theological reflection of the Cartesian emphasis on the individual. As recently as 1967, for example, Karl Rahner reaffirmed that there must be no going back on "the transcendental-anthropological turn in philosophy since Descartes."[11] Modern philosophy has proved bad for rethinking Christian doctrine, so Rahner says, because it has developed into "a philosophy of the autonomous subject," who is closed against the transcendental experience in which dependence on God becomes evident. The modern self is ill at ease in Christianity because of this failure to evince a sense of creatureliness. It appears to be a matter of moral blindness. But what if the difficulty arises at an earlier stage? What if the problem lies with the very idea of the Rahnerian subject? What if the transcendental experience that Rahner wants for the self only obscures and excludes the membership of a community and a—tradition that gives rise to subjectivity in the first place?

My argument is that, far from still having to incorporate Cartesian assumptions about the self, as Rahner supposed, modern theology is already saturated with them.[12] It should be noted at once, however, that Rahner's belief that theology is belatedly still absorbing modern philosophy finds no echo in such an equally distinguished theologian as Eberhard Jüngel. In great detail, he argues that Descartes's attempt to demonstrate the necessity of God's existence by way of establishing the subject's self-certainty has resulted in the "death of God" crisis.[13] Far from having yet to accept the turn to consciousness, theology has already been nearly ruined by it. The perspective within which Jüngel pursues his theological reflections is, then, resolutely anti-Cartesian.

Thus, while both vigorously asserting the importance of the anthropocentric turn, Rahner and Jüngel could not be further apart in their assessment of its significance for theology. Jüngel to my mind, has by far the more plausible story to tell, but he seems to me greatly to exaggerate the ease with which the Cartesian legacy may be detected and disowned. He writes as if *naming* the problem were sufficient to *overcome* it. My contention, by contrast, is that such a powerfully attractive set of metaphysical preconceptions cannot be so swiftly dismissed. Theology (not alone in this) is surely much too deeply colored by varieties of the philosophy of the self-conscious and autonomous individual. Besides, as we shall see, the idea of the detached ego itself springs from decidedly religious roots. In any case, since it is far from being simply wrong, the critique that is required is more a matter of making different connections than one of wholesale rejection.

---

[11] Karl Rahner, *Theological Investigations*, vol. IX, (London: Darton, Longman & Todd, 1972), 38.

[12] A full presentation of this argument can be found in my *Theology After Wittgenstein*.

[13] Eberhard Jüngel, *God as the Mystery of the World* (Edinburgh: T. & T. Clark, 1983), 111–26.

Karl Barth, as one would expect, has provided the most substantial modern critique of theological anthropology.[14] But he had already come to grips in an interesting way with the Cartesian picture of the self.

There are two points to note. First, according to Barth, the Cartesian proof of the existence of God spirals back into the Cartesian metaphysics of the self:

> This idea of divinity is innate in man. Man can produce it at *will* from the treasury or deficiency of his mind. It is made up of a series of preeminent attributes which are relatively and primarily attributes of the human mind, and in which the latter sees its own characteristics—temporality, finitude, limited knowledge and ability and creative power—transcended in the absolute, contemplating itself in the mirror of its possible infinitude, and yet remaining all the time within itself even though allowing its prospect of itself to be infinitely expanded by this speculative extension and deepening. By transcending myself, *I* never come upon an absolute being confronting and transcendent to me, but only again and again upon my own being. And by proving the existence of a being whom *I* have conjured up only by means of my own self-transcendence, *I* shall again and again succeed only in proving my own existence.[15]

This might have been composed as a criticism of Rahner's transcendentalism, although Barth could hardly have heard of it at this time. In the Cartesian proof of God's existence, it is a certain conception of the human being's capacity for self-transcendence that Barth finds endlessly reflected.

Secondly, and even more instructively, Barth finds it necessary to attack the Cartesian emphasis on the thinking self when he discusses the right use of imagination in learning from Scripture.[16] The biblical account of the creation is a saga that has a great deal to teach us:

> We must dismiss and resist to the very last any idea of the inferiority or untrustworthiness or even worthlessness of a "nonhistorical" depiction and narration of history. This is in fact only a ridiculous and middle-class habit of the modern Western mind which is supremely phantastic in its chronic lack of imaginative phantasy, and hopes to rid itself of its complexes through suppression.[17]

As the original practitioner of "narrative theology," Barth denounces the rationalist bias that has affected so much biblical exegesis since the Enlightenment:

> But the human possibility of knowing is not exhausted by the ability to perceive and comprehend. Imagination, too, belongs no less legitimately in its way to the human possibility of knowing. A man without imagination is more of an invalid than one who lacks a leg.[18]

Theologians are thus well aware of the difficulties that the modern philosophy of the self has created. My suspicion, however, is that versions of the mental ego of Cartesianism are ensconced in a great deal of Christian thinking, and that many theologians regard this as inevitable and even desirable. The appeal of some theological writing also seems inexplicable unless it touches crypto-Cartesian assumptions which many readers share.

---

[14] Karl Barth, *Church Dogmatics*, vol. III/2 *The Doctrine of Creation* (Edinburgh: T. &. T. Clark, 1960), 46.

[15] Karl Barth, *Church Dogmatics*, vol. III/1, *The Doctrine of Creation* (Edinburgh: T. &. T. Clark, 1958), 360.

[16] See D.F. Ford, "Barth's interpretation of the Bible," in *Karl Barth: Studies of His Theological Method*, S.W. Sykes, ed. (Oxford: Clarendon Press, 1979), 55–87.

[17] Barth, *Church Dogmatics*, vol. III/1, 81.

[18] Ibid, 91.

3.1 The Deficient Angel

Consider the work of Karl Rahner, whom nobody would dispute is by far the most influential Roman Catholic theologian of the day. The speed with which he charms the reader into his system, and the immediate rewards in theological assurance, conceal, from readers who are philosophically unwary, the problematic character of the first step. The obsession with epistemological preliminaries, which should at once indicate how Cartesian his theological constructions are likely to prove, only persuades students that they are on the right track. The presentation is usually so abstractly "metaphysical," and the sentences so carapaced with qualifications, that the innocent eye runs swiftly over them in order to get to more familiar theological material.

In *Foundations of Christian Faith,* an acknowledged masterpiece of modern theology, Rahner begins by raising basic epistemological problems. To be sure, he expands the discussion to take in other dimensions of human life; but it is always as the cognitive subject that people first appear in Rahner's theology. Students alerted to the bias of the Cartesian legacy would suggest that language or action, conversation or collaboration, are more likely starting points, particularly if, like Rahner, you want to move towards a very strong ecclesiology. Yet consciousness, self-awareness in the cognitive act, is always his favored way into theology.

"Everyone strives to tell someone else, particularly someone he loves, what he is suffering": this innocuous observation is Rahner's example to illustrate the following somewhat more obscure proposition: "The original self-presence of the subject in the actual realization of his existence strives to translate itself more and more into the conceptual, into the objectified, into language, into communication with another."[19] When the sequence is repeated in the next sentence, suspicions are aroused: "Consequently in this tension between original knowledge and the concept which always accompanies it there is a tendency towards greater conceptualization, towards language, towards communication."

Rahner's natural assumption—that communication comes after language, and language comes after having concepts—is precisely what the Cartesian tradition has reinforced. His example suggests that, when I am in pain, I first have the thought that I am in pain, I then put it into words and finally I find someone to whom to communicate it. It is, of course, an entirely natural thing to think: this is often exactly how it happens. But if the picture is that I always, or even normally, have the thought before I put it into words, then something very like the Cartesian vision of my epistemological predicament begins to show through. It looks as if Rahner might be working on the double assumption that I am in a position to identify my sensations prior to my applying the customary labels to them, and secondly that what I reveal of them to someone else remains wholly at my command. The picture would thus be that I enjoy immediate nonlinguistic knowledge of my own inner experiences, while what I am experiencing at any moment necessarily remains hidden from other people unless I deliberately choose to disclose it. On this account, to put it straight into the Wittgensteinian terms which we still have to see, the individual is supposed to be able to locate, by a private mental act of ostensive definition, some item of his own consciousness, while, for knowledge of other people's thoughts or feelings, he is

---

[19] Karl Rahner, *Foundations of Christian Faith: An Introduction to the Idea of Christianity* (London: Darton, Longman & Todd, 1978), 16.

supposed to depend entirely on inferences that he makes from their outward demeanor.[20]

Such suspicions may seem exaggerated, even to readers who have been alerted by Wittgensteinian considerations. It certainly does not follow inescapably, either from his preoccupation with epistemology or from the little remark about having a pain, that Rahner was a firm subscriber to the story of the soul as a solitary individual, with direct acquaintance with his own experiences and command of divulging them. But consider Rahner's account, unfortunately bare of examples, of an ordinary case of knowing something:

> In the simple and original act of knowledge, whose attention is focused upon some object which encounters it, the knowing that is co-known and the knowing subject that is co-known are not the *objects* of the knowledge. Rather the consciousness of the act of knowing something and the subject's consciousness of itself, that is, the subject's presence to itself, are situated so to speak at the other pole of the single relationship between the knowing subject and the known object. This latter pole refers to the luminous realm, as it were, within which the individual object upon which attention is focused in a particular primary act of knowledge can become manifest. This subjective consciousness of the knower always remains unthematic in the primary knowledge of an object presenting itself from without. It is something which goes on, so to speak, behind the back of the knower, who is looking away from himself and at the object.[21]

The subjective consciousness of the knower, which goes on behind his back when he is looking at something, admittedly laced in metaphor, cannot but arouse suspicion, even if the idea makes any sense at all. But it leads immediately to the heart of Rahner's conception of the self:

> If we ask what the *a priori* structures of this self-possession are, then we must say that, without prejudice to the mediation of this self-possession by the experience of sense objects in time and space, this subject is fundamentally and by its very nature pure openness for absolutely everything, for being as such.[22]

Thus, without prejudice to the finitude of time and place, the Rahnerian self turns out to be nothing less than "pure openness for absolutely everything." Even to suspect that our openness is for a good deal less than absolutely everything is already to have passed beyond that suspicion, so Rahner goes on to claim. To recognize your limitations is already to have transcended them. This takes us to Rahner's central notion of "transcendence" and, with some omissions, to this key paragraph:

> [A] subject which knows itself to be finite... has already transcended its finiteness... In so far as he experiences himself as conditioned and limited by sense experience, and all too much conditioned and limited, he has nevertheless already transcended this sense experience. He has posited himself as the subject of a pre-apprehension which has no intrinsic limit, because even the suspicion of such an intrinsic limitation to the subject posits this pre-apprehension itself as going beyond the suspicion.[23]

Obviously much more of Rahner's text would need to be probed; I do not claim to have fully understood even the pages from which I have quoted. But his preoccupation with the cognitive subject is clear: other people remain marginal to his epistemology. The emphasis on the subject's capacity for self-consciousness and

---

[20] For further discussion of this point, see my *Theology After Wittgenstein*, 2nd ed. (London: SPCK, 1997), chap. 4.

[21] Rahner, *Foundations*, 18.

[22] Ibid., 19–20.

[23] Ibid., 20.

self-reflexiveness, and his openness for absolute being, is equally conspicuous. By our insatiable questioning we are, perhaps unwittingly, the products of a dynamic movement of ceaseless self-transcendence towards the steadily receding horizon which is the absolute: in effect, anonymously, the deity. To feel chafed by sense experience is already to be the subject with this capacity for the absolute.

The theological rewards of Rahner's account of the epistemological privileges of the subject are very great. When he comes to discuss arguments for the existence of God, for example, he is able to say that we already have in place all that is required to substantiate them:

> In all the so-called proofs for the existence of God the one and only thing which is being presented and represented in a reflexive and systematic conceptualization is something which has already taken place: in the fact that a person comes to the objective reality of his everyday life both in the involvement of action and in the intellectual activity of thought and comprehension, he is actualizing, as the condition which makes possible such involvement [*Zugriff*] and comprehension [*Begriff*], an unthematic and non-objective pre-apprehension [*Vorgriff*] of the inconceivable and incomprehensible single fullness of reality.[24]

Whenever we have to do with the world to which we belong, whether working on it or thinking about it, we are always carried by that prior grasp *of*, which is simultaneously a being gripped *by*, that absolute which is, however "anonymously," the deity itself. This antecedent invasion by the absolute constitutes our very nature as rational and self-reflexive beings in the first place. Self-presence is necessarily, if often unwittingly, openness to the absolute. We only have to reflect on the subject's constitutive self-presence in any cognitive or volitional act to have proof of the existence of God.

Rahner is also able to put this conception of the subject as essentially openness to the absolute at the heart of his brilliant interpretation of Chalcedonian Christology. The doctrine of the Incarnation almost ceases to be a scandal, so natural does it come to seem that, in one special case, such a self-presence should be transparently, even diaphanously, open to the absolute in what one might call "hypostatic union."

The unattractively abstract jargon, admittedly much less offensive in German and to ears tuned to a different philosophical tradition, easily tempts the reader to skip the epistemological preliminaries to get on to Rahner's theology. To make proofs of God's existence redundant, to offer a coherent account of the doctrine of the Incarnation, and much else, could not fail to draw students into the Rahnerian system—but at what price? And why does it seem so natural to pay it? Much more evidently needs to be said, but there surely is a *prima facie* case for suggesting that Rahner's most characteristic theological profundities are embedded in an extremely mentalist-individualist epistemology of unmistakably Cartesian provenance. Central to his whole theology, that is to say, is the possibility for the individual to occupy a standpoint beyond his immersion in the bodily, the historical, and the institutional. Rahner's consistently individualist presentation of the self emphasizes cognition, self-reflexiveness, and an unrestricted capacity to know. It rapidly leaves time and place behind. It is not surprising if this mentalist-individualist conception of the self seems difficult to reconcile with the insistence on hierarchy and tradition that marks Rahner's Roman Catholic ecclesiology.[25]

---

[24] Ibid., 69.

[25] See Gordon D. Kaufman, "Is there any way from Athens to Jerusalem?" *The Journal of Religion* 59 (1979): 340–46.

We are conditioned and limited by sense experience—"all too much," as Rahner expresses it. But where should we be if we were not so conditioned and limited? What if our relation to our physical and social setting is a matter for gratitude and celebration, rather than resentment and frustration? The idea that it is by leaving the world that one finds oneself is an ancient and a very alluring one, but, without radical reflection on its ambiguities, is it the most productive way of regarding ourselves, particularly from a Christian theological perspective? However interesting the theological developments, what if Rahner's foundation lies in supposing that, as Cornelius Ernst once put it,[26] one is "a more or less deficient angel"?

3.2 The Theistic Gambler

Karl Rahner will never be a popular writer; but much more widely read theologians reveal even more deeply entrenched Cartesian assumptions—Teilhard de Chardin for instance.[27] But consider the following remark in a recent book by Hans Küng, probably the most widely read theological writer in the English-speaking world: "The history of modern epistemology from Descartes, Hume and Kant to Popper and Lorenz has—it seems to me—made clear that the fact of any reality at all independent of our consciousness can be accepted only in an act of trust."[28] He is arguing that, since we have nothing better than such an act of trust at the basis of our belief in the existence of *anything* outside our own minds, it is not very strange for us to have nothing better than an act of trust on which to found our belief in the existence of God. In fact, it is the line that he takes in his earlier massive book on God *Does God Exist?*

Here, after much else, we are offered a brilliant account of the radical scepticism introduced by Descartes and, according to Küng, carried further by Pascal. He might seem to be preparing the way to deflect it. He refers to Wittgenstein and to a galaxy of other philosophers in the Anglo-American tradition, but he shows no sign of understanding the deeply anti-Cartesian campaign which they have been conducting for the past fifty years. Anyone familiar with Wittgenstein's imaginative explorations of the charms of scepticism, as Küng is supposed to be, would expect a rehearsal of the anti-skeptical arguments at this point. But Küng tells us, as the book unfolds, that the alternative to radical scepticism in its most nihilistic Nietzschean form (everything is illusory) is simply the gamble that reality does really exist: "Every human being decides for himself his *fundamental attitude* to reality: that basic approach which embraces, colors, characterizes his whole experience, behavior, action."[29] Innocent of all anti-Cartesian suspicions, he goes for individual decisions as establishing the foundations for belief in the reality of anything outside one's mind: It is up to me to choose the basic attitude I adopt towards this radically dubious reality with which I am surrounded. I simply *decide* to trust the reality of other people and all the rest of the rich tapestry of life.

---

[26] See the pregnant footnote in his introduction to Karl Rahner, *Theological Investigations*, vol. 1 (London: Darton, Longman & Todd, 1961), xiii.

[27] See Anthony Kenny's review, "*The Phenomenon of Man*: The Prototype for Detecting Cartesian Psychology in Modern Theology," reprinted in his *The Legacy of Wittgenstein* (Oxford: Blackwell, 1984), chap. 8.

[28] Hans Küng, *Eternal Life?* (London: Collins, 1984), 275.

[29] Hans Küng, *Does God Exist? An Answer for Today* (London: Collins, 1978), 432.

Küng alludes to Kierkegaard and indeed to Ignatius of Loyola, but this liberty that the individual has to choose his reaction to reality is expounded mainly with reference to recent Anglo-American philosophers, whose writings have encouraged Küng to think that it is all a matter of the individual's decision. From Popper he has learned that "all rational thinking rests on a choice, a resolution, a decision, an attitude." From Carnap he has learned that the principles and rules of argument in an artificial language are a matter of "free choice." From Wittgenstein he has learned, even more amazingly, "that the rules of language may be chosen with complete freedom."[30] This would be laughable if the case were not being built up with such earnestness and apparent learning. All along the line, culling phrases from one supposed authority after another that would knock out the adversary, Küng insists that the individual is at liberty to choose his beliefs about the intelligibility, and even about the reality, of the world around him. The argument comes to a head in these words: "as there is no logically conclusive proof for the reality of reality, neither is there one for the reality of God."[31] Belief in the existence of God is a "basic decision," a creative option one might say, on analogy with the *Urentscheidung* in favor of the reality of the external world. In other words: there is nothing irrational about believing in the existence of God—after all, we have nothing stronger than *belief* in the existence of *anything* outside our own minds.

The self is pictured as having to confront that which surrounds it and then decide whether it is as it seems, or even whether it is there at all. A viewpoint seems to be available from which one is able to survey the passing show and impose a pattern upon it. The individual seems to be free to put what construction he will upon the surrounding world. The supposition is always that one is able to view the world from somewhere else—as if one were God, perhaps.[32]

3.3 A Modern Person

The idea of the self as being in a position to decide how to take the world reappears in the writings of Don Cupitt, another widely read theologian. In *Taking Leave of God,* for example, he writes as follows:

> [T]he principles of spirituality cannot be imposed on us from without and cannot depend at all upon any external circumstances. On the contrary, the principles of spirituality must be fully internalized *a priori* principles, freely adopted and self-imposed. A modern person must not any more surrender the apex of his self-consciousness to a god. It must remain his own.[33]

While one shares Cupitt's detestation of a morality or a religion imposed by intimidation or spiritual terrorism, it looks very much as if we are again presented with the self-conscious individual who is apparently able and willing to provide himself with principles of spirituality from his own inner resources. Indeed, the apex of his self-consciousness, wherever that is supposed to be, has to be entirely independent of all external circumstances as well as of every god. As Cupitt says, in the book's opening words, "Modern people increasingly demand autonomy, the power of legislating for oneself... they want to live their own lives, which means

---

[30] Ibid., 461.

[31] Ibid., 574.

[32] For more detail, see my "Küng's case for God," *Vidyajyoti* XLIX (1985): 118–23.

[33] Don Cupitt, *Taking Leave of God* (London: SCM Press, 1980), 9.

making one's own rules, steering a course through life of one's own choice, thinking for oneself, freely expressing oneself and choosing one's own destiny."

Now, obviously, it is a good thing to think for oneself, to have the freedom to express oneself, and the like. The central values of the Enlightenment, which Cupitt so vigorously defends, have yet to take root in most of the social systems across the world, including many supposedly Christian countries. But does the rhetoric not outrun the case? What is meant by "making one's own rules"? The difficult and precious responsibility of trying to think for oneself, have an informed conscience, etc., is presented as if one had the power to invent one's own standards of right and wrong. Everyone wants to be captain or his or her own soul: "I must be autonomous in the sense of being able to make my own rules and impose them upon myself," as Cupitt says. And, again in the preface to *Taking Leave of God,* "each chooses his own ethic."

There is much more in this vein of heady individualist libertarianism. But there is all the difference in the world between making the rules one's own, critically and responsibly, and making one's own rules: a distinction that Cupitt blurs all the time, apparently without noticing. Much of the time what he says is no more than any sensible parent's advice in a society like ours—attempting to get a child to think for itself, to be reasonably suspicious of the moral standards that it finds around it, and so on. Certainly everybody needs to have the courage to be critical of the decrees of any authority, political or religious, that demands unthinking loyalty and mindless obedience. But again and again, in the midst of much common sense, Cupitt's rhetoric conjures up a more exciting and dramatic portrait of the Modern Person who creates his own moral rules, which, happily—but also just as it happens—coincide with those adopted by like-minded individuals so that it becomes possible, as he says, to form "a liberal democratic republic, the best kind of society."

Don Cupitt's theological writings are haunted by the figure of this self-conscious autonomous individual to whom it falls in radical freedom to choose his own moral world. The Cartesian ego has received a Kantian twist. But this lonely agent of ultimate choice is plainly the alter ego of the epistemological subject who can view everything from a standpoint outside history and community. The solitary individual with the God's eye view turns out, unsurprisingly, to be the creator of his own moral universe. The person at the centre of modern theology has acquired the attributes of the God of classical theology. Far from still having to get theology to make more room for the modern philosophy of the self, it is time to recognize that philosophy today (in Wittgenstein especially) is striving to rid itself of a certain displaced theology. The worldless ego is remarkably like a substitute for the deity in the game of creation. It is, at any rate, bizarre to be writing theology with such deference for the self as deity, transcending and creating the world, just when philosophers have developed strategies to extirpate this alienated theology from philosophy.

Ironically enough, in more recent speculations, Don Cupitt has appealed to Wittgenstein's work in support of his radical voluntarism: "Everywhere he is a thoroughgoing constructivist and voluntarist: logical necessity is created by the rules governing language. If he is a nonrealist about religion, he is also a nonrealist about everything else."[34] The point I wish to make here is that Cupitt's emphasis on the will of the autonomous self is entirely consonant with Küng's picture of the individual who is required to gamble on the reality of the world outside his head.

---

[34] Don Cupitt, *The Sea of Faith* (London: BBC Publications, 1984), 222. I attempt to refute this highly disputable reading of Wittgenstein in *Theology After Wittgenstein,* chap. 5.

3.4 The Hidden Self

What a hydra-headed creature it is! The self who is free to survey the world from no point of view within the world often turns out to be the self who is totally impenetrable to anyone else—once again rather like the hidden God of classical theism.

Consider the following thought from Schubert Ogden's attempt to conceive divine action on analogy with human action:

> Behind all its public acts of word and deed there are the self's own private purposes or projects, which are themselves matters of action or decision. Indeed, it is only because the self first acts to constitute itself, to respond to its world, and to decide its own inner being, that it "acts" at all in the more ordinary meaning of the word; all its outer acts of word and deed are but ways of expressing and implementing the inner decisions whereby it constitutes itself as a self.[35]

To speak of one's private purposes and inner decisions as lying "behind" one's words and deeds is a perfectly good figure of speech. On the other hand, if our words and deeds, thus our speaking and acting, are "but" ways of expressing and implementing the inner goings-on and private states of consciousness by which one is constituted, a dualism of inner and outer, and of private and public, seems to threaten the integrity of the common space constituted by language and our other institutions. It is no great surprise, therefore, to find Ogden continuing as follows:

> [T]he primary meaning of God's action is the act whereby, in each new present, he constitutes himself as God by participating fully and completely in the world of his creatures... his relation to his creatures and theirs to him is direct and immediate. The closest analogy... is our relation to our own bodily states, especially the states of our brains. Whereas we can act on other persons and be acted on by them only through highly indirect means such as spoken words and bodily actions, the interaction that takes place between our selves or minds and our own brain cells is much more intimate and direct.[36]

As a self, then, I can exercise a more direct effect on my brain than I can on other persons. But how does the mind direct the brain? Leaving aside Ogden's non-classical, Whiteheadian doctrine of God, is his conception of the self not very curious? These speculations about the relation of God to the world surely rely on an account of the mind-body question which makes the obscure even more obscure.

It would be difficult to find a better case for Wittgensteinian treatment. We are being asked to think that the self which acts in an "intimate and direct" manner on its brain interacts with other selves "only through highly indirect means such as spoken words and bodily actions." One human being has no access to another except by resorting to the circuitous devices of gesture and speech: a poor relation compared with the immediate access that the self has to its own brain! I have nothing better than language and the expressiveness of my own body with which to communicate with other people, while I have a direct view of my inner experiences, if not even of my brain processes.

3.5 The Dimensionless Pinpoint

Timothy O'Connell, in an important and ambitious recent attempt to reconstruct moral theology as a discipline, obviously requires some working model of the moral agent. Amazingly, he comes out with this beautiful portrait of the self-disembodying Cartesian ego:

---

[35] Schubert Ogden, *The Reality of God and Other Essays* (London: SCM, 1967), 177.

[36] Ibid., 177–78.

In an appropriate if homely image, then, people might be compared to onions... At the outermost layer, as it were, we find their environment, their world, the things they own. Moving inward we find their actions, their behavior, the things they do. And then the body, that which is the "belonging" of a person and yet also is the person. Going deeper we discover moods, emotions, feelings. Deeper still are the convictions by which they define themselves. And at the very center, in that dimensionless pinpoint around which everything else revolves, is the person himself or herself—the I.[37]

At least when Descartes peeled off the layers he discovered a thinking thing, rather than "a dimensionless pinpoint." The essence of the person is now within an ace of vanishing. In the crucible of O'Connell's homely negative anthropology, in fact, we find the residue of the *actus purus* of apophatic theology: God, once again.

Once again, as it turns out, the disembodied self has problems with relating to people (as one onion to another). It is even difficult to say what we mean by the expression "human person":

[P]ersonhood is the one thing about human beings which we cannot actually see. In a process of reflection I seek to discover myself. I hold up to the eye of my mind the experiences that I have. But who looks at these experiences? I do, the person that I am... Repeatedly I attempt to gaze upon the very center of myself. But I always fail. For the real person that I am always remains the viewer, and can never become the viewed.[38]

What I see, in my own case as well as in meeting other people, is never anything but actions behind which I have to "posit" a "moreness" which is "the human person." It certainly seems that, when I meet a human being in ordinary circumstances, I have to hazard the opinion that he has a soul.[39]

Very much the same philosophy of the self reappears in a recent attempt by Peter Chirico to make sense of the notion of the infallibility of the Church. Laying his epistemological foundations necessarily with great care, he comes out with the following two remarks. First: "No man can be more certain of anything than he is of his own self-awareness. The standard or limit of human certitude and human infallibility is that consciousness which one has of oneself."[40] The name of Descartes is never mentioned in Chirico's book, but that first sentence might have been drawn straight from the Second Meditation.

The second remark comes in the context of explaining why Christ, after the resurrection, is "unrecognizable as risen in our experience." The Church, so Chirico says, "cannot identify the risen Christ present in its experience because the risen condition and mode of operation makes Christ inaccessible to human awareness." Making a move very like Hans Küng's, Chirico argues that there is nothing all that strange about our being unable to recognize the risen Christ because, after all,

We never recognize or see another being in itself; we only recognize directly the effects of its activity towards us, activity that occasions the actualization of our experiential continua in a way we can consciously detect and isolate. Hence, for example, we identify

---

[37] Timothy O'Connell, *Principles for a Catholic Morality* (New York: Seabury Press, 1978), 59.

[38] Ibid., 59; see also Stanley Hauerwas, *The Peaceable Kingdom: A Primer in Christian Ethics* (London: SCM Press, 1984), 40–41.

[39] Ludwig Wittgenstein, *Philosophical Investigations*, G.E.M. Anscombe and R. Rhees, eds., G.E.M. Anscombe, trans. (Oxford: Blackwell, 1953), 178.

[40] Peter Chirico, *Infallibility: The Crossroads of Doctrine* (London: Sheed and Ward, 1977), 58.

the change in us as caused by the visible appearance, the barking, and the furry softness of the dog that enters the room.[41]

Thus, when the doggy contour, the canine yelp, the hairy surface and so on variously impinge upon my senses, I put two and two together and judge that it must be that cur from down the street. That is hardly a caricature. Chirico's highly sophisticated attempt to rehabilitate one of the most contentious bits of Roman Catholic dogma is founded on the crudest epistemological dogmas of the Cartesian-empiricist tradition.

3.6 The Godlike Other

Gordon Kaufman, finally, in his attempt to work out a notion of "transcendence without mythology," provides a good example with which to break off this catalogue, for two reasons. His theory also trades on a radically Cartesian conception of the self—but, after criticism, he has disowned it. Thus, he not only shows once again the powerful imaginative hold that the modern philosophy of the subject has in theology, but also that, perhaps a little unusually, a theologian can learn to acknowledge his metaphysical prejudice and disclaim it.

Consider the following account of our relationship to one another:

> What one directly experiences of the other are, strictly speaking, the external physical sights and sounds he makes, not the deciding, acting, purposing centre of the self— though we have no doubt these externalities are not *merely* physical phenomena, but are the outward and visible expression of inner thought, purpose, intention.[42]

So far so good, one may be inclined to say: this is, after all, "strictly speaking." But is it not already a highly artificial sense of "experiencing" someone's presence to suggest that it is "the external physical sights and sounds he makes" that one "directly experiences"? Is it not when someone's behavior is unintelligible that one is left with bare physical sights and sounds? But the passage goes on as follows:

> In our interaction with other persons we presuppose a reality (the active center of the self) *beyond* that which we immediately perceive…It is in the act of communication that we discover that the other is more than merely physical being, is a conscious self; it is in the experience of speaking and hearing that we come to know the *personal* hidden behind and in the merely physical. This is the most powerful experience we have of *transcendence of the given* on the finite level, the awareness of genuine activity and reality *beyond* and *behind* what is directly open to our view.

We "*presuppose*" that there is more to the people around us than there is to rocks and cows, and it is in conversation with them that our hypothesis is tested and, no doubt usually, verified. In later criticism of such remarks as these Kaufman has acknowledged that his conception of transcendence here depended on a picture of the self as a reality that is radically private and normally concealed.[43] He has gone on to develop a conception of the self which involves the possibility of our deceiving one another, and the like, such that one has a certain inaccessibility to others, but, as he now allows, in a great deal of what an individual does and says his or her inner self is perfectly transparent. Accordingly, Kaufman has, set about revising his theory of the hiddenness of God in terms of an avowedly non-Cartesian understanding of the self.

---

[41] Ibid., 76.

[42] Gordon D. Kaufman, *God the Problem* (Cambridge: Harvard Univ. Press, 1972), 63–64.

[43] Ibid., xiii, taking into account criticism by Michael McLain, "On theological models," *Harvard Theological Review* 62 (1969): 155–87.

Of course, in this sample of theologians, drawn from the Roman Catholic, Anglican, Methodist, and Mennonite traditions, the variations on the figure of the mentalist-individualist self would deserve examination. My concern is, however, with the common paradigm. One thing is beyond dispute: in each case the starting point is naturally assumed to be the individual. With Karl Rahner the emphasis is upon the subject in a cognitive situation: the capacity for self-presence in acts of knowledge is tacit apprehension of the absolute. With Hans Küng and Don Cupitt we have a very powerful picture of the self as isolated will and autonomous individual, left in radical freedom to bring a moral universe out of the surrounding chaos either by a gamble of faith or by a God-like act of creation. With Schubert Ogden the self has retreated into the inviolable mystery of a private inner world from which messages emerge obliquely in signs. Timothy O'Connell's "onion peel view of the self," as he calls it, strips off nature, culture, the body, desires and convictions, to lay bare one's identity in a dimensionless pinpoint. Peter Chirico holds out for good old-fashioned Cartesian epistemology in the purest form—nothing is more certainly known to me than the contents of my own mind, nothing else is ever directly accessible. Finally, Gordon Kaufman, until he thought better, supposed that it was only when the other opened his mouth and spoke that one realized that a person lay hidden within the middle-sized, lightly sweating and gently palpitating object on the other side of the dinner table.

In every case, though variously, and sometimes very significantly so, the model of the self is central to some important, sometimes radical and revisionary, theological proposal or program. A certain philosophical psychology is put to work to sustain a theological construction. Time and again, however, the paradigm of the self turns out to have remarkably divine attributes. The philosophy of the self that possesses so many modern theologians is an inverted theology which philosophers today are working bard to destroy.

4 Destroying an Inverted Theology

My claim is that the most illuminating exploration of the continuing power of the myth of the worldless (and often essentially wordless) ego is to be found in the later writings of Ludwig Wittgenstein. It will be helpful to approach them by way of a brief survey of more recent investigations of the question, from somewhat different angles. This will indicate how post-Wittgensteinian philosophers consider the Cartesian legacy.

4.1 The Absolute Conception of Reality

Bernard Williams has suggested that, in its main thrust, the Cartesian project must be intuitively attractive to most people brought up in a culture such as ours, where paradigms of knowledge are tied to ideals of impartiality and objectivity. In the historical context of the rise of modern physics and natural science, Descartes strove to delineate what Williams calls "the absolute conception of reality."[44] The investigator, that is to say, tries to extricate himself from every contingent restriction upon the pursuit of knowledge. He aspires to be so liberated from bias in his outlook that he can attain a representation of the world exactly as it is, uncolored by his prejudices. To enjoy such an objective picture of reality demands the elimination of

---

[44] Bernard Williams, *Descartes: The Project of Pure Enquiry* (London: Penguin Books, 1978), 65.

the observer. The goal of natural science, for people in the Cartesian tradition, is to have a representation of reality that eschews all the properties that things have simply as a result of the presence of human beings among them. To want the absolute conception of reality is to aim at a description of things as they would be in our absence. This or something like it has been the spirit animating science since the seventeenth century. It would be returning to barbarism to surrender or even dilute such ideals of impartiality and objectivity.

For Descartes, however, the absolute conception of reality was tied up with the individual's certainty of his own existence as well as with the idea that God alone guaranteed the existence of reality outside the individual's consciousness. The great modern expression of the ideals of the new science was embedded in the epistemology of the disengaged ego as well as in natural theology. Thus, if certain metaphysical options have swayed theology, they have themselves a theological matrix. The eggshell clings to the thinking.

However well concealed it may be, these metaphysical aspirations still carry a powerful theological charge. As Paul Feyerabend has pointed out, the very idea of a personal experience that would be infallible, self-authenticated, and unprejudiced, etc., because it was the result of sloughing off all received opinion, tradition, and authority, etc.—in fact the very idea with which Descartes opens the *Meditations*—is remarkably reminiscent of certain tendencies in the history of Christianity that would put the individual believer directly and inwardly into a relationship with God which excluded in advance all mediation by an historical community with authoritative tradition, rituals and the like.[45] The desire of philosophers like Bertrand Russell and G.E. Moore, at the dawn of modern analytic philosophy in Cambridge, to get down to the primitive sense-data in order to find a level of experience that would supposedly be free of all interpretation, subjective distortion, etc., is fundamentalism transposed into an adjacent discourse.[46]

The drive for objectivity, then, brought with it the uncanny thought that the only perfect depiction of any reality would have to be from nobody's point of view—or, if there is any difference, from God's. Objectivity requires, as Thomas Nagel argues, "departure from a specifically human or even mammalian viewpoint."[47] A "transcendence of the self"—his phrase—is required for an account of what is of interest simply in itself, independently of anybody's interest in it. As he says, "the power of the impulse to transcend oneself and one's species is so great, and its rewards so substantial" that "detachment from the contingent self" is bound to remain an ethical as well as a scientific obligation—again, if there is any difference. But it is not surprising that there has been an equally powerful drive to protect the inner reality of the subject, which is what makes Cartesian psychology appealing.

4.2 The Drive to Spiritual Reality

Charles Taylor has long been concerned with the same issue. He began with polemics against behaviorism and allied theories (not to mention the practices that

---

[45] Paul Feyerabend, *Problems of Empiricism* (Cambridge: Cambridge University Press, 1981), 19.

[46] I say more about logical atomism in *Theology After Wittgenstein*, chap. 3.

[47] Thomas Nagel, *Mortal Questions* (Cambridge: Cambridge University Press, 1979), chap. 14.

they have legitimated), particularly in psychology.[48] His subsequent essays, recently collected, constitute a general critique of the drive in many (if not all) of the social sciences to exclude subjectivity.[49]

According to the absolute conception of reality, then, the subject is detached from the significance that things have for him personally, so that he attains a certain freedom to objectify, and thus to depict and control, them all, in various ways and with degrees of expected success. The appeal of this ideal, so Taylor says, derives from the obvious benefits of mastering one's natural and historical environment, not to mention one's personality and domestic circumstances. But there is more to it, he suggests. The ideal of the detached self, disengaged from personality and historical contingency, and thus free to see reality objectively, is, so he says, a new variant of that immemorial "aspiration to rise above the merely human, to step outside the prison of the peculiarly human emotions, and to be free of the cares and the demands they make on us"—"a novel variant of this very old aspiration to spiritual freedom."[50] The drive to objectivity is thus "of spiritual origin," or, at any rate, it helps to account for the attractions of the detached self.

It is highly paradoxical, as Taylor goes on to say, because the ideals of our modern scientific culture are supposed to have left religion behind. Of course there is a great difference. In the modern case, it is the natural universe that is to be represented as independently as possible of all human interpretation. In the ancient case, the self wants to lose itself in dispassionate contemplation of the reality that subsists in itself. In both cases, however, the subject is required to transcend human emotions, cultural and historical particularity, and the like, in order to encounter, bare, that which is truly important. What Taylor claims, then, is that the motivating power of the secular and naturalistic drive to extend objective knowledge as far as possible, *and even farther,* receives its deepest energy from "the traditional drive to spiritual purity."[51]

Thus, for several modern philosophers (and my list could easily be extended), one of the deep questions on the agenda is, in one way or another, how to expose and explain the continuing psychological and imaginative appeal of a certain picture of the self's place in the world. Despite what is often supposed, even by themselves, philosophers are, at their best, no more interested in knockdown logical refutations of this or that thesis than they are in spinning airy speculations. They try, rather, to search out the obscure motivations in the recurrence in fresh guises of certain meta-physical prejudices which, although often "refuted," retain their vitality and charm.

What I am suggesting is that Wittgenstein was attempting to free himself from something very like "the absolute conception of reality." Consider only this somewhat gnomic utterance from the *Tractatus*:

> Here it can be seen that solipsism, when its implications are followed out strictly, coincides with pure realism. The self of solipsism shrinks to a point without extension, and there remains the reality co-ordinated with it.[52]

---

[48] Charles Taylor, *The Explanation of Behaviour* (London: Routledge & Kegan Paul, 1964).

[49] Charles Taylor, *Philosophical Papers,* vols. 1–2 (Cambridge: Cambridge University Press, 1985).

[50] Ibid., vol. 1, p. 112.

[51] Ibid., 113.

[52] Ludwig Wittgenstein, *Tractatus Logico-Philosophicus* D.F. Pears and B.F. McGuinness,

As the self withdraws, the world in itself emerges. When subjectivity becomes so perfect that it vanishes into absolute privacy, reality remains in splendid objectivity. In the notes of 1916, from which this remark was extracted, Wittgenstein went on as follows:

> What has history to do with me? Mine is the first and only world! I want to report how *I* found the world.
> What others in the world have told me about the world is a very small and incidental part of my experience of the world.[53]

That sounds very much like an echo of Descartes's voice in the *Meditations:* what I have learned from the community and the tradition, from my companions and precursors, is a tiny and marginal factor in my *Welt-Erfahrung.* But, again in notes that did not survive into the *Tractatus,* Wittgenstein completed his day's reflections thus:

> The human body, however, my body in particular, is a part of the world among others, among beasts, plants, stones, etc., etc.
> Whoever realizes this will not want to procure a preeminent place for his own body or for the human body.
> He will regard humans and beasts quite naively as objects [*Dinge*] which are similar and which belong together.[54]

In the *Investigations,* however, the first move is to secure a focal significance for the human body, and thus to inaugurate a radical critique of "the traditional drive to spiritual purity."

---

trans. (London: Routledge and Kegan Paul, 1961), 5.64.

[53] Ludwig Wittgenstein, *Notebooks 1914–1916*, G.H. von Wright and G.E.M. Anscombe, eds., G.E.M. Anscombe, trans. (Oxford: Blackwell, 1961), 82.

[54] Ibid.

EMOTIONS: HOW I'VE LOOKED FOR THEM IN THE BRAIN

Joseph E. LeDoux

1 Introduction

One of the most exciting things ever to happen in the study of the mind was the emergence of cognitive science. This interdisciplinary approach, based on concepts of information processing derived from computer function, provided a new way of thinking about and doing research on the mind. In particular, it rescued the mind from the shackles of behaviorism, which had eliminated research on mental processes from psychology for much of the first half of the twentieth century.[1] However, the success of cognitive science led to the omission of research on noncognitive topics, especially emotion. Cognitive minds, modeled in computer simulations, could play chess very well. It might even be possible to program them to cheat. But a cheating program feels no guilt or shame, nor does it fear that it will be caught.

Although emotion was left out of cognitive science in the early days, in recent years, there has been an emotional renaissance. Ironically, much of the progress that has been made has come by taking a cognitive-science approach to the study of emotion. Below, I will outline how my own work on emotion has used a cognitive approach to map out brain pathways involved in fear. Further details about my research on emotions and the brain are found in my recent book, *The Emotional Brain.*[2]

2 Emotions and the Brain: A Brief Personal History

In the late 1970s, I was a graduate student in psychology doing research on how cognitive processes work in the human brain. I noticed the lack of concern for emotions, and their underlying brain mechanisms, and began working on this topic. I decided to do this research in an animal model. I chose an animal model over human studies because there were few opportunities at the time for studying the human brain with anatomical precision. I also chose an animal model over a computational model, where emotions are simulated in software, because there were no computational models that could replace actual experimentation on the brain.

The animal model I chose to study was "conditioned emotional responses" in rats, a good choice because so much is known about their anatomy, physiology, and behavior. As a result, new research can build on existing findings. Also, and very important, sufficient similarities exist between the basic anatomy and physiology of the rat brain, on the one hand, and the behavioral expression of emotion in human and rats, on the other, to make the rat a valid model of the human emotional brain.

If you give a rat (or a human) a sound followed by a noxious stimulus (electric shock, loud noise), the sound quickly becomes a warning signal for unpleasant stimulus. As a result, the sound comes to elicit emotional responses that prepare the organism for the impending danger. Included are muscle tension (freezing), changes in blood pressure and heart rate, release of stress hormones, and a host of other

---

[1] For a history of cognitive science, see Howard Gardner, *The Mind's New Sciences: A History of the Cognitive Revolution* (New York: Basic Books, 1987).

[2] Joseph E. LeDoux, *The Emotional Brain* (New York: Simon and Schuster, 1996); see also my companion essay in this volume, "Emotions: A View Through the Brain."

reactions serving protective functions. This is plain old Pavlovian or classical conditioning; the elicited responses are called "conditioned emotional responses."[3]

At first, I thought I was working on the general topic of emotion, but soon I realized I was uncovering mechanisms involved in defensive or fear-related behavior, and that other emotions may have different neural mechanisms. At first, I also thought I was working on subjective emotional experiences, with behavior serving as a measure. Now I believe I was working on how behavioral systems themselves work. My current thinking about these issues can be summarized in two points:

1. Emotion systems (like the defense system) evolved as sensory-motor solutions to survival problems (such as detecting and responding to danger). "Emotion" may not be the best label for these, but the function subserved is the issue, not the label.

2. Feelings (conscious emotions) are not the function that emotion systems were built to perform. The problem of feelings is not a problem about emotions at all, but is simply the mind-body problem.

As a result, studying feelings takes one into philosophical morass of the mind-body problem. In the short run, more progress is likely to be had by focusing on how emotional behavior is regulated and controlled than by focusing on the mind-body problem. For this reason, much of my work has been in the area of emotional behavior rather than emotional feelings—but I have always kept at least one eye open to the question of feelings, for which I suggest a possible mechanism below.[4]

3 Tools of the Trade

How do we actually go about relating emotions to the brain? First we need behavioral tools, ways of reliably activating the system in question. In my studies of fear mentioned above, I started my work using classical conditioning. This turns out to have been an excellent choice. Fear conditioning is incredibly reliable (works in every rat every time), is rapidly learned (requires only one training trial), and lasts for long periods of time (perhaps a lifetime). By simply presenting the unconditioned stimulus (US), which is typically foot-shock, at the end of the conditioned stimulus (CS), usually a tone, the next occurrence of the CS elicits fear responses.

In addition to behavioral tools, we also need neuroscience tools, ways of relating behaviors to brain circuits. One of the oldest and most effective tools is the brain lesion. By damaging part of an animal's brain, or by examining people with brain lesions from neurological disease, one can get clues as to what parts of the brain do. However, the lesion method is flawed and imprecise, telling us as much if not more about the consequences of damaging the brain as it does about how the brain normally works. Therefore, techniques that show normal brain function are also useful. One such approach is to record the brain's electrical activity, especially from single cells, by placing electrodes in discrete brain regions, allowing one to examine the neural coding of stimuli and their associated behaviors. But where to put the electrode? The lesion technique often gives very useful hints. If damage to some area interferes with the behavior, that suggests the area might be part of the mediating circuit. By recording from the area, and determining if the electrical activity of the area is closely associated with the behavior, one can gather additional support for the conclusion that the area is involved. Functional imaging techniques, such as fMRI or PET scans, also reveal the activity of areas in the human brain (albeit indirectly).

---

[3] See LeDoux, *The Emotional Brain*.

[4] See also ibid.

It is important to recognize from the start that brain functions are not, strictly speaking, represented in brain areas, but are represented across brain circuits. So even if damage to some area interferes with the behavior, and the area's electrical activity changes in a way consistent with its involvement in the regulation of the behavior, it is not therefore necessarily the "mediator" of the behavior. Areas contribute to behavior through circuits. Thus, in addition to making lesions and recording activity, it is also often important to trace connections in the brain. Injecting tracer substances into a suspected area (from lesion or recording data) can help to identify the connections of the area. Lesions and recordings can then be made in the connected areas to see if they participate in the behavior. Many other tools are available to neuroscientists. Nevertheless, lesions, recordings, and pathway tracings are some of the basic tools used to explore the circuits mediating a behavior.

4 Component Processes and Emotions

One of the lessons of cognitive science is that to understand the mind we need to figure out how component processes work. This is as true of emotional functions as of cognitive ones. Terms like "cognition" and "emotion" reflect gross abstractions from and generalizations about brain functions; they are *not* functions that are represented in the brain. In other words, I do not think there is an emotion system in the brain. Instead, I believe that the brain has multiple emotion systems. If so, we will need to study individual systems and to dissect the components of the system.

The defense or fear system, for example, is one we have learned a good deal about by applying lesion, recording, and tracing tools. This system is most properly conceived as a set of systems or mechanisms that take care of different aspects of fear, with the aspects in question ranging from the detection and automatic reaction to danger (*reactive mechanisms*), to networks that allow us to act willfully (*action mechanisms*) or habitually (*habit mechanisms*) in the face of danger, to networks that allow us to be consciously aware that we are in danger and feel fear (*feeling mechanisms*). Each of these will be elaborated upon briefly below.

1. Reactive mechanisms: These are circuits "hard-wired" to respond to evolutionarily programed dangers (predators, intense sensory stimulation, etc.) and to produce the same responses to novel stimuli that occur in the presence of and become warning signals of novel dangers. Reactions include somatic motor responses (freezing and other forms of muscle tension) as well as visceral responses (autonomic and endocrine changes). Because of the long duration of visceral (especially endocrine) responses, reactions can lead to enduring states (moods). Reactions are elicited by the initial occurrence of a dangerous stimulus and involve little high-level cognitive processing (long-term memory, working memory, attention, etc).

2. Action mechanisms: In situations of danger, automatic reactions are followed by actions, responses that we willfully emit on the basis of our assessment of the situation, memories of similar situations in the past, expectations, evaluations of possible outcomes, etc. In contrast to reactive mechanisms, which involve minimal interactions with cognitive processes, action mechanisms depend extensively on cognitive processes. Reactions happen to us. We decide to perform actions.

3. Habit mechanisms: Actions can become routines by having been performed many times in similar situations. Habits can be the deliberate outgrowths of actions, or they can be learned implicitly, without much thought, through the mechanisms of reinforcement learning. Behaviors that are reinforced (with rewards or punishments) are either strengthened or weakened, according to the standard behavioral principles outlined by pioneers like Ivan Pavlov, E.L. Thorndike, B.F. Skinner, and C.L. Hull. For example, when rats learn to use conditioned fear stimuli as warning signals to avoid the occurrence of shock, they are essentially learning a habit through the mechanisms of reinforcement learning. The shock is avoided by performing some behavior in the presence of the stimulus, which reinforces the behavior and increases the likelihood that the behavior will occur in the presence of the stimulus that predicts the shock. Habits can be adaptive (don't walk through an intersection when the cross-traffic has a green light) or pathological (don't leave the house because there are many dangers out there). Once formed, habits can be expressed in automatic fashion, regardless of whether they were learned deliberately or implicitly (by reinforcement). Neuroscientists have accumulated evidence that such behavioral changes are mediated by alterations in the strength of synaptic connections in the brain. In other words, behavioral change (learning) is achieved by way of synaptic plasticity within systems that normally control the behavior in question.

4. *Feeling mechanisms:* These are conscious experiences that can accompany reactions, actions, or habits, and possibly elicit (cause) them. An important difference between a conscious emotion and an emotion-less conscious thought (a regular old thought as opposed to one charged with emotional implications) has to do with the fact that emotional states are more intense and enduring, due to the kinds of systems that get activated as part of the reaction, action, or habit—admittedly, though, we still know relatively little about consciousness.

5 How Does the Defense System Work Neurally?

1. *Reactions:* My research for the past fifteen years has focused on the mechanisms underlying fear reactions, which are described in detail in my companion essay in this volume. In brief, inputs to the amygdala from sensory processing areas of the thalamus and cortex are key to the processing of emotional signals, while projections from the amygdala to brainstem areas involved in the control of behavioral, autonomic, and hormonal responses are involved in fear expression.

2. *Actions/habits:* These mechanisms are understood in far less detail. However, it appears that connections from the amygdala to the striatum are involved in linking fear processing with motor control networks organized by the striatum and cortex.

3. *Feelings:* These are the least understood aspect of any emotion. I propose that a conscious feeling, such as being afraid or anxious, results when there is a concurrent representation in working memory of (1) current stimuli, (2) long-term declarative (conscious) memories elicited by those stimuli, (3) arousal from emotional networks like the defense system and its bodily responses, (4) cognitive evaluations of all of the above and resulting thoughts and plans. From what we know about working memory, these circuits involve the dorsal prefrontal cortex together with other prefrontal regions (anterior cingulate cortex, orbital cortex), and their interaction with the amygdala either by way of direct neural connections or by way of bodily feedback occurring during emotional expressions.

6 Conclusion

This outline of my work with emotions is admittedly sketchy—more details are included in my companion essay. It must be emphasized that much of what I have said here applies to the emotion fear. Whether it also applies to other emotions remains to be seen. However, it is clear that the basic strategy used to map the fear pathways could be used to study other emotions as well.

THE UNIQUELY HUMAN CAPACITY FOR
LANGUAGE COMMUNICATION:
FROM *POPE* TO [po:p] IN HALF A SECOND

Peter Hagoort

1 Introduction

No other species besides *Homo sapiens* has developed in the course of its evolutionary history a system of communication in which a finite set of symbols together with a series of principles for their combination allows an infinite set of expressions to be generated. This system of natural language enables members of our species to externalize and exchange thoughts within the social group, and, through the invention of writing systems, within society at large. Speech and language are effective means for the management of social cohesion in societies where the group size no longer allows this to be done by grooming, which is the preferred way of bonding in our genetic neighbors, the old world primates.[1]

The generative power of the human language system rests on its tripartite architecture.[2] In this architecture, language-relevant information is encoded in at least three distinct representational formats: one for meaning, one for syntax, and one for the sound structures of words and utterances. Through the process of mapping these representational structures onto each other, the conceptual structures that specify the content of the speaker's message are expressed as a linear sequence of speech sounds (speaking). Alternatively, during listening to speech conceptual structures are derived from a linear string of speech sounds. In this mapping process, combining units into hierarchical phrase structures (syntax) is the necessary link between conceptual structures and phonological (or sound) structures.

In the division of labor between the sciences investigating the human language faculty, it is the task of the linguist to specify the representational structures involved; and it is the task of the psycholinguist to investigate how these structures are accessed and exploited during listening and speaking. Finally, the cognitive neuroscientist is faced with the challenge of specifying how the brain enables human language, and of determining the spatio-temporal profile of neurophysiological activity underlying speaking (writing) and listening (reading).

In order to get a more precise picture of what these different sciences reveal about the human language faculty, we first have to specify what the overarching term "human language faculty" stands for. It refers to the collection of the following set of complex, related but at the same time distinct skills: speaking and listening, reading and writing, and in communities of the deaf, using sign language. Each of these skills requires that distinct representational structures in memory are accessed and exploited in real time. The cognitive architectures for these skills specify which

---

[1] Robin Dunbar, *Grooming, Gossip, and the Evolution of Language* (London: Faber and Faber, 1996); Willem J.M. Levelt, "Producing Spoken Language: A Blueprint of the Speaker," in *Neurocognition of Language*, Colin M. Brown and Peter Hagoort, eds. (Oxford: Oxford University Press, 1999), 83–122.

[2] Ray Jackendoff, *The Architecture of the Language Faculty* (Cambridge: MIT Press, 1997); idem, "The Representational Structures of the Language Faculty and Their Interactions," in *Neurocognition of Language*, Brown and Hagoort, eds., 37–81.

representational structures are involved and how these are operated on in real time. The neural architectures specify the ways in which these skills are instantiated in the wetware of the human brain. However, one should keep in mind that the distinction between the cognitive architecture and the neural architecture is an idealization. As a first approximation, it is useful to make a distinction between computations in symbolic terms (cognitive architecture) and in neurophysiological terms (neural architecture), but in a complete cognitive neuroscience of language these levels should be brought together.

In the remainder of this essay I will first discuss the cognitive architecture for one of these skills in more detail. I will then discuss the neural architecture. Finally, I draw some implications for a theory of the person.

2 The Cognitive Architecture

A central component of our language skills is the mental lexicon. The mental lexicon is the part of declarative memory specifying the knowledge that a language user has about the words of his or her native language. It is estimated that speakers of a language have an active vocabulary of at least 40,000 words.[3] For these words, speakers know what they mean and how they sound. In addition, they know the syntactic properties of words such as word class (noun, verb, etc.). All this lexical information is retrieved very quickly from memory. On average a speaker produces two or three words per second. This requires not only the retrieval of different sources of word information, but also the coordinated activation of a large ensemble of muscles involved in articulating speech. About 100 muscles are involved in speaking, whose innervation has to be coordinated with millisecond precision. Despite the complexity and speed of this cognitive activity, speakers are very accurate, and, on average, make less than one error in 1000 words. The occasional speech errors are nevertheless very informative about the architecture of the speech process. For instance, sounds can be exchanged between different words as in "*h*eft *l*emisphere" (instead of "left hemisphere"),[4] or can be produced too early as in anticipations ("it's a *m*eal mystery" instead of "it's a real mystery").[5] What these examples of speech errors illustrate is that words are not stored in memory as units, but have to be assembled from the constituting phonemes every time we produce a word. This assembly process occasionally goes wrong, resulting in sounds ending up in the environment of the wrong word.

Figure 1 presents a blueprint for speaking single words. It specifies what happens between the moment that we recognize a particular retinal image as, say, John Paul II, and the actual articulation of the sound stream "pope."

Speaking starts with specifying the conceptual content of the utterance. This specification can be determined by visual input as in our example. But in many cases the conceptual specification is determined by internal input, for example, the speaker's intention to express a certain idea. Whatever triggers the conceptual specification of the utterance, the speaker has to select a particular concept or series of concepts from the knowledge base in memory, and s/he has to select and decide

---

[3] For a general introduction to the mental lexicon, see Jean Aitchison, *Words in the Mind: An Introduction to the Mental Lexicon* (Oxford: Basil Blackwell, 1987).

[4] Victoria A. Fromkin, ed., *Speech Errors as Linguistic Evidence* (The Hague: Mouton, 1973).

[5] Ibid.

about the way of expressing (for example, a message can be expressed as a statement, but also as a question, with or without irony, etc.). For instance, in our example the speaker can decide to say "John Paul the Second" or "the pope," to mention just two possible alternatives.

The conceptual selection and specification process precedes the actual formulation process in which preverbal conceptual structures trigger the retrieval of linguistic structures necessary to express the idea as a series of speech sounds. Here, two completely different types of linguistic information need to be retrieved, one specifying the characteristics of the word sound, the other concerning the grammatical properties of a word.

Each word form in the mental lexicon is associated with syntactic word information.[6] This latter type of information is referred to as lemma information. Lemmas specify the syntactic properties of words, such as their word class (noun, verb, adjective, etc.). For nouns in gender-marked languages their grammatical gender is specified as well (e.g., *horse* in French has masculine gender, in Dutch it has neuter gender). Verb lemmas contain information on syntactic frames (the argument structures), and on the thematic roles of the syntactic arguments (the thematic structure). For instance, the lemma for the verb *donate* specifies that it requires a noun-phrase (NP) as the grammatical subject, an NP as the grammatical object, with the optional addition of a prepositional phrase (PP) as the indirect object (e.g., *John* <subject-NP> *donates a book* <direct object-NP> *to the library* <optional indirect object-PP>). In addition, the mapping of this syntactic frame onto the thematic roles is specified. For *donate* the subject is the *actor*, the direct object the *theme*, and the indirect object the *goal* of the action expressed by the predicate.

In the next phase of the formulation process, the selection of the appropriate lemmas trigger the retrieval of sound pattern of the utterance (see figure 1). During this phase, the speech sounds (phonemes) of the word become available. In addition to the phonemes, a word's metrical information is retrieved, specifying the number of syllables and the stress pattern (not shown in figure 1). In a processing step known as phonological encoding, the phonemes are assigned to their syllable positions in a left-to-right order. The outcome of phonological encoding is a "phonological word," containing the word-sound information as a sequence of syllables with the right stress pattern. Syllables are the codes that form the basis for the articulatory movements of the vocal cords, the velum, the tongue, the jaw and the lips. These are abstract codes, since they are independent of the starting positions of, for instance, the lips and the tongue. Speaking with or without a pipe in the mouth results in different articulation movement trajectories, which are nevertheless instructed by the same abstract syllable codes. The final outcome of this whole cascade of retrieval and activation processes is an acoustic signal that the listener uses to derive the intended message.

Apart from the experimental evidence for a distinction between the retrieval of lemma- and word-form information, we are all familiar with a phenomenon supporting this distinction. This is the so-called tip-of-the-tongue state, referring to the often embarrassing situation in which we know that we know the word, we even

---

[6] Willem J.M. Levelt, *Speaking: From Intention to Articulation* (Cambridge: MIT Press, 1989); idem, "Producing Spoken Language: A Blueprint of the Speaker"; Ardi Roelofs, "A Spreading-activation Theory of Lemma Retrieval in Speaking," *Cognition* 42 (1992): 107–42; idem, "Testing a Non-decompositional Theory of Lemma Retrieval in Speaking: Retrieval of Verbs," *Cognition* 47 (1993): 59–87.

know that it is, say, a noun with a particular grammatical gender (for example, neuter), but for some reason the retrieval of the sound form is hampered. The fact that we can access some aspect of word information but fail to retrieve others illustrates the idea that different aspects of word information are differentially stored

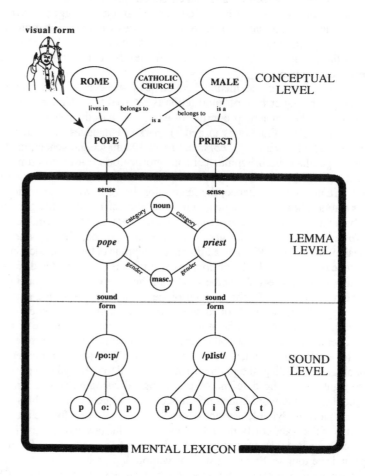

Figure 1. The Levelt and Roelofs model for speaking single words [see Willem J.M. Levelt, *Speaking: From Intention to Articulation* (Cambridge: MIT Press, 1989); Ardi Roelofs, "A Spreading-activation Theory of Lemma Retrieval in Speaking," *Cognition* 42 (1992)]. Concept nodes (POPE) are activated on the basis of sensory and/or conceptual input. Activation from a concept node spreads to its lemma node (*pope*) in the mental lexicon. In addition, activation of POPE results in increased activation of related concepts in semantic memory, such as for instance PRIEST. Each concept node is linked to exactly one lemma in the lexicon. At the lemma level the syntactic word information is specified, such as grammatical gender and word category. For instance, in Italian the gender of the lemma *pope* is masculine (*il papa*), whereas the gender of, for example, the Italian lemma for church is feminine (*la chiesa*). After the lemma has been selected, word form information is retrieved and prepared for articulation.

and retrieved. The simplistic idea that words are units to be found somewhere in the brain is simply wrong. "Word" is just nothing more than a handy catch phrase for an orchestra of information with players of different instruments. Despite differences in the details of various models of speaking, there is general agreement among researchers of language processing that it requires the temporally orchestrated retrieval of the different types of information discussed above.

The way in which I discussed the process of speaking thus far is as a feedforward process from intention to articulation.[7] However, introspectively we often have the feeling that our way of expressing our thoughts sharpens and molds our intentions. That is, speech starts with an intention, but intentions are also (partly) derived from speech. This intuition has led to criticisms of an account that sees speech as the information flow from intention to articulation.[8] However, in my view this criticism can be dealt with easily if we realize that as speakers we are also at the same time listeners. That is, we listen to our own speech, using the same machinery that analyses the speech of others. In the listening process we derive the intention from the speech sounds that hit our ears. Speaking is a highly incremental process, which means that we have not specified all the details of our preverbal message before we start the formulation process. The incremental nature of speech planning opens a window of opportunity in which listening to our own speech can further shape and mold our intentions, and via this route influence the ongoing formulation process. Given that our cognitive machinery of language includes both production and comprehension, it instantiates a continuous internal dialogue between "speaker" and "listener" resulting in the introspective feeling that intentions are not only the source but also the by-product of speaking.

In the example above I have given the rough outlines of the cognitive architecture of speaking, mainly restricted to speaking single words. A full-blown model of speaking specifies additionally how words are combined into longer utterances, how the intonational contours of multi-word utterances are determined, etc. Similar blueprints can be made for listening, reading, and writing.[9] In all these cases, establishing the details of the cognitive architectures for the different language skills is based on a combination of conceptual analysis, computational modeling, and clever experimentation. With the cognitive architectures in hand we can ask sensible questions about the neural instantiation of the different language skills. Without such explicit models, the study of the neural mechanisms of language is doomed to fail.

As an example of science in action, I will discuss one simple experiment in some detail. If lexical concepts such as POPE and PRIEST are stored in a network-like way as shown above in figure 1, and if the activation of a particular concept partly spreads to nearby concept nodes in the network, this would predict certain processing consequences for words that are preceded by semantically related words. This is tested in the following way. Participants in the experiment see word pairs that are flashed on a computer screen. First, one word is flashed on the screen for half a second, followed by a few hundred milliseconds blank screen. Then the target word is flashed on the screen for half a second. The participants are instructed to read aloud this second word as soon as it appears on the screen. The presentation of the

---

[7] Levelt, *Speaking*.

[8] Cf. Daniel C. Dennett, *Consciousness Explained* (Boston, Mass.: Little, Brown and Company, 1991).

[9] For detailed examples, see Brown and Hagoort, eds., *Neurocognition of Language*.

second word starts a clock that is stopped by the verbal response of the participant. As soon as the participant starts to read aloud the word on the screen, the clock stops. This allows the measurement of the participant's reaction time. In one condition of the experiment participants see the target word (for example, "priest") preceded in time by a semantically related word ("pope"). In the other condition the first word is unrelated in meaning to the second word (for example, "horse"). The prediction of the network model is that seeing the word "pope" results in partial activation of the word "priest," through the connection between the concepts POPE and PRIEST. If "priest" appears on the screen immediately after "pope," the reading of "priest" should be faster than in isolation, since it was already partly activated due to the preceding word "pope." However, the word "horse" does not spread part of its activation to "priest," since the concepts HORSE and PRIEST are too far apart in the semantic network space to influence each other. So if "priest" is read immediately after "horse" this should not speed up the reading process. The results of this type of experiment are in line with the predictions from the network model. Subjects are a few tenths of a millisecond faster in reading "priest" preceded by "pope" than in reading "priest" preceded by "horse." This so-called *priming effect* suggests that information about the meaning of words is stored in memory as a network of connected pieces of information, and not as isolated packages of individual word meanings.

3 The Neural Architecture

The neural architecture specifies the spatio-temporal dynamics of the brain processes that convert the retinal image of, say, John Paul II into the speech sound [po:p]. That is, we have to specify which areas of the brain are activated during the processes involved, and how the concomitant activations are temporally orchestrated.

Recently, the recording of the electrical activity of the brain has resulted in a fairly fine-grained estimation of the time course of the different processes involved. Electrical brain activity is usually recorded from a series of electrodes at the scalp. If these recordings are done time-locked to sensory, motor, or cognitive processes, scalp-recorded event-related brain potentials (ERPs) result, reflecting the sum of simultaneous post-synaptic activity of a large ensemble of neurons. The ERPs have a high temporal resolution, in the order of a few milliseconds. Based on the latency of these components, relevant information can be obtained about the time course of the underlying processes (see figure 2).

Although for certain reasons[10] ERPs have been mainly recorded by language researchers in relation to aspects of language comprehension, recent research has applied this techniques successfully in studying the production of speech.[11] Without going into the details, on the basis of these and some other studies an educated guess can be made about the temporal dynamics of converting a visual image into an artic-

---

[10] Cf. Peter Hagoort and M. van Turennout, "The Electrophysiology of Speaking: Possibilities of Event-related Potential Research on Speech Production," in *Speech Production: Motor Control, Brain Research and Fluency Disorders*, Wouter Hulstijn, Herman F.M. Peters, and Pascal H.H.M. Van Lieshout, eds. (Amsterdam: Elsevier, 1997), 351–61.

[11] Miranda van Turennout, P. Hagoort, & C.M. Brown, "Electrophysiological Evidence on the Time Course of Semantic and Phonological Processes in Speech Production," *Journal of Experimental Psychology: Learning, Memory, and Cognition* 23 (1997): 787–806; idem, "Brain Activity During Speaking: From Syntax to Phonology in 40 Milliseconds," *Science* 280 (1998): 572–74.

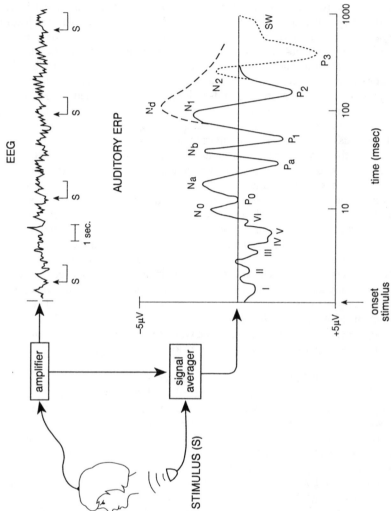

Figure 2. (After Steven A. Hillyard and M. Kutas, "Electrophysiology of Cognitive Processing," *Annual Review of Psychology* 34 [1983]: 33–61.) Idealized waveform of a series of ERP components that become visible after averaging the EEG to repeated presentations of a short auditory stimulus. In this figure, the EEG is recorded from one electrode, placed at a central midline site on the scalp. Usually, averaging over a number of stimulus tokens is required to get an adequate signal-to-noise ratio. Along the logarithmic time axis the early brainstem potentials (Waves I–VI), the midlatency components (N_0, P_0, N_a, P_a, N_b), the largely exogenous components (P_1, N_1, P_2), and the endogenous, cognitive ERP components (N_d, N_2, P_3, Slow Wave) are shown. The components with a negative polarity are plotted upwards; the components with a positive polarity are plotted downwards. The exogenous components mainly reflect the physical stimulus characteristics (e.g., intensity, size, duration). The endogenous components reflect in particular the cognitive information processing consequences of the stimulation of (one of) the sensory systems.

ulated sequence of speech sounds.[12] In our example, it will take at least 150 milliseconds to perceive and categorize the retinal image as John Paul II.[13] The following activation of the concept POPE takes less than 200 milliseconds. It is followed by a cascade of retrieval processes. Activation of the syntactic features of *pope* (the lemma information) precedes the retrieval of the onset phoneme of the word [po:p] by about 40 milliseconds.[14] Importantly, the information about a word's phonological form is not available at once, but accrues in a left-to-right order. For words of 3 phonemes (/p//o://p/), it takes maximally 80 milliseconds to retrieve the remaining segments once the word-initial phoneme is available.[15] Since it takes about 600 milliseconds before articulation of the word [po:p] starts, the remaining time is necessary for preparing (and partly executing) the articulatory motor program on the basis of the phonological information.

Apart from answering the question about the time course, we need to specify the brain areas involved in the cascade of processing operations involved in speaking. For this we have to rely either on evidence from lesion data or on measurements of brain activity with the help of modern brain imaging techniques such as Positron Emission Tomography (PET) and functional Magnetic Resonance Imaging (fMRI). Lesion data come from patients who suffered from a stroke or brain tumor resulting in a language impairment. A precise analysis of the site and the size of the lesion on the one hand, and of the specific nature of the language impairment on the other, are used for making inferences about which areas of the brain subserve a particular aspect of language processing. For instance, lesions in the frontal cortex involving Broca's area often result in an impairment in producing the correct sound pattern for the intended words, which suggests that this brain area is normally involved in, among other things, the assembling of a word's sound pattern. In this way, relating lesion site to impairment symptoms is used for assigning language functions to brain structures.

PET and fMRI measure hemodynamic signals. They enable the detection and visualization of functionally induced local blood flow changes (PET), or changes in blood oxygenation (fMRI), which are assumed to be correlated with the activation of nearby neural tissue.[16] Roughly speaking, in this way the locus of neural activity related to cognitive processes is detected through a vascular filter. This implies that the temporal resolution of these techniques is inherently limited by the temporal dynamics of changes in blood flow or blood oxygenation, which is on the order of hundreds of milliseconds to a few seconds. This contrasts with the electrophysiological recordings that are directly related to neural activity, and have a temporal resolution on the order of milliseconds. However, the ERP measurements suffer from the so-called inverse problem, which makes it difficult to determine the localization of the electrical potentials that are picked up at the scalp. For the time being, only PET and fMRI provide measurements with the required spatial resolution.

---

[12] For more details, see Hagoort and van Turennout, "The Electrophysiology of Speaking."

[13] Simon J. Thorpe, D. Fize, & C. Marlot, "Speed of Processing in the Human Visual System," *Nature* 381 (1996): 520–22.

[14] Van Turennout, Hagoort, & Brown, "Brain Activity During Speaking."

[15] Van Turennout, Hagoort, & Brown, "Electrophysiological Evidence on the Time Course of Semantic and Phonological Processes in Speech Production."

[16] For a general introduction, see Marcus E. Raichle, "Visualizing the Mind," *Scientific American*, April 1994, 36–42.

Recent years have seen an increasing number of PET and fMRI studies on language processing. In the absence of an animal model for language, we are strongly dependent on these new brain imaging techniques to see the brain in action during language tasks. The following logic underlies most brain imaging studies on language. The patterns of brain activation associated with tasks that tap a specific step (for example, the retrieval of lemma information) in the cascade of processes involved in speaking are compared with activation patterns associated with tasks in which this particular process is not involved. Through this comparison one can determine the brain areas that are more strongly activated during this step (lemma retrieval) in the overall process. The areas that are more strongly activated are assumed to be the areas that are particularly involved in this aspect of speaking. For instance, one can ask participants to read aloud words and pseudo-words. The latter are phonotactically legal letter strings, which do not happen to be existing words in, say, English. An example is the letter string *floke*. Everyone can read this word, but no one knows what it means. If one compares the English word *smoke* with the pseudo-word *floke*, the following two differences arise in the processes between seeing it written and saying it aloud. One difference is that for *smoke* the meaning gets accessed in the course of the process, whereas for *floke* we don't have a semantic representation in memory. In addition, we cannot retrieve a phonological code from memory for the pseudo-word *floke*. Instead, we have to assemble such a phonological code by converting the individual graphemes into the corresponding phonemes. If we measure the patterns of brain activity associated with reading aloud words and pseudo-words, the differences between the brain activity associated with words and pseudo-words are due to the retrieval of word meaning and phonology and/or the assembly of the phonological code for pseudo-words. To further segregate the brain activations related to meaning and phonology, other task comparisons are needed. In this way one can figure out which areas of the brain are differentially activated during the different steps in the process of speaking. Of course, there are also areas that are crucial to both tasks, since in addition to the differences there are also commonalities between reading words and pseudo-words. For instance, in both cases one sees activation in the primary visual cortex, since the whole reading process starts with the analysis of the visual patterns that fall on the retina.

In a recent meta-analysis of more than fifty brain imaging studies on single word production, Peter Indefrey and W. Levelt summarized the current understanding of the neural circuitry underlying the cognitive activity that I described above.[17] All core steps in the speaking process are subserved by areas in the left hemisphere, which is the language dominant hemisphere in the large majority of people. Selecting the appropriate concept for speaking (POPE) seems to involve the left middle temporal gyrus (see figure 3 for an overview). From there the activation spreads to Wernicke's area, which is pivotal in retrieving the phonological code of a word stored in memory. Wernicke's area plays a crucial role in the whole network of language processing by linking the lexical aspects of a word form to the widely distributed associations that define its meaning. This role is played by Wernicke's area in both language production and language comprehension.[18] The lexical word form information is relayed to Broca's area in the left frontal cortex and/or the middle part

[17] Peter Indefrey and W.J.M. Levelt, "The Neural Correlates of Language Production," in *The Cognitive Neurosciences*, 2nd edition, M. Gazzaniga, ed. (Cambridge: MIT Press, 2000).

[18] M-Marsel Mesulam, "From Sensation to Cognition," *Brain* 121 (1998): 1013–52.

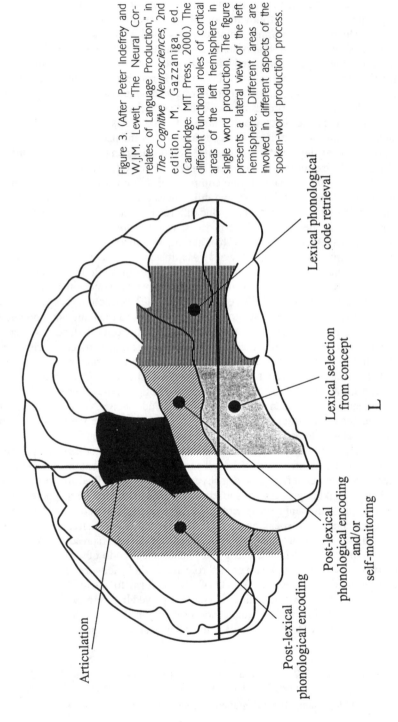

Figure 3. (After Peter Indefrey and W.J.M. Levelt, "The Neural Correlates of Language Production," in *The Cognitive Neurosciences*, 2nd edition, M. Gazzaniga, ed. (Cambridge: MIT Press, 2000.) The different functional roles of cortical areas of the left hemisphere in single word production. The figure presents a lateral view of the left hemisphere. Different areas are involved in different aspects of the spoken-word production process.

Lexical phonological code retrieval

Lexical selection from concept

L

Post-lexical phonological encoding and/or self-monitoring

Articulation

Post-lexical phonological encoding

of the superior temporal lobe in the left hemisphere. These areas play a role in the conversion of the phonological codes in memory into phonological words from which the abstract articulatory program is derived. In the final phase of preparing for articulation and execution of articulation sensorimotor areas become activated, with the possible additional contribution of the supplementary motor area and the cerebellum (the latter two areas are not shown in figure 3).

The conceptual knowledge that we have about the words of our language seems to be distributed more widely than the lexical lemma and form aspects, and is not restricted to the left hemisphere. Moreover, brain imaging and lesion studies on the semantics of concrete nouns indicate that perceptual and functional attributes of word meaning might be accessed through different parts of the brain with the perceptual attributes closer to the primary sensory areas and the functional attributes closer to the motor cortex.[19] The transmodal cortex, the midtemporal cortex, and Wernicke's area are convergence zones or critical gateways for accessing relevant information that is represented in a distributed way.[20] On the basis of the overall organization of the cortex, these areas are well suited to gate our distributed conceptual knowledge into one word form. That all we know about John Paul II converges onto the single word form [po:p] requires the involvement of brain areas that are specialized for binding distributed fragments of knowledge into a single output, in this case a single word form.

Combining knowledge about the overall organization of the brain with information of specific patterns of language-related brain activity and its temporal dynamics allows us to gain insight into the neural organization of the uniquely human capacity for communication by means of natural language. Understanding this highly complex communication system requires a lot of skillful experimental research on detailed issues. Ultimately the understanding that we gain by doing this has wider ramifications for central questions concerning the human person.

4 Language and the Theory of the Person

A full theory of the person requires a specification of the ways in which the signals from such very different functional systems as language, memory, emotion, motor action, etc. with their own dedicated neural circuitry give rise to the sense of self and personhood. How exactly this happens is still largely unknown. How the brain solves the problem of binding the signals of these different systems into the sense of a unified self with continuity from past to future is an almost complete *terra incognita* for current cognitive neuroscience. Despite successful models of different cognitive systems, cognitive neuroscience still lacks an overarching theory of the person. However, even in the absence of a theory of the human person, it is a reasonable guess that such a theory would look quite different if we lacked language. Within the context of his schema theory, Michael Arbib (in this volume) argues that the self is a schema encyclopedia containing hundreds of thousands of schemas that a person uses to interpret and add new information to memory. The schema encyclopedia is used to "tell a story" to fit the new data. Not only are the metaphors used to describe

---

[19] For an overview, see Eleanor M. Saffran and A. Sholl, "Clues to the Functional and Neural Architecture of Word Meaning," in *Neurocognition of Language*, Brown and Hagoort, eds., 241–72.

[20] Mesulam, "From Sensation to Cognition."

the sense of self and person very often derived from the domain of language;[21] it is also clear that language allows us to increase the size of our schema encyclopedia at an amazing rate. Although a cognitive neuroscience of language does not explain the content of our schema encyclopedia, it is indispensable in explaining the machinery that allows us to build up this large schema encyclopedia. No doubt, in the absence of our language capacities our sense of self and person would be substantially more limited and boring. In this regard language is a key component of a theory of the person, for which the input from a cognitive neuroscience of language is thus much needed.

A similar story can be told about the relation between language and awareness. Although consciousness has also not been explained satisfactorily in terms of cognitive neuroscience, despite claims to the contrary,[22] awareness stands a better chance. Awareness is related to our ability to give phenomenal judgments and verbal reports about our sensations. Just as PET and fMRI allow a view of neural activity through a hemodynamic filter, awareness allows a view of consciousness through a linguistic filter. Understanding the characteristics of the filter is also in this case of crucial importance to a better understanding of the central but still mostly evanescent phenomenon of consciousness.

Acknowledgment. I am grateful to Michael Arbib for his comments on an earlier version of this essay.

---

[21] See also the proposal for the interpreter in Michael S. Gazzaniga, *Nature's Mind: The Biological Roots of Thinking, Emotions, Sexuality, Language, and Intelligence* (New York: BasicBooks, 1992).

[22] See David J. Chalmers, *The Conscious Mind: In Search of a Fundamental Theory* (Oxford: Oxford University Press, 1996), for a thought-provoking account.

THE COGNITIVE WAY TO ACTION

Marc Jeannerod

1 Introduction

The generation of voluntary actions is a fascinating problem, because action is bound to the existence of a self. It is the means by which the self expresses itself and inter-acts with the external world. Indeed, action is not limited to the set of muscular contractions which underlie its phenomenal, overt, appearance. Action generation also includes covert aspects, such as the internal representation of the goal and the means to achieve that goal. Covert and overt aspects of an action are parts of a single representation-execution continuum, such that an overt action necessarily involves a covert counterpart, whereas a covert action does not necessarily involve an overt counterpart. This essay is a brief outline of the main concepts and trends in the study of voluntary action.

2 Purposeful Action and Internal Models

The early researchers in this field had noticed (perhaps not innocently) that even the simplest movements produced by the nervous system already appeared to be organized purposefully. Edward Pflüger, for example, claimed that reflex movements elicited by noxious stimuli applied to the leg in the spinal frog were real "intelli-gent"actions, as if the animal tried to escape or to protect itself against undesirable stimuli.[1] Similarly, David Ferrier interpreted the muscular contractions he elicited by applying electrical shocks to the monkey cerebral cortex as coordinated actions: "Many of the movements such as those of the hands, the legs, the facial muscles and the mouth have the aspect of purpose or volition and are of the same nature as those which the animal makes in its ordinary intelligent action."[2] These impressions relied on the fact that muscular contractions in response to stimulation were not localized to single muscles, they were coordinated contractions of several muscles. "Coordina-tion," to use Paul Weiss's terms, "refers to the fact that the central nervous system engages the muscles in such a definite order that... their combined activities result in orderly movements, which, in turn, yield acts of biological adequacy for the whole animal."[3] But, Weiss added, we ignore the principle in operation in the centers to make the appropriate selection. A great deal of current research is devoted to understanding how actions are "represented" in the brain, in other words, how the elementary mechanisms which ultimately control muscle contraction are selected and assembled, and what are the selection and assemblage "principles."

There is a striking disproportion between our knowledge of the physiological properties of the elements and our limited understanding of the operation of the

---

[1] This concept was epitomized by Pflüger as the "Rückenmarkseele," literally, the soul of the spinal cord; see Franklin Fearing, *Reflex Action: A Study in the History of Physiological Psychology*, new edition, (Cambridge: MIT Press, 1970, original ed. 1930, Williams and Wilkins). The brain of a "spinal frog" has been removed, leaving only the spinal cord.

[2] David Ferrier, *The Functions of the Brain* (London: Smith Elder, 1876), 95.

[3] Paul Weiss, 1941, quoted in Charles R. Gallistel, *The Organization of Action: A New Synthesis* (Hillsdale: Erlbaum, 1980), 268.

nervous system as the coordinator of these elemental activities. Charles Sherrington had led the way in searching for the simplest behavioral unit where coordination would already be at work (what he called the "unit of integration"). Simple spinal reflexes, like the extensor or the flexor reflexes, for example, fulfilled his criterion for behavioral elements, as they integrated, in an orderly way, the activity of several motor units. According to his famous keyboard analogy, Sherrington considered that motor units (the motor neuron and the muscle fibers to which it is connected) are played by different reflexes, but in a different order and to a different effect. Coordination, a "co-adjustment" of simple reflexes into more complex ones, was based on mechanisms such as reciprocal facilitation of reflexes having complementary effects, reciprocal inhibition of antagonist reflexes, irradiation for recruiting new reflexes as stimulus strength increased, etc. Another Sherringtonian concept, "sequential combination," accounted for the chaining of reflexes into behavioral sequences.[4] In fact, it now appears that even the most automatized motor sequences, like swallowing or locomoting, are not organized according to the chaining model, but rather depend on built-in "programs" (see below). In addition, the introduction in neuroscience of new concepts (derived from both control theory and cognitive psychology) such as the concept of representation, has changed the views on coordination of complex actions. Although the notion of behavioral elements is still considered as valid, the present view on how these elements are selected and assembled to produce a coordinated action has radically changed.

A major step in changing views on coordination is the notion of "internal models." Organisms not only react to external perturbations or events; they also actively initiate these interactions. This means that representations that account for behavior must not only be reactive, they must be predictive. They must carry internal models of the state of the external world, how this state will be modified by the action of the organism, and how the organism will be modified by that action. An early predecessor to the modern concept of internal models is that of homeostatic regulation, a notion that progressively emerged during the nineteenth century when engineers felt the need to control the motion of machines. Special devices were used for keeping the velocity of the machine as uniform as possible, in spite of variations in the driving force or the resistance. The general principle was to determine a reference value of the parameter to be controlled and to automatically activate the controller when its value departed from the reference.[5] Biological systems also appeared to be liable to the same mode of functioning. Claude Bernard discovered that systemic regulations were circular mechanisms aimed at maintaining "constancy" of the internal milieu. Regulation of blood glucose, for example, was based on constancy of glycemia at a level corresponding to tissular metabolic needs (the reference value). When glycemia dropped below the reference, processes were activated to restore it. Following the work of Claude Bernard, this idea of self-regulation (homeostasis) has received a broad recognition among biologists and has been used to explain many different physiological functions.

Homeostatic systems, however, are closed-loop systems aimed at maintaining the constancy of a fixed inbuilt reference, which can only account for crude interactions between the organism and the external environment. The important point here is that

---

[4] For a review of Sherrington's contribution, see Gallistel, *The Organization of Action*.

[5] This idea was first expressed in 1868 by Maxwell. For references, see Marc Jeannerod, *The Cognitive Neuroscience of Action* (Oxford: Blackwell, 1997).

this notion of a stored reference implies the existence of a certain form of *representation* of the regulated parameter to which incoming signals are compared; it also provides an image of what an internal model could be (figure 1). Kenneth Craik was among the first to assume the existence of such internal models in the brain, a major break from the prevailing emphasis on purely stimulus-response mechanisms.[6] This notion drew attention to the role of endogenous factors such as stored knowledge or mental content, which are now currently considered by cognitive psychologists as causal factors in behavior. Craik's internal models were thought of as analog representations (he died shortly before the first digital computers were built). Later

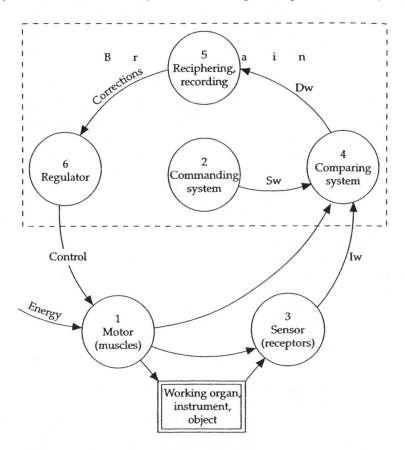

Figure 1. An early example of a motor representation: Bernstein's comparator model. The information (Sw) generated by the commanding system is compared with reafferent input (Iw) resulting from movement execution. The output (Dw) of the comparator determines the global output of the system. The commanding system meets the requirements for a device where action is being represented before execution. From N. Bernstein, *The Coordination and Regulation of Movements* (Oxford: Pergamon, 1967).

---

[6] Kenneth J.W. Craik, *The Nature of Explanation* (Cambridge: Cambridge University Press, 1943).

developments in artificial intelligence emphasized digital representations and hierarchical structures—similar to those of computer programs—and were used to conceptualize how actions could be represented as sequences of steps involving tests and operations.[7]

Internal models also had antecedents in the psychological literature. The popular concept of "schema" that Frederic Bartlett first used to account for the production of behavioral responses is an early formulation of the same idea. According to Bartlett, behavior would be built as a combination of stored elementary reactions, or schemas, ready for use. His legacy can be tracked in modern versions of the same idea, as used by several authors for explaining how the representation of an action can be assembled from stored elements. Modern schemas can be acquired through experience and learning, can be improved, changed, destroyed, etc., and, most importantly, are available as building blocks for creating dynamic representations of actions.[8]

Finally, the existence of internal models became widely recognized in neuroscience. Neurophysiological studies in invertebrates or in lower vertebrates have demonstrated the existence of central discharges which may account for automatic behavior.[9] In the context of ethology, behavioral sequences have been described, which unfold blindly and eventually reach their goal after they have been triggered by external cues (the so-called "trigger features"). Localized brain stimulations can also trigger similarly complex actions which outlast by a considerable amount of time the duration of the stimulus.[10] In the domain of human action, Nikolai Bernstein also expressed the same idea of internal models of action, saying that "There exist in the central nervous system exact formulae of movement or their engrams.... The existence of such engrams is proved... by the very fact of the existence of habits of movements and of automatized movements." According to Bernstein, the engram of an action must contain, "like an embryo in an egg or a track on a gramophone record, the entire scheme of the movement as it is expanded in time. It must also guarantee the order and the rhythm of the realization of this scheme; that is to say, the gramophone record... must have some sort of motor to turn it."[11] The content of these models (the embryo in the egg) is still a matter of intense debate.

---

[7] George A. Miller, Eugene Galanter, and Karl H. Pribram, *Plans and the Structure of Behavior* (New York: Holt, 1960).

[8] Frederic C. Bartlett, "Review of *Aphasia and Kindred Disorders of Speech*, by Henry Head," *Brain* 49 (1926): 581–87; Michael A. Arbib, "Perceptual Structures and Distributed Motor Control," in *Handbook of Physiology, Section I: The Nervous System, Vol. 2: Motor Control*, V.B. Brooks, ed. (Baltimore: Williams and Wilkins, 1981), 1449–80; Ulric Neisser, *Cognition and Reality* (San Francisco: Freeman, 1976); Tim Shallice, *From Neuropsychology to Mental Structure* (Cambridge: Cambridge University Press, 1988).

[9] For example, see Sten Grillner, "Neurobiological Basis of Rhythmic Motor Acts in Vertebrates," *Science* 228 (1985): 143–49.

[10] For a review see Richard Bandler, "Brain Mechanisms of Aggression as Revealed by Electrical and Chemical Stimulation: Suggestion of a Central Role for the Midbrain Periaqueductal Grey Region," *Progress in Psychobiology and Physiological Psychology* 13 (1988): 67–154.

[11] Nikolai Bernstein, *The Coordination and Regulation of Movements* (Oxford: Pergamon Press, 1967), 37–39.

3 Action in the Brain

In the middle of the nineteenth century, converging discoveries in the field of histology, anatomy, and physiology of the nervous system established the role of the cerebral cortex in the organization of actions. As already mentioned, physiologists of the time had discovered the effects of electrical stimulation of parts of the cerebral cortex in producing muscular contractions; they further noted that destroying these excitable zones produced a paralysis of the same muscles. As David Ferrier noticed, "if one attributes a motor function to a nerve because, by irritating it one obtains a muscular contraction and by sectioning it one obtains a paralysis of the same muscle, I do not see why motor functions cannot be similarly attributed to a cortical center, given the fact that the phenomena are essentially the same." However, Ferrier's *motor* theory of the motor cortex was only one among several. The most popular conception was that of a store of motor "images" or motor "ideas" built from the sensations gained during previous movements, and reassembled when a new movement had to be produced. The motor cortex produced voluntary movement because it was the endpoint of the intrapsychic and intracerebral pathways along which information arrived from the sensory systems. Later on, the observation of clinical cases led to a more complicated theory, involving serial steps in the generation of the action. "In the series of physiologic phenomena which the execution of an act presupposes," claimed Hugo Liepmann, "we know sensory excitation, mental representation of the object, representation of the act to be accomplished, the awakening of corresponding motor images in an ideomotor center, excitation leaving this ideomotor center and, finally, muscle contraction."[12]

This classical model of a serial organization of action generation inherited from nineteenth-century associationism is now seriously questioned. New data instead suggest that activation occurs simultaneously at the cortical and subcortical levels of the motor system. Experimentally, unit activity in cortical areas thought to represent levels of organization of a purposeful action (for example, primary motor cortex, dorsal and lateral premotor cortex, cingulate cortex, dorsolateral prefrontal cortex, posterior parietal cortex) occurs more or less synchronously, and the existence of a sequence usually can only be detected statistically. Reciprocal connections between these areas account for this wide distribution of activity within a neural network devoted to generation of actions.

3.1 Is There a Central Coordinator for Complex Actions?

Such a distributed model of action generation raises the issue of the existence of a central coordinator which would determine the temporal structure of the motor output and, ultimately, the sequence of the action (figure 2). Neuropsychological data suggest that the frontal lobes might play the role of such a central coordinator for structuring the pattern of action plans. Patients with prefrontal lesions (lesions located anterior to the primary motor areas), although normally able to perform simple actions, are impaired in decision making, anticipation of action consequences, adaptation to changing rules, and resistance to distracting stimuli. Indeed, one should expect that, if an action is under the control of a goal list, it should continue until the goal is satisfied, and that a failure of this process should trigger the search for

---

[12] See Marc Jeannerod, *The Brain-machine: The Development of Neurophysiological Thought* (Cambridge: Harvard University Press, 1985).

another, more appropriate, action structure. This is exactly what frontal patients cannot do: their performance may be interrupted although the goal remains unsatisfied; or conversely, it may continue in spite of violation of goal satisfaction. Jordan Grafman has used the term "scripts" to designate neurally represented mental

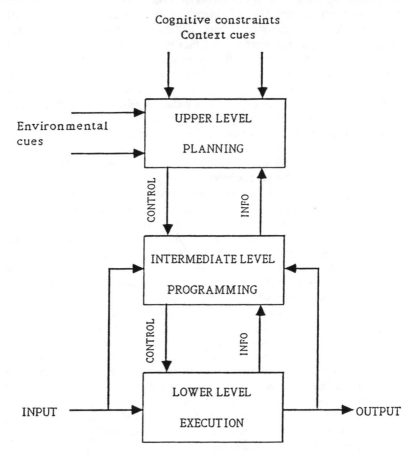

Figure 2. A distributed model for the generation of actions. This simple diagram accounts for three levels of processing action-related information. At the lower level, reflex actions can (in principle) be generated on the basis of direct input-output transformations. This execution level, however, as already thought by Edward Pflüger in the nineteenth century, is under the control of upper levels. The intermediate level also has dynamic properties for generating on-line adaptations, corrections, and coordinations. It can also account for several aspects of learning. Finally, the upper level is immune to direct contact with the environment. It receives higher-level information derived from the state of sensory processing areas (environmental cues), or from the state of cognitive structures. Note that the diagram can account for bottom-up regulation of higher levels (e.g., by sensory cues, ongoing actions) as well as for top-down control of lower levels. Redrawn from Marc Jeannerod, "A Hierarchical model for Voluntary Goal-directed Actions," in *The Principles of Design and Operation of the Brain*, J.C. Eccles and O. Creutzfeldt, eds. (Berlin: Springer-Verlag, 1990), 257–75.

entities that have to be assembled to organize an action sequence.[13] Frontal patients would be impaired in managing the syntax of scripts, that is, in ordering the actions that compose a given script according to the context or to priority rules.[14] By contrast, these same patients may be able automatically to perform relatively complex well-learned routines or stimulus-driven actions. They may even remain stuck in the well-learned set and perseverate into it, or they may imitate actions per-

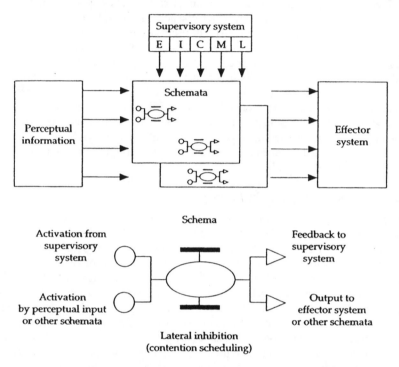

Figure 3. A supervisory system for action planning. The upper diagram shows how perceptual information activates schemas which ultimately produce the action. Schemas are selected under the influence of the supervisory system (supposedly in the frontal lobes) using operations such as energization of schemas (E), inhibition (I), contention scheduling (C), monitoring schema activity (M), and logical control (L). The lower diagram exemplifies one schema with its inputs and outputs. The contention-scheduling action from the supervisory system is responsible for coordination between schemas. From Donald T. Stuss, T. Shallice, M.P. Alexander, and T.W. Picton, "A Multidisciplinary Approach to Anterior Attentional Functions," in *Structure and Functions of the Human Prefrontal Cortex*, J. Grafman, K. Hollyoak, and F. Boller, eds., *Annals of the New York Academy of Sciences* 279 (1995): 191–211.

---

[13] Whether scripts and schemas refer to the same concept is a matter of discussion. See Jordan Grafman, "Plans, Actions and Mental Sets: Managerial Knowledge Units in the Frontal Lobes," in *Integrative Theory and Practice in Clinical Neuropsychology*, E. Perecman, ed. (Hillsdale: Erlbaum, 1989).

[14] See Angela Sirigu, T. Zalla, B. Pillon, J. Grafman, B. Dubois, and Y. Agid, "Planning and Script Analysis Following Prefrontal Lobe Lesions," in *Structure and Functions of the Human Prefrontal Cortex*, J. Grafman, K. Hollyoak, and F. Boller, eds., *Annals of the New York Academy of Sciences* 279 (1995): 277–88.

formed in front of them. This ability to carry on preplanned actions might correspond, in the frontal patients, to the residual function of the basal ganglia and would
be consistent with the role of these structures in timing and checking the sequence
of actions. This hypothesis is consistent with the notion that the basal ganglia may
contribute to the process of habit formation.

Identifying the supervisory system with the frontal lobes, in addition, requires
some revision of the classical view which considered frontal lobe function in terms
of a unitary process (figure 3, overleaf). The prefrontal cortex (the so-called granular
frontal cortex) is known to be a mosaic of subareas, so that there is a possibility that
these subdivisions could match the elementary operations that compose planning. In
addition, the prefrontal cortex is part of several circuits involving not only other
cortical areas such as the cingulate cortex but also the basal ganglia. Experimental
data in monkeys suggest that frontal lobes should in fact be fractionated in several
modules for domain-specific working memories.[15] Donald Stuss and his colleagues
together with Tim Shallice thus suggested that the frontal syndrome should be
divided into several clinical entities, corresponding to an equal number of functional
systems within the frontal lobes.[16]

3.2 Representations within the Brain

The problem of the neural substrates of action generation has recently been revisited
under the light of a classical paradigm in cognitive psychology: mental imagery. This
paradigm has been used extensively in the context of visual perception and has
revealed a potent tool for understanding the cognitive aspects of visual information
processing and their neural substrates.[17] It has now been extended to the study of
mental processes related to action by using motor imagery. Imagining, or mentally
simulating, an action is a particular type of mental imagery, which can be elicited at
will. Mentally simulated actions (as explored by the methods of cognitive psychology, such as mental chronometry) share many properties with actual actions. They
have the same duration and obey the same cognitive constraints, such as (Paul)
Fitts's law. In addition, mentally simulating an action results in a broad subliminal
activation of the motor system, as shown by increase in spinal reflexes, and of the
autonomic system.[18]

---

[15] Patricia S. Goldman-Rakic, "Architecture of the Prefrontal Cortex and the Central
Executive," in *Structure and Functions of the Human Prefrontal Cortex*, Grafman, Hollyoak,
and Boller, eds., 71–83.

[16] Donald T. Stuss, T. Shallice, M.P. Alexander, and T.W. Picton, "A Multidisciplinary
Approach to Anterior Attentional Functions," in *Structure and Functions of the Human
Prefrontal Cortex*, Grafman, Hollyoak, and Boller, eds., 191–211; see also Tim Shallice and
P.W. Burgess, "Deficits in Strategy Application Following Frontal Lobe Damage in Man,"
Brain 114 (1991): 727–41. Note that the concept of a hierarchically higher mechanism is not
in contradiction with that of parallel processing. A hierarchy between levels, although it
implies degrees of specialization for each of these levels, does not imply a sequential order of
activation.

[17] Stephen M. Kosslyn, N.M. Alpert, W.L. Thompson, V. Maljkovic, S.B. Weise, C.F.
Chabris, S.E. Hamilton, S.L. Rauch, and F.S. Buonanno, "Visual Mental Imagery Activates
Topographically Organized Visual Cortex: PET Investigations," *Journal of Cognitive
Neuroscience* 5 (1993): 263–87.

[18] See Jeannerod, *The Cognitive Neuroscience of Action*.

Motor imagery thus offers a good opportunity for studying the covert aspects of action generation, uncontaminated by execution. More specifically, the study of brain activity during mental simulation of action may reveal the functional anatomy of the involved neural network. This study was first attempted by David Ingvar and L. Philipsson in 1977 and by Per Roland and his coworkers in 1980,[19] and was the subject of a series of recent experiments using neuroimaging techniques.[20] In the study by Jean Decety and his coworkers, subjects imagined themselves grasping virtual objects with their right hand. Activity increased in several areas concerned with motor behavior. At the cortical level, area 6 in the inferior part of the frontal gyrus was strongly activated on both sides, as was area 40 in the left inferior parietal lobule. Subcortically, the caudate nucleus was found to be activated on both sides and the cerebellum on the left side only. Another focus of activity was observed in left prefrontal areas, extending to the dorsolateral frontal cortex (areas 9 and 46). Finally, the anterior cingulate cortex (areas 24 and 32) was bilaterally activated. In addition, Klaus-Martin Stephan and his colleagues, in comparing the effects of execution, mental simulation, and preparation to move, found a prominent involvement of supplementary motor area (SMA). Whereas the rostral part of SMA was preferentially activated during simulated movements, its caudal part was activated during execution. The question of whether the primary motor cortex is also active during mental simulation (that is, when no execution occurs) remains a controversial issue. Activation of the primary motor cortex in this condition, however, should not be considered as a surprising finding: as shown by Apostolos Georgopoulos and his coworkers in monkey experiments, the directional activity of cortical cells in M1 is clearly modified during a task which involves mentally rotating the direction of a movement prior to executing it, a situation close to motor imagery.[21]

Forming the mental image of an action thus involves a pattern of cortical activation that resembles that of an intentionally executed action.[22] Other modalities of action representation, like that elicited by the observation of actions performed by other people, also involve activation of a similar cortical network. The specificity of the network for each modality of representation is still the subject of intensive research. This point is developed in my companion essay in this volume.

---

[19] David Ingvar and L. Philipsson, "Distribution of the Cerebral Blood Flow in the Dominant Hemisphere During Motor Ideation and Motor Performance," *Annals of Neurology* 2 (1977): 230–37; Per E. Roland, E. Skinhoj, N.A. Lassen, and B. Larsen, "Different Cortical Areas in Man in Organization of Voluntary Movements in Extrapersonal Space," *Journal of Neurophysiology* 43 (1980): 137–50.

[20] Modern neuroimaging techniques use Positron Emission Tomography (PET) or functional Magnetic Resonance Imaging (fMRI). These techniques allow mapping of brain (mostly cortical) activity in normally behaving subjects. See Jean Decety, D. Perani, M. Jeannerod, V. Bettinardi, B. Tadary, R. Woods, J.C. Mazziotta, and F. Fazio, "Mapping Motor Representations with PET," *Nature* 371 (1994): 600–2; Klaus-Martin Stephan, G.R. Fink, R.E. Passingham, D. Silbersweig, A.O. Ceballos-Baumann, C.D. Frith, and R.S.J. Frackowiak, "Functional Anatomy of the Mental Representation of Upper Extremity Movements in Healthy Subjects," *Journal of Neurophysiology* 73 (1995): 373–86.

[21] Apostolos P. Georgopoulos, J.T. Lurito, M. Petrides, A.B. Schwartz, and J.T. Massey, "Mental Rotation of the Neuronal Population Vector," *Science* 243 (1989): 234–36.

[22] Christopher D. Frith, K. Friston, P.F. Liddle, and R.S.J. Frackowiak, "Willed Action and the Prefrontal Cortex in Man: A Study with PET," *Proceedings of the Royal Society*, 244 (1991): 241–46; see Jeannerod, *The Cognitive Neuroscience of Action*, for discussion.

4 A Moral Conclusion

This review of some of the conceptions of how action is organized makes it clear that the vast territory of research concerning the covert aspects of action generation has only been partly explored. The conjunction of the cognitive approach and the modern techniques for mapping brain activity should progressively unmask the biological underpinnings of such central aspects of the self as intentionality and self-consciousness. To what extent can intentions be understood by external agents? What are the signals that are used for differentiating one's actions from those of other agents? These and other questions need to be asked, as a first step towards the naturalization of mental mechanisms.

In other words, an understanding of the neural mechanisms of action generation is critical for elaborating a scientific theory of the person.[23] As was stated at the beginning of this essay, action is an expression of the self and, for this reason, has to be (and always has been) considered from a moral point of view. Intention is viewed as a kind of action by many prominent religious traditions. This agrees with the notion developed here in light of the neurosciences that the mental representation of an action is also an expression of the self. Insofar as the action is directed toward other selves, one has to consider that an action's effects are not constrained to the realm of its outward execution. The process of interacting with others begins inside the body with the generation of internal states that govern the expression of actions.

---

[23] Nicolas Georgieff and M. Jeannerod, "Beyond Consciousness of External Reality: A 'Who' System for Consciousness of Action and Self-consciousness," *Consciousness and Cognition* 7 (1998): 465–77.

A NEUROSCIENTIFIC PERSPECTIVE ON HUMAN SOCIALITY

Leslie A. Brothers

1 The Social Brains of Primates

The evolution of modern primates bespeaks the importance of sociality in our history. A four-stranded evolutionary thread emerges when we compare the traits of modern primates that retain ancestral features to the traits of monkeys, apes, and humans. First, there has been a transition from a primarily nocturnal to a primarily diurnal mode of living. Second, accompanying this transition, there has been a switch from dependence on the sense of smell for conveying social information to a reliance on vision. Third, the switch to vision has been accompanied by rearrangements of the nerves and muscles of the face, to yield faces capable of rich and subtle social signals. Finally, social groups and relationships have become more complex.[1]

As a result of these changes, the primate face transmits a stream of signals needed to negotiate demanding and ever-changing social situations. (Vocal signals, both linguistic and paralinguistic, are highly integrated with facial signals and other bodily gestures into a total signaling package.) Transmitters imply receivers: primates also had to develop the ability to decode social signals carried by faces and voices. Here, a fifth strand appears, for comparative anatomical studies of the primate brain indicate that certain primitive structures that had been associated with interpreting olfactory (smell-related) information gradually changed their connections so as to receive much more information from the expanding visual areas of the brain. These structures appear to be crucial links in the chain that connects incoming social information, as processed by the visual system for example, to appropriate bodily responses, ranging from the dramatically observable to the nearly covert.

Of particular interest in this regard is the amygdala, a distinctive cluster of neurons found deep in the anterior temporal lobe of each hemisphere, and the nearby temporal cortex with which it has immediate connections. For most of this century, the amygdala has been thought to be involved in assigning emotional significance to items and events in the environment: discussions of the emotional systems of the brain have invariably highlighted the role of the amygdala. In recent decades, however, some neuroscientists interested in the social functions of the primate brain have demonstrated a more specifically social role for this structure. Their demonstrations began with observing the social behavior of monkeys with experimental lesions of the anterior temporal lobes. Such animals appeared to have particular difficulty responding to the social signals of other monkeys. In recent years, the importance of these earlier observations for understanding human social cognition has been confirmed by detailed analysis of patients with circumscribed amygdala lesions. It has now been shown that certain of these individuals have difficulty interpreting facial expressions, direction of eye gaze, and tone of voice. To what extent the amygdala and surrounding anterior temporal lobes may be implicated in clinical disorders typified by social deficits, such as schizophrenia and autism, is a matter of considerable research interest.

---

[1] This essay summarizes material in Leslie A. Brothers, *Friday's Footprint: How Society Shapes the Human Mind* (New York: Oxford University Press, 1997), which contains detailed references for material discussed here.

A separate line of evidence for the role of the anterior temporal lobes in social cognition derives from studies that stimulate areas of the brain or record the spontaneous activity of neurons in response to social stimuli. For ethical reasons, studies involving penetration into the brains of humans can only be carried out in neurosurgical settings where the investigation can be made part of diagnostic and treatment activities. One example is the study done by the neurosurgeons Itzhak Fried, George Ojemann, and their colleagues.[2] They showed that a person's ability to identify pictured facial expressions could be disrupted by applying electrical stimulation to a specific region of the right temporal cortex, without disrupting his or her ability to make other judgments about other visual stimuli. Researchers who employ noninvasive techniques such as functional imaging have also been able to show that some temporal lobe regions are particularly responsive to the sight of faces, and even the movements of facial features such as the eyes and mouth.[3]

Turning to monkeys, in the late 1960s and early 1970s researchers studying the neural basis of vision encountered temporal lobe neurons that seemed to be responsive only to social visual stimuli such as faces. These neurons were found in regions of the temporal cortex known to have connections with the amygdala. In short order, researchers studying neural activity in the amygdala as it related to motivation also discovered neurons responsive to the sight of faces. Their reports suggested to me that neurons in the amygdala and surrounding regions might respond to many other visual stimuli of social significance, including bodily movements of other animals and the sight of interactions taking place between other animals, for example. I therefore compiled a large number of short clips of natural social scenes by videotaping social colonies of macaque monkeys. I selected and presented these clips, each about two seconds in length, to single macaque monkeys that had been trained to sit quietly and watch a television monitor while undergoing recording from single neurons in the region of the amygdala. By showing large numbers of social scenes, I was able to correlate the firing of single neurons with particular features of the presented scenes.

The figures show some examples. Figure 1 depicts a stumptail macaque monkey with an open mouth. Macaques may open their mouths for a number of reasons, including eating. One facial expression that involves an open mouth is the yawn, often used by males to signal dominance, since it displays their large and potentially dangerous canines. Figure 2 shows the response of an amygdala neuron to many moving clips showing monkeys with open mouths. In some of the clips, the open mouth was part of a yawn sequence. As the figure shows, the response of the neuron was much greater to yawn sequences than to other sequences that included open mouths, suggesting that this cell was attuned to a feature specific to yawns—perhaps their dynamic structure—rather than mouth features such as teeth and tongues. Figure 3 shows a neuron's response to another kind of motion, that of walking, trotting, or running. This neuron responded briskly to all sequences that included alternating movements of the legs, regardless of direction of motion or gait. Other kinds of movement in stationary animals, such as movements of the head, arms, or legs, produced no response.

---

[2] Itzhak Fried, C. Mateer, G. Ojemann, R. Wohns, and P. Fedio, "Organization of Visuospatial Functions in Human Cortex," *Brain* 105 (1982): 349–71.

[3] Aina Puce, T. Allison, S. Bentin, J. Gore, and G. McCarthy, "Temporal Cortex Activation in Humans Viewing Eye and Mouth Movements," *Journal of Neuroscience* 18 (1998): 2188–99.

Figure 1. A female stumptail macaque monkey during an exchange of threat displays.

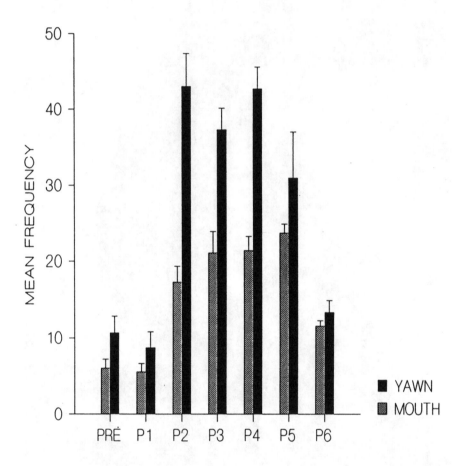

Figure 2. Response (spikes per second) of a neuron in the anterior entorhinal cortex to various scenes depicting open mouths and yawn expressions.[4] Epochs are 500 ms in length. "Pre" is the average baseline firing rate prior to stimulus presentation. The stimulus segments appeared during P1 to P4 to moving clips of monkeys with open mouths.

---

[4] This cell was described in Leslie Brothers and B. Ring, "Mesial Temporal Neurons In the Macaque Monkey with Responses Selective for Aspects of Social Stimuli," *Behavioural Brain Research* 57 (1993): 53–61.

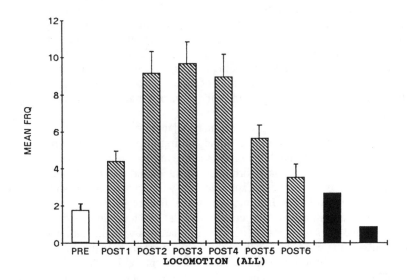

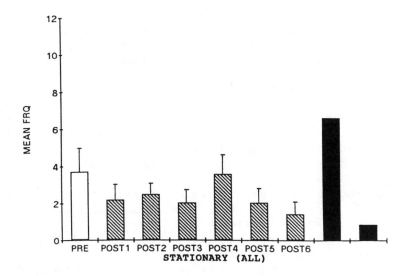

Figure 3. Response of a neuron in the medial amygdala to scenes containing alternating limb movements in the context of locomotion (top) and other views of animals whose movements did not involve locomotion (bottom).[5]

---

[5] The cell was described in Leslie Brothers and B. King, "Response of neurons in the Macaque Amygdala to Complex Social Stimuli," *Behavioural Brain Research* 41 (1990): 199–213.

These and other results indicate that neurons in the amygdala and nearby cortex of macaque monkeys respond to such features as significant motion, identity of individuals, and particular kinds of interactions taking place between individuals. Since the responses persisted over repeated presentations of stimuli, and were not elicited by sequences that contained material the subjects found disturbing (depictions of animals making threatening lunges toward the camera), I concluded that these cells were not simply encoding gradients of emotional responses. Rather, the responses are consistent with the kind of sophisticated encoding that social brain systems must be able to carry out in order to interpret and respond to a complex social environment.

The picture that is emerging from human and monkey studies is that representations of features of the outside social world are first assembled in the temporal lobe cortices of the primate brain. Meaningful social events are registered when a host of signals and relevant contextual information are integrated. Our brains need to tell us the difference between someone approaching with friendly intent and someone whose aims are hostile, for example. The visual features of a face have to be put together to yield an image of a particular individual who has a unique identity—*who* is approaching gives us access to our past interactions with this individual. Next, movements of the eyes and mouth indicate the person's disposition. Similarly, information from head position and body movement tell us where this person is looking or going, providing raw material for the representation of a mental state such as his or her goal or desire.

As these processes are taking place, the neural representations of others' social intentions must be linked to an appropriate, responsive behavior in the perceiver—otherwise the gears of society could not mesh and turn. Social representations should thus be intimately tied to response dispositions that can range from overt muscle activity, such as running away, to subtle visceral and hormonal reactions of anxiety, attraction, or tenderness. Response dispositions should be set into play "downstream" from the temporal cortices where face-responsive neurons have been found, in structures such as the amygdala. The amygdala, together with several other interconnected structures that have been classically designated "limbic," receive sensory information and in turn project directly to somatic effector structures such as the hypothalamus, brainstem, and primitive motor centers, making it a candidate for the link between social perception and response.

Confirmation of this role for the amygdala is provided by the research of the neurologist Pierre Gloor, of the Montreal Neurological Institute.[6] In the process of preparing patients for subsequent neurosurgery, Gloor was able to run low-level electrical current into the amygdala and record the accounts patients gave of their experiences. Strikingly, a number of accounts contained "social feelings," that is, the feelings one might have in a particular social situation. One of these was apprehension at the displeasure of an authority figure; another was the feeling of being shunned or unwelcome at a social gathering. The patients did not always have labels for naming their feelings, so they described the situation for which the feeling would be appropriate in order to communicate their experiences to Gloor: "It's as if I am at a party where I wasn't invited and am not welcome." These data strongly suggest

---

[6] See, for example, Pierre Gloor, "The Role of the Human Limbic System in Perception, Memory and Affect: Lessons from Temporal Lobe Epilepsy," in *The Limbic System: Functional Organization and Clinical Disorders*, B.K. Doane, K.E. Livingston, eds. (New York: Raven Press, 1986), 159–69.

that the amygdala is a critical link between the representations of others' attitudes and dispositions and the subject's own somatic dispositions to react—in the preceding example, either by retreating from the group or by seeking out an ally so that one can feel part of the group. Both the representation and the response would be conditioned by previous experience and cultural beliefs, but it is also possible that there are cultural universals such as submission to dominance and sensitivity to social rejection.

2 Persons and Their Relations

We assign unique identities to individuals. To be more precise, we assign identities to a mind-body unit called "person." Persons have two dimensions at the same time. First, they are loci of subjectivity, which is central to mind. Second, persons are assigned attributes that mark their position in a larger social order, an order which has publicly recognized—and often contested—moral aspects.

Attributing mind implies attributing mental states such as beliefs and goals. Although there are debates as to nonhuman primates' ability to attribute mental states to others, clearly their abilities in this regard are more limited than our own.[7] It appears that there has been a cognitive watershed in the evolution of human beings: under the pressure of an ever-increasing load of social signals, a higher-level representational capacity has emerged—"person." The person-system is like a language whose grammar is realized through the behavior of the individuals who speak it. The grammar of person has rules such as "one mind per body" and "one identity per mind-body unit." In this language, persons are related to one another through moral accountabilities.

What studies of the social brain in primates suggest is that the evolution of our brains made it possible for us to construct and participate in the semantics of personhood. These semantics are both culturally external and dependent on internal, individual brains—just as language is both cultural and neural. This dependence on brain function is illustrated by the fact that some individuals become unable to use person rules properly when they have brain lesions: patients suffering from misidentification syndromes may attribute one mind to several bodies, or perceive an identifiable body as having been taken over by an alien mind. Disorders of brain development such as autism and its variants also result in an inability to make proper attributions of mentality, as well as difficulties engaging in the lower-level social routines through which personhood is exercised.

Contributors to this volume struggled at length with how to assess the reality of levels of phenomena "above" the elements described by physics. The research on primate sociality presented in this short essay, with its focus on the emergence of the person concept in human thought, shows that the world of human reasons emerges from the operations of brains specially equipped for social participation. It raises the question of a reality that exists at the level of relations between subjects; the laws of this reality probably are best approached through traditional disciplines such as ethics. Some would hold that we must reach to the neuronal level to explain moral behavior completely. Indeed, since subjectivity and its social embeddedness are made possible by our physical beings, including especially our brains, it would follow that any orientation of the individual to the moral-social order would

---

[7] Daniel Povinelli, T. Eddy, "Factors Influencing Young Chimpanzees' (Pan Troglodytes) Recognition of Attention," *Journal of Comparative Psychology* 110 (1996): 336–45.

encompass him or her at all levels including the neural. However, neural events take place at a lower level, selected by but not determining moral content—just as an account in terms of ion channels would be even more trivial. I believe our brains make personhood possible, but the explanatory limits of neuroscience are reached where the person dimension begins.

II. FROM NEUROSCIENCE TO PHILOSOPHY

Scientific:

Michael A. Arbib
Joseph E. LeDoux
Marc Jeannerod

Philosophical:

William R. Stoeger, S.J.
Nancey Murphy
Theo C. Meyering

TOWARDS A NEUROSCIENCE OF THE PERSON

Michael A. Arbib

1 Neurology and the Person

The bulk of this essay (and my companion essay in this volume, "Crusoe's Brain") provides a framework for a discussion by neuroscientists, philosophers, and theologians of the relation of a science of the person to neuroscience on the one hand, and the issues raised by belief in God on the other. The present section is by way of preamble: it provides examples of ways in which neuroscientists and cognitive scientists have shown that many aspects of the person, far from being ineffable, can strongly be linked to neurological processes as seen at the level of clinical data on disease, lesions, and the effects of drugs.

Blindsight: Following damage to the primary visual cortex, humans can perform certain visual discriminations despite reporting that they are blind in corresponding parts of their visual fields.[1] Indeed, the subjective "experience" that one possesses a "top-level executor" does not imply that one is neurally defined. For example, our current understanding of vision is that many different "features" and "affordances" of the visual input are extracted in many different brain regions[2] and that there is no single place where all the features are integrated. (And how small must such a place be to accommodate experiences of the unity of experience? One neuron? A cubic millimeter? One nucleus?) For example, the identity of a face and its emotional import seem to be processed somewhat separately, yet they are normally experienced as a unity. Such neurological data can help us probe the phenomenon of consciousness as a systems property.[3]

Episodic Memory: Lesions in areas such as the hippocampus can result in profound disruption of episodic memory—the "narrative" memory of places and events that provides context for our experiences and a sense of continuing existence. A recent study suggests that the hippocampus may be especially important for event memory, while the surrounding cerebral cortex contributes both to fact memory and to event memory.[4] Can one really be a person without being able to construct narratives from past experience, or relate new episodes to those which have gone before? If there is no memory, every event is totally new.

Emotion and Motivation: Antonio Damasio's account of Phineas Gage's accident shows how loss of affect and a radical change in personality can result from a large lesion disconnecting much of the frontal lobes. Gage's intelligence level was

---

[1] Lawrence Weiskrantz, E.K. Warrington, M.D. Sanders, and J. Marshall, "Visual Capacity in the Hemianopic Field Following a Restricted Occipital Ablation," *Brain* 97 (1974): 709–28.

[2] David C. Van Essen and C.H. Anderson, "Information Processing Strategies and Pathways in the Primate Retina and Visual Cortex," in *An Introduction to Neural and Electronic Networks*, Steven F. Zornetzer, Joel L. Davis, and Clifford Lau, eds. (San Diego, Calif.: Academic Press, 1990), 43–72.

[3] See, for example, section 8.3 of Michael A. Arbib, *The Metaphorical Brain 2: Neural Networks and Beyond* (New York: Wiley-Interscience, 1989).

[4] Mortimer Mishkin, W.A. Suzuki, D.G. Gadian, and F. Varga-Khadem, "Hierarchical Organization of Cognitive Memory," *Philosophical Transactions of the Royal Society of London B* 352 (1997): 1461–67.

almost unaffected, and his memory had momentary lapses, but the most important effect was a loss of affectivity. His attitude was transformed by the accident into a relaxed one, characterized by carelessness.[5]

Joseph LeDoux (in this volume) develops his view that emotion systems evolved as sensory-motor solutions to problems of survival; he thus distinguishes emotions from "feelings," which as "conscious emotions" are not, he suggests, the function that emotion systems evolved to perform. This is helpful in showing how neuroscientists can focus on a well-defined subproblem to define meaningful animal experiments—in this case, on fear-conditioning in the rat[6]—but also serves to emphasize how wide is the gap between most neuroscience and the integrated study of the person. How does "behavioral fear" relate to the feeling of fear or anxiety? How general are the implications of studies linking fear to the amygdala for our understanding of emotion and feeling more generally. The behavioral role of emotion may have arisen independently of feelings, but feelings enable humans to rethink priorities, for example, and thus are not redundant in relation to the behavioral base. LeDoux concedes this, noting that connections of the amygdalar circuit with the hippocampus and cerebral cortex provide the possible basis for enrichment of "behavioral emotion" by episodic memory and cognitive state.

Consciousness: Theo Meyering notes that the philosophical distinction between intentional and phenomenal aspects of mind is clearly important with regard to the distinction between the behavioral role of emotion and feelings. He goes on to say that:

> Philosophers have lately grown quite concerned about the ramifications of these [intentional and phenomenal] aspects of mental, life which strike them as rather heterogeneous. Since the rise of functionalist and, in general, of nonreductive materialist accounts of mind a few decades ago there was a widespread expectation that in principle an account along functionalist lines could be given to cover all aspects of mentality. Yet consciousness, or the phenomenal aspect of mind, turns out to generate very special problems of its own not encountered in the area of the intentional aspect of mind.... [It] looks as though, generally speaking, structure and function can always be equivalently emulated or reproduced in a zombie world, that is, in a world functionally equivalent to ours but altogether devoid of consciousness. If so, a ghastly dilemma may arise. For either consciousness consists in certain functions it may serve or it doesn't. If it does, those very functions may be subserved by structures devoid of consciousness. Or else it cannot be reduced to the functions the having of relevant feelings may subserve, in which case these feelings may be ineliminable, yet they would also be rendered epiphenomenal.[7]

Mary Hesse and I have given an evolutionary account of how communication might have coevolved with consciousness in the sense of a unitary integration of the multitudinous patterns of competition and cooperation between brain regions (structural) and schemas (functional), and this account has recently been revisited by Giacomo Rizzolatti and myself in our account of the evolution of brain mechanisms of language.[8] It is, I suppose, a functionalist account, for it concedes that neurosci

[5] Antonio R. Damasio, *Descartes' Error: Emotion, Reason, and the Human Brain* (New York: Putnam Books, 1994).

[6] Joseph LeDoux, *The Emotional Brain: The Mysterious Underpinnings of Emotional Life* (New York: Simon & Schuster, 1996).

[7] Personal communication.

[8] Michael A. Arbib and Mary B. Hesse, *The Construction of Reality* (Cambridge: Cambridge University Press, 1986), section 4.2; Michael A. Arbib and G. Rizzolatti, "Neural Expectations: A Possible Evolutionary Path from Manual Skills to Language," *Communica-*

ence may aspire to show how the evolution of, for example, cortical systems which modulate and extend the function of "older" systems[9] can explain all externally observable, measurable aspects of consciousness, but cannot "prove" that an embodied neural network with all those properties must "feel conscious." The classic argument is that neuroscience can explain more and more of the known color phenomena (for example, how a patch may appear to have a different color as the background changes), but cannot resolve the question "How do you know that when I see green, I have the same phenomenal experience as when you see green?"[10]

Prozac: For thousands of years humans have known and used chemicals that alter subjective experience—materials that affect the mind. One of the newer substances that change the emotional state of a person is Prozac. It can transform a depressive person into a happy one. The action of Prozac takes place "below" the synaptic level, affecting the uptake of serotonin in the synaptic cleft (the small space between a tip of one neuron and the "receptor site" on another) and the effect is a complete change in personality.[11] How can changes at the molecular level yield global changes in personality? Leslie Brothers notes that such serotonin re-uptake inhibitors have positive effects not only on mood but on sociality as well. In humans, impulsive aggression and low serotonin levels are positively correlated.[12] Brothers notes that, conversely, high serotonin activity in monkey cerebrospinal fluid was correlated with affiliative sociality.

Neglect: Unilateral lesions of the parietal cortex can yield a neglect syndrome in which the person neglects the existence of any object situated in the contralateral side.[13] Patients with lesions on one side of the parietal cortex were unable to describe the part of the street that was on the contralateral side. When they were turned around, the same phenomenon appeared, but now the side that previously was neglected was normally described. This means that the information about the street was stored but the person was unable to recall it in a contralateral spatial frame.

Social Behavior: Social interactions between people, while largely neglected in neuroscience, are crucial for the development of the mind.[14] It has been argued that specific brain structures evolved for the purpose of social behavior. Study of social interactions may provide the key to understanding the connections between mind and spirit. For example, ethics might be related to the biology of social interaction, with our understanding then refined through studies in historical and cultural anthropol-

---

tion and Cognition 29 (1997): 393–424. Note, too, the earlier discussion of blindsight.

[9] See chap. 3 of Michael A. Arbib, Péter Érdi, and János Szentágothai, *Neural Organization: Structure, Function, and Dynamics* (Cambridge: MIT Press, 1997).

[10] Actually, given my own experience of life with and without sunglasses, I suspect that "you" and "I" do not have the same phenomenal experience, but are nonetheless able (generally) to agree on the state of a traffic light, or the ripeness of a plum.

[11] B.L. Jacobs, "Serotonin, Motor Activity and Depression-Related Disorders," *American Scientist* 82 (1994): 456–63.

[12] Leslie Brothers, *Friday's Footprint: How Society Shapes the Human Mind* (New York: Oxford University Press, 1997).

[13] Edoardo Bisiach and G. Vallar, "Hemineglect in Humans," in *Handbook of Neuropsychology*, vol. 1, François Boller and Jordan Grafman, eds. (Amsterdam: Elsevier, 1988), 195–222. For a link to the earlier topic of blindsight, see John C. Marshall and P. Halligan, "Blindsight and Insight in Visuo-spatial Neglect," *Nature* 336 (1988): 766–67.

[14] See Brothers, *Friday's Footprint*, and discussion of this book in my companion essay in this volume.

ogy. Brothers (personal communication) has emphasized the space between what the brain does and what we construct through our narratives, collective performances, etc.—a space between biological underpinnings and what is constructed through our social systems. Cognitive neuroscience assumes we will find these things in the brain, but Brothers disagrees.

Language: Only humans have the complex lexicon and grammar of language, which, presumably, evolved to increase the effectiveness of communication. Different lesions yield strikingly different aphasias. Did evolution favor the development of communication systems which then led to changes in brain structures or were there more general pressures to use complex, sequential structures in our interactions with the environment yielding brain mechanisms that then permitted the cultural evolution of language?[15]

Mental Disorders: Extensive research is under way on the anatomical, biochemical, and genetic bases of mental disorders, such as Alzheimer's disease, schizophrenia, Huntington's disease, and mania, leading to the promise of therapies for some of these diseases.[16]

2 Science, Religion and the Understanding of Personhood

William Stoeger, speaking both as a cosmologist who uses physics to study such phenomena as black holes and the Big Bang and as a Jesuit priest, notes that:

> [science has shown the universe to be] on every level more vast, more intricate in its structure and development, more amazing in its evolution, in its variety flowing from fundamental levels of unity, and in its balance of functions, than we could have imagined without the contribution of the sciences. Certainly, at least in some way, such a perspective and such understanding enriches theological reflection, and provides some of the detailed experiential points of reference from which we consider who God is, and who He is not, and who we are in relation to Him, to one another, and to our world.[17]

As for cosmology, so for neuroscience.

---

[15] See Giacomo Rizzolatti and M.A. Arbib, "Language Within Our Grasp," *Trends in Neurosciences* 21 (1998): 188–94, for discussion and further references.

[16] See, for example: Huntington's Disease Collaborative Research Group, "A Novel Gene Containing a Trinucleotide Repeat That Is Expanded and Unstable in Huntington's Disease Chromosomes," *Cell* 72 (1993): 971–83; Ming T. Tsuang, G. Winokur, and R.R. Crowe, "Morbidity Risks of Schizophrenia and Affective Disorders among First Degree Relatives of Patients with Schizophrenia, Mania, Depression and Surgical Conditions," *British Journal of Psychiatry* 137 (1980): 497–504; James C. Coyne and G. Downey, "Social Factors and Psychopathology: Stress, Social Support and Coping Processes," *Annual Review of Psychology* 42 (1991): 401–25; M. Flint Beal, "Neurochemistry and Toxin Models in Huntington's Disease," *Current Opinion in Neurology* 7 (1994): 542–47; William J. Jagust, "Functional Imaging in Dementia: An Overview," *Journal of Clinical Psychiatry* 55 (1994, suppl.): 5–11; Donald L. Price, C.H. Kawas, S.S. Sisodia, "Aging of the Brain and Dementia of the Alzheimer's Type," in *Principles of Neural Science*, 4th ed. Eric R. Kandel, James H. Schwartz, Thomas M. Jessell, eds. (New York: Elsevier, 1995); and Peter R. Rapp and D.G. Amaral, "Individual Differences in the Cognitive and Neurobiological Consequences of Normal Aging," *Trends in Neurosciences* 15 (1992): 340–45.

[17] William R. Stoeger, "Contemporary Cosmology and Its Implications for the Science-Religion Dialogue," in *Physics, Philosophy and Theology: A Common Quest for Understanding*, Robert J. Russell, William R. Stoeger, S.J., and George V. Coyne, S.J., eds. (Vatican City State: Vatican Observatory, 1988), 240.

2.1 Theology and a Science of the Person

But first, what is theology, and is it a science? Is it the study of God (in which case a nonbeliever like myself would say it cannot be a science—unless we accept it as a science of non-existent things) or is it the study of human belief in God (in which case it is part philosophy, part a fascinating chapter in anthropology)? The latter would include the former if God exists. Part of theology recounts and analyzes myths and legends, whether to extract eternal Truths or to chart shifting patterns of belief. As such, it has one foot in the humanities, the other in anthropology of religion and culture. Other parts are "pure" philosophy, while yet others may be related to the psychology of "altered states" or to the sociology of tribal ritual (whether in isolated tribes, modern cities, or Internet communities). Thus, I do not see theology itself as a science, but I do see many specific parts of it either posing challenges to, or themselves becoming part of, an evolving "science of the person." To further open the discussion to the secularist as well as the theist, I would broaden the definition of theology even further from the study of God to the study of those aspects of the human condition for which many have found God to be the answer. In other words, we may accept the importance of enduring questions about the nature of person-hood—about *love, faith, and commitment*—whether we are theists, atheists, or something in between.

I argue that we cannot approach theology (in the narrow or broad sense) without some sense of the intricacy of the human brain—a sense which the previous section sought to suggest.

Brain talk speaks of lesion data, anatomy, neurophysiology, and neurochemistry. *Mind talk* speaks of intention, action, perception, consciousness, and responsibility. Together they can be viewed as neuroscience embedded within cognitive science. *Spirit talk* may be construed as mind talk *or* God talk: as a variant of mind talk, or as something that regards our identity as rooted in our relation to God. Mind has properties (self-consciousness, wonder, emotion, reason) that make it *seem* more than "merely material." There is much work to be done towards a *neuroscience of the person* no matter which of these beliefs one holds. Yes, people have religious longing; yes, they have a sense of soul. Nonetheless, I believe that all of this can be explained in terms of the physical properties of the brain. We need to restructure and refocus the concepts of the mind and the soul to make them amenable to scientific inquiry. Neuroscience does not address the *concept* of "God" directly, but it can make progress towards theological questions (in the generalized sense above) if it focuses on the "science of the person"—the biological, psychological, and ultimately social aspects of humanness. We have already seen that, although the case is not closed, there is a respectable case for an approach to neuroscience predicated on the view that all of this can be explained *eventually* by the physical properties of the brain—within bodies constituting persons within societies, expressing individual and cultural histories as well as the effects of biological evolution.[18]

While science seems generally to reduce the world into analyzable parts in order to understand these parts, and ideally to be able to reconstruct them and thus understand the whole, religion seems to minister directly to the whole person. A

---

[18] This view is what Philip Clayton (in this volume) refers to as the Arbib Credo. He attacks it vigorously, but I have not found his arguments sufficiently persuasive to change my "faith."

science of the person should be concerned with the interaction of nature with nurture, with understanding the needs of individuals across the environmental landscape, and with cultural and biological coevolution. What must be added to our current science to understand how life becomes lyrical and meaningful? As neuroscience is aided by interdisciplinary perspectives, so must a complete science of the person be aided by an interdisciplinary perspective. Each discipline—philosophy, biology, political science, psychology, sociology, evolutionary theory, computer science, chemistry, etc.—brings its own biases, foci, conceptual level, and tools for understanding.

Do a sense of and need for spirituality and religion exist as a consequence of the evolution of the human brain and the properties it projects in virtue of its being a complex survival machine? The answer may well be "Yes" whether we explain religious sentiments in physical or psychological terms —as a human "invention" to meet a need —or see religion as an appreciation of a Divine Reality, elucidating the universal feeling of spirituality and the role it has played and continues to play for humanity. To close this section, here are a number of questions to be addressed by a science of the person:

- What makes an individual person distinctive?
- What makes for "free will" beyond the determination of genetics and environment?
- What is the meaning of morality as the foundation for virtuous human behavior?
- Is there continuity of life beyond death, immortality?
- What is the source for our sense of awe and wonder?
- What is the purpose and meaning of our lives and the cosmos?

I believe that all of this can be addressed in a framework rooted in the physical properties of the brain—but not without changing the terms of the debate. For example, the "purpose and meaning of our lives" will, I believe, come to be seen as rooted in personal and social contingencies as conditioned by biological and environmental realities, rather than as reflecting, say, God's will. In particular, we must, I suggest, seek bridging concepts that analyze and dissect the various different qualities and processes that currently are mishmashed together when people discuss mind and soul. Understanding these differences will provide a first step toward seeing which aspects can indeed be understood in terms of concepts (themselves evolving) from the social and biological sciences.

Can we hope for a "complete" science of the person? A complete science is at best a horizon goal, at worst a chimera. Just consider the state of physics as the prime example: A complete science of matter? No. A complete science of matter and energy? No. A complete science of matter and energy and information?... However, my point is that we can increasingly link neuroscientific research (aided immensely by new techniques for human brain imaging) to studies of phenomena "of the person" that once seemed outside the reach of laboratory science. It is not that I seek a single science to which all others may be reduced, but rather that I seek patterns of communication which allow paths from one science to another, enabling them to contribute together to a coherent worldview. I reject any form of dualism of mind and body, but I do not imply that there is in current neuroscience a complete grounding for a reduction of the cognitive and social sciences, let alone of "theology." Rather, I regard each level—neural, cognitive, individual, and social—as possessed of many important truths and yet also incomplete. It is a process of *mutual reduction* that we need, as insights at any one level propagate through to change the science at other levels.[19]

For example, it is not very helpful to link psychology to physics directly, but the path psychology↔neuroscience↔biology↔chemistry↔physics is plausible. It is not that we seek to reduce mental operations to quantum field theory (though stranger things may happen before we are done!) but rather that, while each science develops its own vocabulary, laws, and studies, it will accept in principle that its concepts can be integrated with those of neighboring sciences. I say "in principle" because the conversion may be too complex in many cases to carry this through explicitly (though increasing computing power will push the boundaries of explicit analysis further and further).

2.2 Morality

> In the beginning God created the heavens and the earth. The earth was without form and void, and darkness was upon the face of the deep; and the Spirit of God was moving over the face of the waters.... So God created man in his own image, in the image of God he created him; male and female he created them. (Genesis 1:1, 27 RSV)

Linking morality with God, the theist believes that religion brings all questions and problems about values to an "extranatural relation of person to God."[20] The Torah of Judaism and the Old Testament of Christians holds that humans are made in the divine image, so that human and God share the same character. From this, it has been argued, follow principles of morality—if all people are made in the image of the creator, then all have something of the divine within them, and all are worthy of dignity and respect. Yet the notion that "all men are created equal" was a radical cry of eighteenth-century Enlightenment, and devout people of all religions still find it all too easy to brand groups of humans as "the other" for whom death by warfare or ethnic cleansing is a moral imperative. Atheists can be moral, while immense inequalities have been tolerated in religious societies up to the present day.

The certainty that for many comes with religious belief can lead to intolerance and inflexibility in the name of morality. A religion with a healthy dose of liberality can avoid intolerance. But: (1) Many look to religious faith for a certainty that a liberal view would weaken. (2) How to reconcile such relativism with a critical search for truth? (3) How to reconcile it with a meaningful adherence to ritual?

Whether or not we are formed in God's image and likeness, we seek not only to understand what we share as human beings but also the basis for the individuality that distinguishes us one from another. Must these be rooted in a faith in God's existence and in there being something better in the world-to-come through eternity?

What makes religion distinct from the scientific approach in helping us to understand ourselves? Religion highlights what is distinct and irreducible in all of

---

[19] In *The Construction of Reality*, Hesse and I discuss at some length the benefits of "two-way reduction" rather than reduction *simpliciter*. The two-way reduction of personal experience and neuroscience can act reciprocally to enrich our understanding of both the mind and the brain. This is in contrast to the view that psychology, linguistics, anthropology, and social sciences could be reduced to the scientific vocabulary of an existing science, such as neuroscience, in the way that chemical phenomena can be explained by the formalism of physics. Even in the latter case, note that we do *not* use physical terminology to discuss chemical processes in general; we only use it when seeking explanations of generally important chemical processes.

[20] Ibid.

us. Religion is beyond objectivity and pragmatism, it brings ties with the world and Cosmos, morality, and the continuity beyond life. Religion brings "proofs" about the existence of a structure above all the unifying principles: God. But proofs of what kind? Religious experience adds new dimensions to human experience, beyond the pragmatic boundaries of the scientific realm, including the personal, emotional, social, and moral.

Leaving aside issues of human equality and justice, we may still use "soul"—even while denying theistic significance to the term—for the personal inner calling, a feeling of significance of both the self and humanity generally, of lofty aspirations and openness to change, of freedom of the will and responsibility, and of immortality. Consider the magic of the mind, the beauty of the earth, the sense of spiritual significance, and the human need for these; consider attributes like creativity, imagination, love, free will, and spiritual reflection that distinguish human from animal. One may seek to understand the source of these human feelings whether or not they are "true," and whether or not the soul provides a local habitation for the divine image.

I do not believe that neuroscience can ever tell us what to do in terms of morality and ethics, but, nonetheless, it may teach us much about reasoning and decision-making, logic and irrationality, the neural basis of compassion and empathy and certain facets of social reality. What are the implications for a science of the person?

3 Bridging the Levels with Computational Neuroscience

3.1 Computational Models

Will we ever understand the brain well enough to predict behavior? Just as the prime parable of chaos theory states that the flapping of a butterfly's wings can produce changes in weather patterns across the ocean that make weather unpredictable, so might the mechanisms of behavior be completely understood, while the actions of an individual may remain generally unpredictable. Weather is unpredictable, but the weather does not have a nonphysical "spirit." This example implies that we cannot conclude that very complex systems must be supernatural things, nor that scientific explanation must involve complete predictability. The program for the future of computational neuroscience is to provide the theoretical and computational framework which will integrate the different viewpoints on brain and mind, human being and world. This program should include elucidating the phenomena related to learning and integration, as well as storage and recall of information in cortical and subcortical areas, in order to provide information about the interactions and specific functions of different neural structures implied in behavioral actions.[21]

There is a paradox at the core of neuroscience. Neuroscience should consider the brain as a whole, yet neuroscience cannot succeed without subdivision of the problem into a multiplicity of subgoals whose very detail obscures the overall quest.

---

[21] Three books of mine bear on this program: Michael A. Arbib, *The Metaphorical Brain 2: Neural Networks and Beyond* provides a personal overview of how to model the brain; Michael A. Arbib, ed., *The Handbook of Brain Theory and Neural Networks* (Cambridge: MIT Press, 1995) is a massive compendium embracing studies in detailed neuronal function, system models of brain regions, connectionist models of psychology and linguistics, mathematical and biological studies of learning, and technological applications of artificial neural networks; while Michael A. Arbib, Péter Érdi, and János Szentágothai, *Neural Organization: Structure, Function, and Dynamics* integrates modeling, anatomy, and physiology in a comprehensive view of neural organization.

The issue, then, is not to avoid subdivision, but rather to chart the bridging disciplines that can link the subspecialties with each other as well as with "sciences of the person" like anthropology and sociology, for which neuroscience has little current or obvious relevance.

For example, we have a wealth of experimental data from pharmacology and molecular biology on the functioning of individual ion channels and receptors of a cell. However, how these diverse factors interact to give rise to the physiology of the cell is poorly understood. At the next higher level, we have traced outlines and interconnections of various neuronal circuits in the brain. But the extensive description at the circuit level is lacking in the details of how these circuit properties are determined by characteristics of individual cell groups. Finally, we have the results of decades of careful behavioral experiments, but apart from a lesion here and a spritz of neurotransmitter there, the causative circuit-level interactions that underlie the behavior are still poorly understood. What is required is a systematic linking of these various levels of analysis (and their mountains of data) into a tight framework that is testable. That is where computational neuroscience can come to the rescue. Computational models provide the most efficient way (perhaps sometimes the only way) that the complex, nonlinear interactions between different elements of one biological "level" can be made to fit the constraints and observations of the next higher level. For instance, we have increasingly accurate models of molecular events taking into account the biophysics of channels and the kinetics of intermolecular interactions. At the next higher level, we have models that fit observations of synaptic terminal activity of a single neuron based on its cable properties, as well as inputs upon it. If these two computational levels with their biological accuracy could be combined into a more holistic model, new predictions might arise regarding the nature of the underlying interactions. In other words, "bottom-up" and "top-down" models need to be combined at the different levels of biology. In turn, these different levels need to be combined to give a panoramic view of structure and function, which can then be rigorously tested. However, the mental level may require more "abstract" models that are more psychological—or even sociological and less neural (thus the development of schema theory in the fashion described below). The neuroscientist's explanation of what the mind is and how it works is only one of many. In issues of cognitive neuroscience, the humanistic and social understanding of our experiences can complement our lesion experiments and pharmacology. The challenge is to interface these many approaches for a fuller understanding and appreciation of our selves.[22]

The approach to cognitive science espoused here distinguishes two modes of functioning: (1) a coarse-grain style of computation distributed across a network of relatively large subsystems, and exemplified by the *schema theory* developed in the next section; and (2) a fine-grain style of computation involving parallel processing by an array of neurons implementing similar processes on an array of data. The key

---

[22] Theo Meyering (personal communication) comments: "If what you envisage is a pluralistic multilevel analysis of the neuroscience of the person, where more basic levels may provide explanatory extensions of theories relevant at the next higher level of phenomena without any implication of wholesale reducibility of higher levels to more basic levels down to some privileged neuroscientific level of analysis deemed to be fundamental, I believe that would certainly constitute a wholesome program for computational neuroscience. However, if your hopes are fastened on the emergence of some ultimate overarching monoconceptual theory 'in which ultimately everything gets explained in terms of virtually nothing' (Sperry), I believe such a vision is bound to remain just a gleam in your eye."

question for the integration of processes at either level is that of *cooperative computation*: How is it that the local interaction of a number of systems can be integrated to yield some overall result without imposing explicit executive control? The study of cooperative phenomena has its roots in physics, as when we seek to understand how atomic magnets can "cooperate" to yield global magnetism through the mass effects of local interactions. Our current approach to cooperative computation is distinguished by an interest in how pattern is defined across the subsystems, rather than on the attainment of a global state.

3.2 Self-Organization and the Nature of the Individual

An analysis of what constitutes the science of the person is equivalent to an examination of the sum total of all of the factors that compose individuality. For each person one would need first to consider the relative impact of individual anatomy. Next, one must consider the body of experiences that each of us brings to every situation, making every reaction unique not only between people but within a person. Intangible but very critical components such as religion and faith (in God or even in self-ability) must be evaluated in an analysis of the science of the person.

Current neurophysiology relates much of learning to a simple set of rules that modify the strength of connections between neurons in the brain, and these changes play an important role in making each individual unique. The analysis of learning may therefore provide insights into the molecular mechanisms underlying a mental process and so begin to build a bridge between cognitive psychology and molecular biology. However, it is one thing to analyze the tuning of specific neural networks; it is quite another to explain how the many, many networks in one head *cohere* to constitute a personality.

And then there is the debate over nature (the result of evolution) and nurture. How much of our neurobiology (and ultimately behavior) is determined by innate/genetic mechanisms, how much by environmental factors, and how much by complex interactions between the two? In practically all developing nervous systems, there is an activity-independent phase of growth, innervation, and synapse formation that relies on "guidepost" molecules and gradients of chemoattractants and repellents; then there is an activity-dependent phase that modifies these early patterns. The role of activity—due to external stimuli as well as to spontaneous rhythms—has in recent years been found to be more vital than previously thought. Neuroscience now studies the formation of the nervous system not as the simple unfolding of a genetic blueprint, but rather as the way in which the sensory and motor experience of the individual define constantly changing patterns of neural activity which affect the ongoing "self-organization" based on that blueprint of the nervous system. This self-organization, while most dramatic in the child, may continue in the adult, being involved in both learning and in compensation for disease.

4 Basic Notions of Schema Theory[23]

Many workers in cognitive science have little interest in brain or action, and much of their work focuses on linking Artificial Intelligence (AI) and cognition to symbol manipulation in general and to linguistics in particular. My own work, on the

---

[23] An abridged version of this section occurs in my companion essay. On the other hand, that essay develops far more fully the notion of "social schemas."

contrary, tries to see our cognitive abilities as rooted in our more basic capabilities to perceive and interact with the world. What, then, is this schema theory in which we are to give an account of the embodied mind, an account which is to transcend mind-body dualism by integrating an account of our mental representations with an account of the way in which we interact with the world?

I use the term "schema theory" to designate an approach to cognitive neuroscience which explains behavior in terms of the concurrent interaction of many functional units called *schemas* (composable units of action, thought, and perception).[24] There are schemas for recognition of different objects, for the planning and control of different activities, and for more abstract operations as well. Schema theory now combines three distinct levels of theorizing:

Basic Schema Theory: Schema theory *simpliciter* provides a basic language which matches well with the "mental." It has its basic definition at a functional level which associates schemas with specific perceptual, motor, and cognitive abilities and other complex dispositions—and then stresses how our mental life results from the dynamic interaction, the competition and cooperation, of many schema instances. For example, one *perceptual* schema would let you recognize that a large structure is a house; in doing so, it might provide strategies for locating the front door. The recognition of the door (activation of the perceptual schema for door) is not an abstract end in itself—it helps activate, and supplies appropriate inputs to, *motor* schemas for approaching the door and for opening it.[25] However, even at this functional level, a "computationally complete" explanation may involve schemas which are quite different from those that are suggested by introspection from conscious mental behavior.

Neural Schema Theory: Just as much human behavior can be explained by psychology without recourse to neurology, so can much successful schema theory proceed at a purely functional level. However, if we are to understand phenomena described earlier—blindsight, episodic memory, emotion, consciousness, Prozac, neglect, mental disorders, etc.—it is clear that the details of schema function must make contact with data on brain localization and even with neurochemistry. This motivates the "downward" extension of schema theory to form *neural schema theory*, in which we move from psychology and cognitive science as classically conceived (viewing the mind "from the outside") to cognitive neuroscience. The description of a schema can often be refined into a network of more detailed schemas.[26] For a psychologist looking at overt behavior, the lowest-level schemas employed may be relatively molar, themselves relatable to the subject's introspection

---

[24] Of course there are many other approaches to schema theory, a few of which will be mentioned in the section "A Historical Sketch" below. The roots of my own approach may be traced in two papers: Michael A. Arbib, "Artificial Intelligence and Brain Theory: Unities and Diversities," *Annals of Biomedical Engineering* 3 (1975): 238–74; and idem, "Perceptual Structures and Distributed Motor Control," in *Handbook of Physiology, Section 2: The Nervous System, Vol. II, Motor Control, Part 1*, Vernon B. Brooks, ed. (Bethesda, Md.: American Physiological Society, 1981), 1449–80.

[25] See sec. 2.2 of Arbib, *The Metaphorical Brain 2*, for further examples.

[26] Theo Meyering (personal communication) comments: "This sounds a bit like homuncular decomposition as in [Daniel] Dennett's... homuncular functionalism. The ultimate homunculi's operations are so stupid and mechanical that our fellow neuroscientists may recognize them as simple routine response actions on the part of individual neurons."

or to the gross regional analysis of brain activity afforded by current human brain imaging techniques. For the neurophysiologist, further decomposition may be required until the schemas so defined are sufficiently fine-grained that their function may be played out across the detailed structures of specific neural networks of the brain. *Neural* schema theory provides a language for neuroscience appropriate to the analysis of data at the level of neuropsychology and human brain imaging, while at the same time showing that this "molar language" for neuroscience in no way precludes the relevance of finer-grained analysis in terms of neural circuitry and neurochemistry. Note that that our functional definition of a schema may change as we work out its implementation, revealing details that escaped our attention on superficial examination.

A schema in *basic schema theory* is a functional notion (emphasizing its causal role, regardless of what implements it). It is only when we turn to neural schema theory that we seek to go the further step of studying the neural implementation of the schemas—thus linking schemas to the structural entities (brain regions or neural circuits, for example) which implement them. Interestingly, the language of schema theory is little used by neuroscientists. This is not because schema theory is irrelevant to neuroscience, but rather because few neuroscientists study large systems. Instead, they focus on one specific schema (for example, depth perception), the response of one neural circuit to specific patterns of stimulation, or fine details of neurochemistry and biophysics. I believe schema theory will become more widely accepted as more neuroscientists seek to link these details with larger cognitive systems, and relate them to the results of human brain imaging (which forces a more global view of interacting brain regions).

Neural schemas, then, are intermediate between behavior and neurons. How does a schema differ from an Edelman group, a Hebb assembly, or other notions that are similarly intermediate?[27] They occupy the same "ecological niche," but my theory offers explicit analyses of perceptual-motor linkages and of the formation of assemblages/coordinated control programs that go beyond their theories. Below, I will strongly distinguish "schemas" from "modules" in the sense of Jerry Fodor's *Modularity of Mind*.[28]

Social Schema Theory: In seeking to reconcile the "collective representations" of a community with the thought processes of individuals—creating an epistemology that integrates a sociology of knowledge with a psychology of knowledge—Mary Hesse and I in *The Construction of Reality* extended basic schema theory "upward" to develop *social schema theory*. This theory shows how "social schemas" constituted by collective patterns of behavior in a society may provide an external reality for a person's acquisition of schemas "in the head," in the sense of basic schema theory; conversely, it is the collective effect of behavior expressing schemas within the heads of many individuals that constitutes, and changes, this social reality. To understand the human individual one studies the coherence and conflicts within a schema network that constitutes a personality, with all its contradictions. Social schema theory extends this to the holistic nets of social reality, custom, language, and religion. I say more about social schema theory in my companion essay, while here I concentrate on schema theory at the psychological and neural levels.

---

[27] Gerald M. Edelman, *Neural Darwinism: The Theory of Neuronal Group Selection* (Basic Books, 1987); Donald O. Hebb, *The Organization of Behavior* (New York: John Wiley & Sons, 1949).

[28] Jerry Fodor, *The Modularity of Mind* (Cambridge: MIT Press, 1983).

4.1 From Action-Oriented Perception to Knowledge

A schema is both a store of knowledge and the description of a process for applying that knowledge. As such, a schema may be instantiated to form multiple active copies called *schema instances*. For example, given a schema that represents generic knowledge about a chair, we may need several active instances of the chair schema, each suitably tuned, to subserve our perception of a scene containing several chairs. A schema is more like a molecule than an atom in that schema instances may well be linked to others to form *schema assemblages* which provide yet more comprehensive schemas.

Schema theory provides, *inter alia*, a language for the study of *action-oriented perception*[29] in which the organism's perception is in the service of current and intended action rather than (though not exclusive of) providing stimuli to which the organism provides unintended responses. According to schema theory, our minds comprise a richly interconnected network of schemas. Schema theory can also express models of language and other cognitive functions.[30]

An assemblage of some instances of these schemas represents our current situation. A crucial notion is that of *dynamic planning*: the organism is continually making and remaking plans—in the form of schema assemblages called *coordinated control programs* which combine perceptual, motor, and coordinating schemas—but these are subject to constant updating as perception signals obstacles or novel opportunities. In particular, action-oriented perception involves passing parameters from perceptual to motor schemas. For example, perceiving a ball instructs the hand how to grasp it; perceiving obstacles adjusts one's navigation. However, schema assemblages and dynamic planning ensure that behavior seldom involves direct relationships of a behaviorist, stimulus-response simplicity; rather, context and plans help determine which perceptual clues will be sought and acted upon.

Schemas are modular entities whose instances can become *activated* in response to certain patterns of input from sensory stimuli or other schema instances that are already active. The *activity level* of an instance of a perceptual schema represents a "confidence level" that the object represented by the schema is indeed present; while that of a motor schema may signal its "degree of readiness" to control some course of action. The activity level of a schema may be but one of many parameters that characterize it. Thus a schema for "ball" might include parameters for its size, color, and velocity—properties we might notice when we see a ball or play with it, or put differently, properties that would be observed at a level of detail appropriate to our skill and interest rather than at the level of highly precise measurements.

To make sense of any given situation we call upon hundreds of schemas in our current schema assemblage. Our lifetime of experience might be encoded in a personal "encyclopedia" of hundreds of thousands of schemas. As we act, we perceive; as we perceive, so we act. Perception is not passive, like a photograph. It is active, as our current schemas determine what we take from the environment.

---

[29] Ulric Neisser, *Cognition and Reality: Principles and Implications of Cognitive Psychology* (San Francisco, Calif.: W.H. Freeman, 1976); Michael A. Arbib, *The Metaphorical Brain: An Introduction to Cybernetics as Artificial Intelligence and Brain Theory* (New York: Wiley-Interscience, 1972).

[30] Michael A. Arbib, E. Jeffrey Conklin, and Jane C. Hill, *From Schema Theory to Language* (Oxford: Oxford University Press, 1987).

The essence of schema theory goes beyond the fact that we have concepts, for example, of a ball, because it makes explicit aspects of "concepts" that might be lost in other accounts. We first need to distinguish the "concepts" from the "schema." Are whales mammals? Science says "Yes"—but "Do whales activate the mammal schema?" would be answered "No" for many individuals. Further, schema theory integrates perceptual schemas (for example, how to recognize a ball) with motor schemas (such as what to do with a ball) through the parameter-passing mechanism, but also expresses likely and unlikely patterns of co-occurrence through the patterns of competition and cooperation that develop within the schema network.[31] I discuss learning below—where perceptions lead (in an ongoing action-perception cycle) to actions with attendant expectations; failure of these expectations can lead to modifications of perceptual, motor, and other schemas.[32]

One would like to have criteria (whether functional, neurological, phenomenological, conceptual, or behavioral) to individuate, or pick out, distinct schemas but none such exists at present. A schema analysis will often start with some overall function or phenomenon of interest and then refine the definition of the schema and its decomposition into other schemas in such a way as to match data on speed and error of behavior, or (if one studies schemas at the level of brain theory) the effects of lesions and other neural measurements and perturbations.[33]

4.2 Schema Instances and Cooperative Computation

Schema theory sees behavior as based *not* on inferences from axioms nor on the operation of an inference engine on a passive store of knowledge. This moves us from the domain of serial computation to an understanding of how behavior results from *competition* and *cooperation* between schema instances (that is, interactions which, respectively, decrease and increase the activity levels of these instances) which, due to the limitations of experience, cannot constitute a completely consistent axiom-based logical system.

Schema theory thus offers a new paradigm of computation, with "schemas" as the programs, and *cooperative computation*—a shorthand for "computation based on the competition and cooperation of concurrently active agents"—as their style of interaction. Cooperation yields "strengthened alliances" between mutually consistent schema instances, allowing them to achieve high activity levels to constitute the overall solution of a problem (as perceptual schemas become part of the current short-term model of the environment, or motor schemas contribute to the current course of action). It is as a result of competition that instances which do not meet the evolving (data-guided) consensus lose activity, and thus are not part of this solution (though their continuing subthreshold activity may well affect later behavior).

---

[31] It would be interesting to discuss whether my notion of a "schema network" might relate to the notion of "semiotic network" as used by Wesley Wildman and Leslie Brothers (in this volume)—but extended by an indissoluble link with the pragmatic network.

[32] Admittedly, this is an inadequate classification of schemas. Specific models introduce schemas whose role is to coordinate other schemas; for example, Bruce Hoff and M.A. Arbib, "Simulation of Interaction of Hand Transport and Preshape During Visually Guided Reaching to Perturbed Targets," *Journal of Motor Behavior* 25 (1993): 175–92. Moreover, as assemblages or coordinated control programs are built up, they constitute compound schemas which are primarily neither perceptual nor motor.

[33] For more details see, for example, Arbib, *The Metaphorical Brain 2*, or chap. 3 of Arbib, Érdi, and Szentágothai, *Neural Organization*.

A schema network does not, in general, need a top-level executor since schema instances can combine their effects by distributed processes. This may lead to apparently emergent behavior,[34] due to the absence of global control. As a very simple example, take my model of the frog, *Rana computatrix*.[35] The decision on whether to feed or flee results from the interaction of schemas related to these two behaviors, not from explicit analysis of the relative merits of these two courses of action by higher level schemas. But the process does not stop there. A schema for hunger can shift the balance from "flee" to "feed" not by top-down control but by lowering the threshold for the "feed schema" to initiate behavior; schemas for recognition of obstacles can bias the chosen behavior to yield an appropriate detour, and this is expandable by learning.

To further see why a schema network may not need a top-level executor, think of schemas as linked in a network with two kinds of links. One kind passes data. For example, the ball-schema might pass time-until-contact information to the catch-schema. The other kind of link passes activity levels so that, for example, perceptual schemas for two regions of an image may excite each other if the objects they represent are likely to occur in that spatial relationship, or they might inhibit each other if such a juxtaposition is unlikely, as in seeing a snowball atop a fire. Since a surrealist painting *could* be seen to depict a snowball atop a fire, it is clear that these activity-links bias a dynamic process of interpretation rather than determining what can and cannot be seen. Similar considerations apply to other forms of integration of action, perception, and thought. Elsewhere, I provide a more fully developed example of cooperative computation in recognition in visual scene perception, which involves the continued interaction of bottom-up (more data-driven) and top-down (more hypothesis-driven) schemas.[36]

4.3 Learning

Schema theory is a learning theory too. A schema provides us not only with abilities for recognition and guides to action, but also with expectations about what will happen. These may be wrong. We sometimes learn from our mistakes. Our schemas, and their connections within the schema network, change. In a general setting, there is no fixed repertoire of basic schemas. Rather, new schemas may be formed as assemblages of old schemas; but once formed a schema may be tuned by some adaptive mechanism. This tunability of schema-assemblages allows them to start as composite but emerge as primitive, much as a skill is honed into a unified whole from constituent pieces. My approach to schema theory thus adopts the idea of Jean Piaget, the Swiss developmental psychologist and genetic epistemologist, that the child has certain basic schemas and basic ways of *assimilating* knowledge to schemas, and that the child will find at times a discrepancy between what it experiences and what it needs or anticipates.[37] On this basis, its schemas will change,

---

[34] I use the term "apparently emergent" here as a shorthand for rejecting claims like Nancey Murphy's (in this volume) of "supervenience without reduction"—I do not want my use of the term "emergent" to be understood as implying that the specific phenomenon cannot be explained in lower-level terms.

[35] For more details, see chap. 3 of Arbib, Érdi, and Szentágothai, *Neural Organization*.

[36] Arbib, *The Metaphorical Brain 2*, sec. 5.3.

[37] See, for example, Jean Piaget, *Biology and Knowledge* (Edinburgh: Edinburgh University Press, 1971).

accommodation will take place. It is an active research question as to what constitutes the initial stock of schemas. Much of Piaget's writing emphasizes the initial primacy of sensorimotor schemas, where other scientists study the interactions between mother and child to stress social and interpersonal schemas as part of the basic repertoire on which the child builds.

Another important concept in Piaget's work is that of *reflective abstraction*.[38] Piaget emphasizes that we do not respond to unanalyzed patterns of stimulation from the world. Rather, current stimuli are analyzed in terms of our current stock of schemas. It is the interaction between the stimulation—which provides variety and the unexpected—and the schemas already in place that provides patterns from which we can then begin to extract new operational relationships. These relationships can now be reflected into new schemas which form, as it were, a new plane of thought. And then—and this is the crucial point—since schemas form a network, these new operations not only abstract from what has gone before, but now provide an environment in which old schemas can be restructured. To the extent that we can form a general concept of an object, our earlier knowledge of a dog, a ball, and so on, become enriched.

4.4 An Historical Sketch

Schema theory is designed to give an account of the embodied mind, an account which is to transcend mind-body dualism by integrating an account of our mental representations with an account of the way in which we interact with the world. To enrich the discussion of schemas, this section offers a brief historical review.[39] The history of schemas goes back to Immanuel Kant and beyond, but its links to neuroscience start with the work of the neurologists Henry Head and Gordon Holmes[40] who discussed the notion of the *body schema*.[41] A person with damage to one parietal lobe of the brain may lose all sense of the opposite side of his body (see our earlier discussion of "neglect"), not only ignoring painful stimuli but even neglecting to dress that half of the body; conversely, a person with an amputated limb but with the corresponding part of the brain intact may experience a wide range of sensation from the "phantom limb." Even at this most basic level of our personal

---

[38] The idea of reflective abstraction is developed by Evert W. Beth and Jean Piaget, *Mathematical Epistemology and Psychology*, trans. W. Mays, (Dordrecht: Reidel, 1966); I argue, in "A Piagetian Perspective on Mathematical Construction," *Synthese* 84 (1990): 43–58, that Piaget pays insufficient attention to the role of social structures, including formal instruction, in the child's construction of logic and mathematics.

[39] Far more information is provided in Michael A. Arbib, "Schema Theory: From Kant to McCulloch and Beyond," in *Brain Processes, Theories and Models: An International Conference in Honor of W.S. McCulloch 25 Years After His Death*, Robert Moreno-Diaz and José Mira-Mira, eds. (Cambridge: MIT Press, 1995), 11–23, wherein I also acknowledge my intellectual debt to Warren McCulloch.

[40] Henry Head and G. Holmes, "Sensory Disturbances from Cerebral Lesions," *Brain* 34 (1911): 102–254.

[41] Stephen Happel (personal communication) states that my view of schemas will need to be explored for "its relationship to earlier philosophical traditions on schemas of the understanding and innate ideas.... As a 'store' of knowledge and a 'process', they differ from Kantian schemata of understanding which are purely formal, but which nonetheless make all sensory perception interpretive." I agree that it would be most valuable to have a critique of my version of schema theory by someone more versed in these traditions.

reality—our knowledge of the structure of our own body—our brain is responsible for constructing that reality for us. Our growing scientific understanding of knowledge takes us far from what "common sense" will tell us is obvious. One of Head's students was Frederick Bartlett, who noted that people's retelling of a story is based not on word-by-word recollection, but rather on remembering the story in terms of their own internal schemas, and then finding words in which to express this schema assemblage.[42]

Such ideas prepare us for the work of Kenneth Craik, who understood the brain to "model" the world, so that when you recognize something, you "see" in it things that will guide your interaction with it.[43] There is no claim of infallibility, no claim that the interactions will always proceed as expected. But the point is that you recognize things not as a linguistic animal, merely to name them, but as an embodied animal. I use the term "schema" for the building blocks of these models that guide our interactions with the world. To the extent that our expectations are false, our schemas can change; we learn. Many writers in the 1960s and 70s built upon this notion of an internal model of the world—first in the cybernetic tradition—in order to develop the concept of representation so central to work in AI today.[44]

One of the best-known users of the term "schema" is Piaget, whose *Biology and Knowledge* gives an overview of his "genetic epistemology," which develops an embryological metaphor for the growth of a human's, and of human, knowledge. Piaget defines a schema as the structure of interaction, the underlying form of a repeated activity pattern that can transcend the particular physical objects it acts on and can become capable of generalization to other contexts. He traces the cognitive development of children, starting from basic schemas that guide their motoric interactions with the world, through stages of increasing abstraction that lead to language and logic, to abstract thought. We have already noted the importance of Piaget's concepts of *assimilation*, the ability to make sense of a situation in terms of the current stocks of schemas, and of *accommodation*, the way in which the stock of schemas may change over time as the expectations based on assimilation to current schemas are not met. These processes within the individual are reminiscent of the way in which a scientific community is guided by the *pragmatic criterion* of successful prediction and control.[45] We keep updating our scientific theories as we try to extend the range of phenomena they can help us understand. It is worth noting, however, that the increasing range of successful prediction may be accompanied by revolutions in ontology, in our understanding of what is real, as when we shift from the inherently deterministic reality of Newtonian mechanics to the inherently probabilistic reality of quantum mechanics.

---

[42] Frederic C. Bartlett, *Remembering* (Cambridge: Cambridge University Press, 1932).

[43] Kenneth J.W. Craik, *The Nature of Explanation* (Cambridge: Cambridge University Press, 1943).

[44] Richard L. Gregory, "On How so Little Information Controls so much Behavior," in *Towards a Theoretical Biology, no. 2, Sketches,* Conrad H. Waddington, ed. (Edinburgh: Edinburgh University Press, 1969); Donald M. MacKay, "Cerebral Organization and the Conscious Control of Action," in *Brain and Conscious Experience,* John C. Eccles, ed. (New York: Springer-Verlag, 1966), 422–40; Marvin L. Minsky, "Matter, Mind and Models," in *Information Processing 1965, Proceedings of International Federation of Information Processing Societies Congress 65*, vol. 1, (New York: Spartan Books, 1965), 45–59.

[45] Mary B. Hesse, *Revolutions and Reconstructions in the Philosophy of Science* (Bloomington: Indiana University Press, 1980).

Much work in brain theory and artificial intelligence contributes to schema theory, even though the scientists involved do not use this term. Schema theory provides a knowledge-representation protocol which is part of the same theory-building enterprise as frames and scripts[46] but is distinguished in that, for example, schema theory has a grain size smaller than frames and scripts, but larger than neural models. Schema theory stresses the building up of new schemas; script theory stresses overarching organizational schemas for some family of behaviors. In its emphasis on the interaction of active computing agents (the schema instances), schema theory is related to studies in distributed artificial intelligence rooted in work on actors,[47] the HEARSAY speech understanding system,[48] and distributed problem solving.[49] Since each schema combines knowledge with the processes for using it, schemas are more like actors than like frames or systems with unitary blackboards. Marvin Minsky espouses a *society of mind* analogy in which "members of society," the agents, are analogous to schemas.[50] Rodney Brooks controls robots with layers of asynchronous modules that can be taken as a version of schemas.[51] This work shares with schema theory the mediation of action through a network of schemas; no single, central, logical representation of the world need link perception and action.[52] It shares with Grey Walter and Valentino Braitenberg the study of the "evolution" of simple "creatures" with increasingly sophisticated sensorimotor capacities.[53]

The term "schema theory," then, does not refer to one polished and widely accepted formalism.[54] For work within artificial intelligence, including work in machine vision and robotics, one seeks to define schemas as program units for cooperative computation that meet criteria for ease of implementation or for

[46] Marvin L. Minsky, "A Framework for Representing Knowledge," in *The Psychology of Computer Vision*, Patrick H. Winston, ed. (New York: McGraw-Hill, 1975), 211–77; Roger Schank and Robert Abelson, *Scripts, Plans, Goals and Understanding: An Inquiry into Human Knowledge Structures* (New York: Erlbaum, 1977).

[47] Carl E. Hewitt, "Viewing Control Structures as Patterns of Passing Messages," *Artificial Intelligence* 8 (1977): 323–64.

[48] Lee D. Erman, F.A. Hayes-Roth, V.R. Lesser, and D.R. Reddy, "The Hearsay-II Speech-Understanding System: Integrating Knowledge to Resolve Uncertainty," *Computing Surveys* 12 (1980): 213–53.

[49] Randall Davis and R.G. Smith, "Negotiation as a Metaphor for Distributed Problem Solving," *Artificial Intelligence* 20 (1983): 63–109.

[50] This analogy seems to subsume rather than replace his frame theory of 1977; see Marvin L. Minsky, *The Society of Mind* (New York: Simon and Schuster, 1985), 244–50.

[51] Rodney A. Brooks, "A Robust Layered Control System for a Mobile Robot," *IEEE Journal of Robotics and Automation* RA-2 (1986): 14–23.

[52] Arbib, *The Metaphorical Brain*, 168.

[53] Valentino Braitenberg, "Taxis, Kinesis, Decussation," *Progress in Brain Research* 17 (1965): 210–22; idem, *Vehicles: Experiments in Synthetic Psychology* (Cambridge: MIT Press, 1984); W.Grey Walter, *The Living Brain* (New York: Penguin Books, 1953).

[54] For other approaches to schema theory as developed within cognitive psychology see George Mandler, *Cognitive Psychology: An Essay in Cognitive Science* (Hillsdale, N.J.: Lawrence Erlbaum Associates, 1985), and Tim Shallice, *From Neuropsychology to Mental Structure* (Cambridge: Cambridge University Press, 1988). For a related schema-based approach to motor control, see Richard A. Schmidt, "The Schema as a Solution to Some Persistent Problems in Motor Learning Theory," in *Motor Control: Issues and Trends*, George E. Stelmach, ed. (New York: Academic Press, 1976), 41–65.

computational efficiency. For work within brain theory and cognitive psychology, schemas are designed to serve as units of complexity intermediate between behavior and neuron and to help us "decompose" overall behavior in a fashion that gives us insight into the data of psychology and neuroscience. While my schema theory has been informed by that of Piaget—especially in its emphasis on the sensorimotor basis for mental development—and other work reviewed above, it is distinguished by the following four points:

1. It emphasizes the fact that our experience is usually mediated by an assemblage of schemas rather than a single schema. A situation is represented (consciously or unconsciously, repressed or not) by the activation of a network of schemas that embody the significant aspects of a situation for the organism. Then, schemas determine a course of action by a process of analogy formation, planning, and schema interaction, in which formal deduction is not necessary implicated. Moreover, memory of a schema assemblage may be tuned to create a new schema.

2. It relates perception to action within a unified representational framework.[55]

3. It presents a view of adaptation with links to Piaget's assimilation and accommodation, but sees developmental stages as "emergent" rather than genetically prespecified.

4. Related to *1* above, it introduces cooperative computation as a unifying style for cognitive science and neuroscience.

My approach is also distinguished by being structured in such a way that a schema may either be viewed purely as a functional unit in a network of interacting schemas (*basic* schema theory), further analyzed in terms of its neural underpinnings (*neural* schema theory), or further developed by linking individual schemas to social schemas (*social* schema theory).

4.5 Linking Schemas to the Brain

In brain theory (see footnote 20), the analysis of schema instances is intermediate between the overall specification of some behavior and the neural networks that subserve it. A given schema, defined functionally, may be distributed across more than one brain region; conversely, a given brain region may be involved in many schemas. A top-down analysis may advance specific hypotheses about the localiza-tion of (sub)schemas in the brain, and these may be tested by lesion experiments, with possible modification of the model (for example, replacing one schema by several interacting schemas with different localizations) and further testing. Once a schema-theoretic model of some animal behavior has been refined to the point of hypotheses about the localization of schemas, we may then model a brain region by seeing if its known neural circuitry can indeed be shown to implement the posited schema. In some cases the model will involve properties of the circuitry that have not yet been tested, thus laying the ground for new experiments.

Schemas as "functional units" may be contrasted with the "structural units" of neuroanatomy and neurophysiology. The work of the nineteenth-century neurologists led us to think of the brain in terms of large interacting regions each with a more or less specified function, and this localization was reinforced by the work of the

---

[55] See also Marc Jeannerod, *The Cognitive Neuroscience of Action* (Oxford: Blackwell Publishers, 1997).

anatomists at the turn of the century who were able to subdivide the cerebral cortex on the basis of cell characteristics, cytoarchitectonics. It was at this same time that the discoveries of the neuroanatomist Santiago Ramón y Cajal and the neurophysiologist Charles Sherrington[56] helped establish the neuron doctrine, leading us to view the function of the brain in terms of the interaction of discrete units, the neurons. The issue for the brain theorist, then, is to map complex functions, behaviors, and patterns of thought onto the interactions of these rather large entities, anatomically defined brain regions, or these very small and numerous components, the neurons. This has led many neuroscientists to look for structures intermediate in size and complexity between brain regions and neurons to provide stepping stones in an analysis of how neural structures subserve various functions. One early example was the Scheibels' suggestion that the reticular formation could be approximated by a stack of "poker chips" each incorporating a large number of neurons receiving roughly the same input and providing roughly the same output to their environments. This modular decomposition provided the basis for William Kilmer and Warren McCulloch's model of the reticular formation.[57]

In another direction, the theoretical ideas of Walter Pitts and McCulloch combined with the empirical observations of Jerry Lettvin and Humberto Maturana on the frog's visual system to suggest that one might think of important portions of the brain in terms of interacting layers of neurons, with each layer being retinotopic in that the position of neurons in the layer was correlated with position on the retina, and thus in the visual field.[58] A neuron may participate in the implementation of multiple schemas. For example, in the toad brain we find that certain neurons in the pretectum, whose activity correlates with that of the perceptual schema for predators, will also contribute via an inhibitory pathway to the tectum to the perceptual schema for prey (this is an explicit example of "cooperative computation"). A representation of some overtly defined concept or behavioral parameter will in general involve temporally coordinated activity of a multitude of neurons distributed over multiple brain regions. Moreover, each region will in general exhibit *coarse coding* of parameters: it is not the firing of a single cell that codes a value, but rather the averaged activity of a whole set of neurons that is crucial. In any case, the brain embodies many different schemas, some based on circuitry evolved for that purpose, others developed on the basis of experience with both social and nonsocial interactions.

Vernon Mountcastle and T.P.S. Powell working on the somatosensory cortex, followed by David Hubel and Torsten Wiesel working on the visual cortex, established the notion of the column as a "vertical" aggregate of cells in the cerebral

---

[56] Santiago Ramón y Cajal, "Histologie du systeme nerveux," English trans. by Neely and Larry Swanson, (Paris: A. Maloine, 1911/Oxford: Oxford University Press, 1995); Charles S. Sherrington, *The Integrative Action of the Nervous System* (New Haven, Conn.: Yale University Press, 1906).

[57] M.E. Scheibel and Arnold B. Scheibel, "Structural Substrates for Integrative Patterns in the Brain Stem Reticular Core," in *Reticular Formation of the Brain*, Herbert H. Jasper *et al.*, eds. (Boston: Little, Brown and Co., 1958), 31–68; William L. Kilmer, W.S. McCulloch, and J. Blum, "A Model of the Vertebrate Central Command System," *International Journal Man-Machine Studies* 1 (1958): 279–309.

[58] Walter H. Pitts, and W.S. McCulloch, "How We Know Universals, the Perception of Auditory and Visual Forms," *Bulletin of Mathematical Biophysics* 9 (1947): 127–47; Jerry Y. Lettvin, H. Maturana, W.S. McCulloch, and W.H. Pitts, "What the Frog's Eye Tells the Frog's Brain," *International Journal of the Institute of Radio Engineers* 47 (1959): 1940–51.

cortex, again working on a common set of inputs to provide a well-defined set of outputs.[59]

With this research, the notion of the brain as an interconnected set of "modules"—intermediate in complexity between neurons and brain regions—was well established within neuroscience,[60] but here it may be useful to distinguish "neural modules" and "schemas" from "modules" in the sense of Fodor's *Modularity of Mind*,[61] a sense which has been excessively influential in recent cognitive science and related philosophizing. Rather than go into details, I simply list the key points from an earlier critique:[62] The fundamental point is that Fodor's modules—such as "language" or "vision"—are too large. It is clear from schema analyses of visual perception or motor control[63] that a computational theory of cognition must use a far finer grain of analysis than that offered by Fodor. Fodor offers big modules (for example, one for all of language), arguing vociferously that they are computationally autonomous, and despairs at the problem of explaining the central processes, since they are not informationally encapsulated. By contrast, my approach is to analyze the brain in terms of those smaller *functional* units called schemas, while stressing that each schema may involve the cooperative computation of many *structural* units ("modules" in the classical, medium-grained sense of neuroscience outlined above). Since the interactions between these schemas play a vital role in my models, the case for autonomy of large modules becomes less plausible. As a result, schema theory offers a continuity of theorizing between, say, vision and action and central processes, rather than recognizing the reality of the divide posited by Fodor.

George Ellis (personal communication) stresses that the notion raised in the section on "Learning" that "schema assemblages may start as composite but emerge as primitive" [that is, functionally grouped schemas can be described and/or activated by a single label] underlies the essential feature of hierarchical structuring.[64] In fact, the issue of hierarchical structuring has been a central concern since, for example, the publication of Donald Hebb's *The Organization of Behavior* to present day study of neural networks. The main ingredients are patterns of neural activity—which become established ("attractors") as quasi-stable (namely, until there is a certain amount of change of input activity)—and excitatory and inhibitory links, which encourage coactivation of several such patterns together (Hebb's "assemblies") or the activation of such patterns in some order (Hebb's "phase

---

[59] Vernon B. Mountcastle and T.P.S. Powell, "Neural Mechanisms Subserving Cutaneous Sensibility, with Special Reference to the Role of Afferent Inhibition in Sensory Perception and Discrimination," *Bulletin of Johns Hopkins Hospital* 105 (1959): 201–32; David H. Hubel and T.N. Wiesel, "Sequence Regularity and Geometry of Orientation Columns in the Monkey Striate Cortex," *Journal of Comparative Neurology* 158 (1974): 267–94.

[60] For a recent overview, see Szentágothai's discussion of "modular architectonics of the brain" in chap. 2 of Arbib, Érdi, and Szentágothai, *Neural Organization*.

[61] Fodor, *The Modularity of Mind*.

[62] Michael A. Arbib, "Modularity and Interaction of Brain Regions Underlying Visuomotor Coordination," in *Modularity in Knowledge Representation and Natural Language Understanding*, Jay L. Garfield, ed. (Cambridge: MIT Press, 1987), 333–63.

[63] Detailed expositions may be seen in Arbib, *The Metaphorical Brain 2*.

[64] Ellis (personal communication) notes that this structuring is given a central place in Stafford Beer's illuminating book *Brain of the Firm*, 2nd ed. (Chichester, U.K.: John Wiley and Sons, 1981), and in quite a different way in Alwyn Scott's book *Stairway to the Mind: The Controversial New Science of Consciousness* (New York: Springer-Verlag, 1995).

sequences"). However, many processes that we can now describe at the abstract schema level (as in visual scene perception, mentioned above) still pose unanswered questions about whether and how they are realized in the brain's circuitry.

4.6 The Self as a "Schema Encyclopedia"

People carry in their heads what appears to them to be a total model of the universe; what they experience must be assimilated to their current stock of schemas, though it may also change them. The child or adult is an active participant in his or her world; when confronted with something new, each tries to "tell a story" to fit the new data, starting from the current stock of schemas. This new story becomes a contribution to the extended stock of schemas that constitutes their personhood within space and time. What they already know provides the machinery through which they incorporate new data. But what is crucial is that this highly personal network has a coherence. It is this coherence of the network that we call the *self*. A self seems to be something that develops in a sentient being able to reflect on his or her experience. Not all of us have equally coherent selves any more than we are equally sensitive to experience, but everyone is conscious of being and having a self.

Schemas constitute a network that brings together our notions of reality at many different levels. Schemas and their connections within the schema network change through a process of accommodation—learning is necessary because schemas are fallible. In schema theory we characterize much of the decision making that we engage in, whether or not it is conscious, as being *a time-limited process of analogical reasoning* based on a vast array of examples which by the nature of our limited experience do not constitute a completely consistent logical system of certain deduction from consistent axioms. But even a computer program, which is completely explicit, may take many hours or even days of computing to deliver an adequate answer. And so we may find ourselves forced to make a decision knowing that we have imperfect data, knowing that we have too little time, knowing that the outcome of the decision is of immense importance to us.

I have already suggested that to make sense of any given situation we call upon hundreds of schemas in our current "schema assemblage" and that our lifetime of experience—our skills, our general knowledge, our recollection of specific episodes—is encoded in a personal "encyclopedia" of hundreds of thousands of schemas, enriched from the verbal domain to incorporate the representations of action and perception, of motive and emotion, of personal and social interactions, of the embodied self. It is in terms of these hundreds of thousands of schemas that I would offer a naturalistic account of the self, embodied in space and time. Nonetheless, for many people this raises the question: could hundreds of thousands of schemas, or billions of neurons, cohere to constitute a single personality, a self, a personal consciousness?

In any particular study of brain and mind, schema theory separates a part of the network from its connections inside and outside and thus must do damage to the plenitude of lived experience. However, our job as cognitive scientists is only to *chart* the territory of mental life to establish the major phenomena and their relationships; not to provide the full-scale map, not to replace a life richly lived by the running of some computer program. Schema theory seeks the benefits of holism while avoiding the disadvantages attendant upon any approach that precludes analysis of processes into constituents with distinct contributions to the whole.

4.7 Can Schema Theory be Falsified?

There are two facets to schema theory: (1) as a language for cognitive theories, and (2) as a specific theory of intelligent behavior. Schema theory provides the beginnings of a language for distributed systems at a level abstract enough to convey real understanding of complex problem-solving behaviors and yet precise enough that we can refine the specification to some concrete implementation. Discussion of specific brain models leads us to the other facet of schema theory: a model of perceptuo-motor integration and cognition that uses schema theory *qua* language for *expressing* such models. In the language sense, schema theory is more like group theory than general relativity. Whereas the latter is a model of the physical world—it can be falsified or revised on the basis of physical experiments—group theory stands or falls for the scientist seeking to explain the world (as distinct from the mathematician proving theorems) not by whether it is true or false, but rather on whether its terminology and theorems aid the expression of successful models. Schema theory as an abstract model of computation does not yet have the rigor or stock of theorems of group theory, but the success of models using the language of instantiation, modulation, activity levels, etc., strengthens its claim to be a valuable tool in the development of brain theory, cognitive science, and philosophy of mind. However, the language of schema theory has developed in tandem with a schema-based theory of human and artificial intelligence. Specific schema models can be falsified. If they are usually replaced by more effective models which do not use schema language, then schema theory in general becomes weakened. If, however, the result effectively employs schemas, then the case for schema theory is strengthened.[65]

5 Implications

The brief examples presented in the section 1—blindsight, episodic memory, Prozac, etc.—show that more and more aspects of the person can be shown to depend on brain mechanisms, but that many open questions remain. It is now clear that the unity of self is an intricate construction that can be fractured in surprising ways by neurological damage (with hemi-neglect the standard example); thus "mental causation" as a level whose laws form a closed system must be subjected to renewed critique. I have offered schema theory, with its schema instances and cooperative computation, as a computational approach to mind-talk, while making clear that, when linking schemas to the brain, an initial schema-theoretic model based on introspection might be subject to drastic revision in the face of neuroscientific data. Earlier sections charted ways of bridging the levels with computational neuroscience.

My discussion in section 2, with its all too brief comments on morality, made it clear that the challenges for a "science of the person" are manifold—feelings, moral choice, social constructions, experiences of ultimacy, an individually held "theory of mind" that relates self and others, etc. There is as yet no consensus as to whether neuroscience can indeed contribute to all these questions. I suggested that the self could, in some sense, be viewed as a "schema encyclopedia," and I offered a preview of the theory of social schemas set forth in my companion essay. Thus there is much to encourage the monist who wishes to see personhood firmly rooted in brains and bodies in social interaction, but too many questions remain for such a monist to insist that the case is closed.

---

[65] Such successes are reported in chap. 3 of Arbib, et al., *Neural Organization*.

Dualism sees the brain as the seat of processes that control our body and its interactions with the world, and possibly even those giving rise to thoughts and emotions, but ultimately regards the core of what gives us our identity as rooted in something "not of the flesh." One religious claim which is irreconcilable with the idea that the mind *and* the spirit are merely the brain in action would seem to be the immortality of the soul. On the other hand, *aspirations* to immortality might be explained as yet another expression of the activity of the human brain. The idea of immortality is a very powerful one; many religions state that after this life the soul will continue its existence. Believers feel that they participate in the Divine, this world being only a temporary station and God the fundamental reality. For many Christians, the human soul is immortal—human life transcends death of the material body, and does so in a way that preserves the community; not only do we live forever, but we live forever with our friends, kin, and fellow believers.

Faith in resurrection would seem to be a place where the Christian believer parts company with "science as we know it"—this despite the fact that the Christians contributing to this volume are extremely reluctant to allow God to transgress the laws of physics. For them, God created the universe along with the laws of physics and is ever present in maintaining these laws (ensuring, for example, that $E = mc^2$ at all places and times), but their God is noninterventionist—he does not capriciously change the laws of physics in response to prayer or to work a miracle. Nonetheless, many of them would see the resurrection of Jesus as a genuine exception. However, the crucial viewpoint for most of them is that God acts, in general, through the nature of the world, a world so designed by God as to ensure the evolution of humans as caring beings for whom the life and resurrection of Jesus may serve as inspiration. For some Christians, prayer is more a form of self- or group-inspiration than it is a means of encouraging a beneficial intervention by God.

Christian theology is diverse: even Catholic theologians may have views differing widely from each other's and the Pope's. While many Christians believe in an immaterial soul, most of the Christians writing in this volume do not, and see themselves as being in the mainstream of current Christian theology in rejecting dualism and embracing a belief in the resurrection of the body. They are therefore able to accept with equanimity the efforts of neuroscience to see mind as a product of brain (but with many caveats about reduction), since the New Creation would see the body resurrected (in a more perfect form) by God, with the mind thus recreated "automatically" without need for a disembodied spirit.[66] Of course, this avoidance of the problem of dualism raises another problem: How does God hold "the pattern of the body" when the body decomposes in this creation?[67] Neuroscience offers no support for reintegration of the nervous system after its decomposition following death. For this and further reasons, many other monists do not believe in immortality—the material of the body might be eternal in different forms of matter and energy, but the form of that material that gives rise to mind is temporary and loses its function at death.

---

[66] For some, the resurrection of Jesus is a "guarantee" of bodily resurrection, but it is unclear how the resurrection of Jesus within this world after three days and with the stigmata upon him is related to the idea of a New Creation in which each of us will be bodily resurrected in a more perfect form. Other discussions of resurrection can be found in the essays by Green, Barbour, and Peters in this volume.

[67] Ian Barbour's essay in this volume provides one of several examples of a monist Christianity that escapes dualism by belief in the resurrection of the body.

EMOTIONS: A VIEW THROUGH THE BRAIN

Joseph E. LeDoux

1 Emotions, Feelings, and the Brain: An Overview

So what is an emotion? Many have asked and answered this question, but little agreement exists about what the correct answer is. Some even say that emotions are natural categories and that we should not ask about their representation in the brain.[1] In part, the controversy is due to the lack of agreement over how terms like "emotion" and "emotional" are used. When brain researchers study emotions, they typically study neural systems involved in the control of emotional behavior or associated physiological responses, often in experimental animals. But when psychological theorists talk about "emotions" they are usually referring to subjective feelings experienced by humans. Everyone recognizes that these differences exist. However, the differences and their implications are not always attended to when emotions are discussed. In particular, there is a strong tendency to start with our own subjective experiences of an emotion and then ask, through studies of the brains of experimental animals, how that experience is mediated by the brain. The problem with this approach is that it assumes that feelings are the functions of emotional systems in the brain. This, as it turns out, is probably wrong.

Emotion systems in the brain evolved as behavioral (essentially sensory-motor) solutions to problems of survival. Feelings can (but do not necessarily) occur when one of these systems is active. Emotional responses, in other words, do not require feelings. Put differently, feelings are not the functions that emotion systems were designed to perform.

For example, all animals have to be able to detect danger in their world and respond in ways that will keep them alive. This is as true of a bug and a slug as it is of a fish, frog, rat, and human. There are fairly universal ways of responding to danger. When a sudden danger occurs many animals freeze (even if only momentarily). Predators respond to movement and freezing is usually a better bet than moving (though it breaks down sometimes, as when a deer is frozen in the headlights of an oncoming car; but then again, evolution does not know about cars). This behavioral response is accompanied by a variety of physiological changes, some of which occur to support the behavior, some of which are consequences of the behavior, and some of which are in anticipation of having to defend or run away. These kinds of responses occur in humans and most of our mammalian cousins, as well as in many other vertebrates. Although humans experience fear when responding to danger, the same (or similar) responses occur in other animals that are unlikely have the capacity for subjective experiences. Conscious fear does not always accompany the expression fear responses, and it is certainly not a requirement for these expressions. Most crucial, the capacity to express behaviors that we call "emotional" in all likelihood existed prior to (in an evolutionary sense) the capacity to be consciously aware of emotional states. Emotion systems did not evolve to produce feelings.

---

[1] Leslie A. Brothers, *Friday's Footprint: How Society Shapes the Human Mind* (New York: Oxford University Press, 1997).

Part of the confusion here is clearly related to fact that the distinction between "emotions" and "feelings" that I am driving at is not captured by these terms. I am obviously using "emotion" to refer to objectively measurable bodily responses and the brain systems that control them and "feelings" to refer to associated subjective experiences. However, if we had a different word for the responses, a word not derived from subjective feelings, things would be much clearer. Since I do not yet have an alternative word, I will continue using "emotions" to refer to responses and the brain systems underlying them.

If the generation of subjective feelings is not the function of what we have come to think of as emotion systems, then what system is responsible for feelings? Subjective feelings, like being afraid, occur when an emotion system, like the fear system, is active in a brain with the capacity to be aware of its own activities. Emotional feelings are thus what happens when emotion systems are present in brains that are conscious. The problem of feelings is, in other words, the mind-body problem. Find consciousness in the brain and an understanding of feelings is sure to follow.

In many other areas of cognitive science and psychology progress has been made in understanding how fundamental processes work without solving the problem of how experiences occur. For example, we have learned a great deal about how the brain processes color or plans movements without having first figured out how the brain experiences the intention to move or the color of objects. In the study of emotion, though, researchers have often acted as though they first needed to solve the problem of feelings—the mind-body problem—before doing anything else.

This is attributable in large part to William James's famous question about whether we run from the bear because we are afraid or whether we are afraid because we run. Although psychologists have not accepted James's claim that we are afraid because we run, they have accepted his framing of the problem: that the study of emotion should be about figuring out where feelings, like being afraid, come from. As a result, little attention has been paid to the more fundamental question of how the sensory registration of the bear is processed by the brain in such a way as to trigger defensive reactions in us.

How emotional stimuli are processed by the brain is a more tractable problem than how feelings come out of the brain. Just as color processing can be understood without first understanding how color qualia are experienced, emotional processing can be studied without first understanding how emotional feelings are experienced. In the case of color perception, it is believed that the experience of color must be based on the underlying processes. Surely something similar goes on for an emotion as well—the brain system that detects danger and controls defense responses is likely to contribute the raw data that become the experience of fear.

In this view, the mechanism that allows us to experience the color of an apple is *not* different from the mechanism that allows us to experience emotions. What differs is the system that provides the data that become the content of consciousness. When consciousness is filled with inputs from an emotion system, we are aware of (feel) an emotion. Working backwards from our subjective experiences of emotions will probably tell us more about consciousness than emotions. This approach can tell us about how emotional content is represented in consciousness, but not about the pre- or unconscious processes that provide the information that becomes the feeling.

What then is the difference between an emotional state of consciousness and a non-emotional one? One difference has already been stated: different systems supply the raw data that become the content. Two other differences are also obvious: emotional states of consciousness tend to be more intense and longer lasting;

emotions engage brain areas and systems that are not greatly involved in non-emotional states. For example, fear-arousing and otherwise stressful events tend to lead to the release of hormones from the adrenal cortex and medulla.[2] These circulate back to the brain either directly or indirectly and have profound consequences on forebrain networks. These and other "arousal" functions of hormones and other modulators intensify and "lock in" brain states that are occurring, thus allowing the organism to remain focused, attentive, and aroused in challenging situations. A wandering mind during an attack would not be a good idea. At the same time, this intensity and duration of emotional brain states can have detrimental effects when they occur during undesirable emotions (anxiety, depression).

Below, I will present the basic organization of the brain system involved in the control of fear responses. Toward the end, I will touch on how this information may also be useful in understanding the origin of fearful feelings. I used fear in the discussion above, and will focus on the fear system below because we understand it better than any other emotion system. First, however, I want to justify focusing on one emotion by critiquing the notion that the brain has a general purpose emotion system.

2 Emotion and the Brain: One System or Many?

For decades now, the limbic system theory has functioned as an explanation of how our brains make emotions.[3] However, there are numerous problems with this view.[4] Two are particularly important. First, the limbic system itself is poorly defined, if not undefinable. Second, some standard limbic areas, like the hippocampus, are relatively uninvolved in emotional behaviors, whereas others, like the amygdala, are strongly implicated in some but not necessarily all emotional behaviors. As a result, the concept of the limbic system is not very useful in making predictions about how a particular emotion might be represented in the brain other than saying that some brain areas located in the no-man's land between the neocortex and the hypothalamus may play some role.

The limbic-system theory was a seductive concept because it seemed to explain all emotions at once. This was convenient during the early days of the cognitive revolution in psychology and neuroscience, a time when researchers were happy not to have to think about emotions. But now the tides have changed and there is

---

[2] Bruce McEwen and R. Sapolsky, "Stress and Cognitive Functioning," *Current Opinion in Neurobiology* 5 (1995): 205–16; Robert M. Sapolsky, "Why Stress is Bad for Your Brain," *Science* 273 (1996): 749–50; James L. McGaugh, M.H. Mesches, L. Cahill, M.B. Parent, K. Coleman-Mesches, and J.A. Salinas, "Involvement of the Amygdala in the Regulation of Memory Storage," in *Plasticity in the Central Nervous System*, J. L. McGaugh, F. Bermudez-Rattoni, and R. A. Prado-Alcala, eds. (Mahwah, New Jersey: Lawrence Erlbaum Associates, 1995), 18–39.

[3] Paul D. MacLean, "Psychosomatic Disease and the 'Visceral Brain': Recent Developments Bearing on the Papez Theory of Emotion," *Psychosomatic Medicine* 11 (1949): 338–53; idem, "Some Psychiatric Implications of Physiological Studies on Frontotemporal Portion of Limbic System (Visceral Brain)," *Electroencephalography and Clinical Neurophysiology* 4 (1952): 407–18.

[4] Alf Brodal, *Neurological Anatomy* (New York: Oxford University Press, 1982); Joseph E. LeDoux, "Emotion and the Limbic System Concept," *Concepts in Neuroscience* 2 (1991): 169–99; Rolf Kotter and N. Meyer, "The Limbic System: a Review of its Empirical Foundation," *Behavioural Brain Research* 52 (1992): 105–27.

renewed interest in emotions. And since the limbic system is not the answer, a different approach is needed.

The approach I have followed is to assume nothing, and instead to let the brain tell us how it does its emotional job. To do this, I have studied one emotion, fear, and have looked for its representation in neural circuits. If it turns out that the system uncovered for fear is involved in other emotions, then so be it. But we need to get to that point by actually studying different emotions rather than by ordaining an emotional system on the basis of a purely anatomical conceptualization.

Recently, Leslie Brothers has questioned the value the terms "emotion" and "fear" on the basis of her claim that they are not "natural kinds."[5] I use the term "fear" to refer to brain networks that detect and respond to danger, and learn about new dangers. These networks are real (they are natural kinds), and their functions are essential to life.

3 What is Fear and Why Should We Care About Its Organization in the Brain?

Defensive behaviors and associated bodily responses are mediated by the brain networks referred to as the fear system. Activation of this system is a normal response to danger and occurs often in daily life. When the degree of fear expressed is greater than that called for in a particular situation, or when fear responses occur often in inappropriate situations, a fear or anxiety disorder exists.[6] Excluding substance abuse, so-called anxiety disorders account for roughly half of all psychiatric illnesses reported each year.[7] The fear system of the brain is likely to be involved in at least some anxiety disorders, and it is important that we understand in as much detail as possible how the fear system works. This information may lead to a better understanding of how anxiety disorders arise and how they might be prevented or controlled. If through studies of fear we were only to learn about fear-related processes, we would still have accomplished quite a lot.

There are a number of experimental tools for studying fear in animals (including people), but one of the simplest and most straightforward is fear conditioning. With this procedure, meaningless stimuli acquire affective properties when they occur in conjunction with a biologically significant event. The initially neutral stimulus is called a conditional (or conditioned) stimulus (CS) and the biologically significant one an unconditional (or unconditioned) stimulus (US). Through CS-US associations, innate physiological and behavioral responses come under the control of the CS. For example, if a rat is given a tone CS followed by an electric shock US, after a few tone-shock pairings it will exhibit a complex set of conditioned fear responses to the tone.[8] These include direct alterations in the activity of autonomic systems (for

---

[5] Brothers, *Friday's Footprint*.

[6] E.g., Stephen Maren and M.S. Fanselow, "The Amygdala and Fear Conditioning: Has the Nut Been Cracked?" *Neuron* 16 (1996): 237–40; Arne Öhman, "Fear and Anxiety as Emotional Phenomena: Clinical, Phenomenological, Evolutionary Perspectives, and Information-processing Mechanisms," in *Handbook of the Emotions*, M. Lewis and J. M. Haviland, eds. (New York: Guilford, 1992), 511–36.

[7] Ronald W. Manderscheid and M.A. Sonnenschein, *Mental Health, United States 1994* (Rockville, Md.: U.S. Dept . Public Health and Human Services, 1994).

[8] Robert J. Blanchard and D.C. Blanchard, "Passive and Active Reactions to Fear-Eliciting Stimuli," *Journal of Comparative Physiological Psychology* 68 (1969): 129–35; Mark E. Bouton and R.C. Bolles, "Conditioned Fear Assessed by Freezing and by the Suppression of

example, heart rate, blood pressure), endocrine systems (hormone release), and skeletal systems (conditioned immobility, or "freezing"), as well as modulations of pain sensitivity (analgesia) and somatic reflexes (fear-potentiated startle, fear-potentiated eyeblink responses). This form of conditioning works throughout the phyla, having been studied experimentally in flies, worms, snails, fish, pigeons, rabbits, rats, cats, dogs, monkeys, baboons, and humans, to name some.[9]

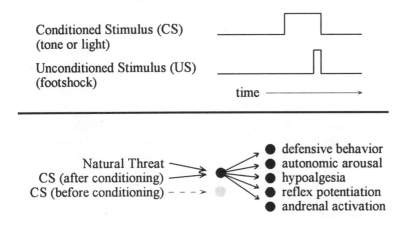

Figure 1: How fear conditioning works. Fear conditioning occurs when a relatively innocuous stimulus (the conditioned stimulus, CS) occurs in conjunction with some aversive event (the unconditioned stimulus, US). Once this happens, the CS acquires the capacity to activate a variety of behavioral and visceral defense responses similar to those activated by natural dangers. A rat, for example, will express fear responses to a CS but also to a cat, even though it has never had previous experiences with cats.

Fear conditioning is a very efficient form of learning. It takes very few encounters with a predator—often one is sufficient—to create long-lasting memories that are useful in guiding behavior in future situations where the sights, sounds, or smells of the predator are later detected. But the price we pay for this efficiency is substantial. We often develop fears and anxieties that are not very useful. And these, like their more beneficial counterparts, are very difficult to get rid of. When we pair a tone with a shock, we are tapping into this evolutionarily old learning system that underlies both adaptive and maladaptive aspects of human behavior.

Three Different Baselines," *Animal Learning and Behavior* 8 (1980): 429–34; Robert C. Bolles and M.S. Fanselow, "A Perceptual-defensive-recuperative Model of Fear and Pain," *Behavioral and Brain Sciences* 3 (1980): 291–323; William K. Estes and B.F. Skinner, "Some Quantitative Properties of Anxiety," *Journal of Experimental Psychology* 29 (1941): 390–400; W.R. and D.E. McAllister, "Behavioral Measurement of Conditioned Fear," in *Aversive Conditioning and Learning*, F.R. Brush, ed. (New York: Academic Press, 1971), 105–79; Joseph E. LeDoux, *The Emotional Brain* (New York: Simon and Schuster, 1996).

[9] See Joseph E. LeDoux, "Emotion, Memory and the Brain," *Scientific American* 270 (1994): 32–39.

Fear conditioning may not tell us all we need to know about all aspects of fear, or all aspects of fear or anxiety disorders, but it is an excellent starting point. Furthermore, many of the other fear assessment procedures, such as the various forms of avoidance conditioning, crucially involve an initial phase of fear conditioning that then provides motivational impetus for the later stages of instrumental avoidance learning.[10] There are some fear assessment procedures that do not require learning (for example, open field, the elevated maze, or light avoidance tasks), but these are somewhat less amenable to a neural systems analysis than fear conditioning, mainly due to the fact that the stimulus situation is often poorly defined in these.

4 Basic Neural Systems Involved in Fear Conditioning

Significant progress has been made over the past two decades in elucidating the neural circuits underlying fear conditioning.[11] Most of this work has been performed in rats, and has focused on the processing of a simple auditory CS, such as a pure tone (that is, a single-frequency acoustic signal), that is paired with a US consisting of a mild electric shock to the feet. For this reason, I will focus below on the circuits in the rat brain involved in auditory fear conditioning. Nonetheless, many of the auditory findings also apply to other sensory modalities, and the results from rat studies apply to other animals, including birds, dogs, cats, mice, monkeys, and humans.[12] The basic circuits and mechanisms thus seem to have been conserved throughout much of the evolutionary history of vertebrates.

The centerpiece of the fear conditioning system is the amygdala, a small region located in the temporal lobe. It is essential for the acquisition of conditioned fear as well as for the expression of innate and learned fear responses.[13] It is generally believed that through the process of conditioning, the CS acquires the capacity to access and trigger the hard-wired fear response network organized by the amygdala and its output connections to the brainstem, producing a cascade of innate defensive reactions. It is thus important to understand how information about the CS is transmitted to, and processed by, the amygdala, and to determine the cellular mechanisms in the amygdala that underlie the changes in processing during learning.

Auditory CS information is transmitted from the ear, through auditory pathways of the brainstem, to the auditory relay nucleus in the thalamus, the medial geniculate

---

[10] E.g., Orval H. Mowrer, "A Stimulus-response Analysis of Anxiety and its Role as a Reinforcing Agent," *Psychological Review* 46 (1939): 553–65; John C. Dollard and Neal E. Miller, *Personality and Psychotherapy* (New York: McGraw-Hill, 1950).

[11] For reviews, see Michael Davis, "The Role of the Amygdala in Conditioned Fear," in *The Amygdala: Neurobiological Aspects of Emotion, Memory, and Mental Dysfunction*, J.P. Aggleton, ed. (New York: Wiley-Liss, Inc., 1992), 255–306; Bruce S. Kapp, P.J. Whalen, W.F. Supple, and J.P. Pascoe, "Amygdaloid Contributions to Conditioned Arousal and Sensory Information Processing," in *The Amygdala*, Aggleton, ed.; Joseph E. LeDoux, "Emotion: Clues from the Brain," *Annual Review of Psychology* 46 (1995): 209–35; idem, *The Emotional Brain*; Maren and Fanselow, "The Amygdala and Fear Conditioning."

[12] For a summary, see LeDoux, *The Emotional Brain*.

[13] See D. Caroline Blanchard and R.J. Blanchard, "Innate and Conditioned Reactions to Threat in Rats with Amygdaloid Lesions," *Journal of Comparative Physiological Psychology* 81 (1972): 281–90; LeDoux, *The Emotional Brain*; Kapp, et al., "Amygdaloid Contributions to Conditioned Arousal and Sensory Information Processing"; Davis, "The Role of the Amygdala in Conditioned Fear."

body (MGB).[14] The signal is then relayed to the amygdala by way of two parallel pathways. A direct monosynaptic projection originates in a particular subset of nuclei of the auditory thalamus.[15] A second, indirect, pathway conveys information from all areas of the auditory thalamus to the auditory cortex. Several cortico-cortical pathways then transmit auditory information to the amygdala.[16] Both the direct and indirect pathways terminate in the lateral nucleus of the amygdala (LA),[17] often converging onto single neurons.[18] Experimental manipulations that disrupt normal functioning of LA—such as permanent lesions,[19] temporary inactivation of the region,[20] or pharmacological blockade of excitatory amino-acid receptors[21]— interfere with fear conditioning.

---

[14] Joseph E. LeDoux, A. Sakaguchi, and D.J. Reis, "Subcortical Efferent Projections of the Medial Geniculate Nucleus Mediate Emotional Responses Conditioned by Acoustic Stimuli," *Journal of Neuroscience* 4 (1984): 683–98.

[15] Joseph E. LeDoux, C.F. Farb, and D.A. Ruggiero, "Topographic Organization of Neurons in the Acoustic Thalamus That Project to the Amygdala," *Journal of Neuroscience* 10 (1990): 1043–54.

[16] F. Mascagni, A.J. McDonald, and J.R. Coleman, "Corticoamygdaloid and Corticocortical Projections of the Rat Temporal Cortex: a Phaseolus Vulgaris Leucoagglutinin Study," *Neuroscience* 57 (1993): 697–715; Lizabeth M. Romanski and J.E. LeDoux, "Information Cascade from Primary Auditory Cortex to the Amygdala: Corticocortical and Corticoamygdaloid Projections of Temporal Cortex in the Rat," *Cerebral Cortex* 3 (1993): 515–32.

[17] B. Turner and M. Herkenham, "Thalamoamygdaloid Projections in the Rat: a Test of the Amygdala's Role in Sensory Processing," *Journal of Comparative Neurology* 313 (1991): 295–325; Joseph E. LeDoux, P. Cicchetti, A. Xagoraris, and L.M. Romanski, "The Lateral Amygdaloid Nucleus: Sensory Interface of the Amygdala in Fear Conditioning," *Journal of Neuroscience* 10 (1990): 1062–69; Romanski and LeDoux, "Information Cascade from Primary Auditory Cortex to the Amygdala"; Mascagni, et al., "Corticoamygdaloid and Corticocortical Projections of the Rat Temporal Cortex."

[18] Xingfang Li, G.E. Stutzmann, and J.L. LeDoux, "Convergent but Temporally Separated Inputs to Lateral Amygdala Neurons from the Auditory Thalamus and Auditory Cortex Use Different Postsynaptic Receptors: *In Vivo* Intracellular and Extracellular Recordings in Fear Conditioning Pathways," *Learning and Memory* 3 (1996): 229–42.

[19] LeDoux, et al., "The Lateral Amygdaloid Nucleus."

[20] Fred J. Helmstetter and P.S. Bellgowan, "Effects of Muscimol Applied to the Basolateral Amygdala on Acquisition and Expression of Contextual Fear Conditioning in Rats," *Behavioral Neuroscience* 108 (1994): 1005–9; Jeff Muller, K.P. Corodimas, Z. Fridel, and J.E. LeDoux, "Functional Inactivation of the Lateral and Basal Nuclei of the Amygdala by Muscimol Infusion Prevents Fear Conditioning to an Explicit Cs and to Contextual Stimuli," *Behavioral Neuroscience* 111 (1997): 683–91.

[21] Mindy J.D. Miserendino, C.B. Sananes, K.R. Melia, and M. Davis, "Blocking of Acquisition but Not Expression of Conditioned Fear-potentiated Startle by Nmda Antagonists in the Amygdala," *Nature* 345 (1990): 716–18; Michael S. Fanselow and Jeansok J. Kim, "Acquisition of Contextual Pavlovian Fear Conditioning Is Blocked by Application of an Nmda Receptor Antagonist D,l-2-amino-5-phosphonovaleric Acid to the Basolateral Amygdala," *Behavioral Neuroscience* 108 (1994): 210–12; Stephen Maren, G. Aharonov, D.L. Stote, and M.S. Fanselow, "N-methyl-d-aspartate Receptors in the Basolateral Amygdala Are Required for Both Acquisition and Expression of the Conditional Fear in Rats," *Behavioral Neuroscience* 110 (1996): 1365–74.

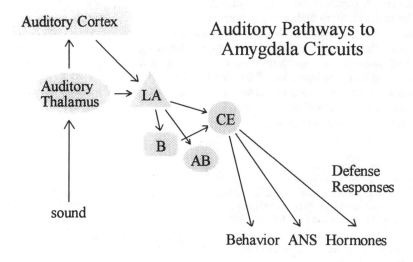

Figure 2: Fear conditioning pathways. Auditory conditioned stimuli reach the amygdala from the thalamus and cortex. The lateral amygdala (LA) receives the inputs and then projects to the central amygdala (CE) directly and by way of the basal (B) and accessory basal (AB) nuclei. CE then controls the expression of defense responses by way of projections to the brainstem.

Although the auditory cortex is not required for the acquisition of conditioned fear to a simple auditory stimulus,[22] processing of the CS by cells in the auditory cortex is modified as a result its pairing with the US.[23] In situations involving more complex stimuli that must be discriminated, recognized, and/or categorized, the auditory cortex may be an essential link to the amygdala.[24]

What are the advantages of the parallel processing capabilities of this system? First, the existence of a subcortical pathway allows the amygdala to detect threatening stimuli in the environment quickly, in the absence of a complete and time-consuming analysis of the stimulus. The speed of this processing route may confer an evolutionary advantage to the species. Second, the rapid subcortical

[22] Lizabeth M. Romanski and J.E. LeDoux, "Equipotentiality of Thalamo-amygdala and Thalamo-cortico-amygdala Projections as Auditory Conditioned Stimulus Pathways," *Journal of Neuroscience* 12 (1992): 4501–9; Jorge L. Armony, D. Servan-Schreiber, L.M. Romanski, J.D. Cohen, and J.E. LeDoux, "Stimulus Acquisition of Fear Responses: Effects of Auditory Cortex Lesions in a Computational Model and in Rats," *Cerebral Cortex* 7 (1997): 157–65.

[23] Norman M. Weinberger, "Retuning the Brain by Fear Conditioning," in *The Cognitive Neurosciences*, M.S. Gazzaniga, ed. (Cambridge: MIT Press, 1995), 1071–90; Greg J. Quirk, J.L. Armony, and J.E. LeDoux, "Fear Conditioning Enhances Different Temporal Components of Tone-Evoked Spike Trains in Auditory Cortex and Lateral Amygdala," *Neuron* 19 (1997): 613–624.

[24] E.g., Jerry L. Cranford and M. Igarashi, "Effects of Auditory Cortex Lesions on Temporal Summation in Cats," *Brain Research* 136 (1977): 559–64; Ian C. Whitfield, "Auditory Cortex and the Pitch of Complex Tones," *Journal of the Acoustical Society of America* 67 (1980): 644–47.

pathway may function to "prime" the amygdala to evaluate subsequent information received along the cortical pathway.[25] For example, a loud noise may be sufficient to alert the amygdala, at the cellular level, to prepare to respond to a dangerous predator lurking nearby, but defensive reactions may not be fully mobilized until the auditory cortex analyzes the location, frequency, and intensity of the noise to determine specifically the nature and extent of this potentially threatening auditory signal. The convergence of the subcortical and cortical pathways onto single neurons in the lateral nucleus provides a means by which the integration could take place.[26] Third, recent computational modeling studies show that the subcortical pathway can function as an interrupt device[27] enabling the cortex, by way of amygdalo-cortical projections, to shift attention to dangerous stimuli that occur outside the focus of attention.[28]

Information processed by the lateral nucleus is then transmitted via intra-amygdala connections to the basal and accessory basal nuclei, where it is integrated with other incoming information and further transmitted to the central nucleus.[29] The central nucleus is the main output system of the amygdala. Damage to the central amygdala or to the structures that project to it interferes with the acquisition and expression of all conditioned responses, whereas lesions of areas to which the central amygdala projects interferes with individual responses, such as blood pressure changes, freezing behavior, or hormone release.[30]

Most of the work described so far has involved rodents. However, recent studies have shown that damage to the human amygdala interferes with fear conditioning[31]

---

[25] Joseph E. LeDoux, "Neurobiology of Emotion," in *Mind and Brain*, J. E. LeDoux and W. Hirst, eds. (New York: Cambridge University Press, 1986); Li, et al., "Convergent but Temporally Separated Inputs to Lateral Amygdala Neurons."

[26] Li, et al., "Convergent but Temporally Separated Inputs to Lateral Amygdala Neurons."

[27] Herbert A. Simon, "Motivational and Emotional Controls of Cognition," *Psychological Review* 74 (1967): 29–39.

[28] See Jorge L. Armony, D. Servan-Schreiber, J.D. Cohen, and J.E. LeDoux, "Computational Modeling of Emotion: Explorations Through the Anatomy and Physiology of Fear Conditioning," *Trends in Cognitive Sciences* 1 (1997): 28–34; idem, "Arousal of Fear by Unattended Stimuli: Exploring Interactions Between Cognition and Emotion Through a Neurally-Based Computational Model," *Journal of Cognitive Neuroscience* (submitted).

[29] Asla Pitkänen, L. Stefanacci, C.R. Farb, C.-G. Go, J.E. LeDoux, and D.G. Amaral, "Intrinsic Connections of the Rat Amygdaloid Complex: Projections Originating in the Lateral Nucleus," *Journal of Comparative Neurology* 356 (1995): 288–310; Asla Pitkänen, V. Savander, and J.L. LeDoux, "Organization of Intra-amygdaloid Circuitries: an Emerging Framework for Understanding Functions of the Amygdala," *Trends in Neurosciences* (submitted); V. Savander, J.E. LeDoux, and A. Pitkänen, "Interamygdala Projections of the Basal and Accessory Basal Nucleus of the Rat Amygdaloid Complex," *Neuroscience* 76 (1996): 725–35; idem, "Lateral Nucleus of the Rat Amygdala Is Reciprocally Connected with Basal and Accessory Basal Nuclei: a Light and Electron Microscopic Study," *Neuroscience* 77 (1997): 767–81.

[30] E.g., Joseph E. LeDoux, J. Iwata, P. Cicchetti, and D.J. Reis, "Different Projections of the Central Amygdaloid Nucleus Mediate Autonomic and Behavioral Correlates of Conditioned Fear," *Journal of Neuroscience* 8 (1988): 2517–29; Louis D. van de Kar, R.A. Piechowski, P.A. Rittenhouse, and T.S. Gray, "Amygdaloid Lesions: Differential Effect on Conditioned Stress and Immobilization-Induced Increases in Corticosterone and Renin Secretion," *Neuroendocrinology* 54 (1991): 89–95.

[31] Antoine Bechara, D. Tranel, H. Damasio, R. Adolphs, C. Rockland, and A.R. Damasio, "Double Dissociation of Conditioning and Declarative Knowledge Relative to the Amygdala

and that activation of the human amygdala occurs during conditioning, as demonstrated with fMRI.[32]

The amygdala is thus the key to the fear conditioning system. It receives information about external stimuli, interprets the significance of the stimulus, and initiates defensive responses in a wide variety of species, including ours.

5 Fearful Situations

Whether a stimulus signals danger, and thus elicits fear reactions, may depend on the situation (context) in which it occurs. For example, the sight of a bear in the zoo poses little threat, but the same bear seen while on a walk in the woods would make us run away in fear. Furthermore, contexts may themselves acquire aversive value through prior experiences. If we are mugged, we will most likely feel "uneasy" when we return to the scene of the crime.

The relationship between environmental situations and fear responses has been investigated in the laboratory through contextual fear conditioning: when a rat is conditioned to expect a foot-shock in the presence of a tone CS, it will also exhibit fear reactions to the chamber where the conditioning took place, even in the absence of the CS.[33] Recent studies have shown that the formation and consolidation of contextual fear associations depend on the hippocampus. Lesions of the hippocampus made prior to training interfere with the acquisition of conditioned responses to the context without having any effect on the conditioning to the CS.[34] Furthermore, hippocampal lesions made after training interfere with the retention of contextual fear association.[35] This selective retrograde amnesia for contextual fear, however, is temporally graded, as lesions made more than two weeks after conditioning have no effect.[36] These findings are consistent with human studies[37] and computational models[38] that suggest a time-limited contribution of the hippocampus in the

and Hippocampus in Humans," *Science* 269 (1995): 1115–58; Kevin S. LaBar, J.E. LeDoux, D.D. Spencer, and E.A. Phelps, "Impaired Fear Conditioning Following Unilateral Temporal Lobectomy in Humans," *Journal of Neuroscience* 15 (1995): 6846–55.

[32] LaBar, et al., "Impaired Fear Conditioning Following Unilateral Temporal Lobectomy in Humans."

[33] Jeansok J. Kim and M.S. Fanselow, "Modality-Specific Retrograde Amnesia of Fear," *Science* 256 (1992): 675–77; Russell G. Phillips and J.E. LeDoux, "Differential Contribution of Amygdala and Hippocampus to Cued and Contextual Fear Conditioning," *Behavioral Neuroscience* 106 (1992): 274–85.

[34] Phillips and LeDoux, "Differential Contribution of Amygdala and Hippocampus to Cued and Contextual Fear Conditioning"; idem, "Lesions of the Dorsal Hippocampal Formation Interfere with Background but Not Foreground Contextual Fear Conditioning," *Learning and Memory* 1 (1994): 34–44; Nathan R.W. Selden, B.J. Everitt, L.E. Jarrard, and T.W. Robbins, "Complementary Roles for the Amygdala and Hippocampus in Aversive Conditioning to Explicit and Contextual Cues," *Neuroscience* 42 (1991): 335–50.

[35] Kim and Fanselow, "Modality-Specific Retrograde Amnesia of Fear."

[36] Ibid.

[37] Larry R. Squire, P.C. Slater, and P.M. Chace, "Retrograde Amnesia: Temporal Gradient in Very Long Term Memory Following Electroconvulsive Therapy," *Science* 187 (1975): 77–79.

[38] Pablo Alvarez and L.R. Squire, "Memory Consolidation and the Medial Temporal Lobe: a Simple Network Model," *Proceedings of the National Academy of Sciences, USA* 91 (1994): 7041–45; Mark A. Gluck and C.E. Myers, "Hippocampal Mediation of Stimulus

formation and consolidation of explicit memories. The role of the hippocampus in the evaluation of contextual cues in fear conditioning is also consistent with current theories of spatial, configural, and/or relational processing in the hippocampus.[39]

Tones and Contextual Stimuli
Enter the Amygdala Differently

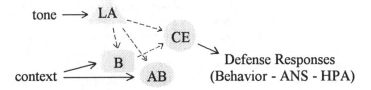

Figure 3: Pathways for conditioning to specific cues and contexts differ. For a simple conditioned stimulus, like a tone or a light, conditioning is mediated by projections to the lateral amygdala (LA) which then projects directly to the central amygdala (CE). For conditioning to a global context, though, projections to the basal (B) and accessory basal (AB) amygdala from the hippocampus (not shown) appear to be required.

Exactly how contextual information coded in the hippocampus interacts with the emotional system is still unclear. Bidirectional projections between the hippocampal formation and the amygdala provide anatomical channels through which the attachment of emotional value to context may take place.[40] The fibers from the hippocampus to the amygdala terminate extensively in the basal and accessory basal nuclei, and to a much lesser extent in the lateral nucleus, suggesting why lesions of the lateral nucleus have little effect on context conditioning, but lesions of the basal and/or accessory basal nuclei are disruptive.[41]

Representation: a Computational Theory," *Hippocampus* 3 (1993): 491–516; James L. McClelland, B.L. McNaughton, and R.C. O'Reilly, "Why There Are Complementary Learning Systems in the Hippocampus and Neocortex: Insights from the Successes and Failures of Connectionist Models of Learning and Memory," *Psychological Review* 102 (1995): 419–57.

[39] Neal J. Cohen and Howard Eichenbaum, *Memory, Amnesia, and the Hippocampal System* (Cambridge: MIT Press, 1993); John O'Keefe and Lynn Nadel, *The Hippocampus as a Cognitive Map* (Oxford: Clarendon Press, 1978); Robert J. Sutherland and J.W. Rudy, "Configural Association Theory: the Role of the Hippocampal Formation in Learning, Memory, and Amnesia," *Psychobiology* 17 (1989): 129–44.

[40] David G. Amaral, J.L. Price, A. Pitkänen, and S.T. Carmichael, "Anatomical Organization of the Primate Amygdaloid Complex," in *The Amygdala*, Aggleton, ed., 1–66; Newton S. Canteras and L.W. Swanson, "Projections of the Ventral Subiculum to the Amygdala, Septum, and Hypothalamus: a PHAL Anterograde Tract-Tracing Study in the Rat," *Journal of Comparative Neurology* 324 (1992): 180–94; Ole P. Ottersen, "The Afferent Connections of the Amygdala of the Rat as Studied with Retrograde Transport of Horseradish Peroxidase," in *The Amygdaloid Complex*, Y. Ben-Ari, ed. (New York: Elsevier/North-Holland Biomedical Press, 1981), 91–104.

[41] Stephen Maren, G. Aharonov, and M.S. Fanselow, "Retrograde Abolition of Conditional Fear after Excitotoxic Lesions in the Basolateral Amygdala of Rats," *Behavioral Neuroscience* 110 (1996): 718–26; P. Majidishad, D.G. Pelli, and J.E. LeDoux, "Disruption of Fear

6 Getting Rid of Fear

Fear responses tend to be very persistent. This can be extremely helpful for survival, as it allows us to keep a record of all previously encountered threatening experiences, and thus to respond quickly to similar future situations. Nonetheless, it is also important to be able to learn that a stimulus no longer signals danger. Otherwise, unnecessary fear responses will be elicited by innocuous stimuli and may potentially become a liability, interfering with other important routine tasks. In humans, the inability to inhibit unwarranted fear responses can be devastating, as in phobias, post-traumatic stress disorder, generalized anxiety disorder, etc.

In laboratory experiments, learned fear responses can be reduced (extinguished) by repeatedly presenting the CS without the US. Note, however, that extinction of conditioned fear responses is not a passive forgetting of the CS-US association, but an active process, possibly involving a new learning.[42] In fact, CS-elicited responses can be spontaneously reinstated following an unrelated traumatic experience.[43]

Experimental observations in fear conditioning studies suggest that neocortical areas, particularly the prefrontal cortex, are involved in the extinction process. Lesions of the ventromedial and medial orbital prefrontal cortex retard the extinction of behavioral responses to the CS.[44] Such findings complement electrophysiological studies showing that neurons within the orbitofrontal cortex are particularly sensitive to changes in stimulus-reward associations.[45] Lesions of sensory areas of the cortex also retard extinction[46] and neurons in the auditory cortex exhibit extinction-resistant changes to an auditory CS.[47] Thus, the medial prefrontal cortex, possibly in conjunction with other neocortical regions, may help regulate amygdala responses to stimuli

---

Conditioning to Contextual Stimuli but Not to a Tone by Lesions of the Accessory Basal Nucleus of the Amygdala," *Society for Neuroscience Abstracts* 22 (1996): 1116.

[42] Mark E. Bouton and D. Swartzentruber, "Sources of Relapse after Extinction in Pavlovian and Instrumental Learning," *Clinical Psychology Review* 11 (1991): 123–40.

[43] Ivan P. Pavlov, *Conditioned Reflexes* (New York: Dover, 1927); W.J. Jacobs and L. Nadel, "Stress-Induced Recovery of Fears and Phobias," *Psychological Review* 92 (1985): 512–31; Robert A. Rescorla and C.D. Heth, "Reinstatement of Fear to an Extinguished Conditioned Stimulus," *Journal of Experimental Psychology: Animal Behavior Processes* 104 (1975): 88–96.

[44] Maria A. Morgan, L.M. Romanski, and J.E. LeDoux, "Extinction of Emotional Learning: Contribution of Medial Prefrontal Cortex," *Neuroscience Letters* 163 (1993): 109–13; Maria Morgan and J.E. LeDoux, "Differential Contribution of Dorsal and Ventral Medial Prefrontal Cortex to the Acquisition and Extinction of Conditioned Fear," *Behavioral Neuroscience* 109 (1995): 681–88; but see J.C. Gerwitz and M. Davis, "Second-Order Fear Conditioning Prevented by Blocking NMDA Receptors in Amygdala," *Nature* 388 (1997): 471–73.

[45] Simon J. Thorpe, E.T. Rolls, and S. Maddison, "The Orbitofrontal Cortex: Neuronal Activity in the Behaving Monkey," *Experimental Brain Research* 49 (1983): 93–115; Edmund T. Rolls, "The Orbitofrontal Cortex," *Philosophical Transactions of the Royal Society B* 351 (1996): 1433–44.

[46] Joseph E. LeDoux, L.M. Romanski, and A.E. Xagoraris, "Indelibility of Subcortical Emotional Memories," *Journal of Cognitive Neuroscience* 1 (1989): 238–43; Albert H. Teich, P.M. McCabe, C.C. Gentile, L.S. Schneiderman, R.W. Winters, D.R. Liskowsky, and N. Schneiderman, "Auditory Cortex Lesions Prevent the Extinction of Pavlovian Differential Heart Rate Conditioning to Tonal Stimuli in Rabbits," *Brain Research* 480 (1989): 210–18.

[47] Quirk, et al., "Fear Conditioning Enhances Different Temporal Components of Tone-Evoked Spike Trains in Auditory Cortex and Lateral Amygdala."

based on their current affective value. This suggest that fear disorders may be related to a malfunction of the prefrontal cortex that make it difficult for patients to extinguish fears they acquired.[48] Recent studies have shown that stress has the same effects as lesions of the medial prefrontal cortex (fear exaggeration).[49] Given that stress is a common occurrence in psychiatric patients and can induce functional changes in the prefrontal cortex, it is possible that the exaggeration of fear in anxiety disorders results from stress-induced alterations in the medial prefrontal region.

7 Will and Emotion: From Reaction to Action

The defensive responses we have considered so far are hard-wired reactions to danger signals. These are evolution's gifts to us. They provide a first line of defense against danger. Some animals rely mainly on these. But mammals, especially humans, can do much more. We are able to take charge. Once we find ourselves in a dangerous situation, we can think, plan, and make decisions. We make the transition from reaction to action. How does this occur?

Considerably less is understood about the brain mechanisms of emotional action than is understood about reaction, due in part to the fact that emotional actions come in many varieties and are limited only by the ingenuity of the actor. For example, once we freeze and express physiological responses to a dangerous stimulus, the rest is up to us. On the basis of our expectations about what is likely to happen next and

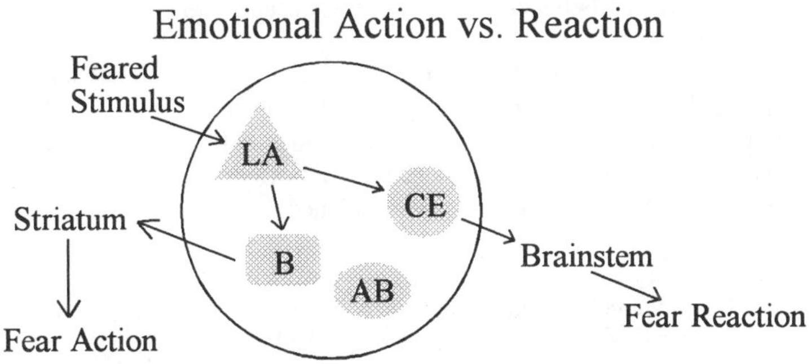

Figure 4: Emotion reaction vs. action. In situations of danger, we react; then we act to protect ourselves (see text). Reactions involve connections from the lateral amygdala (LA) to the central amygdala (CE). Actions, though, involve connections from LA to the basal nucleus (B), which then connects with the striatum.

---

[48] Jorge L. Armony and J.E. LeDoux, "How the Brain Processes Emotional Information," *Annals of the New York Academy of Sciences* 821 (1997): 259–70; Morgan, Romanski, and LeDoux, "Extinction of Emotional Learning"; Morgan and LeDoux, "Differential Contribution of Dorsal and Ventral Medial Prefrontal Cortex to the Acquisition and Extinction of Conditioned Fear."

[49] Keith P. Corodimas, J.E. LeDoux, P.W. Gold, and J. Schulkin, "Corticosterone Potentiation of Learned Fear," *Annals of the New York Academy of Sciences* 746 (1994): 392–93; Cheryl D. Conrad, A.M. Magariños, J.E. LeDoux, and B.S. McEwen, "Chronic Restraint Stress Enhanced Contextual and Cued Fear Conditioning in Rats," *Society for Neuroscience Abstracts* 718.4 (1997).

our past experiences in similar situations, we make a plan about what to do. We become instruments of action.

Instrumental responses in situations of danger are often studied using avoidance conditioning procedures. Avoidance is a multistage learning process.[50] First, conditioned fear responses are acquired. Then, the CS becomes a signal that is used to initiate responses that prevent encounters with the US. Finally, once avoidance responses are learned, animals no longer show the characteristic signs of fear.[51] They know what to do to avoid the danger and simply perform the response in a habitual way. Consistent with this is the fact that the amygdala is required for avoidance learning (for the fear conditioning part) but not for the expression of well-trained avoidance responses (the instrumental part).[52] The involvement of an instrumental component in some aversive learning tasks may explain why these are not dependent on the amygdala for long-term storage.[53]

Because avoidance learning involves fear conditioning, at least initially, it will be subject to all the factors that influence fear conditioning and conditioned fear response. However, because avoidance learning involves more than simple fear conditioning, it is to be expected that avoidance will be subject to influences that have little or no effect on conditioned fear. Much more work is needed to understand how fear and avoidance interact and thus how emotional actions emerge out of emotional reactions. From what we know so far, it appears that as in other habit systems,[54] interactions between the amygdala, basal ganglia, and neocortex are important in avoidance.[55]

The avoidance circuits may be the means through which emotional behaviors are performed initially as voluntary responses and then converted into habits. Emotional habits can be useful, but also quite detrimental. Successful avoidance is known to prevent extinction of conditioned fear since the opportunity to experience the CS in the absence of the US is eliminated by the avoidance. In real life, this can perpetuate

---

[50] Orval H. Mowrer and R.R. Lamoreaux, "Fear as an Intervening Variable in Avoidance Conditioning," *Journal of Comparative Psychology* 39 (1946): 29–50.

[51] Robert A. Rescorla and R.L. Solomon, "Two-Process Learning Theory: Relationships Between Pavlovian Conditioning and Instrumental Learning," *Psychological Review* 74 (1967): 151–82.

[52] Marise B. Parent, C. Tomaz, and J.L. McGaugh, "Increased Training in an Aversively Motivated Task Attenuates the Memory-Impairing Effects of Posttraining N-methyl-d-aspartate-Induced Amygdala Lesions," *Behavioral Neuroscience* 106.5 (1992): 789–98.

[53] Mark G. Packard, C.L. Williams, L. Cahill, and J.L. McGaugh, "The Anatomy of a Memory Modulatory System: from Periphery to Brain," in *Neurobehavioral Plasticity: Learning, development, and response to brain insults*, N.E. Spear, L.P. Spear, and M.L. Woodruff, eds. (New Jersey: Lawrence Erlbaum Assoc., 1995), 149–50; McGaugh, et al., "Involvement of the Amygdala in the Regulation of Memory Storage."

[54] Mortimer Mishkin, B. Malamut, and J. Bachevalier, "Memories and Habits: Two neural Systems," in *The Neurobiology of Learning and Memory*, J.L. McGaugh, G. Lynch and N.M. Weinberger, eds. (New York: Guilford, 1984).

[55] Barry J. Everitt and T.W. Robbins, "Amygdala-Ventral Striatal Interactions and Reward-Related Processes," in *The Amygdala*, Aggleton, ed., 401–29; Jeffrey A. Gray, *The Psychology of Fear and Stress*, vol. 2, (New York: Cambridge University Press, 1987); S. Killcross, T.W. Robbins, and B.J. Everitt, "Different Types of Fear-Conditioned Behavior Mediated by Separate Nuclei Within Amygdala," *Nature* 388 (1997): 377–80.

anxiety states. The panic patient who never leaves home as a means of avoiding having a panic attack is but one example.

8 Other Emotions?

The term "other emotions" has two meanings that we should consider. The first is the difference between fear and other so-called basic emotions, such as disgust, anger, or joy.[56] In contrast to fear, other basic emotions have been studied far less and are generally less well understood at the neural level. The one exception is anger or aggression, for which there is a large body of work on brain mechanisms.[57] The circuits involved in anger expression seem to overlap with the fear expression circuits. It is not known whether this is interesting (that is, indicative of a universal emotional system) or trivial (that is, whether this is just a function of how to get muscles and organs wired up), as the processing of anger by the brain is not understood at all. Further, since the same basic kinds of signals (threats) can elicit aggression or fear, it may be that the overlap reflects commonalities between these two systems rather than a general rule that will apply to other systems. While there is some evidence that the amygdala may also be involved in positive or appetitive emotions,[58] much less is known about how these other emotions work. The amygdala could be involved in both fear and positive emotions because it is a general purpose emotional computer, or it could be involved in each emotion for a different reason. The other sense of the term "other emotions" has to do with the difference between basic and higher order emotions.[59] Essentially nothing is known about the neural basis of jealousy, envy, or trust, but at the psychological level these are believed to involve cognitive systems to a greater extent than basic emotions. Only future research that focuses on other emotions—in both senses—will tell us the answer.

9 So What About Feelings?

Consciousness is an important part of the study of emotion and other mental processes. My hypothesis is that the mechanism of consciousness is the same for emotional and nonemotional subjective states; what distinguishes these states is the brain system that is present to consciousness at the time.

We are far from solving the problem of how consciousness arises, but a number of theorists have proposed that it has something to do with working memory, a serially organized mental work-space where things can be compared and contrasted

---

[56] See Paul Ekman, "An Argument for Basic Emotions," *Cognition and Emotion* 6 (1992): 169–200.

[57] See Joseph LeDoux, "Emotion," in *Handbook of Physiology*, F. Plum, ed. (Bethesda, Md.: American Physiological Society, 1987), 419–60.

[58] For example, see T. Ono and H. Nishijo, "Neurophysiological Basis of the Kluver-Bucy Syndrome: Responses of Monkey Amygdaloid Neurons to Biologically Significant Objects," in *The Amygdala*, Aggleton, ed., 167–90; Edmund T. Rolls, "Neurophysiology and Functions of the Primate Amygdala," in *The Amygdala*, Aggleton, ed., 143–65; Everitt and Robbins, "Amygdala-Ventral Striatal Interactions and Reward-Related Processes"; Michela Gallagher, P.W. Graham, et al., "The Amygdala Central Nucleus and Appetitive Pavlovian Conditioning: Lesions Impair One Class of Conditioned Behavior," *Journal of Neuroscience* 105 (1990): 1906–11.

[59] E.g., Ekman, "An argument for Basic Emotions."

and mentally manipulated.[60] Working memory allows us, for example, to compare an immediately present visual stimulus with information stored in long-term (explicit) memory about stimuli with similar shapes and colors or stimuli found in similar locations.

A variety of studies of humans and nonhuman primates suggest that the prefrontal cortex, and especially the dorsolateral prefrontal areas, is involved in working memory processes.[61] Immediately present stimuli and stored representations are integrated in working memory by way of interactions between prefrontal areas, sensory processing systems (which serve as short-term memory buffers as well as perceptual processors), and the long-term explicit (declarative) memory system involving the hippocampus and related areas of the temporal lobe. Recently, the notion has arisen that working memory may involve interactions between several prefrontal areas, including the anterior cingulate and orbital cortical regions, as well as the dorsolateral prefrontal cortex.[62]

Now suppose that the stimulus is affectively charged, say a trigger of fear. The same sorts of processes will be called upon as for stimuli without emotional implications, but in addition, working memory will become aware of the fact that the fear system of the brain has been activated. I propose that this additional information, when added to perceptual and mnemonic information about the object or event, is the condition for the subjective experience of an emotional state of fear.

But what is the additional information that is added to working memory when the fear system is activated? The amygdala projects to many cortical areas, even some from which it does not receive inputs.[63] It can thus influence the operation of perceptual and short-term memory processes, as well as processes in higher-order areas. Although the amygdala does not have extensive connections with the dorsolateral prefrontal cortex, it does communicate with the anterior cingulate and orbital cortex, two other components of the working memory network. But in addition, the amygdala projects to nonspecific systems involved in the regulation of cortical arousal. And the amygdala controls bodily responses (behavioral, autonomic,

---

[60] Philip N. Johnson-Laird, *The computer and the mind: An Introduction to Cognitive Science* (Cambridge: Harvard University Press, 1988); John F. Kihlstrom, "Conscious, Subconscious, Unconscious: A Cognitive Perspective," in *The Unconscious Reconsidered*, K. S. Bowers and D. Meichenbaum, eds. (New York: John Wiley and Sons, 1984), 149–211; Daniel L. Schacter, "On the Relation Between Memory and Consciousness: Dissociable Interactions and Conscious Experience," in *Varieties of Memory and Consciousness: Essays in Honour of Endel Tulving*, H.L.I. Roediger and F.I.M. Craik, eds. (Hillsdale, New Jersey: Lawrence Erlbaum Associates, 1989), 355–89; Tim Shallice, "Information Processing Models of Consciousness," in *Consciousness in Contemporary Science*, A. Marcel and E. Bisiach, eds. (Oxford: Oxford University Press, 1988).

[61] Alan Baddeley and S. Della Sala, "Working Memory and Executive Control," *Philosophical Transactions of the Royal Society B* 351 (1996): 1397–1404; Joaquín M. Fuster, *The Prefrontal Cortex* (New York: Raven, 1989); Patricia S. Goldman-Rakic, "Circuitry of Primate Prefrontal Cortex and Regulation of Behavior by Representational Memory," in *Handbook of Physiology: The Nervous System V*, F. Plum, ed. (Bethesda, Md.: American Physiological Society, 1987), 373–417.

[62] Mark D'Esposito, J. Detre, D. Alsop, R. Shin, S. Atlas, and M. Grossman, "The Neural Basis of the Central Executive System of Working Memory," *Nature* 378 (1995): 279–81; David Gaffan, E.A. Murry, and M. Fabre-Thorpe, "Interaction of the Amygdala with the Frontal Lobe in Reward Memory," *European Journal of Neuroscience* 5 (1993): 968–75.

[63] Amaral, et al., "Anatomical Organization of the Primate Amygdaloid Complex."

endocrine), which then provide feedback that can influence cortical processing indirectly. Thus, working memory receives a greater number of inputs, and receives inputs of a greater variety, in the presence of an emotional stimulus than it does in the presence of other stimuli. These extra inputs may just be what is required to add affective charge to working-memory representations, and thus to turn subjective experiences into emotional ones.

How Fearful Feelings Come About

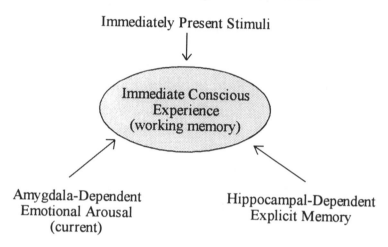

Figure 5: Fearful feelings. The hypothesis proposed for fearful feelings is that they involve the representation in working memory of immediately present stimuli, past memories about similar stimuli or situations, and amygdala-mediated arousal, which is also triggered by the stimulus (see text for details).

Though described in terms of the fear system, the present hypothesis about feelings is a general one that applies to any emotion. That is, an emotional feeling results when the general purpose system of working memory is occupied with the fact that a specific emotion system is active. The difference between emotional states and other states of consciousness, then, is not due to different underlying mechanisms that give rise to the qualitatively different subjective experiences. Instead, there is one mechanism of consciousness, and it can be occupied either by mundane events or by emotionally charged ones. If this view is correct, we might expect to make progress in understanding how subjective states come from the brain by figuring out which brain systems are feeding working memory when people are in particular states. While neurobiology has not solved the problem of feelings, it has at least given us a way to think about how our most personal states might emerge from the organization of synaptic pathways in the brain.

Acknowledgment. This work supported by PHS Grants, MH46516, MH38774, and MH00956, and by a grant from the W.M. Keck Foundation to NYU.

ARE THERE LIMITS TO THE NATURALIZATION OF MENTAL STATES?

Marc Jeannerod

1 Introduction

Humans are social beings. Any theory of the self must take into account its inherently social nature, which makes understanding other selves and communicating with them a basic feature of human behavior. In the realm of psychology, this means that the ensemble of functions we use to communicate with other people (our mind and our behavior) must be tuned to other minds in such a way that our behavioral productions can be understood by others. Most psychological theories have used a solipsist approach to describing behavior. They concentrate on functions such as self-consciousness (how I feel myself as a distinct entity) and consciousness of external events (how I achieve goals by interacting with the environment), although they tend to overlook the basic fact that these functions are two-way processes. Our mental states are directed to other selves which, in turn, generate mental states directed toward us.[1]

The above considerations add another dimension to the current objective of cognitive neuroscience, namely, the search for the neural underpinnings of mental states. Mental states should no longer be considered only from the standpoint of the self where they originate: rather, attention should be focused on the properties of mental states that make them "social." New questions arise: Is social ability an independent function which requires specific neural mechanisms? Is it possible, on physiological grounds, to draw a distinction between private mental states, to which others have no access, and public mental states which can be shared with other people? In raising these questions, this essay intends to make a contribution to what could be called "social neuroscience," by establishing the existence of brain states underlying representations (about actions, external events, etc.) that can be shared by several people and that could be the substrate of intersubjectivity.[2]

The basic tenets of cognitive neuroscience, namely, that mental states have a causal role on behavior, and that this role is a consequence of their embodiment in brain states, will be taken for granted. A few points of clarification are in order, however, before moving to the core of the problem. First, it must be stated explicitly that mental functioning does not imply that mental states have to be conscious. In the domain of the generation of actions—which will be the main theme of this essay—the definition of a mental representation extends from a declarative desire to do something with a conscious image of the goal, to an implicit intention, the content of which completely escapes subjective awareness, or will only be known after the generation of the corresponding action. It could be argued that, according to this definition, a mental state could simply be a computational state implemented on any system with the appropriate physical properties (for example, a brain or a computer) and therefore be devoid of any specifically human content. In this essay, however,

---

[1] The psychoanalytic theory is a notable exception to psychological solipsism. It has clearly claimed that construction of the self is based on interactions with the affective milieu.

[2] For a review of recent literature, see Marc Jeannerod, "To Act or Not to Act: Perspectives on the Representation of Actions," *Quarterly Journal of Experimental Psychology* 52A (1999): 1–29.

mental representations will be considered specifically in the context of social communication between individuals, a dimension that computers do not have.

Second, the relationship of mental states to brain states raises the question of multiple realizability: different brains may produce a similar mental state and yet this mental state may correspond to quite different neural states in each of these brains (as a matter of fact, it could be produced equally well by quite different information-processing systems, be it a brain or a computer, for example). Obviously, the concept of multiple realizability stands as a potential difficulty for assigning mental states to specific brain states and for comparing brain states during different mental states, as will be attempted below. In fact, establishing correlations between a mental state and a brain state does not imply a full compatibility between the two. Brain states are far more complex than mental states: they involve interactions between a great number of elements (cells, synapses, etc.), which are not individually represented at the mental level. One has to assume that productions arising as a consequence of brain functioning (as are mental states) integrate many of the elementary aspects of brain states and do not reflect the details of these underlying mechanisms. To give an analogy at a different level, a limb movement is the integration of the activity of many neurons and muscle fibers, and yet it is currently accepted to be coextensive with the ensemble of these elementary activities: the movement is the way the elementary activities appear when the motor system is examined at the effector level. Similarly, mental states are the way certain brain states appear when the system is examined at the cognitive level.

2 The Relationship of Mental States to Brain States in the Social Context

Modern cognitive psychology tends to follow one of two ideas about how people understand other minds, the ability referred to as mind-reading.[3] According to one of these ideas, human beings are born with elementary rules for building and operating with their mental content. Such rules have been identified for, among other domains, language (how to parse speech sounds), or perception (how to organize a chaotic sensory world). They are activated at an early age by interaction with the external world, so that each individual can form representations of reality. In the domain of social communication, it is posited that humans are born with similar rules to attribute mental states to other individuals and with elementary skills for deciphering these states. According to the alternative idea, mind-reading is not based on organizing the social world according to rules or theories (even implicitly); instead, having mental states (intentions, etc.) is the precondition for reading and understanding the mental states of others, in such a way that the system we use for detecting intentions of other people is part of the system that generates our own intentions. We can know what people think (or believe, or intend) because we are able internally to simulate the states of their minds.[4] In this essay, the consequence of this second idea will be explored. The problem, therefore, will be to understand

---

[3] The term "mind-reading" is now used in the context of the Theory of Mind, to indicate the ability of people to understand the mental states of other individuals. Mind-reading becomes efficient in young children around 3–4 years of age. It seems to be deficient in a number of pathological developmental conditions, including infantile autism. See Simon Baron-Cohen, *Mindblindness: An Essay on Autism and Theory of Mind* (Cambridge: MIT Press, 1995).

[4] For references, see Alvin Goldman, "Interpretation Psychologized," *Mind and Language* 4 (1989): 161–85.

how mental states are generated, how they can cause behavior, how they can be understood by others, and how the self can make a distinction between its own mental states and those of others.

Let us assume that the postulate "mental states are coextensive with brain states" is correct. Some of the logical consequences of this assumption when applied to the context of communication between individuals will be examined first. Social interaction during communication involves mental states being shared by several individuals; the people who interact must have a common code for their intentions, beliefs, or desires in order to understand each other. In line with the above propositions, it follows that the same intention, for example, may be simultaneously (though not identically) represented in several brains. This possibility raises the obvious problem of how the intention can be attributed to its proper source, and suggests the existence of specific signals, the presence or absence of which will determine attribution of the intention to the self or to another source, respectively. Although this seems to be an easy task, as we normally have no difficulty disentangling our thoughts from those of others, pathological examples illustrate the fact that subjects can decipher thoughts they have produced and, at the same time, misattribute them to another person (see below). Producing a mental state and attributing this state to oneself are therefore two different processes—and they are not automatically linked. The brain has to inform itself about what it does; if this information is not present, or not used, or improperly interpreted, the mental state can still be recognized but will be misattributed.

An essential aspect of social communication is to understand the meaning of overt behavior, which is the first step to accessing underlying mental states. For this reason, a large amount of our daily life is spent watching and interpreting the actions of others. Recognition of actions of conspecifics is a genuine ability, which seems to be highly developed in humans and nonhuman primates.[5] Humans can easily distinguish biological motion from impulsions produced by mechanical devices, even when only a limited number of cues are available.[6] Each individual builds up from this observation his/her own theory of the mind of others.[7] In addition, observation of action is the first step of imitation, a powerful means of establishing contact with other individuals and acquiring new skills from them, which starts at birth and continues throughout life.[8]

These issues raise key questions about the type of representation that is built while observing an action. As stated above, the idea that an observed action can be understood and imitated whenever it becomes the source of a representation of the same action by the observer implies that it is transformed into a potential action in order to be understood. A few experiments in human subjects support this idea. These experiments were prompted by results obtained in monkeys, showing the existence of neuronal populations in several brain areas that selectively encode

[5] David Premack and G. Woodruff, "Does the Chimpanzee Have a Theory of Mind?" *Behavioral and Brain Sciences* 4 (1978): 515–26.

[6] Gunnar Johansson, "Studies on Visual Perception of Locomotion," *Perception* 6 (1977): 365–76. Verena Dasser, I. Ulbaek, and D. Premack, "The Perception of Intention," *Science* 243 (1989): 365–67.

[7] Alan M. Leslie, "Pretense and Representation: The Origins of Theory of Mind," *Psychological Review* 94 (1987): 412–26.

[8] Andrew N. Meltzoff, "Understanding the Intentions of Others: Re-enactment of Intended Acts by 18-month-old Children," *Developmental Psychology* 31 (1995): 838–50.

postures and movements performed by conspecifics. Such neurons were found in an associative visual area in the inferotemporal cortex by David Perrett and his colleagues[9] and in the premotor cortex. In the latter region, where neurons are known to be selective for the monkey's active performance of a particular type of movement (for example, a hand movement), a significant proportion of them are also selective for the monkey's observation of the same hand movement made by an experimenter or by another monkey (the so-called mirror neurons of Giacomo Rizzolatti and his colleagues[10]). This finding directly supports the above assumption that the same intention may be simultaneously represented in several brains.

A series of experiments using PET scans were recently performed in humans to explore brain activity during observation of actions performed by others. In these experiments, subjects are typically instructed to remain immobile and carefully observe simple actions displayed in front of them while their brain metabolism is monitored. Areas mostly located in the parietal and the frontal lobes, and clearly related to motor functions, were found to be activated.[11] Further experiments by Jean Decety and his colleagues showed that the subjects' cognitive strategy during observation influenced the distribution of cerebral activation. When the subjects had received the instruction to memorize the actions with the purpose of later imitation, structures involved in motor functions (the dorsolateral prefrontal cortex on both sides, dorsal and medial premotor cortex) were involved. By contrast, when the instruction was to observe the actions with the purpose of simply recognizing them afterwards, only the parahippocampal gyrus in the temporal lobe was activated. The pattern of activation also differed according to whether the observed actions were meaningful (that is, they referred to a recognizable goal), or meaningless sequences of movements. Observing a meaningful action activated areas which were mainly confined to the left hemisphere. The main structures involved were the inferior frontal (area 45), middle temporal (area 21) and parahippocampal regions.[12] Activation of area 45 is of a particular interest because it might be functionally similar to the monkey premotor ventral area 6 where mirror neurons are recorded. Indeed, the same region has been found to be activated during mental simulation of hand actions and even during the recognition of man-made tools presented visually.[13]

---

[9] David I. Perrett, M.H. Harris, R. Bevan, S. Thomas, P.J. Benson, A.J. Mistlin, A.J. Citty, J.K. Hietanen, and J.E. Ortega, "Framework of Analysis for the Neural Representation of Animate Objects and Actions," *Journal of Experimental Biology* 146 (1989): 87–113; David P. Carey, D.I. Perrett, and M.W. Oram, "Recognizing, Understanding and Reproducing Action," in *Handbook of Neuropsychology*, F. Boller and J. Grafman, series eds., vol. 11, *Action and Cognition*, M. Jeannerod, ed. (Amsterdam: Elsevier, 1997), 111–29.

[10] Giacomo Rizzolatti, L. Fadiga, V. Gallese, and L. Fogassi, "Premotor Cortex and the Recognition of Motor Actions," *Cognitive Brain Research* 3 (1996): 131–41.

[11] Scott T. Grafton, M.A. Arbib, L. Fadiga, and G. Rizzolatti, "Localization of Grasp Representations in Humans by Positron Emission Tomography, 2: Observation Compared with Imagination," *Experimental Brain Research* 112 (1996): 103–11.

[12] Jean Decety, J. Grezes, N. Costes, D. Perani, M. Jeannerod, E. Procyk, F. Grassi, and F. Fazio, "Brain Activity During Observation of Action: Influence of Action Content and Subject's Strategy," *Brain* 120 (1997): 1763–77.

[13] Jean Decety, D. Perani, M. Jeannerod, V. Bettinardi, B. Tadary, R. Woods, J.C. Mazziotta, and F. Fazio, "Mapping Motor Representations with PET," *Nature* 371 (1994): 600–2; Grafton et al., "Localization of Grasp Representations in Humans by Positron Emission Tomography, 2"; Daniela Perani, S.F. Cappa, V. Bettinardi, S. Bressi, M. Gorno-Tempini, M. Matarrese, and F. Fazio, "Different Neural Systems for the Recognition of

Finally, in contrast to observation of meaningful actions, observation of meaningless sequences of movements primarily engaged the right hemisphere. Areas in the occipito-parietal region and the inferior temporal gyrus (areas 19–37) were activated.

These results reveal that in many situations involving observation of an action (for example, observing meaningful movements to learn and replicate them), cortical areas that would be involved if one intended that same action become active. It is tempting to generalize this finding to other situations where observation of conspecifics is a source of inter-individual communication. The fact that several individuals can share similar representations (in neural terms) might provide a rationale for such phenomena as empathy, contagion of emotions, etc.

3 Shared Representations and the Agency Problem: About Distinguishing One's Own Intentions from Those of Other People

This notion of shared representations raises in turn two main questions: (1) How can one so easily distinguish one's actions from those of other people? (2) How does one become aware of one's own actions? This section examines the idea that, because understanding actions implies representations which are common to several persons, the possibility of attributing these actions to their real agent (the agency problem) requires the presence of specific signals at the level of these representations.

John Barresi and C. Moore have attempted to specify the difference between conditions when an action is observed and when it is self-generated. When the action is observed, the information available to the subject would carry a "third person" knowledge, based on visual analysis of the movements of the agent toward objects, his gaze orientation, facial expression, etc. When an action is self-generated, by contrast, the available information would be of the "first person" type, that is, mainly based on self-produced signals, such as proprioceptive signals, for example. It is the presence or the absence of the latter signals which would allow making a distinction between self-caused and world-caused effects on external objects, respectively. In addition, Barresi and Moore suggest, the fact that an action is taking place activates an "intentional schema," a structure internal to each agent with the capacity of coordinating first and third person information. According to the input signals available to the intentional schema, the action will be attributed to the self or to the other person.[14] This mechanism may become critical in situations where the two types of information about the action are available at the same time, that is, when two agents are involved in situations like joint attention, matched actions, mutual imitation, etc. A similar theory was used by physiologists to account for how the central nervous system can distinguish between internally generated and externally generated changes of the external world.[15] According to this theory, a specialized

---

Animals and Man-made Tools," *NeuroReport* 6 (1995): 1637–41.

[14] John Barresi and C. Moore, "Intentional Relations and Social Understanding," *Behavioral and Brain Sciences* 19 (1996): 107–54.

[15] There are several versions of this model: the corollary discharge model (Roger W. Sperry, "Neural Basis of the Spontaneous Optokinetic Response Produced by Visual Inversion," *Journal of Comparative and Physiological Psychology* 43 [1950]: 482–89), and the efference copy model (Erich von Holst, "Relations Between the Central Nervous System and the Peripheral Organs," *British Journal of Animal Behavior* 2 [1954]: 89–94). For a modern version, see David M. Wolpert, Z. Ghahramani, and M.I. Jordan, "An Internal Model for Sensorimotor Integration," *Science* 269 (1995): 1880–82.

structure, the comparator, receives action-related signals from internal and external (sensory) sources. During a self-generated action, the internal signals, which copy the commands sent to the effectors (and which therefore reflect the desired action), create within the comparator an anticipation for the consequences of the action. When the action is effectively executed, sensory signals related to changes in the external world also reach the comparator. If these sensory signals match the anticipation of the comparator, the desired action is registered by the system; if they do not, a mismatch is registered between the desired action and the action that has been produced; finally, if sensory signals arrive in the absence of internal signals, a change in the external world independent from the agent is registered.

One way to deal with the problem of agency is to explore the degree of accuracy of subjects when they consciously have to determine the origin of an action. There are very few studies dealing with the subjective experience related to the execution of voluntary movements. In one of these studies, performed by Tozsten Nielsen in 1963, a situation was created where subjects were presented with movements of an uncertain origin: they were shown the image of an alien hand visually superimposed on (and undistinguishable from) their own hand. Movements performed by the alien hand could be either in concordance or in discordance with the subjects' own movements. Even in the latter case, subjects experienced the alien hand as theirs, without regard for obvious discrepancies between the self-generated and the seen movements; they simply reported feelings of strangeness or, on some occasions, the impression of having their hand pushed by some external force, or having lost control of their movements.[16] Normal subjects are thus poorly aware of their own movements. When placed in an ambiguous situation, they tend to experience movements of an alien hand as their own. In addition, when required to make an agency judgment, they tend to privilege movement-related visual information over kinesthetic information.

This problem was systematically reexamined in a new experiment by Elena Daprati and her colleagues. Using a closed TV circuit, the subject's hand and the experimenter's hand could be displayed briefly on the video screen seen by the subject. The two hands looked alike as they were covered with similar gloves. In each trial, both the experimenter and the subject had to perform a given hand movement on command (for example, stretch thumb, stretch fingers 1 and 2, etc). On some trials, however, the experimenter's movement departed from the instruction. As a result of this experimental arrangement, the subject was randomly shown either his own hand, or the experimenter's hand performing the same movement as his, or a different movement. At the end of each trial, a verbal agency judgment was recorded: the subject had to say whether the hand he had seen was his hand or another hand.[17] Normal subjects were able unambiguously to determine whether the moving hand seen on the screen was their own or not in the two "easy" conditions. When they saw their own hand, they correctly attributed the movement to themselves; and when they saw the experimenter's hand performing a movement which departed from the instruction they had received, they correctly denied seeing

---

[16] Tozsten I. Nielsen, "Volition: A New Experimental Approach," *Scandinavian Journal of Psychology* 4 (1963): 225–30.

[17] Elena Daprati, N. Franck, N. Georgieff, J. Proust, E. Pacherie, J. Dalery, and M. Jeannerod, "Looking for the Agent: An Investigation into Consciousness of Action and Self-consciousness in Schizophrenic Patients," *Cognition* 65 (1997): 71–86. This experiment is also described in Jeannerod "To Act or Not to Act."

their own hand. By contrast, their performance degraded in the "difficult" trials where they saw the experimenter's hand performing the same movement as required by the instruction. In this condition, they misjudged the alien hand as their own about thirty percent of the time. The pattern of responses that Daprati and her colleagues recorded in their difficult condition can be better understood if one assumes that agency judgments made by the subjects were based on the state of a comparator similar to that postulated above. In the difficult condition, no obvious mismatch was likely to occur between the anticipated and the perceived final hand postures because the subject's (invisible) hand and the experimenter's (visible) hand both executed very similar movements. Only slight differences in timing and kinematic pattern between the internal signals and the sensory signals arising from the visual and kinesthetic receptors were available to the comparator to give the correct agency response. This explanation also fits into the above framework of Barresi and Moore. Indeed, in the ambiguous condition, the intentional schema was fed simultaneously with third-person signals (those from the experimenter's seen hand) and first-person signals (from the subject's unseen hand), and these signals, particularly those of visual origin, were partly confounded.

This result emphasizes the weakness of signals arising from the execution of one's own movements, with its obvious potential consequence on the perceived agency. Although the above mechanism may be sufficient for correct attribution of actions in everyday life, it is less efficient in ambiguous situations. In addition, pathological conditions may produce specific alterations at one or the other level of the mechanism: the nervous system may fail to generate the proper signals, or they may be produced but remain unperceived. Using the same paradigm as in normal controls, Daprati and her colleagues examined agency responses in schizophrenic patients. Although their responses were close to normal in the easy conditions of the task, they failed to give correct responses in the difficult condition. In this condition, they misattributed the experimenter's hand to themselves more than eighty percent of the time. This striking behavior is likely to reflect a dysfunction of the representations generated while observing actions. It is consistent with clinical reports showing that most hallucinating schizophrenic patients show a tendency to incorporate external events into their own experience, or to interpret environmental cues as specifically directed to themselves.

4 Personal Identity: A Limitation to Mind-Reading?

In positing that mental states can be the cause of behavior because they are brain states, it was assumed that the mechanism for translating an intention into a behavior is inherent in the physical nature of the mental state. In addition, this property was considered a prerequisite for understanding mental states of others, due to the construction of shared representations. These assumptions raise a critical point. How far can this attempt at naturalizing the mental content go? Trying to answer this question might reveal limitations in the amount of mental content that can effectively be shared between individuals. Consider for example the mental state of intending. Assuming that this state corresponds to the activation of a given set of brain areas (as illustrated in the previous section), to what extent can this pattern of activation reveal what the "intention" is about? This distinction between the brain mechanism for intending and the semantics of the intention is reminiscent of similar distinctions in other contexts, such as that between the syntax and the semantic content of a sentence in language studies.

This distinction, however, requires a number of specifications. Because, as I have suggested, an intention cannot have a causal effect on behavior without being implemented by brain mechanisms, the neural representation should include both the mechanism for intending and the content of the intention. The existence of a neural network for intending seems to be demonstrated by PET studies. These studies show that willed action is associated with activation of the prefrontal cortex on the left side, the premotor cortex, and the cingulate cortex.[18] In addition, for the intention to carry out the desired effect, the network for intending must be penetrated by a specific content which pertains to the author of the intended action. This requires further activation of neural structures which monitor information such as memories of previous experiences, expectations in relation to the present situation, knowledge about the properties of objects, etc. It is likely that, although the network for intending is common to everyone (as revealed by the above PET studies), the neural structures which relate to the significance and the goal of that particular action will be unique to the subject and to the situation.

This point can be illustrated by an experimental result described in 1987 by R.S. Marteniuk and his colleagues. Imagine a situation where a subject is required to reach by hand and grasp a wooden disk located at 25 cm from his body. Before starting the reach and grasp action, the subject receives an instruction by a colored light. If the light is, say, red, the disk has to be placed carefully on a small box; if the light is green, the disk has to be thrown in a larger container. An optoelectronic device is used to record the kinematics of the reach movement. Note that this movement (moving the hand from the starting position to the disk) is common to all trials, whatever the instruction about what to do with the disk after it has been grasped. The main finding of the experiment was that the reaching movement was affected by the instruction: when the instruction was to throw the disk, the peak velocity was higher, the deceleration phase was shorter, and the grip size formed by the fingers was larger than when the instruction was to place it in the small box.[19] This finding raises the following question: Could an external observer, who ignored the final goal of the movement given by the instruction, monitor this subtle difference in kinematics and decipher the intention represented in the subject's mind? In the context of this experiment, this question may seem an easy one. If an external observer could monitor with great accuracy the behavior of the subject, he should have access to his intention. However, what initially seems plausible may become more difficult when fewer cues are available. Researchers have stressed the role of such cues as the direction of gaze, facial expressions, body posture, etc., which are more difficult to decipher. Context cues—knowledge about the subject's desires and beliefs—may provide additional information. The ensemble of these cues, those that can be extracted from the subject's behavior and those that are present in the environment, constitute a network of neural events that is unique to a given actor in a given situation.[20] The above example stresses the hierarchical nature of the

---

[18] Christopher D. Frith, K. Friston, P.F. Liddle, and R.S.J. Frackowiak, "Willed Action and the Prefrontal Cortex in Man," *Proceedings of the Royal Society* 244 (1991): 241–46.

[19] Ronald S. Marteniuk, C.L. MacKenzie, M. Jeannerod, S. Athenes, and C. Dugas, "Functional Relationships Between Grasp and Transport Components in a Prehension Task," *Canadian Journal of Psychology* 41 (1987): 365–78.

[20] Note that the semantic content of an intention may also remain outside subjective awareness of its own author. Experimental or pathological situations even raise the question of whether "intentions" devoid of any semantic content can be generated. Electrical

mechanisms that generate mental states in the context of social communication. The public side of the mental state is the side that can be accessed and understood by other selves because it is part of experience common to all individuals. By contrast, the private side cannot be accessed by other selves because it relates to the history of a particular self. I can share the sorrow of someone who lost a beloved person because I know what sorrow is. I know that it creates sadness and is often manifested by crying, etc. But the limitation to my sharing the sorrow of that person is that I cannot build exactly the same representation because I did not ever experience the same relationship with the person who died. It cannot be the case that sorrow is the same for all individuals who lose a beloved person, and that there is an invariant pattern of sorrow for losing a mother, a wife, or a son. This stresses the limitation of empathy and sympathy.

The reason why the private part of the mental state of a person cannot be experienced by other persons, therefore, is that it is impossible to build an appropriate neural representation, which would have to match that part of their mental state. Would this become possible if all the information were available? Imagine a situation where two persons have exactly similar brains and experience exactly the same events in their personal history (an extreme variety of monozygotic twins). Would the content of their mental states be exactly the same? If the answer to this question were positive, then the public and the private sides of the mental state would be confounded, and the (private) contents of each mental state should be fully accessible to each of the two persons. A neurobiologist's critique of this example is that it is not possible that two brains will ever be the same because of ontogenetic constraints.[21] Similarly, it should not be possible that two individuals will ever share the same experience because they will necessarily have different standpoints when facing the same external events. If the neurobiologist's critiques are correct, then different persons' mental states relative to the same event will necessarily carry different contents, with the consequence that these contents will be inaccessible to external observers who try to monitor the mental states of these persons. Note that if the answer given to the above question turns out to be negative, as seems to be the case, this would not be because of some in-principle incommensurability between content and vehicle, but because of the biological and the physical constraints pertaining to individual development and interactions with the external world.

The fact that mental states have to be brain states, and that part of them remain incommunicable, generates a false impression of dualism between what can be shared and what cannot. Ipseity would therefore reside in that part of the mental/neural content for which representations cannot be built from observation. The

---

stimulation of the premotor cortex, for example, triggers a strong experience of "intending" or "wanting to move." Subjects feel the urge to raise their arm, for example, without being able to give any justification for this movement; see Marc Jeannerod, *The Cognitive Neuroscience of Action* (Oxford: Blackwell, 1997). In a similar vein, a single case was recently reported where a stimulation of the anterior cingular gyrus produced a pleasant experience with laughing, yet the subject felt unable to explain the reason for this feeling and could only find ad hoc justifications; Itzhak Fried, C.L. Wilson, K.A. MacDonald, and E.J. Behnke, "Electric Current Stimulates Laughter," *Nature* 391 (1998): 650.

[21] Gerald M. Edelman, "Group Selection and Phasic Reentrant Signalling: A Theory of Higher Brain Function," in *The Mindful Brain: Cortical Organization and the Group-Selective Theory of Higher Brain Function*, G.M. Edelman and V.B. Mountcastle, eds. (Cambridge: MIT Press, 1978), 51–100.

paradox is that personal identity, although it is clearly in the realm of physics and biology, pertains to a category of facts that escape description and, because of that, may remain outside the aim of scientific inquiry. It is not that there is no way of understanding how meaning is embodied; rather, it is that knowing how meaning is embodied does not entail access to it.

Acknowledgment. I am pleased to thank Michael Arbib and Nancey Murphy for their thoughtful comments on a previous version of this essay.

THE MIND-BRAIN PROBLEM, THE LAWS OF NATURE, AND CONSTITUTIVE RELATIONSHIPS

William R. Stoeger, S.J.

1 The Mind-Brain Problem

The mind-brain, or mind-body, problem and issues associated with it continue to resist satisfactory resolution despite the concerted assault by cognitive scientists and philosophers of mind. What is the relationship between what we refer to as "mind" and the brain—between mental states and brain states? How is the mind related to the brain? How do mind-states arise from brain states? Or do they? Are mind-states just certain special types of highly complex brain states experienced holistically "from the inside?" But, if so, what makes such brain states special, different from those which are not mind-states? And how does such a continuous unified subjective experience, from the inside, arise?

From a more scientific point of view, it is clear that the mind is closely related to the brain, and to its function as a physical organ. From all the neurophysiological, psychological, and medical evidence we now have, we can say provisionally but very securely that without a properly functioning human brain there is no human mind. Thus, the mind, as we know it, seems to be strongly dependent on, and perhaps even *in some sense* reducible to, the physical or the material. But here another ambiguity arises: What are the boundaries of the physical or the material? And how are the physical or the material to be distinguished from what is not physical or material? This is an important question that we shall address in section 3.

From a more philosophical or phenomenological point of view, however, our experience of mind seems to transcend what is purely physical or material, in our knowing and understanding, in our communication via language, all of which involves symbolic reference,[1] and in our instinctive quest for the true, the good, and the beautiful. Our subjective, or "inner," experience of these mental functions, along with our awareness of them, and of ourselves as a unified center of consciousness,[2] seem to require an explanation which transcends the purely physical functioning of the brain, as we usually conceive it, even though these all depend upon the brain and are expressed through it and through the body. Thus, it seems on the one hand that the mental and personal must be completely physically or materially based (what else could there be?), and on the other hand that a purely physical or material explanation is seriously inadequate, given what we are able to say reliably about physical, chemical, and biological systems, even very complex ones. At least at our level of understanding of the brain and its processes—extensive and detailed as it is—we do not yet know how mind and brain are related.[3] And we have very little idea of what an adequate account of this relationship would even look like.

---

[1] Terrence W. Deacon, *The Symbolic Species* (New York: Norton, 1997), 43–59, 83–99.

[2] Colin McGinn, *The Character of Mind*, 2nd ed. (Oxford: Oxford University Press, 1997), 142–43.

[3] McGinn, *The Character of Mind*, 17–39, gives a brief and very clear explication of how neither the purely physical or monistic account nor the dualistic account is satisfactory, thus underscoring the extreme difficulty of the mind-brain problem.

The mind-brain problem does not stand alone. It is intimately connected with an account of the unity of the self, and the meaning of the human person. Conscious experience, mental acts, memory, intentional behavior, and language—these are all somehow essential to our being persons or selves.

In this essay I shall not attack the mind-brain problem head-on, nor attempt to provide a solution to it. Instead I shall discuss some of the concepts and presuppositions which are important for disentangling elements of the problem. The central concepts are "reducibility" and its opposite "irreducibility," "emergence," "supervenience," "the physical," "the material," "the immaterial," "the mental," and "the spiritual." The precise content of these concepts, and the precise nature of the presuppositions involved in using them for studying the mind-body problem, depend to a large extent, as I shall show, upon how we conceive "the laws of nature," particularly upon how we conceive the relationship between the laws of nature as we know them and the regularities, processes, and relationships which really obtain in nature. The first four or five of these concepts are not specific to the mind-brain problem, but are used in important ways to elaborate it.

2 The Laws of Nature and Mental Phenomena

The principal point to be developed here is that when we speak of "the laws of nature" we can mean two rather different things. We may mean the regularities, relationships, processes, and structures in nature: (1) *as we know, understand, and model them*; or (2) *as they actually function in reality*, which is much, much more than we know, understand, or have adequately modeled. Thus, the distinction is between epistemology and ontology—between our descriptions of the regularities and relationships we discover and the *actual regularities and relationships* in nature.[4] When it is not otherwise clear which meaning I intend, I shall use "the laws of nature" for the regularities, etc., as they actually function, and "our laws of nature" for our descriptions or models of them. It is essentially the old distinction between our models and understandings of some reality and that reality itself. I have insisted on this distinction in order to maintain a balanced epistemological and metaphysical perspective on a variety of interdisciplinary issues.[5] And I am convinced this distinction is very helpful in dealing with the mind-brain problem.

---

[4] I shall sometimes speak of these actual regularities and relationships in nature as "the laws of nature as they actually function" or even "the laws of nature as perfectly known," or "known by God," in order to emphasize that "our laws of nature" are incomplete and imperfect compared not only to reality, but also to what could be known about them. This is in light of the improvement of our scientific knowledge, and of our tendency to compare the limited knowledge we have with our imagined perfect knowledge of nature. This is necessary because in section 5 I discuss reducibility, particularly epistemological reducibility, and in so doing need to envision epistemological reducibility in the limit of perfect, or divine, knowledge. In using these expressions I am presuming that a perfect, or divine, knowledge and description of the regularities and relationships of nature would be isomorphic with the actual regularities and relationships themselves. This may or may not be true. But that is of secondary importance—it is a way of speaking that is very common among scientists. The fundamental distinction, however, is between the ontological level of the regularities and relationships themselves and the epistemological level of our description of them.

[5] William R. Stoeger, "The Ontological Status of the Laws of Nature" in *Quantum Cosmology and Laws of Nature: Scientific Perspectives on Divine Action*, Robert J. Russell, Nancey Murphy, and Christopher J. Isham, eds. (Vatican City State: Vatican Observatory; Berkeley, Calif.: Center for Theology and the Natural Sciences, 1993), 209–34.

In insisting on this distinction I am not espousing or implying an unbridgeable Kantian divide between the knowable phenomena and radically unknowable noumena. I hold that the natural sciences are able to discover essential facts about the way reality is in itself. The recalcitrance of reality as we encounter it is one of the strongest reasons for believing this—reality refuses to be easily molded to our imaginations and desires. That is why careful observation and experiment so often force modification and abandonment of theories. However, the descriptions and theories of the natural sciences are at the same time provisional and always incomplete. Though they provide marvelously adequate descriptions of the regularities, processes, and relationships of nature relative to some questions and for some applications—particularly in cases in which analytic methods have succeeded in revealing the essential details of components and their relationships with one another—they can never provide a complete or fully adequate and ultimate description and explanation. And yet people oftentimes presume that they are capable of doing so—or even are close to doing so. This is one of the reasons why the distinction is important to emphasize.

If we confuse these two different senses of "the laws of nature," we can make serious mistakes in using concepts which directly or indirectly rely upon it or upon derived notions—such as "reducibility," "emergence," and "supervenience." The laws of nature as we understand and have modeled them are provisional, imperfect, and very limited.[6] In fact, in almost all cases they were derived or discovered by studying very simple, ideal situations, in which complicating factors could be excluded, and/or by abstracting from the relationships, regularities, processes, and characteristics—including ultimate ones—which were irrelevant to the focus of the particular study (for example, gravitational phenomena or the behavior of a charged particle in an electromagnetic field).

Thus, for some situations the laws of nature as we know them are adequate and very well confirmed. But for other situations they are completely inadequate and unsubstantiated. In some cases we have yet to formulate them (for example, with regard to human consciousness). However, this obviously does not mean that there are no regularities, relationships, or processes which obtain. It just means that we do not yet know enough about them or have not adequately modeled them. This is true even at the level of fundamental physics. For example, we know the laws of nature as they pertain to explaining how planets orbit stars; but we know very little about why or how mass-energy generates a gravitational field in the first place—we just know that it does. But that does not mean that there is no reason, or network of relationships, which would explain it. We just have not discovered it yet.[7] Some of these relationships may, in fact, take us out of the realm of what we traditionally call physics and demand metascientific or philosophical considerations. We just do not know enough yet about mass-energy and gravity, and how they fit into the big picture of why reality is the way it is. We know a great deal, thanks to physics, astronomy,

---

[6] In "The Ontological Status of the Laws of Nature," ibid., I also show that "our laws of nature" are only descriptive, not prescriptive. This distinction is not directly important for the principal argument of the present essay.

[7] I am presuming here that everything has a sufficient reason. From a philosophical point of view, this is a huge assumption. However, it is one which is almost always made in the natural sciences and in daily life. We pervasively have the experience of seeking a reason or an explanation for an event or a state of affairs and eventually finding one. We have strong indications that reality is profoundly intelligible, though we cannot rigorously justify this.

and mathematics, but we do not have a satisfyingly complete understanding. To that extent we do not know the laws of nature as they actually function.

Moving to the case at hand, we know very little about how mind and brain are related. They are intimately related, but we do not have as yet anything even approaching a provisionally adequate account or model of that complex relationship. Thus, given the inadequacy of our understanding of the full network of interrelationships and processes governing highly organized biological and neurological systems, it is certainly premature and counterproductive to assume that we shall be able to understand the conscious and mental characteristics and behaviors of higher animals and human beings solely on the basis of—that we shall be able to "reduce" them to—what we now know of physics, chemistry, and biology. As a matter of fact, what is being done in neuroscientific investigation—and in such fields as the physics of complex systems—is pushing physics, chemistry, and biology into radically new territory and opening avenues which may lead to a deeper understanding of the crucial connections.[8] This may, in turn, radically change what we regard as "physical" or "material," and "nonphysical" or "nonmaterial" (see below). In other words, we seem to be discovering that the material is much richer in potentiality and possibility than what we give it credit for on the basis of traditional physics, chemistry, and biology. The very bases of our understanding and appreciation of physical reality are changing and being enriched in the process.

These seem like obvious points. Why should they be stressed? First of all, the phenomena we are considering here—consciousness, abstract thought, perceptual synthesis, symbolic communication (language)—seem to be different in kind from the phenomena that the natural sciences have had success in modeling and explaining. They have both an apparently irreducible subjective, or "inner," aspect and a radically synthetic character. (By their radically synthetic character I mean that they are given in such a way that it is not possible to break them down into experienced components in any satisfactory way. Furthermore, from the completely different neural point of view, we have no handle on the constitutive relationships[9] among basic neural events, which constitutive relationships instantiate these phenomena, nor even a clue as to what would count as a provisionally adequate description of these interneural relationships [see below].) Though it is relatively easy to model these phenomena, *assuming* their inner and radically synthetic character, as Michael Arbib (using his schema theory), Terrence Deacon, and others have so impressively done, it is not at all easy—nor has it been possible so far—to account adequately for such synthetic, "transcendent," subjective mental phenomena or to explain *how* they arise from interacting patterns of neural events.

Put differently, the methods of the natural sciences alone, as presently conceived and practiced, are not equipped to deal with this range of phenomena and experience. They do not seem to be able to "penetrate the barrier of interiority" to reveal

---

[8] This point has been recently made by Ernan McMullin in his essay "Biology and the Theology of Human Nature," in *Controlling Our Destinies: Historical, Philosophical and Ethical Perspectives on the Human Genome Project*, Phillip Sloan, ed. (Notre Dame, Ind.: University of Notre Dame Press, forthcoming 1999). Michael Arbib refers to the same idea in his essay in this volume, "Towards a Neuroscience of the Person," fn. 18, calling it "two-way reductionism."

[9] The concept of "constitutive relationship" is central to the argument of this essay—constitutive relationships are principal components of the laws of nature, both as they function and as we model them. This concept will be defined and discussed in section 4.

the links between the external phenomena and the internal phenomena in a detailed and compelling way. This is absolutely essential for an adequate explanation.

There is, as I have said, an irreducible interiority, or subjectivity here, which is essential to the experiences themselves. But it is also an interiority which is linked to the "outside"—and this in two different senses. First, the interiority is revealed and functions in perceptive and intentional relationships with objects "outside" the self —over against what is outside—but drawing what is outside into its interior conscious experience and acting from the inside towards what is outside in patterns of behavior which are accessible to measurement and observation.[10] Second, this interiority is necessarily supported by (some say it is identical with) certain complex patterns of neural activity of the brain and the body, which can be studied, measured, and modeled from the "outside." From the perspective of the natural sciences, we can correlate our inner experiences with the actuation of certain regions of the brain—even relating certain areas of our experience with one another in new ways on the basis of overlapping regions of brain activity. And we can confirm these findings by studying the loss of perceptual, mental, and conscious function due to injury to specific parts of the brain.

But, as I have emphasized above, what we cannot yet do is explain or describe exactly how interacting neural events, or patterns of neural events, yield or instantiate the inner integrated experiences of perception and consciousness, linking them intimately with the outer world. Nor can we yet explain how the inner experiences of intention and reference arise and are realized or expressed in a new series of interacting neural events. This is probably why the best we can do philosophically at this stage is to postulate an irreducible (in the sense delineated below) psychosomatic unity.[11]

Additionally, so far as I am aware (if someone can provide adequate evidence to the contrary, please do so!), there is, as I have indicated above, no heuristic which gives us the essential characteristics such an explanatory model should have. What type of model would count as an explanation of these mental phenomena? How do we show that a candidate model is a good candidate for explaining the observed integration and interiority? There are as yet no criteria for determining this. Feedback loops, self-organizing maps, schemas, etc., certainly provide some of the necessary underpinnings. But they do not as yet, as far as I can see, touch the fundamental issues presented by the binding problem, subjectivity, and consciousness. So serious is this gap that some eminent philosophers of mind have suggested either that consciousness and subjectivity are fundamental, irreducible givens in reality, like mass-energy,[12] or that our mental capacity as human beings is too limited to understand the essential connection and relationships involved in these phenomena.[13]

3 Distinguishing "the Physical" from "the Nonphysical"

Articulating the distinction between "the physical" and "the nonphysical," "the material" and "the nonmaterial," is a key example of where the difference between

---

[10] David Braine *The Human Person, Animal and Spirit* (Notre Dame, Ind.: University of Notre Dame Press, 1992), 42–57, 96ff.

[11] Ibid., passim.

[12] David Chalmers, *The Conscious Mind* (Oxford: Oxford University Press, 1996), 3–28, 276–308.

[13] McGinn, *The Character of Mind*, 42–48.

"our laws of nature" and the inter-relationships, regularities, and processes as they actually function in nature comes into play in a crucial way. It is a complicated example, but an important one. Since it sheds some light on why the terms "mental" and "spiritual" are so ambiguous, it is directly relevant to the mind-brain problem.

To begin with, from both an historical and a philosophical point of view, the concept of "matter" is not a well-defined scientific concept. Only the concept of "mass" is, and it is a fundamental or irreducible scientific concept, now conceived as equivalent to "energy." Michael Heller has written on this,[14] as has Ernan McMullin.[15] "Matter" is an ambiguous term. From an Aristotelian and Thomistic point of view, it is closely related to "potentiality," and thus cannot simply be equated with "mass-energy." One might be able to make a case for saying that "mass-energy" is a form or realization of "matter," conceived as potentiality, but certainly not all "matter" is simply mass-energy.

Furthermore, if we ask what characteristics matter, or, even more narrowly, entities possessing mass-energy, can have, what restrictions should we place on answers to those questions? We really have to include life, consciousness, and mental capacity as possible properties of matter, or of entities with mass-energy, even though we do not yet understand the laws of nature that relate to these remarkable characteristics. Of course, we include these characteristics not simply in virtue of their possessing mass-energy (not all such systems manifest these characteristics) but rather in virtue of their being highly organized in very definite ways. All of the issues with which we are concerned here have at their basis the intimate relationships between the manifestations of what we refer to as "spirit," "mind," and the properties of highly neurologically organized "matter." Though we acknowledge this profound connection, we also must acknowledge, as I have already stressed, that what we understand about entities with mass-energy in these highly organized configurations is not sufficient to account for these characteristics. This does not mean that some other completely extraneous principle is needed—it *may be*—but only that we lack essential knowledge and information concerning such complex neurological structures and their many levels of function. Thus, from a scientific—and even a philosophical—point of view, referring to the human person as a psychosomatic unity is a preliminary intuitive conclusion based on our experience of personal integrity, rather than on the results of careful analysis.

Thus, we begin to see that speaking about "matter," or what is "physical," involves deep ambiguities, both scientifically and philosophically. Obviously then, "the nonphysical" and "the immaterial" are even more ambiguous designations. In this context it is difficult to specify to what these predicates refer. And when we move to "the mental" and "the spiritual," the difficulties escalate. As we have just seen, a strong case can be made for maintaining that "the mental" and "the spiritual," as we experience and know them, are intimately connected with highly neurologically organized matter—they are *not* "immaterial" in the sense that they are separate from matter, or independent of matter. They possibly could be considered "immaterial" or "unphysical" in the sense that they involve characteristics of matter which go beyond what we can model and understand. In this sense, the designations draw from the way these terms seem to be used in ordinary language.

---

[14] Michael Heller, "Adventures of the Concepts of Mass and Matter," *Philosophy in Science* 3 (1988): 15–35.

[15] McMullin, "Biology and the Theology of Human Nature," 11–12 (typescript), and references therein.

Thus, to appreciate the ambiguities that arise, we need to reflect on how the words "physical" and "material" are employed in common parlance—not just on what they mean, or should mean, scientifically and philosophically. If we talk about "the physical" and "the nonphysical," or "the material" and "the immaterial," what do we usually mean? I submit that we normally mean by "the physical" and "the material" that category of reality whose structure and behavior we can model, describe and understand with provisional accuracy using the natural sciences. That is, "the physical" and "the material" refer to those aspects of reality which are well-modeled and understood using "our laws of nature." These roughly coincide with what is accessible to controlled observation and experiment and adequately describable by general qualitative and quantitative law-like formulations.

Correlatively, the "nonphysical" or "immaterial," has come to refer to those phenomena, behavioral characteristics, and capacities which give no secure evidence of being described or accounted for in this way and which transcend the regularities, processes, and relationships we presently understand and have adequately modeled. Thus, for example, we speak of a personal relationship as having an essentially immaterial, or even a spiritual, component, and of certain values as being spiritual, irreducible to the regularities, processes, and relationships describable by our laws of nature. But that does *not* mean that they are not related to physical or material entities in a profound and intimate way. It just means that our laws of nature do not adequately describe that relationship, and almost always abstract from those features of reality which fall outside the narrow focus of scientific inquiry, or are not easily generalized or subsumed under some easily describable regularity or "law." Furthermore, what presents itself as "nonphysical" and "immaterial" almost always has a synthetic or holistic character which defies scientific analysis, as we have already indicated. The very fact that our laws of nature are unable to integrate these aspects of reality into their purview is an indication that they are incomplete and imperfect, powerful as they may be in a variety of other situations—not that such phenomena are illusory or unfounded. In other words, our laws of nature—the natural sciences as we know and practice them—do not describe or model all important aspects of reality. This is obvious with regard to physical, chemical, and biological phenomena themselves. It is even more obvious with respect to phenomena which fall outside those categories.

The category of "the mental" is similar. We know it is closely related to what is physical—the brain—and to physical processes and events, as we understand them. But we do not know exactly how. There are relevant relationships we do not yet know or understand. As I have already emphasized, what we must admit is that we really do not understand the capacities of the physical and the material when it is neurologically organized. Nor do we really know how such neurological organization is effected and rendered active as a whole. We know that matter is necessary for the mental and spiritual we experience, but we also know that what *we understand and know* about neurologically organized matter is not sufficient for explaining the manifestations of "the mental" or "the spiritual." In the way we usually use these terms, they are sometimes identified with "the immaterial" and "the nonphysical." However, this is not because they are unconnected with matter, but simply because the phenomena they designate fall outside our adequate scientific understanding and explanation. We do not know the laws of nature as they actually function at this level. But obviously such laws of nature—relationships and regularities—are in force. Otherwise, we would not be asking these questions.

4 The Hierarchical Structuring of Reality and Constitutive Relationships

We can turn the question around: Instead of asking how such complex, highly neurologically organized systems can exhibit conscious mental behavior, including subjectivity—the question we have been discussing above—we can ask what is the principle of such organization, what causes there to be such a neurologically rich entity in the first place? What are the laws of nature effecting such unified, sophisticated organization? This question leads directly to the scientific/philosophical question we shall address in the last section of this essay: What makes a thing be what it is, endowing it with a definite unity of structure and behavior, persistence, and consistency of action? The Aristotelean answer to this question was "substantial form," or in the case of human beings, the "soul." We shall discuss this question briefly in the last section, taking into account what we know from the natural sciences and using the notion of constitutive relationships, which we shall define more carefully below.

As a prelude to that discussion, and as an elaboration of our comments about the extremely complex organization of which matter is capable, it is crucial to raise the point explicitly which George Ellis and others have frequently emphasized.[16] The reality of which we are a part is hierarchically structured in innumerable levels—from the levels of fundamental particles and atoms, through the levels of molecules and supermolecules and subcellular structures, to the levels of cells, multicellular entities, eventually those with highly differentiated cellular structures, and finally those with highly developed brains. The brain itself, as we well know, exhibits many levels of structure and activity, even within its different components. There are regularities, processes, and structures at every level which provide a necessary foundation for those on higher, more complex levels. These are part of "the laws of nature" relevant to that level.

But there are also the relationships between the entities on a given level which constitute the entities on the next level—enabling them to interact and combine. These concrete relationships linking entities on a given level are essential to the realization of the entities on the next level, and render the more complex entities much more than simple aggregations of their components. Thus, for instance, the properties of a molecule cannot be understood simply on the basis of the properties of its atomic components, but only in terms of the very specific interrelationships among them—its primary structure, its secondary structure, and its tertiary structure (that is, how the molecule—such as a protein—folds). This is true at every level of hierarchical structuring. At elementary levels, where we believe we understand the essence of these constitutive relationships in terms of the four fundamental physical interactions, it is often possible to specify in some detail how concrete properties arise from underlying properties of the components and the relationships among them. But we rarely achieve a complete explanation. As we get to higher and higher levels, it becomes more and more difficult to understand and model adequately these constitutive relationships themselves, let alone to specify how the characteristics of the more complex entity are to be understood in terms of the inter-relationships among its components and with its environment. When we come to the higher animals, and particularly to human beings, and consider the behavioral characteristics and the experiential richness they manifest, their self-consciousness and intentional

---

[16] George F.R. Ellis, essay in this volume, and references therein.

orientation, their linguistic, conceptual, and cultural capabilities, we are so far unable to explain how these remarkable capacities are generated by the inter-relationships among the components and the multi-level patterns of activity of the brain. Nor are we even aware of all the inter-relationships which are necessary for these capacities to be realized. In fact, the brain alone is not enough. The body is also essential, as are relationships among persons, which stimulate the development and proper programing and functioning of the brain relative to its bodily and environmental context.

What then are "constitutive relationships"? The constitutive relationships of a certain complex whole are all of those connections, relationships, and interactions which either incorporate its lower-level components into that more complex whole, relate that complex whole to higher-level unities in such a way as to contribute essentially to its character, or maintain its connection with the Ground of its being and existence. Depending on the levels of organization involved, these constitutive relationships may be physical, metaphysical, biological, or social in character. An example would be the relationships which make the liver what it is—some of these involve its constituent cells, some of them involve its relationship to the rest of the body, and some of them involve the whole physical and metaphysical context which supports the body and its organs in existence. Another example would be the human person, and the various levels of relationships and interconnections—to lower-level components and to higher-level unities (for example, family, society, etc.). Essentially, then, constitutive relationships are the foundation for the complex unity an entity or organism manifests and for the functions it fulfils. This concept is correlative to the hierarchical structuring that is such a universal feature of reality. Things are what they are because of the relationships they have with one another, and with the Ground of Being and order.

Thus, constitutive relationships are not simply "compositional," nor simply "functional." They are, rather, those inter-connections among components and with the larger context which jointly effect the composition of a given system and establish its functional characteristics within the larger whole of which it is a part, and thereby enable it to manifest the particular properties and behavior it does. A unified whole at any level is not explained simply by its components, but by the components *together with* the relationships among them, *and* by the relationships it has with its surrounding evolutionary matrix (for example, the body is the liver's surrounding evolutionary matrix). Similarly, the functional organization is not explained simply by the functions themselves, but by the infra-structural and supra-structural inter-relationships which enable the performance of the functions. These key constitutive relationships are obviously essential to the hierarchical structuring of reality, and are "the laws of nature."

5 Reducibility, Emergence, and Supervenience

We have already discussed how our operative concept of what is "physical" and "nonphysical"—"material" and "nonmaterial"—is affected, if not determined, by our knowledge, or lack of knowledge, of the laws of nature as they actually function. Let us now briefly examine three concepts, "reducibility," "emergence," and "supervenience"—all of which play an important role in cognitive science and philosophy of mind, not to mention philosophy of science in general—to see how they may be similarly affected. There have been extensive discussions concerning all of these notions, which I shall not review here. My intention is to highlight how the meaning, and therefore the usefulness, of these concepts depends on the concept of the laws of nature we use in defining them—that is, whether or not we define them in terms of "our laws of nature," our provisional, incomplete, and imperfect descriptions of the regularities, processes, and relationships in reality, or rather in terms of those regularities, processes, and relationships as they actually function. I shall show that the meaning of these terms is significantly affected by this distinction.

In discussing reducibility we must first indicate with which forms of reduction we are primarily concerned. *Methodological* reduction, which refers to the research strategy characteristic of the sciences,[17] is not one of them. It has been highly successful, and does not imply ontological, epistemological, or others forms of reductionism. *Ontological* reducibility holds that higher-level entities are complex organizations of lower-level entities, and that nothing else is needed besides those lower-level entities and the relationships or interactions which they have with one another and with their environment.[18] If we analyze all the things that make up the world, we will not find any sorts of things making it up other than these lower-level entities. This form of reductionism, as I have just characterized it, will provide an important reference point but will not directly concern us in the end, because we are not advocating either vitalism or matter-spirit dualism. Ontological reducibility is supported by strong indications from both science and philosophy, and is therefore a characteristic of our description of reality we wish to maintain.[19]

It is worth pointing out at this stage that our ability to explain and model the constitutive relationships of an entity or system—thus giving an account of its behavior, its properties, and its relationships with other entities or systems—has no bearing on the reality of its properties, activities, and relationships. Whether they are described as "reducible" or as "irreducible"—depending on how we define these terms—is beside the point.[20] If the higher-level properties and behavior are observed and have effects, then they are real. Furthermore, the constitutive relationships which mold the lower-level entities into higher-level ones radically change what the lower-level entities are.

The forms of reducibility which do concern us more directly are those which are often referred to as *epistemological* or *theoretical* reducibility, *mereological* reducibility, and *causal* reducibility. "Epistemological reduction" means that higher-level laws of nature—those which describe the behavior of more complex entities and situations—can be identified with descriptions of special cases of lower-level regularities and relationships among more fundamental entities through bridge rules.[21] For this to be possible, the concepts of the higher-level theory must be connectable to those of the lower-level description, and the theoretical statements of the former must be derivable from the latter.[22]

By "mereological reducibility" I mean that the characteristics and behavior of the whole are reducible to the characteristics and behavior of its parts. A mereologically reducible entity does not manifest characteristics or behavior different from that of

---

[17] Ian G. Barbour, *Religion and Science*, (San Francisco, Calif.: HarperSanFrancisco, 1997), 230–31.

[18] See Nancey Murphy, *Anglo-American Postmodernity: Philosophical Perspectives on Science, Religion, and Ethics* (Boulder, Co.: Westview Press, 1997), 14.

[19] Murphy, in this volume, p. 153.

[20] McMullin emphasizes this in "Biology and the Theology of Human Nature," 6.

[21] See Murphy, in this volume, p. 153, and references therein.

[22] Barbour, *Religion and Science*, 231–32.

its parts or the simple sum or aggregation of its parts. An example is a pile of wood or of logs, or a pile of sand, with regard to chemical and physical characteristics. Mereological reducibility, or irreducibility—along with causal reducibility, or irreducibility—will be most helpful in speaking about the importance of constitutive relationships and in providing epistemologically neutral conceptual tools for describing and exploring "emergence," "supervenience," and related issues in the philosophy of mind. "Causal reducibility" indicates that the fundamental causes function in the realm where the lowest-level entities reside, and that higher-level causes are ultimately determined by those at the lowest level, and not the other way around.[23] There will be some situations where causal reducibility will hold and others in which higher-level causes will be determined by more than the causes effective on lower, subvenient, levels—as Nancey Murphy stresses.[24]

Whether some characteristic, or body of knowledge, is epistemologically reducible or not depends on our understanding of "the laws of nature"—in other words, on "our laws of nature." If we have a thorough understanding of physics and chemistry, for instance, we can say that the chemical properties of water are, in a specific sense, reducible to physical properties—to the constitutive relationships which are given in the specific chemical bonds between the two atoms of hydrogen and the single atom of oxygen, and to their relative orientation within the water molecule. These relationships in turn can be understood to a considerable degree in terms of atomic theory and the detailed characteristics of the electromagnetic inter-action. In other words we can establish provisionally adequate bridge laws, in this case between the chemical properties of water and that physical realities underlying them. However, if we do not understand "the laws of nature" and therefore cannot specify such bridge laws, we often say that there is no reducibility.

Obviously then, a particular regularity or behavior may not be epistemologically reducible at one stage of scientific development but may become so at a later stage. Though we do not know the regularities, relationships, and processes of nature as they actually function, our laws of nature reveal something of how they do so. But we want to speak of how they do so on the basis not only of the detailed but imperfect models we have of them, but also on the basis of the general pattern that emerges in the relationships between entities and regularities on one level and those on higher levels as scientific knowledge grows. That general pattern is the hierarchical structuring of which we spoke earlier. This nested hierarchical organization is realized through the connections among lower-level entities *together with* whole-part relationships which establish the boundary conditions by determining the distribution and collective function of groups of lower-level components.

With that general pattern which we observe from nucleons all the way up to the most complex of all organs, the human brain, it is tempting to say that all of reality as it functions in itself is organized that way—the more we discover, the more hierarchical organization will be reinforced as a general principle. On this basis it is often presumed that perfect knowledge of higher-level phenomena, such as inten-tional mobility, imagining, thinking, and telling stories, would give us a complete and perfect specification of all the vertical and horizontal relationships constituting them, in their generality and in their particularity. That may or may not be the case. If it is not, that would seem to require some key feature of reality being unrelated to

---

[23] Murphy, *Anglo-American Postmodernity*, 14.

[24] See her essay in this volume, sec. 4, pp. 154–163.

anything else, and would thus break the pattern of nested hierarchical structuring which so far seems to hold. So far no such unrelated feature has been uncovered.

The very fact that we wish to speak not about our theories but about reality as it is being revealed to us through our theories indicates that we do not want a concept of reducibility that depends on the level of our detailed knowledge, but on what we can securely but provisionally say about reality in itself, from its general characteristics, such as its nested hierarchical structuring. This is what science is about. At the same time, we also do not want a notion of reducibility which applies to all cases. Rather, we want one which is capable of distinguishing entities whose essential characteristics are given by their components and the causes associated with their components from those whose essential characteristics are not given simply by the characteristics of their components and by the causes associated with them. Mereological reducibility together with causal reducibility provide the desired conceptual framework. These notions of "reducibility" do not vary as our knowledge grows; and this is what we require. Furthermore, they reflect how reality seems to be constructed, and at the same time do not apply to all entities, regularities, and processes.

As I have stressed above, all characteristics and features of reality seem to be reducible in the ontological sense that they can ultimately be described in terms of relationships of fundamental entities with one another and with the history and context of which they are a part—that is, in terms of their network of constitutive relationships. So far as we know there is nothing which needs to be injected besides these entities and their constitutive relationships. True, we cannot completely rule this out, but we have no inkling of it either from the sciences or from philosophy.[25] Thus, from the point of view of regularities, processes, and relationships as they actually function and are potentially knowable, everything seems to be reducible in this sense. Such reducibility is important to recognize but not particularly helpful in classifying different sorts of entities and phenomena.

For this classification I suggest, then, that we use both "mereological irreducibility," and "causal irreducibility." The former means that the characteristics of an entity or a system are essentially different from the characteristics of the components which make it up. In such a case it functions as a separate whole, and not just as an aggregation of its parts, within the network of its relationships with its context or environment. This criterion does not depend crucially upon our knowledge or understanding of its constitutive relationships, but rather upon its own unified capacity for relationships, activity, and behavior. At the same time, the entity's irreducible character is due precisely to the importance of the constitutive relationships among its components, and with its history and context. Its components are basic to it, but its distinguishing characteristics are given and determined, not by its components alone but also by its constitutive relationships. We can say that such an entity is "mereologically irreducible." It cannot be reduced to the simple functioning of its components. Thus, an object, a system, or a property is irreducible in this sense if the constitutive relationships among its components, or with other systems or objects outside itself, render its properties, behavior, or function essentially different from that of its components, and endow it with functional unity

---

[25] Of course, one needs a ground of existence and order—from a theistic perspective, God. There are essential constitutive relationships of every being with God, I would maintain. But, at the level of the sciences, which presume both order and existence, this ground is not evident. God is not an object or entity to be found within the universe. I am speaking here only of objects or entities which are found within the universe.

and integrity so that some of the characteristics of its components are no longer manifest separately, apart from the whole in which they now reside. Characteristics, objects, and systems are "mereologically reducible" when this is not the case.

Examples of such irreducibility are common: water is not mereologically reducible to hydrogen and oxygen, nor is salt to chlorine and sodium; a bound atom is not mereologically reducible to the behavior of its protons, neutrons, and electrons; nor a cell to its chemical and biological components. The constitutive relationships of all these objects are essential to its characteristics. In fact, another indication of such mereological irreducibility is that some of the characteristic properties of the components themselves become different when they are incorporated into a mereologically irreducible entity. For example, though a free neutron decays with a half-life of about ten minutes, a bound neutron is stable against decay. A free electron interacts very strongly with photons—a bound electron does so, but differently and less strongly. Examples of mereological reducibility would include a pile of rocks, a cloud of hydrogen gas, a glacier, or a mountain, relative to the physics and chemistry of their physical components—hydrogen molecules, water ice crystals, or rocks.

"Causal irreducibility" refers to the case where the higher-level causes within a system or sequence are not determined solely by the causes operating at more fundamental levels of organization in that system—there are external causal factors impinging on the system from equivalent or higher levels. Top-down causality of any sort renders a system or sequence causally irreducible. If, however, the higher-level causes within a system, entity, or sequence are determined solely by those operating at more fundamental levels, then the system is "causally reducible." One can make a very good case that the behavior of a water molecule, though not mereologically reducible, is causally reducible. Murphy shows that sequences of higher mental functions such as deciding and analyzing are not causally reducible.[26] Another example would be a wedding ring—as such it requires not only the internal causes that make it this physical object—the ring—but also the relationships which link it to the two people whose commitment it symbolizes and celebrates.

We can now easily see that there is a parallel problem with the concept of "emergence." This term is often used, it seems, to connote the appearance of a novel property which we cannot explain in terms of "our laws of nature." If we can explain the property, many people tend not to refer to it as "emergent." This is again somewhat unhelpful, for once we understand certain processes and relationships more fully, what was once considered an "emergent property" is no longer described that way. I suggest, consistent with my suggestion with regard to "reducibility," that any novel property which arises, whether it can be explained or not by what we know of the constitutive relationships at the moment—and in the limit they all may be explainable!—be considered "emergent." That is, any property which is "mereologically irreducible" in the sense that I suggested above, should be considered as "emergent" as one goes from a lower level of organization to a higher level.[27] This is, in fact, completely consonant with how most scientists use the term and therefore does not pertain solely to biological and psychological properties.[28]

---

[26] Nancey Murphy, in this volume, sec. 4.

[27] This concept of emergence which I am proposing seems to be very similar, if not equivalent, to John Searle's notions of emergence. See Searle, *The Rediscovery of the Mind* (Cambridge: MIT Press, 1992), 111–12.

[28] Cf. Bernd-Olaf Küppers, "Understanding Complexity," in *Chaos and Complexity: Scientific Perspectives on Divine Action*, Robert J. Russell, Nancey Murphy, and Arthur

Another concept which is often employed in trying to unravel the mind-brain problem is that of "supervenience." There have been many treatments of what this concept means and what it should mean.[29] It is intended to designate a dependent but generally irreducible relationship that higher-level properties or states have with lower-level properties or states.[30] Chemical properties are "supervenient"—or "supervene"—on physical properties, and mental states are "supervenient" on brain states. This dependency is not causation, strictly speaking. The basic requirement is that higher-level states (mental states) can change only if the lower-level states (brain states) also change. (It is not required that this hold the other way around—we can have supervenience even if lower-level brain states change and higher-level mental states remain unchanged. Clearly, the higher-level states would be identical with the lower-level states only if, along with the supervenience relation, varying lower-level states implies variation in the higher-level states as well.) Thus, this relationship of supervenience itself does not involve either identity, as I have already said, nor necessarily the reducibility, of the higher-level states to the lower-level ones, but simply dependence.

Now, there are several different ways in which this concept of supervenience is affected by the concepts of laws of nature we employ—whether we use "the laws of nature" or "our laws of nature." The first and most obvious way is through the concept of irreducibility. By using our suggested notions of "mereological irreducibility" and "causal irreducibility" the distinctiveness of the higher-level (mental) states is kept very clear in the general case, despite their utter dependence on the lower-level (brain, or physical) states. Thus, when mereological or causal irreducibility obtains, as we have defined it above, it is clear that the supervenient states are essentially different from the subvenient states, though radically dependent upon them. This is because there are relationships constituting the supervenient states, linking them with one another and with the surrounding environment, which have no counterpart on the subvenient level. Each subvenient state partially determines a supervenient state, but complex supervenient states, or particular clusters or causal sequences of supervenient states are not mereologically or causally reducible to their determining subvenient states.[31] What is most important to notice is that this irreducibility is not epistemological—that is, it does not depend on the adequacy of our knowledge or our theories, and it is not simply ontological. Mereological and causal irreducibility are perfectly consistent with ontological reducibility.

Again, we appeal to the wonderfully hierarchically organized structure of the reality of which we are a part. But the wonder is not primarily in the hundreds and thousands and even millions of different microlevels of organization, but rather in the highly differentiated constitutive relationships which marshal more fundamental entities into innumerable kinds of more complex entities with characteristics, behaviors, and capabilities which cannot be predicted from even a detailed knowledge of the more fundamental entities themselves. The constitutive relationships involved in each case make the novel higher-level systems, entities, and properties which emerge to be what they are. Except for mere aggregates of lower-

---

Peacocke, eds. (Vatican City State: Vatican Observatory; Berkeley, Calif.: Center for Theology and the Natural Sciences, 1994), 93–105.

[29] For further discussions of this term, see Murphy and Theo Meyering in this volume.

[30] See also McGinn, *The Character of Mind*, 28–33.

[31] See Murphy's essay in this volume, sec. 4, pp. 154–163.

level entities they are generally mereologically irreducible. And sometimes, because some of their constitutive relationships are not bottom-up—that is, associated with their components—but rather top-down, linking them to their larger wholes and to their environment, they are causally irreducible as well. And thus a thorough, detailed, and complete understanding of *all* the constitutive relationships in a given situation is necessary for an understanding of the character and nature of the higher-level states, systems, and properties. Even in the simplest cases (for example, in physics) "our laws of nature" do not give us this, and never will. They give us a limited and approximately accurate description of some of the principal constitutive relationships as they pertain to a certain narrow range of properties or characteristics. So, for example, though we have very reliable theories of electromagnetism, the weak and strong nuclear interactions, and gravity, which enable us to predict how slightly more complicated systems of basic components such as a hydrogen or helium atom, a medium-sized atomic nucleus, or the earth-moon system behave, we do not really know the underlying constitutive relationships which tell us how these basic interactions have arisen with the relative strengths they have, how they are related to one another, and how they are related to the fundamental properties of the particles whose behavior and coupling they "govern." Our understanding at higher-levels of organization is even worse than this.

This discussion leads us to the second way our operative concept of the laws of nature influences "supervenience" and our application of it to the mind-brain problem. Recalling our discussion of the "physical" and the "nonphysical"—the "material" and the "nonmaterial"—we see that we cannot simply refer to the lower brain state level as "physical" or "material" and the upper mind-state level as "nonphysical" or "nonmaterial," if this distinction is made on the basis of what is scientifically knowable. Strictly speaking, if we intend this distinction to be based on the degree of our knowledge of the relevant phenomena, then both levels contain "material" and "nonmaterial" aspects. If we intend by this distinction more traditional philosophical categories, as in, for example, the writings of Thomas Aquinas, then again both levels contain the "material" and the "nonmaterial," but with a very different meaning: pure potentiality and substantial form. In either case, we realize that the essential issue does not rest on these distinctions but rather on how the determination of the higher-level by the lower-level is brought about.[32] And then, in light of the discussion above, we realize that the answer itself is not simply in terms of the lower-level states themselves, but in terms of the levels of complex organization and patterns among these lower-level states and the information about the systems, subsystems, and their relationships with the outside world that these intermediate and higher-states and patterns encode.[33] This essentially concerns the character of the constitutive relationships which establish the organization and empower its capabilities.

Mental states supervene on brain states. However, a particular mental state, or a particular sequence of brain states, each of which is a necessary condition for a certain mental state, will not in general be completely determined by the brain states themselves but by constitutive relationships at the level of the mental states—their relationships with one another and with their environment and context.[34] It is the

---

[32] McGinn, *The Character of Mind*, 32ff.

[33] Both Murphy and Meyering (in this volume) make this point very compellingly.

[34] See again Murphy's essay in this volume.

brain states *together with certain types of constitutive relationships, not just among themselves but also relating the mental states they determine with one another and with other historical and environmental conditions* which determine the sequences and clustering of mental states, leading to subjectivity, abstract thought, etc. What neuroscience and philosophy of mind are searching for is a detailed knowledge and appreciation of those constitutive relationships! In speaking about the fundamental disparity between consciousness and brain activity, McGinn puzzles: "How could something inherently unobservable (namely, consciousness) be just a combination of items (brain states) whose essence is to be observable?"[35] The answer, though we do not know it in detail, is that consciousness is not "just a combination" of brain states, but rather involves a complex evolving pattern of brain states constituted in some very special way by certain relationships, including those with the outside world and with previous states of itself, which we need to discover and understand. Furthermore, it is likely that there is considerably more to the brain states themselves than meets the eye. They themselves are probably constituted by a fuller range of relationships than we have been able to recognize. Many researchers in philosophy of mind and in neurophysiology have been "locked in" to a "bottom-up" orientation. The full picture—concerning these crucial constitutive relationships to which we have been appealing—must also essentially involve the "top down" orientation. This means that the character of the brain states themselves is strongly influenced by the mind and by consciousness, as well as by the body and its components. However, exactly how this happens is not yet known and is a matter of research and discussion.

From this discussion we can see how a careful consideration of what we know and do not know about the laws of nature—the regularities, processes, structures, and relationships within the world on different levels—affects the meaning of the key notions of "reducibility," "emergence," and "supervenience" as well as the way those concepts are applied, and the consequences of their application. Much more detailed study of this needs to be pursued, but the principal points have been sketched here. We can also see a little more clearly on the basis of this analysis what the mind-brain problem involves, and some of the key concepts and assumptions relevant to it which must be carefully examined.

6 The Nature and the Essence of Entities and Selves

I conclude with a brief discussion of the closely related problem I mentioned earlier (at the beginning of section 4): What makes a thing what it is, endowing it with a definite unity of structure, persistence, and consistency of action?

In Aristotelian and Thomistic terms the answer to this question is "its substantial form" or "its soul," which is not conceived as a substance separate from the matter, existing independently, but rather as the entity's principle of unity and activity.[36] From a scientific and a contemporary philosophical point of view this is seen as very unhelpful—though I cannot see that it is in any way contradicted by the natural sciences. In order for this model to establish contact with what the natural sciences reveal to us about reality, there must be scientifically accessible correlates of the "substantial form" or "soul," whether or not our present level of scientific knowledge and understanding enables us to identify them securely in a given case. This does not mean that everything about the "substantial form" or "soul" must be scientifically

---

[35] McGinn, *The Character of Mind*, 47.

[36] See Stephen Happel's essay in this volume.

accessible, for, as we have already seen, the natural sciences are limited, as are all disciplines, by their particular focus or interest, by their methods, and by the evidence made available by those methods. But there must be manifestations of the "soul" or "substantial form" which are unambiguously revealed by the sciences.

From our discussions so far in this essay, I think it is fairly clear that the scientifically accessible correlates we are looking for are the constitutive relationships—or the manifestations of constitutive relationships—which are in principle accessible to scientific observation, experiment, modeling, and analysis. They are a subset of the laws of nature, as they are potentially knowable using the natural sciences. The "substantial form" or "soul" itself is then the complete network of constitutive relationships for a system or an entity as they actually exist and function, which "our laws of nature" only imperfectly describe and partially indicate.[37] Some components of this complete network of constitutive relationships are accessible to the natural sciences; other components will never be. In some cases, as in the example of water given above, we have a provisionally adequate (at a certain level), but by no means complete, understanding of these constitutive relationships in terms of physics, chemistry, and biology. In many cases which are much more complicated, we simply do not have an adequate understanding of them—for instance, in the case of what makes a dog or a human being what it is as a separate, unified, active entity with properties which are not mereologically or causally reducible to its components.

As I have pointed out before, not all of the potentially scientifically accessible constitutive relationships will be "bottom-up"—effecting new and more complex entities from more fundamental ones. Some will be "whole-part" or "top-down" constitutive relationships. And some may very well link the entity in question, or its enveloping whole, with key objects in the environment in an essential way. For instance, if we take the example of the human brain, it is certainly obvious that some of its constitutive relationships specify how it is constructed from individual neurons into certain types of neuron bundles, which in turn are parts of larger, highly differentiated neuron groups or brain areas, such as the cerebral cortex, the amygdala, the hippocampus, etc. However, what also essentially specifies that it is a human brain is its relationship with the rest of the human body, not only at the present moment but also at previous moments in the body's history, including its conception from a particular egg and a particular sperm cell, its fetal development, and its infancy. This automatically involves the fact that this body is or was a living human person interacting with his or her environment, with other persons, and with society as a whole. Thus, the brain is the brain of particular person; its capabilities—in terms of its brain states and the bodily, personal, and mental behavior they support—depend on an enormous variety of relationships. Some are constitutive of the brain's being what it is; and others, while not being constitutive of the brain

---

[37] This concept of soul is very similar to that proposed by Eleonore Stump in "Non-Cartesian Substance Dualism and Materialism without Reductionism," *Faith and Philosophy* 12.4 (October 1995): 510. I am indebted to Murphy for pointing out this reference to me. Stump conceives of the soul or form as a configurational state of matter, determined by the interactions and relationships among its components. Where my proposal may differ from Stump's is that the network of constitutive relationships includes not only those relating the components of the higher-level whole but also those which relate the complex whole to other higher-level unities in an essential way, as well as those which maintain its connection with the Ground of its being (see p. 137, above). Stump's concept does not seem to include these—at least they are not emphasized in her definition.

itself, are constitutive of the brain states being what they are, and of the mental states at a given moment which supervene upon them, being what they are.

Finally, not all of the constitutive relationships of an entity or a system or a state will necessarily be subject to scientific analysis and discovery. One reason for this is the essentially analytic or reductionist methodology employed by the natural sciences. Another is that some of the constitutive relationships are not accessible to the sciences in principle. For instance, mental states or states of consciousness as such seem to involve relationships of this sort.

It is precisely in attempting to understand these essential constitutive relationships in the case of living organisms and conscious, reasoning organisms that the natural sciences encounter their present limitations in this regard—which may or may not be essential limitations. To highlight this, I ask a very simple set of questions, which I recently posed to a biologist (Marty Hewlett, Department of Molecular and Cellular Biology, University of Arizona): How can we best describe life and death from a strictly biological point of view? That is, when an organism is living, how do we describe that state? And how do we describe what causes it to be living? And when it is dead, how do we biologically describe that state in an adequate way? Obviously, it must be in terms of the proper functioning of the organism as whole. But how is this done? Hewlett replied:

> Indeed we are at the limits of biology when we speak of this.... Modern biology, with its very reductionist approach cannot answer this question, since the properties of molecules which make up a living system do not predict life *per se*. I think that the answer lies in a kind of "metabiology" which asks the larger questions that biology itself is not equipped to address. One approach to an answer is to consider life as an emergent property of the system. Unfortunately, this begs the question in many ways. It also smacks of vitalism and leads to fears that we are returning to a nineteenth-century view.... Right now there is really no good answer. Chris Langton and Stuart Kauffman at the Santa Fe Institute are going in the direction of systemics and what Stuart calls "spontaneous order." But neither of these answers your question.

If we return to the question with which we began this section, "What makes a given entity what it is?" and ask it of a living dog, a cat, or a human being, we realize that there is no simple scientific answer. And yet, as we have discussed, the natural sciences have a great deal that can be useful in trying to formulate or search for an answer. In particular, they reveal a great deal of information concerning the constitutive relationships which obtain in terms of, for instance, the animal's genome and the evolution of its expression in this concrete dog, cat, or human. But they do not supply an adequate synthesized answer.

We can translate the key question into another form, which we also posed in section 4: What is the principle of organization of such a neurologically complex living entity? What are the laws of nature—what are the constitutive relationships— effecting such a unified, sophisticated, and dynamic kind of organization? This is the question I pose for further discussion, clarification, and hopefully an eventual answer from the natural sciences and from philosophy.

SUPERVENIENCE AND THE DOWNWARD EFFICACY OF THE MENTAL: A NONREDUCTIVE PHYSICALIST ACCOUNT OF HUMAN ACTION

Nancey Murphy

1 Introduction

The purpose of this essay is to contribute to a nonreductive physicalist account of human nature consistent with views advocated by others in this volume: Ian Barbour, Philip Clayton, George Ellis, Stephen Happel, Arthur Peacocke, Ted Peters, Fraser Watts.[1] My own position is consistent with Barbour's in seeing the human being as a multilevel psychosomatic unity, who is both a biological organism and a responsible self. This view is to be contrasted with both (reductive) materialism and body-soul (or mind-body) dualism. Barbour and Peacocke both emphasize the importance of understanding human life and functioning in terms of a hierarchy of levels of complexity or organization.

The most pressing problem for such an account of human nature, I believe, is a more compelling argument against the *total* reduction of the mental (and moral and spiritual[2]) levels of human functioning to the level of neurobiology. That is, I intend to answer the following question: If mental events are intrinsically related to (supervene on) neural events, how can it *not* be the case that the contents of mental events are ultimately governed by the laws of neurobiology? If neurobiological determinism is true then it would appear that there is no freedom of the will, that moral responsibility is in jeopardy, and, indeed, that our talk about the role of reasons in any intellectual discipline is misguided. Thus, the main goal of this essay will be to show why, in certain sorts of cases, complete *causal* reduction of the mental to the neurobiological *fails*. I shall attempt to do this, first, by clarifying the concept of *supervenience*, which is used by philosophers of mind to give an account of the relations between the mental and the physical. My redefinition of "supervenience" will make it clear that in many cases, supervenient properties are *functional* properties.

I then turn to Donald Campbell's account of downward causation, which also trades on the functional character of the state of affairs being explained. Then, employing accounts of cognition and neurobiology that emphasize functionality and

---

[1] I am assuming that nonreductive physicalism and emergent monism are essentially the same positions. However, I prefer the term "nonreductive physicalism" because it is more precise. "Monism" is vague in that it means only that the person is composed of only one kind of substance but does not distinguish among idealist, physicalist, and dual-aspect monist accounts. "Emergent" has been used in a variety of ways since the late nineteenth century. This term has two drawbacks: one is that it has become emotionally charged as a result of debates in philosophy of biology. The other is that whereas there has been a significant amount of progress in distinguishing among different forms of reductionism (see, for instance, Theo Meyering's essay in this volume), there is no comparable set of distinctions among different sorts of emergentist theses.

[2] I deal with religious experience in "Nonreductive Physicalism: Philosophical Issues," in *Whatever Happened to the Soul?: Scientific and Theological Portraits of Human Nature*, Warren S. Brown, Nancey Murphy, and H. Newton Malony, eds. (Minneapolis: Fortress Press, 1998), chap. 6. There I describe religious experience as supervenient on ordinary cognitive and emotional experiences.

feedback loops, I show that downward causation consistent with Campbell's account occurs in a variety of cognitive processes. It turns out that rational and moral principles can be seen to exert a top-down effect on the formation and functioning of neural assemblies.

Thus, I hope to contribute to the argument that Christian theology has nothing to lose in substituting a nonreductive-physicalist account of human nature for the various forms of body-soul dualism that have appeared in Christian history.

2 Supervenience

The concept of supervenience is now used extensively in philosophy of mind. Many suppose that, in contrast to mind-brain *identity* theses, it allows for a purely physicalist (that is, ontologically reductionist) account of the human person without entailing the explanatory or causal reduction of the mental. In other words, it leaves room for the causal efficacy of the mental. However, it is not clear to me that typical approaches to constructing a formal definition of "supervenience" have this consequence.[3] In this section I first review some of the history of the development of this concept in philosophy. I then offer an alternative characterization of the supervenience relation and attempt to motivate this alternative by showing that it fits examples from domains in which the concept of supervenience is most often used—ethics, biology, and philosophy of mind. I claim that my definition sheds light on the question of causal reductionism, helping us indeed to account for instances of mental causation without giving up the dependence of the mental on the neurobiological.

Accounts of the development of the concept of supervenience in philosophy generally mention Richard M. Hare's use in ethics and Donald Davidson's in philosophy of mind. Hare used "supervenience" as a technical term to describe the relation of evaluative judgments (including ethical judgments) to descriptive judgments. Hare says:

> ...let us take that characteristic of "good" which has been called its supervenience. Suppose that we say "St. Francis was a good man." It is logically impossible to say this and to maintain at the same time that there might have been another man placed in precisely the same circumstances as St. Francis, and who behaved in them in exactly the same way, but who differed from St. Francis in this respect only, that he was not a good man.[4]

In 1970 Davidson used the concept to describe the relation between mental and physical characteristics. He describes the relation as follows:

> mental characteristics are in some sense dependent, or supervenient, on physical characteristics. Such supervenience might be taken to mean that there cannot be two events alike in all physical respects but differing in some mental respect, or that an object cannot alter in some mental respect without altering in some physical respect. Dependence or supervenience of this kind does not entail reducibility through law or definition...[5]

---

[3] Jaegwon Kim now argues that nonreductive physicalism is a myth; that is, it is an unstable position that tends toward outright eliminativism or some form of dualism; see "The Myth of Nonreductive Materialism," in *The Mind-Body Problem*, Richard Warren and Tadeusz Szubka, eds. (Oxford: Blackwell, 1994), 242–60. I agree, but only so long as one sticks to Kim's own definition of "supervenience."

[4] Richard M. Hare, *The Language of Morals* (New York: Oxford University Press, 1966, orig. 1952), 145.

[5] Donald Davidson, *Essays on Actions and Events* (Oxford: Clarendon Press, 1980), 214. Reprinted from *Experience and Theory*, ed. Lawrence Foster and J.W. Swanson (University

David Lewis characterizes the intuition that definitions of "supervenience" are meant to capture: "The idea is simple and easy: we have supervenience when [and only when] there could be no difference of one sort without difference of another sort."[6]

Jaegwon Kim has been influential in the development of formal definitions of supervenience. It is now common to distinguish three types: weak, strong, and global. Kim has defined these as follows, where A and B are two nonempty families of properties:

- A *weakly supervenes* on B if and only if necessarily for any property F in A, if an object x has F, then there exists a property G in B such that x has G, and if any y has G it has F.[7]
- A *strongly supervenes* on B just in case, necessarily, for each x and each property F in A, if x has F, then there is a property G in B such that x has G, and *necessarily* if any y has G, it has F.[8]
- A *globally supervenes* on B just in case worlds that are indiscernible with respect to B ("B-indiscernible," for short) are also A-indiscernible.[9]

In short, supervenience is now widely understood as an asymmetrical relation of property covariation, and the interesting questions of definition are taken to turn on the placement of modal operators.

However, consider again Hare's original *use* of "supervenience" (see above). I believe that the qualification "placed in precisely the same circumstances" is an important one. St. Francis' behavior (for example, giving away all his possessions) would be evaluated quite differently were he in different circumstances (for example, married and with children to support). If this is the case, then the standard definitions of supervenience are not only in need of qualification but are entirely wrongheaded. On the standard account the fact that G supervenes on F means that F materially implies G. But if circumstances make a difference it may well be that F implies G under circumstance c, but that F implies not-G under c'. Thus, as Theo Meyering has pointed out, there is room for "multiple supervenience" as well as multiple realizability.[10]

A second reason for redefining supervenience is that property covariation is too weak a relation to capture some notions of supervenience. A number of authors argue that what supervenience is really about is a relationship between properties such that the individual has the supervening property *in virtue of* having the subvenient property or properties.

---

of Massachusetts Press and Duckworth, 1970).

[6] David Lewis, *On the Plurality of Worlds* (Oxford: Basil Blackwell, 1986), 14. Quoted by Brian McLaughlin, "Varieties of Supervenience," in *Supervenience: New Essays*, Elias E. Savellos and Ümit D. Yalçin, eds. (Cambridge: Cambridge University Press, 1995), 16–59; quotation on p. 17. McLaughlin added "and only when," claiming that it is clear from the context that this is what Lewis meant.

[7] Jaegwon Kim, "Concepts of Supervenience," *Philosophy and Phenomenological Research* XIV, no. 2 (December 1984): 153–76; quotation on p. 163.

[8] Ibid, 165.

[9] Ibid, 168.

[10] See Meyering, in this volume.

Thus, I propose the following as a more adequate characterization of supervenience:

- Property G in A supervenes on property F in B if and only if x's instantiating G is in virtue of x's instantiating F under circumstance c.[11]

A number of authors call attention to the sorts of factors that I mean to highlight by making the supervenience relation relative to circumstances. Externalists in philosophy of mind argue that relevant features of the way the world is are crucial for determining what intentional state supervenes on a given brain state. My definition makes it possible to say that mental properties supervene on brain properties and at the same time to recognize that some mental properties are co-determined by the way the world is. Another relevant case is Thomas Grimes's example of the economic properties of currency.[12] To put his example in my terms, the property of being, say, a U.S. penny supervenes on its being a copper disk with Lincoln's head stamped on one side, and so forth, only under the circumstances of its having been made at a U.S. mint, and under a vast number of other, more complex, circumstances having to do with the federal government, its powers, and its economic practices.[13]

Berent Enç claims that there is a *species* of supervenient properties that have causal efficacy that is not fully accounted for by the causal role played by the microbase properties. The properties he has in mind are,

> locally supervenient properties of an individual that are associated with certain globally supervenient properties. These globally supervenient properties will have their base restricted to a region outside the individual in which properties causally interact with the properties of the individual in question. I do not know how to give a general formula that captures all of these globally supervenient properties. But some examples will illustrate the idea....
> 1. Properties that are defined "causally," for example, being a skin condition that is caused by excessive exposure to sun rays, that is, being a sunburn. . . .
> 2. Properties that are defined in terms of what distal properties they have, for example, fitness in biology....
> 3. Properties that are defined in terms of what would have caused them under a set of specifiable conditions, like being a representation of some state of affairs.[14]

In my terminology, the property of being a sunburn supervenes on a microcondition of the skin cells under the circumstance of its having been brought about by over-exposure to the sun. Fitness supervenes on any particular configuration of biological characteristics only within certain environmental circumstances.

---

[11] See also my "Supervenience and the Nonreducibility of Ethics to Biology," in *Evolutionary and Molecular Biology: Scientific Perspectives on Divine Action*, R.J. Russell, W.R. Stoeger, S.J., and F.J. Ayala, eds. (Vatican City State: Vatican Observatory; Berkeley, Calif.: Center for Theology and the Natural Sciences), 463–89.

[12] Thomas R. Grimes, "The Tweedledum and Tweedledee of Supervenience," in *Supervenience*, Savellos and Yalçin, eds., 110–123; 117.

[13] George Ellis suggests modifying my definition to read "Property G in A supervenes on property F in B if and only if x's instantiating G is in virtue of x's instantiating F under circumstance c *belonging to the set C*," where the choice of C specifies what issues are relevant to this supervenience. This suggestion deserves further thought, but at present I suspect that the sorts of circumstances that are relevant cannot ordinarily be specified apart from an examination of particular cases.

[14] Berent Enç, "Nonreducible Supervenient Causation," in, *Supervenience*, Savellos and Yalçin eds., 169–80; 175.

Paul Teller makes similar points:

Let us restrict attention to properties that reduce in the sense of having a physical realization, as in the cases of being a calculator, having a certain temperature, and being a piece of money. Whether or not an object counts as having properties such as these will depend, not only on the physical properties of that object, but on various circumstances of the context. Intentions of relevant language users constitute a plausible candidate for relevant circumstances. In at least many cases, dependence on context arises because the property constitutes a functional property, where the relevant functional system (calculational practices, heat transfer, monetary systems) are much larger than the property-bearing object in question. These examples raise the question of whether many and perhaps all mental properties depend ineliminably on relations to things outside the organisms that have the mental properties.[15]

So the moves I am making are not unheard of in the supervenience literature. Furthermore, I claim that my definition does in fact meet the desiderata for the concept of supervenience in that it reflects both dependence and nonreducibility. However, my claim for nonreducibility is circumscribed—I claim that it gives us nonreducibility only where and in the sense in which we should want it. That is, it gives us a way of talking about the genuine dependence of human characteristics on the brain, but leaves room for the codetermination of *some* of those characteristics by the external world, especially by culture.[16]

There seem to be at least three types of dependence or in-virtue-of relations that deserve to be counted as supervenience relations. On some accounts, the supervenience of moral on nonmoral properties involves what we may call "conceptual supervenience." St. Francis' goodness supervenes on his generosity because being generous is part of what we *mean* by goodness. Notice, though, that the kind of supervenience in question depends on one's moral theory. For example, if rule utilitarianism is true, then we need an intervening level of description between the property of being a generous act and the description of it as good—namely, the property of being a pleasure-enhancing action. The relation between the goodness and the pleasure enhancement, again, is conceptual. But the relation between generosity and pleasure-enhancement is functional.

"Supervenience" is often used to characterize the relation between the temperature of a gas and the mean kinetic energy of its molecules: the gas has the temperature it does *in virtue of* the kinetic energy of its particles. This is an instance of microdetermination.

The type of supervenience that may repay the most attention is functional. The sort of relation envisioned here is that between a supervenient functional property and its realizand. Examples are plentiful in biology: the relation between fitness and an organism's physical properties, the relation between being the gene for red eye pigmentation and being a particular sequence of base pairs in a strand of DNA, and (from another domain) the relation between being a coin and being a metal disk with certain characteristics.

Functional terms describe causal roles in larger systems. Reference to these larger systems is one way of making more specific the circumstances that I highlight in my definition, and thus shows in a vivid manner the *codetermination* of the

---

[15] Paul Teller, "Reduction," in *The Cambridge Dictionary of Philosophy*, Robert Audi, ed. (Cambridge: Cambridge University Press, 1995), 679–80; 680.

[16] This desideratum is comparable to Michael Arbib's notion of two-way reducibility: neuroscience is necessary for understanding mental processes and sufficient for some purposes, but for others, culture is necessary as well.

supervenient property by the subvenient property or properties and the circumstances. For example, fitness is codetermined by the environment; being currency is
codetermined by the system of economic exchange.[17]

So one advantage of a stronger definition of supervenience in terms of the in-
virtue-of relation is that the dependence of supervenient on subvenient properties is
built in. However, this dependence is (usefully) moderated by the recognition that
in some cases, due to circumstances of various sorts, the subvenient properties are
not sufficient determinants of the supervenient property.

The second advantage of defining supervenience in terms of the in-virtue-of
relation is its heuristic value. It leads us to look at actual (purported) cases of
supervenience and ask *how it is* that properties at one level exist in virtue of
properties described at another level. In the process, we learn more about the
complex "layered" reality in which we live. There is a great deal of room here for
further exploration—exploration of types of supervenience based on different types
of dependence relations. I project that such exploration would result in a body of
literature more interesting than the current one focusing on strong, weak, and global
supervenience—more interesting, at least, to philosophers of science and ethicists,
if not to logicians.

3 Nonreducibility

The supervenience relation, as I have already stated, is supposed by some to be a
nonreducible relation. A number of authors complain that the question of reduction
is complicated by the fact that there is no accepted account of reduction.[18] In this
section I distinguish a variety of kinds of reduction and claim that we should want a
definition of supervenience that avoids only certain of these kinds. The most
significant open issue, I believe, involves causal reduction. However, it is only in
limited instances that we should want to argue for a definition of supervenience that
entails causal nonreducibility. My primary concern in this essay is to preserve the
distinction between reasons and causes. However, free will and moral responsibility
are also at issue.[19] Thus, the desideratum for a definition of supervenience is that the
relation turn out to be nonreducible in only the right way and for only the right kinds
of cases.

Francisco Ayala has usefully distinguished three forms of reduction: methodological, ontological, and epistemological.[20] Methodological reductionism is a research
strategy, seeking explanations by investigation of lower levels of complexity.

---

[17] I have argued that a speech act or illocutionary act in John L. Austin's terms supervenes
on a locutionary act; see my *Anglo-American Postmodernity: Philosophical Perspectives on
Science, Religion, and Ethics* (Boulder: Westview Press, 1997), 24–25. An excellent example
of the role of circumstances in such cases is Arbib's example of the same locutionary act—
saying "How are you?"—functioning as either a greeting or a request for medical information
depending on the "frame" of the conversation.

[18] McLaughlin, "Varieties of Supervenience," 45–46; Paul K. Moser and J.D. Trout,
"Physicalism, Supervenience, and Dependence," in *Supervenience*, Savellos and Yalçin, eds.,
187–217; 190–91.

[19] Warren Brown and I intend to deal with these further issues in *An Essay in
Neurophilosophy: Neuroscience, Mental Causation, Free Will, and Morality* (forthcoming).

[20] Francisco J. Ayala, "Introduction," in *Studies in the Philosophy of Biology: Reduction
and Related Problems*, F.J. Ayala and T. Dobzhansky, eds. (Berkeley and Los Angeles:
University of California Press, 1974).

Ontological reductionism, as Ayala uses it, is the denial of nonmaterial ontological entities such as vital forces and souls. Ayala's epistemological reduction corresponds to the theoretical reduction that has been the focus in much of the supervenience literature to date; that is, translation of higher-level laws or theories into special cases of lower-level laws via bridge rules. There are at least two other possible sorts of reduction: semantic and causal. Semantic reduction, of course, is related to epistemological or theoretical reduction in that bridge laws or definitions are needed to relate the vocabulary of the reduced theory to that of the reducing theory. Causal reduction is also related to theoretical reduction if one takes law-like regularity to be the criterion for causal relations. Note that Barbour (in this volume) is using "ontological reduction" to cover what I mean by both ontological and causal reductionism.

Let us consider now, in the particular case of philosophy of mind, where these various reductionist theses stand. Methodological reductionism in philosophy of mind is the thesis that one should pursue explanation of mental phenomena by investigating the underlying neural mechanisms. It cannot be denied that the recent marriage between the cognitive and neurosciences has been fruitful, so methodological reductionism here seems plainly unobjectionable, so long as it is not taken to the extreme of saying that cognition can be approached only in this way.

I believe it is safe to say that most philosophers of mind and neuroscientists now hold to an ontologically reductionist view of the relation between the mental and the physical. In fact, one of the primary uses of the concept of supervenience is to enable us to say that while complex living organisms have mental properties, there is no need to postulate a substantial mind to account for those properties.[21] So most would say that ontological reductionism in this domain is both true and desirable, and any good account of supervenience ought to help us see why it is true.

Most of the discussions of supervenience and reductionism in the literature have focused on semantic and theoretical reduction. In philosophy of mind this pertains to whether the laws of psychology are practically or in principle derivable from the laws of neurobiology. This is the kind of reductionism that an understanding of supervenience is intended (by many) to thwart. I believe it is still an open question whether multiple realizability blocks semantic and therefore epistemological reduction.[22] However, I shall not attempt to contribute to this discussion, since I see as more interesting and important the question of causal reduction.

Some would argue that, causation being understood in terms of lawful regularity, the fate of causal reductionism hangs exactly on the issue of epistemological reduction. However, if we consider a more primitive notion of causality, that of "making something happen," then the failure of epistemological reductionism does not entail the absence of causal determination of the mental by the neurobiological.[23]

Note that the question here is not whether the subvenient property causes the supervenient, but rather whether casual *relations* described at the subvenient level are sufficient to determine the sequence of events as described at the supervenient level. Bear in mind that the best account we can give of the causal efficacy of mental

---

[21] "Substantial" is meant here in a Cartesian rather than Thomist sense. Note that this is different from the eliminative materialist position, which also denies mental properties or events.

[22] See Andrew Melnyk, "Two Cheers for Reductionism: Or, the Dim Prospects for Non-Reductive Materialism," *Philosophy of Science* 62 (1995): 370–88; especially 379; see also Theo Meyering in this volume.

[23] See the essay by Meyering in this volume.

states is to assume that they supervene on physical states, and it is the physical states that are causally efficacious. That is, it is because my thought "I should take an umbrella" is realized neurobiologically that it is capable of initiating the chain of physical events resulting in my picking up the umbrella. But if all mental states are physically realized, and if we assume causal closure at the physical level, then what are we to make of the supposed reasoned, as opposed to causal, relations among intentional states? That is, how can we avoid having to assume that what are usually taken for reasoned connections among intentional states are not in fact determined by the laws of neurobiology? A central goal of this essay is to show that the reasoned connections *do* matter.

Consider this example: let M_1 be the thought, "it is cloudy and the wind is out of the southwest," and M_2 be the thought "it will probably rain today." By hypothesis, M_1 supervenes on some brain state, B_1, and M_2 on B_2. If we assume in addition that B_1 is an adequate physical cause of B_2, then how can it *not* be the case that M_2 is merely *caused* by B_1 rather than it being a reasoned judgment based on M_1?[24]

So the question can be expressed symbolically as follows: Let the dollar sign represent the supervenience relation, the solid arrow represent the assumed causal relation between brain states (which we need in order to give an account of the causal efficacy of mental states), and the outlined arrow represent the supposed reasoned connection between mental states (judgments). We can then draw the following picture:

$$M_1 \Rightarrow M_2$$
$$\$ \qquad \$$$
$$B_1 \rightarrow B_2$$

The question now is the relation between the two arrows. Causal reductionism here means that the bottom arrow is the significant one; the top arrow is dependent on or determined by it. Yet we seem to need to show, in order to maintain the meaningfulness and efficacy of the mental *qua* mental, that the order of dependence is reversed. This is not to say that M_1 *causes* B_2 in a straightforward manner—I reject all moves to make supervenience or realization a causal relation. Rather, it is to say that the reasoned relation between M_1 and M_2 plays an indispensable role in the neurobiological processes. I shall show how this may be the case below.

I think that several attempts in the literature to argue for the causal relevance of intentional states are on the right track but not fully adequate.[25] However, I shall not take time here to comment on them; rather I shall attempt to show the value of my own account of supervenience for explaining the causal relevance of *some* intentional states (which involves the limitation of causal reductionism in certain cases).

4 Downward Causation

Let me emphasize that I want no general argument against the causal reducibility of the supervenient to the subvenient, or even of the mental in general to the neurobiological. If so, we would lose the benefits of methodological reductionism and

---

[24] I shall in the end (try to) show how M_2 can be both reasoned and caused.

[25] Cynthia Macdonald, "Psychophysical Supervenience, Dependency, and Reduction" in *Supervenience*, Savellos and Yalçin, eds., 140–57; and Enç, "Nonreducible Supervenient Causation."

would be in danger of losing ontological reductionism as well. That is, we *want* to be able to give an account in neurobiological terms of the necessary and sufficient conditions of, say, feeling pain or seeing a patch of blue. We can assume that perceptions and other low-level conscious states[26] are realized by neurobiological events or properties and that these sorts of conscious states are perfectly determined by the neurobiological (or other physical) antecedents of the neural events that realize them. For example, a pin prick is followed by a deterministic chain of physical events, one of which realizes the experience of a stabbing pain.

Where underdetermination at the neurobiological level becomes important is primarily with regard to "higher-level" mental events: deciding, judging, reasoning. To see how it might be possible to argue against causal reductionism in certain instances[27] I turn to Donald Campbell's account of downward causation in the evolutionary process. We shall see that Campbell's analysis involves what I have called functional supervenience and that nonreducible circumstances play an essential role.

Here is Campbell's familiar account of downward causation:

> Consider the anatomy of the jaws of a worker termite or ant. The hinge surfaces and the muscle attachments agree with Archimedes' laws of levers, that is, with macromechanics. They are optimally designed to apply maximum force at a useful distance from the hinge. A modern engineer could make little if any improvement on their design for the uses of gnawing wood, picking up seeds, etc., given the structural materials at hand. This is a kind of conformity to physics, but a different kind than is involved in the molecular, atomic, strong and weak coupling processes underlying the formation of the particular proteins of the muscle and shell of which the system is constructed. The laws of levers are one part of the complex selective system operating at the level of whole organisms. Selection at that level has optimized viability, and has thus optimized the form of parts of organisms, for the worker termite and ant and for their solitary ancestors. We need the laws of levers, *and organism-level selection…* to explain the particular distribution of proteins found in the jaw and *hence* the DNA templates guiding their production…. Even the *hence* of the previous sentence implies a reverse-directional "cause" in that, by natural selection, it is protein efficacy that determines which DNA templates are present, even though the immediate microdetermination is from DNA to protein.[28]

This example is meant to illustrate the following set of theses:

> 1. All processes at the higher levels are restrained by and act in conformity to the laws of lower levels, including the levels of subatomic physics.
> 2. The teleonomic achievements at higher levels require for their implementation specific lower-level mechanisms and processes. Explanation is not complete until these micromechanisms have been specified.

But in addition:

> 3. *The emergentist principle:* Biological evolution in its meandering exploration of segments of the universe encounters laws, operating as selective systems, which are not described by the laws of physics and inorganic chemistry, and which will not be described

---

[26] By "low-level conscious states" I mean, roughly, the sorts of conscious states that we presume we share with the higher animals: the deliverances of the senses, proprioception, etc. My point here is that nothing is lost in terms of our sense of our humanness if these turn out to be strictly determined by the laws of neurobiology. However, I believe that downward causation is as operative in many of these instances as it is in reasoning, judging, etc. For example, "mental set" (i.e., expectation) affects sensory perception.

[27] I say "in certain instances" because I recognize that not all of what we take to be reasoned action is. Sometimes the reasoning is rationalization of behavior determined by other causes.

[28] Donald T. Campbell, "'Downward Causation' in Hierarchically Organised Biological Systems," in *Studies in the Philosophy of Biology*, Ayala and Dobzhansky, eds., 181.

by the future substitutes for the present approximations of physics and inorganic chemistry.

4. *Downward causation:* Where natural selection operates through life and death at a higher level of organization, the laws of the higher-level selective system determine in part the distribution of lower-level events and substances. Description of an intermediate-level phenomenon is not completed by describing its possibility and implementation in lower-level terms. Its presence, prevalence, or distribution (all needed for a complete explanation of biological phenomena) will often require reference to laws at a higher level of organization as well. Paraphrasing Point 1, all processes at the lower levels of a hierarchy are restrained by and act in conformity to the laws of the higher levels.[29]

Campbell uses the term "downward causation" reluctantly, he says, "because of the shambles that philosophical analysis has revealed in our common sense meanings of 'cause'." If it is causation, he says, "it is the back-handed variety of natural selection and cybernetics, causation by a selective system which edits the products of direct physical causation."[30]

We can represent the bottom-up aspect of the causation as follows:

jaw structure

$

DNA → protein structures

That is, the information encoded in the DNA contributes to the production of certain proteins upon which the structure of the termite jaw supervenes.

However, to represent the top-down aspect of causation we need a more complex diagram as in figure 1. Here the dashed lines represent the top-down aspects and solid lines represent bottom-up causation.

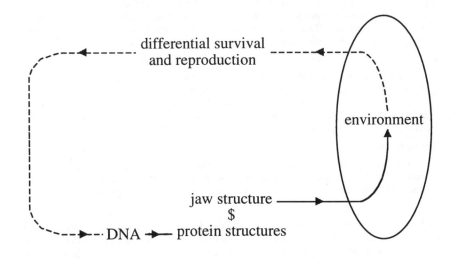

Figure 1.

---

[29] Campbell, "Downward Causation," 180.

[30] Ibid., 180–81.

This "back-handed" variety of causation is all that is needed for the purposes of blocking *total* causal reduction of the mental to the neurobiological.[31] My argument has to be somewhat speculative because it depends on accounts of neurobiology and of cognition that are still contentious. The key to my argument is the fact that some neuroscientists and cognitive scientists describe mental processes such as concept formation, memory, and learning in terms formally identical to the evolutionary process in biology. In short, because of the organism's actions and interaction with the environment, cognitive processes can be understood to develop by means of feedback systems that suppress some responses and enhance others. Gerald Edelman describes neuron growth and the development of synaptic connections as random, rather than instructed, processes. These processes provide the somatic diversity upon which "selection" can occur.[32] Learning is selection whereby some groups of neurons and their connections survive and thrive at the expense of others.[33] Jean-Pierre Changeaux uses the term "selective stabilization" to refer to this process.[34]

4.1 Tuning Neural Networks

The distinction I made above between lower and higher mental functions may correspond roughly to different levels of "functional validation"; this refers to differences among cognitive processes regarding the extent to which they are "wired in" for the species versus needing to be shaped in each individual by interaction with the environment. Actually, three levels are recognized here. There are cognitive processes whose character is entirely independent of the environment, even though the organism needs environmental stimulation to activate them. Other processes are partially independent of the environment, but inappropriate stimulation may change their character (for example, kittens raised in visual environments with nothing but vertical lines will later respond only to vertical, not horizontal lines). Finally, there are cognitive processes whose development is entirely dependent upon the character of the environmental stimulation.[35] Notice that even in the case of states such as pain or color sensation only part of the explanation of why, say, hot surfaces produce this *kind* of feeling can be given by tracing the neural realization of pain back to its physical antecedents, since selection surely helps account for the fact that a burn feels *like that* rather than feeling pleasurable.

While color perception may be built into the species, learning to associate the visual experience of blue with the word "blue" has to be explained in terms of the history of the individual. In a present instance the tendency to think "blue" in the presence of a certain visual stimulus may be causally explained by response thresholds of a given set of neurons. So in this instance the physical realization of the

---

[31] For a more detailed treatment, see my "Downward Causation and Why the Mental Matters," *CTNS Bulletin* 19.1 (Winter 1999): 13–21.

[32] Michael Arbib points out that a better expression than "selection" would be "search in the space of synaptic weights."

[33] Gerald M. Edelman, *Bright Air, Brilliant Fire: On the Matter of the Mind* (New York: BasicBooks, 1992).

[34] Jean-Pierre Changeaux, *Neuronal Man: The Biology of the Mind*, trans. by L. Garey (New York: Pantheon, 1985).

[35] Bryan Kolb and Ian Q. Whishaw, *Fundamentals of Human Neuropsychology*, 4th ed. (New York: W.C. Freeman, 1996), 500. See also Arbib, "Toward a Neuroscience of the Person," sec. 3.2.

visual sensation causes in a fairly straightforward manner the physical realization of the thought "that's blue."[36] But the complete causal account requires, in exact parallel with Campbell's example of downward causation, reference to history—to the individual's having been taught the word and the concept, with appropriate associations of stimulus and response strengthened and others being eliminated. The social environment here provides the selective pressure.

For a slightly more complex example I draw upon an account by Paul Churchland of experiments in computer modeling of recognition tasks. We can hypothesize that the processes in human learning are formally identical. His simplest example is training a target cell to detect a "T" shape when it is projected onto a "retinal" grid with nine squares:

A_1	A_2	A_3
B_1	B_2	B_3
C_1	C_2	C_3

Each square is connected to the target cell. The target cell has been successfully trained to recognize a "T" when it responds maximally to the combined input from A_1, A_2, A_3, B_2, and C_2, but does not respond if it receives signals from B_1, B_3, C_1, or C_3. It may also respond weakly to patterns similar to its "preferred stimulus." Training of a receptor cell requires feedback. At first the receptor cell's firing is random with relation to the incoming stimuli. Feedback from the environment weeds out inappropriate responses and strengthens appropriate ones.[37]

I find Donald MacKay's treatment of cognitive processes particularly useful. He diagrams cognitive processes as feedback systems, as in figure 2. This sort of feedback loop is exemplified in simple self-governing systems such as a thermostat.

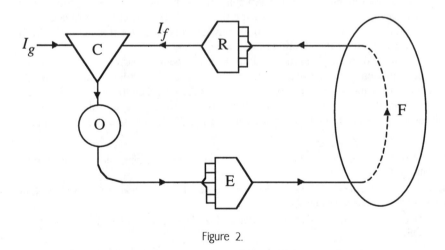

Figure 2.

---

[36] I say "fairly" because of the complications of analyzing probabilistic causes.

[37] Paul M. Churchland, *The Engine of Reason, the Seat of the Soul: A Philosophical Journey into the Brain* (Cambridge: MIT Press, 1995), 36–39.

Here the action of the effector system, E, in the field, F, is monitored by the receptor system, R, which provides an indication I_f of the state of F. This indication is compared with the goal criterion I_g in the comparator, C, which informs the organizing system, O, of any mismatch. O selects from the repertoire of E action calculated to reduce the mismatch.[38]

If we assume that something comparable to what Churchland describes is actually going on in the brain of a child learning to recognize letters (and to name them), then E effects the child's response to the visual stimulation and to the question "Is this a T?" R is the reception of feedback from the teacher, which is recognized by C as either matching or mismatching the goal of "getting it right." In a mismatch the organizing system, in this simple case, would simply generate randomly a different response to the next set of stimuli. However, a "match" would result in strengthening the receptor cell's response to the correct pattern of input and would strengthen the connection between this and the verbal representation.

Once the receptor cell is trained there is bottom-up causation whenever the child recognizes a T—the connection between the (imaginary) nine receptors in the visual field and the T-receptor cell is now "hard wired." However, there is top-down causation in that interaction with the social environment needs to be invoked to explain why that particular connection has come to exist. This is formally identical to the case Campbell describes, where the existence of the DNA that causes, in bottom-up fashion, the protein structures that constitute the termite jaw needs to be explained by means of a history of feedback from the environment (top-down).

4.2 Schemas and Downward Causation

A schema, according to Michael Arbib, is a basic functional unit of action, thought, and perception. Schemas, in the first instance, are realized by means of neural processes in an individual brain. So here is an instance of *functional supervenience*. A schema *is* nothing but a pattern of neural activity, but to understand the schema *qua* schema it is necessary to know what it does; that is, what function it serves in cognition and action.

Arbib's account of *tuning* schemas via interaction with the environment is an account at a higher level of description that parallels and partially overlaps the foregoing account of tuning neural networks. Here the unit upon which selection and downward causation operates is not individual neural connections but schemas, which are realized by medium-scale neural structures and distributed systems. Apparently some schemas are innate yet subject to restructuring, while others are learned. In addition to tuning as a result of action and interaction with the environment, schemas are enriched as a result of interaction with other schemas.

Arbib's emphasis on the action-oriented character of schemas, not merely their representational functions, is crucial for my account. This creates the feedback loops that allow for downward causation from the environment, including the socio-cultural environment. Another crucial ingredient is the element of competition among schemas and schema instances. "It is as a result of competition that instances which do not meet the evolving (data-guided) consensus lose activity…"[39]

---

[38] Donald M. MacKay, *Behind the Eye*, The Gifford Lectures, ed. Valerie MacKay (Oxford: Basil Blackwell, 1991), 43–44.

[39] Arbib, "Towards a Neuroscience of the Person," in this volume, p. 90.

4.3 The Downward Efficacy of Reason

Colin McGinn points out that one of the most significant problems in philosophy of mind is to explain how a physical organism can be subject to the norms of rationality: "How, for example, does *modus ponens* get its grip on the causal transitions between mental states?"[40] Here I employ the concept of downward causation to begin to develop an account of the top-down efficacy of the *intellectual* environment. In the following examples I intend to show that causal closure at the neurobiological level does not conflict with giving an account at the mental level in terms of reasons. To the question, "What is 5×7?" most of us respond automatically with "35." Let us presume that understanding the question (M_1) and thinking the answer (M_2) supervene upon (or are realized neurologically by) B_1 and B_2 respectively. Then we can ask the question whether M_2 occurs because "35" is the *correct* answer or because B_1 *caused* B_2 and M_2 supervenes on B_2. However, considering again the role of feedback loops, we can see that it is clearly both. In short, the laws of arithmetic are needed to account for the prior development of the causal link between B_1 and B_2. In this case, using again the diagram in figure 2, the organizing system can be thought of as first producing random responses to the question, "What is 5×7?" The effector writes or speaks the answer. The field of operation, again, is the classroom; the receptor system receives feedback from the teacher, and the comparator registers "right" or "wrong." If wrong, the organizing system selects a different response. In time, wrong guesses are extinguished, and the causal connection from B_1 to B_2 is strengthened. Then, of course, the response "35" is caused (bottom up) by the connection between B_1 and B_2. The complete explanation requires a top-down account of interaction with the teacher. But this in turn raises the question of why the teacher reinforces one answer rather than another, which can only be answered, ultimately, by an account of the *truth* of "$5\times7 = 35$."[41]

It is clear, I believe, from Arbib's work as well as that of Edelman and his colleagues that analysis of the mental in terms of selection can be extended to all higher mental functions. Higher-order consciousness involves the ability to categorize lower-order conscious functions (schemas) and to evaluate them. So, just as the concept "blue" or the connection of "5×7" with "35" can be learned, *schemas having to do with the evaluation of thought itself can be learned and internalized.* Thereafter they function to select, among lower-order thought processes, those that are and are not "fit."

We can see this by turning again to the learning of arithmetic. Instruction does not aim simply at rote learning but at teaching both skills and evaluation procedures that ultimately allow for internal correction of mental operations. Consider the slightly more complex case of learning to multiply. Here a problem is posed (say, 55×77). The organizing system now involves not a random generator of responses but a *skill* or *operation* that produces answers.[42] The feedback system, then, not only

---

[40] Colin McGinn, "Consciousness and Content," in *The Nature of Consciousness: Philosophical Debates*, Ned Block, Owen Flanagan, and Güven Güzeldere, eds. (Cambridge: MIT Press, 1997), 295–307; quotation on p. 305, n. 2.

[41] In a previous version my example was "$2+2 = 4$." Some objected that this did not need to be taught. Here I have tried to choose a case where the answer is typically taught, in the first instance, rather than intuited or calculated.

[42] See MacKay, *Behind the Eye*, 144, where he conceives of the typical organizing system

corrects for wrong answers but also, via a supervisory system (see figure 3), corrects flaws in the operation itself. The student learns *how* to multiply.

Figure 3 represents a slightly more complex information-flow system that involves internal evaluation of cognitive processes. This is a map of a system capable of modifying its own goals, including: SS, supervisory system; C, comparator; O, organizing system; F, field of action; E, effectors; R, Receptors; I_f, indicated state of field; and I_g, indicated goal-state or criterion of evaluation. The organizing system, O, organizes the repertoire of possible activity, which may be prewired or built up in response to the regularities of its field of action. The supervisory system selects from O parts of the repertoire appropriate to incoming mismatch signals from the comparator and supervises the development, trial, and updating of the repertoire.[43]

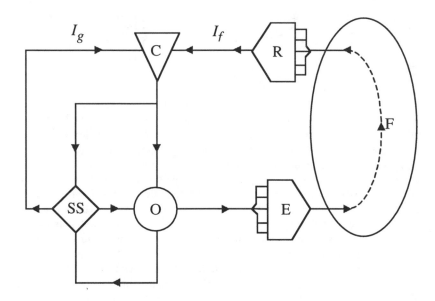

Figure 3.

With this diagram we can represent cognitive processes in which the criteria for correct performance are internalized. The supervisory system now allows for another feedback loop, internal to the information-flow system, and thus allows for more sophisticated instances of downward causation.

MacKay provides a more complex information-flow map both to represent the hierarchical nature of internal evaluation of cognitive processes and also to make it possible to represent a system capable of self-organized goal-directed activity (see figure 4 below). Here the supervisory system of figure 3 is represented by two components, the meta-comparator, MC, and the meta-organizing system, MO. Heavy lines represent the supervisory or meta-organizing level. (FF represents a feed-forward system with feature filters.)

---

as involving norms, maps, and skills.

[43] Ibid., 51.

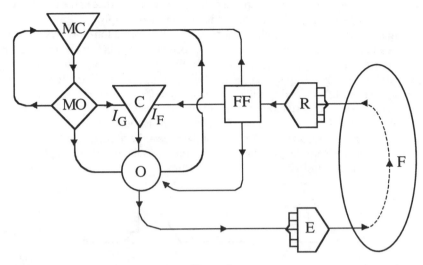

Figure 4.

In figure 2 the goal state of the system, I_g, is set by some criterion outside the system. For example, the goal of getting right answers in class is (initially, at least) set by the teacher. In figure 4 the goal state itself is set by higher-level processes within the system. MacKay says: "we have drawn the meta-organizing system with a meta-evaluative procedure to take stock of how things are going. It adjusts and reselects the current goal in the light of its evaluation of the success of ongoing agency, and it also keeps up-to-date the organizing system to match the state of the world indicated by the feed forward from the receptor system.... So there is a meta-organizing system, which organizes the activity of the organizing system in the way that the organizing system organizes the activity in the outside world."[44]

A simple example of the role of the meta-organizing system, building again on the examples from mathematics classes, would be a student who decides to rebel and rejects the goal of getting right answers. This change in priorities completely changes the operation of the entire system, but may itself be revised if the environmental consequences make the new strategy too costly. And here we are imagining yet a higher level of evaluation, which is removed from the sphere of algorithmic cognitive processes and instead falls within the sphere of practical reasoning.

This puts us in position to consider an example of moral reasoning. In this case, just as in the simpler cases of arithmetic, the reasons (here moral reasons) can have top-down efficacy despite the presumed causal closure at the neurobiological level.[45]

The fight or flight response is pre-wired in humans, as in other animals. The ability to recognize threatening behavior is apparently built up easily from pre-existing perceptual capacities.[46] Thus, a typical series of events can be represented by figure 2, involving, say, perception of the behavior of another person (R); evaluation of the behavior as threatening (C); selection of response—fleeing,

---

[44] Ibid., 141 and 142.

[45] This is not to say that moral reasons are always effective; there are also rationalizations.

[46] Leslie Brothers, *Friday's Footprint: How Society Shapes the Human Mind* (Oxford and New York: Oxford University Press, 1997), 28–29.

fighting, conciliation—by the organizing system (O); and effecting the response (E). Feedback from the field of operation (F) will refine the actor's ability to choose responses that have the best survival value.

Now we complicate the picture by adding a moral supervisory system. Let us say that our agent is a pacifist. Now we need figure 4. The meta-comparator now places a higher value on nonviolent conflict resolution than on survival. The meta-organizing system then adjusts C's priorities accordingly. C's job will now be to evaluate threatening behavior not in terms of threats to survival, but in terms of threats to the peace of the community. A different repertoire of skills, maps, and norms will have to be developed in O. As this system develops, the FF path, which selects relevant features of sensory input, will be affected by action and reactions of the environment. G. Simon Harak points out that virtuous behavior effects changes in the agent's perceptions. As an example he recounts an incident from a time in his life when he was practicing martial arts. He commented to a companion on the threatening demeanor of a man they had just encountered. However, the companion, a seminarian dedicated to pacifist principles, had not perceived the stranger as threatening. So patterns of action, and thus patterns of readiness to act, gradually shape one's perceptions of reality.[47] This is a particular instance of Arbib's more general claim that action shapes one's schemas, which in turn shape expectations and perceptions.[48]

Notice that in MacKay's diagram (figure 4) there is feedback from the field of operation to the meta-comparator. This represents the fact that, in the case in question, the moral principle is subject to readjustment in light of the effects produced by acting in accordance with it. For example, it is often supposed that pacifist responses increase others' aggression. The pacifist might re-evaluate her commitment to this principle if this turned out to be true. So here we have a representation of the top-down efficacy of moral principles (as well as the environment) in shaping human behavior. This discussion is not intended to be an adequate account of moral reasoning, but only to show the possible downward efficacy of moral principles.[49]

5 Conclusion

Let me now try to tie all of this together. I began with Barbour's and Peacocke's account of the human being as a multilevel psychosomatic unity, but claimed that the threat of reductionism has not yet been adequately met. To pursue this goal, I suggested, first, that we redefine the concept of supervenience, which is used by philosophers of mind to give an account of how the mental relates to the neurobiological, and I emphasized two necessary ingredients. One is that supervenient properties be taken to obtain *in virtue of* subvenient properties. This builds into the very meaning of "supervenience" the sort of asymmetric property dependence that is wanted but not, apparently, explicable when the relation is defined in terms of material implication and modal qualifiers. Second, I suggested that *circumstances*

---

[47] G. Simon Harak, *Virtuous Passions: The Formation of Christian Character* (New York: Paulist Press, 1993), 34.

[48] Arbib, "Towards a Neuroscience of the Person," sec. 4.1.

[49] An earlier version of this essay contained a section on free will, in which I argued for a conception of freedom based on the possibility of self-transcendence, that is, evaluation of one's own cognitive processes. I have omitted the section here but will pursue it at greater length elsewhere (see fn. 19 above). I thank Michael Arbib and Thomas Tracy for their useful comments on it.

pertaining to the supervenient level of description will often codetermine the presence or absence of the supervenient property—this qualification thus needs to be reflected in the definition.

Concentration on the in-virtue-of relation has heuristic value in leading us to ask, of real properties in the real world, what kinds of dependence relations there are. I suggested that there are at least three that deserve attention: conceptual, micro-determinational (to coin a term), and functional. The functional sort turns out to be particularly useful, in part because it draws our attention to the additional circumstances at the supervenient or functional level that have to be taken into account—namely the rest of the causal system into which this part fits. So concern with the functional relation has also led, in this essay, to examination of cases widely recognized as instances of supervenience in which downward causation obtains.

In philosophy of mind and cognitive science, then, we can take mental events or schemas to supervene, in a functional sense, on neural processes. I used Donald MacKay's understanding of cognitive processes as information-flow feedback systems, and spelled out increasingly complex cases where we can see the top-down efficacy of epistemological (mathematical) and moral principles. That is, evaluative standards of an epistemological sort shape, by means of downward causation, the very structure of the individual's neural networks.

In her comments on an earlier draft of this essay, Leslie Brothers summarized my position so well that I end with a quotation from her remarks. She begins with an example that well illustrates my concept of supervenience and then shows how it works to account, first, for the nonreducibility of the social to the individual level and, second, for the top-down efficacy of higher-level mental events:

> Take the sound of a single musical note. (1) It can be reduced to sound-wave frequencies that causally give pitch. (2) If a component of a melody, the sound of the note will also have relations with the notes around it determined by the musical system in which the melody was composed. Standard supervenience applies to the note in isolation (so-and-so many cycles per second) and Murphy-supervenience to the note as a diminished seventh, which it is in virtue both of the sound-waves and the musical context....
>
> Think of persons as having two dimensions that intersect. The first is that they are loci of experience—for example, they can "have" pain. The second (I owe this to Rom Harré) is that they are locations in a moral-social order, an order woven together of shoulds and reasons and villains and *mensches* and so on. The actions of persons are accounted for, by themselves and others, in the context of this order. Now, if people like Francis Crick and Christof Koch are right, in the fullness of time we will causally reduce qualia to a feature of collective neuronal activity (personally I suspend judgment on this but let's give it to them for the sake of argument). This will be like reducing the note to its sound-wave frequencies. On the other hand, what Murphy refers to as "higher-level" mental events (deciding, judging, reasoning) will not be able to be extricated from the moral-social order to which they belong. Furthermore, I see no problem at all with downward neural causation here: if I choose to practice the golden rule by returning a lost wallet, I activate neural assemblies that represent the worried state of someone whom I might imagine to be like a friend or relative; I activate assemblies representing the person's joy when her wallet is returned, etc. By themselves, the neurons can only encode various states of the world, not a moral dimension. But moral behavior will select neural activity having to do with scenarios of giving something to someone, as opposed to scenarios having to do with spending the cash myself. It is a backhand, selective system of downward causation, but...[it] is what Murphy has in mind as "all we need or want for the purposes of blocking total causal reduction of the mental to the neurobiological."[50]

---

[50] I thank the conference participants for helpful comments on this essay, particularly Leslie Brothers, Theo Meyering, and Warren Warren Brown, my colleague at Fuller.

MIND MATTERS:
PHYSICALISM AND THE AUTONOMY OF THE PERSON

Theo C. Meyering

1 Introduction

Physicalism—or roughly the view that the stuff that physics talks about is all the stuff there is—has had a popular press in philosophical circles during the twentieth century. And yet, at the same time, it has become quite fashionable lately to believe that the mind matters in this world after all and that psychology is an autonomous science irreducible to physics. However, if (true, downward) mental causation implies nonreducibility and physicalism implies the converse, it is hard to see how these two views could be compatible. In this essay I review some classical arguments purportedly showing how the autonomy of the special sciences can be upheld without violating the laws of physics or the principle that physics constitutes a complete and closed system. I present these arguments in order of increasing strength, indicating how the more popular arguments in fact fall short of establishing anti-reductionism of the intended kind. I go on to add new arguments which I take to demonstrate quite effectively how downward causation is compatible with the reign of physics. To set the stage I begin my essay with a section in which I distinguish among various kinds of reductionism.

2 Reduction is Not a Unitary Concept

The standard textbook example of successful reduction is no doubt the nineteenth-century reduction of heat to motion, the reduction, that is, of phenomenological to statistical thermodynamics, and thus the reduction of thermodynamics to mechanics. Up to the nineteenth century thermodynamics attempted to describe the behavior of what was then thought of as a substance, heat (or caloric). Its principles accounted for such facts as heat transfer from hotter to colder bodies, the amount of transfer being dependent on the material constitution of the relevant bodies, etc. Once the atomic-molecular structure of matter was recognized towards the end of the nineteenth century, heat (or temperature) could be identified as the degree to which the constitutive particles of matter move or vibrate, that is, to the mean molecular kinetic energy of the molecules of the relevant substance. It thus became possible to unify what were formerly known as two separate disciplines and to reduce the "special science" of (phenomenological) thermodynamics to mechanics, giving rise to what was henceforth known as statistical thermodynamics.

During the logical-positivist era the ideal of the unity of science caught hold of the philosophical imagination: physics was considered the basic science, with the other sciences (chemistry, biology, psychology, and the social sciences) stacked up on top, each one dealing with mereologically more complicated fusions of physical events than its predecessor.

But such are the whims and vicissitudes of philosophical fashion that, having seen a climate of austere and bald reductionism in the era of positivist and analytical philosophy, current philosophical literature is now being swept by a wave of anti-reductionism, asserting no more than a very weak form of ontologically reductionist claims. In philosophy of mind, in particular, we have first seen purported *analytical* reductions proposed by phenomenalism and behaviorism, to be succeeded by the weaker *theoretical* reductions of the later generations. Then, starting with sporadic suggestions in the 1960s and 1970s the pendulum has now gone back virtually full swing, it seems, with claims of (radical or relative) autonomy of the special sciences in confident ascendancy. To paraphrase Frank Zappa: reductionism isn't dead, it just smells funny; such seems to be the philosophical mood at the present day.

In the following I will discuss a number of anti-reductionist arguments. However, reduction is hardly a unitary concept. It may be wise, therefore, to take stock and distinguish various kinds of reductionist positions in order of decreasing strength.

First, what the positivists labeled "physicalism" is in fact a radically reductionist view according to which all the sciences are ultimately reducible to physics. In the classical account of reduction proposed by Ernest Nagel, so-called "bridge laws" are required to connect higher- to lower-level laws, thus enabling the derivation of higher- from lower-level theories.[1] I will designate this logical positivist conception of reductionism *radical* (or *industrial strength*) *physicalism*, to be contrasted with two distinct kinds of milder versions of physicalism, *ideal* (or *regular strength*) *physicalism* versus *mild* or *token physicalism* (pun intended).

Ideal physicalism, as opposed to *radical physicalism*, allows irreducibility only for pragmatic or epistemological reasons, insisting meanwhile that in principle every event or fact that has a scientific explanation can be explained within some theory of an ideally complete physics. Thus, ideal physicalists maintain that even though reduction may be only in the mind of God, it does not cease to be honest-to-God reducibility for all that. As we will see, to make room for this version of physicalism—and to see it as distinct from both *radical physicalism* on the one hand and *token physicalism* on the other—may bring serious critical force to bear on those overly confident *non*reductionists who base their argument solely on the fact that special science properties are multiply realizable by "wildly disjunctive"[2] physical properties lacking sufficient perspicuity or explanatory coherence. While these so-called "nonreductive materialists" surely *expect* their argument to establish a robust form of anti-reductionism, their position may in fact be a mere planet within the strong gravitational field of radical physicalism. That is to say, nonreductive materialism may be just a little removed from the mainstream of traditional reductionism, with no more than a slight epistemological twist added on. Put briefly, their argument runs as follows: While every particular mental *event* may be (*token-*) identical to a particular physical *event* (thus allowing an ontological physicalism of sorts), yet mental *properties* need not be (*type-*) identical to physical *properties* (thus thwarting reductionism, or *radical physicalism*). No doubt the argument aspires to establish its own position as a species of objective anti-reductionism. Yet, as I will argue below, multiple realizability alone supports no more than a very tenuous form of anti-reductionism better classified as *regular strength physicalism* than as a genuinely anti-reductive species of *token physicalism*.

Mild or *token physicalism* then, in its turn, constitutes a radically weaker form of reductionism than *ideal* or *regular strength physicalism*, inasmuch as it allows events to have no *physical* explanation at all, not even within some ideally complete physics. Yet, *token physicalism* is nevertheless physicalist and (ontologically)

---

[1] Ernest Nagel, *The Structure of Science* (London: Routledge and Kegan Paul, 1961), 364.

[2] Jerry A. Fodor, "Special Sciences," *Synthese* 28 (1974): 77–115.

reductionist insofar as it holds every event to be token-identical with some physical event or other, as I have indicated above.

Finally, then, this token-identity thesis may be weakened even further when it is allowed that a given (higher-order) token-event may be differently *composed* out of (token-different) physical particulars. Consequently the above thesis of token-token identity of the events subsumed in the explanations of the various (higher- and lower-level) sciences can no longer be maintained. Thus, it is "immaterial" (to mix metaphors) to the integrity of a particular mental-state-token as being the mental-state-token that it is (or as being the mental state that I am in now) whether or not it is realized by, for example, a particular sodium ion crossing a certain synaptic cleft somewhere in my parietal cortex here and now, even though that very same mental-state-*token* would have been *physically* differently composed had it not. Or, to give a slightly different example, a physicalist would no doubt have to hold that, say, a stock market crash is in point of fact *constituted* by innumerable physical particulars in all sorts of states and interactions. Yet he is thereby not at all committed to the implausible view that this gigantic collection of wildly heterogeneous physical particulars has enough physical integrity to count as a *physical* event, which also happens to be describable as an economic event (of rather disastrous proportions). That is to say, it is an economic event *composed* of physical particulars, but it is not a physical event in any meaningful sense of the word. Thus *compositional physicalism* or *milder than mild materialism* holds that even though all the entities subsumed in the explanations of the various special sciences are physically *constituted*, no single pre-existing set of physical events is identically available for all of the sciences: different special sciences carve up the world in a way sufficiently different from that of physics for there to be room for *ontological pluralism*. Even so, for *compositional physicalism* the tie with the ideal of the unity of science—though delicate—is still intact, for the ontological and causal independence of the nonphysical facts is still constrained by their physical constitution.[3]

In the next section I will propose and discuss several arguments that may be put forward against reductionism and in favor of the relative autonomy of the special sciences, in particular of psychology as the science of the person. Another way to express the same point is in terms of the notion of macro-explanation. The issue at stake, then, is whether macro-explanations as furnished by the special sciences have a distinctive and irreducible role to play, or whether they are of no more than merely instrumental value, reflecting at best our epistemological needs or our ignorance, but not some objective causal order in the nature of things.

3 The Legitimacy of Macro-Explanations and the Relative Autonomy of the Special Sciences

There are, I believe, at least five reasons to retain and to positively value macro-explanations in general. I will present them below in order of increasing strength and effectiveness, the earlier ones falling short of the anti-reductionist ambitions they are meant to satisfy—or so I will argue.

---

[3] Cf. Graham MacDonald, "Reduction and the Unity of Science," paper presented at the OSW Conference on Reductionism, Oisterwijk, The Netherlands, 1998.

3.1 The Argument from Ignorance of Ultimate Causes

The first argument I wish to consider in support of the need and the positive value of macro-explanations points to the fact that such explanations are obviously useful in cases where we want to explain and predict, but have no more than a rough indication of where to locate the relevant cause and how to specify it. In some such cases our explanations may turn out to be genuinely causal after all, but perhaps only lacking in the fineness of their grain. These are philosophically innocent cases. Thus brewers in the nineteenth century used yeast without the foggiest notion of enzymes being the true causal mechanisms responsible for fermentation. Or sailors would consume lime juice and brown rice to prevent scurvy or beriberi, with no idea whatsoever of vitamins being the actual causal agents sustaining their good health.

However, such explanatory practices are not without risk. In the absence of detailed knowledge of the true causal agents we might be latching on to spurious correlations. Thus the proffered macro-explanation may not be genuinely causal at all, but may turn out to be based on a pseudo-process, as if we were to explain the current position of the shadow of a car on the shoulder of the road by reference to its earlier position and its velocity. Similarly perhaps—or so it has been argued— mental causation, for all we know, may be nothing but the shadow of physiology (in ignorance we are all captives in Plato's cave, mistakenly taking shadows to be true causal agents). Thus, Jaegwon Kim, for one, does believe that all *macro*-causation is epiphenomenal, including in particular mental causation. This, in a nutshell, is the upshot of Kim's doctrine of "supervenient causation."[4] In Kim's view there exists a macro-causal relation between two events only in the case where there is a micro-causal relation between the two events upon which they supervene. Clearly such a definition of supervenient causation renders all macro-causation epiphenomenal. And since mental causation is a species of macro-causation, the way the mind matters in this world is epiphenomenal just as well.

3.2 The Argument from Multiple Realizability—Or the Autonomy of the Special Sciences

While the argument from ignorance of ultimate causes appears to have some force *vis-à-vis* (radical) *reductionism*, it does not have nearly enough force to establish the *autonomy of the special sciences*. By contrast, the next argument to be considered is animated by the ambition to accomplish just that philosophical feat. This rather influential argument, made popular by early functionalists such as Hilary Putnam and Jerry Fodor, argues in favor of the autonomy of the special sciences and of the legitimacy, indeed indispensability, of macro-explanations in science on the basis of the so-called *multiple realizability* of special science properties. In particular, there cannot be any type-type identities between mental and physical properties, nor is there room for any strict psycho-physical laws so as to reduce psychology to neuro-physiology and ultimately to physics.

Thus, in a very general argument Fodor has argued that there cannot be any so-called bridge laws of the sort required by Nagel's classical account to bring about the reduction of any special science to physics, since the predicates of a given special science will only map onto predicates of physics that are at best *"wildly disjunc-*

---

[4] Jaegwon Kim, "Epiphenomenal and Supervenient Causation," *Midwest Studies in Philosophy*. 9 (1984): 257–70.

tive."[5] Take for instance the concept of monetary exchange, which figures in laws of economics such as Gresham's "law." Let us assume, for the sake of argument, that any instance of a monetary exchange is also an event truly describable in the vocabulary of physics and in virtue of which it falls under the laws of physics. Yet commonplace considerations suggest that a physical description which covers all such events will be wildly disjunctive. As Fodor points out, some monetary exchanges involve strings of wampum, some involve dollar bills, and some involve signing a check or (extending Fodor's examples) striking keys for a PIN code.

> What are the chances that a disjunction of physical predicates which covers all these events (i.e., a disjunctive predicate which can form the right hand side of a bridge law of the form "x is a monetary exchange ↔ ... ") expresses a physical kind? In particular, what are the chances that such a predicate forms the antecedent or consequent of some proper *law* of physics? The point is that monetary exchanges have interesting things in common; Gresham's law, if true, says what one of these interesting things is. But what is interesting about monetary exchanges is surely not their commonalities under *physical* description. A kind like a monetary exchange *could* turn out to be coextensive with a physical kind; but if it did, that would be an accident on a cosmic scale.[6]

However, who says (apart from the logical empiricists) that the ideal of the unity of science may only be realized by exceptionless mappings of the natural-kind predicates of the special sciences onto those of physics in terms of biconditional bridge laws? Surely there are more ways of achieving intelligibility besides austere semantic connectivities as in the classical construal. Thus, on Fodor's view, all we expect

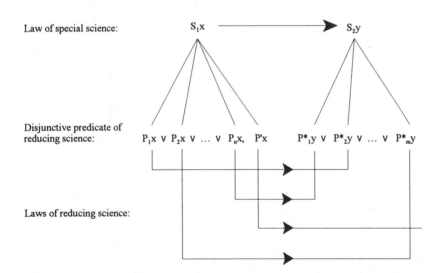

Figure 1. Schematic representation of the proposed relations between the reduced and the reducing science on a revised account of the unity of science. If any S_i events are of the type P', they will be exceptions to the law $S_1x \rightarrow S_2y$.

---

[5] Fodor, "Special Sciences."

[6] Ibid. Reprinted in Jerry A. Fodor, *Representations: Philosophical Essays on the Foundations of Cognitive Science* (Cambridge: MIT Press, 1981), 134.

from reduction is "to explicate the physical mechanisms whereby events conform to the laws of the special science."[7] There is no reason why heterogeneous physical mechanisms which happen to play a similar causal role could not furnish that explication from case to case, thus removing mystery by providing understanding in physical terms, which is the real goal of scientific reduction. Thus Fodor suggests a liberalized version of the relation between physics and the special sciences, in which bridge statements only express token event-identities but no correspondences between the kind-predicates of the reduced and the reducing science (see figure 1).[8] This view, though weaker than what standard reductionism requires, squares well with the ontological position of *token physicalism*, which is all that is required for the unity of science, according to Fodor.

Yet, however cogent this influential argument may seem to be, it is quite possible to shoot holes into its seemingly impenetrable walls. First of all, the argument may be taken to undermine no more than the *epistemological* case for reductionism, but not the *metaphysical* case. After all, the argument has only pointed to the fact that the natural-kind predicates of the special sciences are most likely to map onto predicates of physics that are "wildly disjunctive" and thus *humanly* unmanagable and unperspicuous. But this is far from showing that there really are no type-type identities, or that the special sciences cannot be exhaustively reducible in the mind of God, so to speak. As Jaegwon Kim points out, the argument allows that every mental property M, say, has a (possibly infinite) series of physical properties $P_1, P_2, \ldots$, each of which is sufficient for M. Consider then the union (disjunction) of these physical properties, $\cup P_i$. It is easily seen that $\cup P_i$ is a necessary coextension of M. Clearly, there always exists the possibility that the $\cup P_i$ may some day be represented by a perspicuous description in an appropriate scientific theory.

> Even if this never happens and, indeed, cannot happen given our cognitive capacities or inclinations, why isn't the demonstrated existence of $\cup P_i$ for M, and similar physical coextensions for all other mental properties, sufficient to show the *metaphysical reducibility* of mental properties?[9]

Referring back to the taxonomy of versions of physicalism of different strength proposed above, the argument from multiple realizability may thus be said to have established no more than the failure of *industrial strength physicalism* along the lines set forth by the logical positivists. However, that kind of radical physicalism would have struck most philosophers as unrealistically strong to begin with. What the argument from multiple realizability has failed to do, however, is deliver a robust version of anti-reductionism conferring truly objective autonomy to the special sciences. Instead it has yielded what I have labeled *regular strength physicalism*, which is nonreductionist only in a rather lame sense of the word.

That all of this is not just idle philosophical speculation may be shown by pointing to the ongoing dispute started by Philip Kitcher and Alexander Rosenberg over the reducibility of classical genetics to molecular genetics. Thus Kitcher has argued that classical genetics does not reduce to molecular genetics via bridge laws in cytology in view of the immensely heterogeneous character of the molecular

---

[7] Fodor, *Representations*, 138.

[8] Ibid., 139.

[9] Jaegwon Kim, "Supervenience," in *A Companion to the Philosophy of Mind*, S. Guttenplan, ed. (Cambridge: Blackwell, 1994), 575–83, esp. 580.

realizations of cytological entities and processes such as genes or meiosis.[10] The conclusion Kitcher has been eager to reach is that the autonomy of classical genetics is not a merely temporary reflection of our scientific imperfection, but rather a reflection of different levels of organization objectively existing in nature. Against this Rosenberg has maintained that Kitcher's arguments for the purported non-transitivity of explanation between classical and molecular genetics via cytology fall far short of entailing the intended conclusion. Kitcher's arguments are pragmatic arguments at best, arguments which may therefore support no more than an instrumentalist conclusion.[11] Thus, in Rosenberg's view, biology is of merely practical or instrumental value, and may be subsumed in some ideally complete physics by a purely molecular theory of life.

Finally, I wish to point out that even the parade case of scientific reduction I have cited at the beginning of this essay—the reduction of heat to motion—ironically turns out to be a prime example of a phenomenon that is multiply realized! Thus temperature in gases differs in physically important ways from that in solid bodies (where the interconnected molecules are confined to a variety of vibrational motions), which is different again from heat in a plasma (which has no constituent molecules), or in a vacuum (which has a so-called "blackbody" temperature in the distribution of electromagnetic waves propagated through it). Accordingly, multiple realizability as such is no proof of nonreducibility. What it may show at best is that reduction must be domain-specific, not that it is impossible.

3.3 Functional Explanation in View of Multiple Realizability

As we have just seen, the functionalists' orthodox argument from multiple realizability—which received virtually universal acclaim among philosophers, as it seemed to open a middle position of nonreductive materialism, perched neatly in between the equally untenable horns of Cartesian dualism on the one hand and radical reductionism on the other—may at the end of the day turn out to have only limited validity as an argument for nonreducibility. However, the functionalists did more than just point to the *multiple realizability* of mental states. They also stressed the related point that mental states are essentially *functional* states. That is, mental states are relational states defined by their typical causal role rather than being defined by the states that happen to realize that role. Originally most functionalists tended to be so-called "narrow" functionalists (or individualists), seeing mental states as relations between inputs, other mental states, and outputs *within the head* of the subject. More recently so-called "externalists" have urged that mental states are in fact codetermined by contextual relations to the (social or physical) environment or to the history of the subject.[12]

It is here, where our explanatory preference is driven not so much by ignorance of ultimate causes (as in the first argument considered), nor by the apparent

[10] Philip Kitcher, "1953 and All That: A Tale of Two Sciences," *Philosophical Review* 93 (1984): 335–73.

[11] Alexander Rosenberg, *The Structure of Biological Science* (New York: Cambridge University press, 1985); idem, "From Reductionism to Instrumentalism?" in *What the Philosophy of Biology Is: Essays dedicated to David Hull*, M. Ruse, ed. (Dordrecht: Kluwer, 1989), 245–62.

[12] Nancey Murphy builds parts of the argument in her essay in this volume on the externalist character of *moral* properties.

unmanageability of realizer-state predicates corresponding to the relevant special science kind predicates (as in the second argument considered), but rather by a well-motivated positive need for *functional abstraction* and for *relational explanation in terms of typical causal role* that we may succeed in finding much better and more robust reasons for upholding the indispensability and (relative) autonomy of macro-explanations. Typically, in the case of such functional macro-explanations, the real causal agent may be a member of a disjunctive class, each of which would have had the same result and all of which share a second-order functional property, which is thus rightly singled out as the causally relevant factor. Thus if you have a headache, you are well advised to take *some* familiar analgesic (it doesn't matter very much which one) rather than to look specifically for acetylsalicylic acid (the stuff contained in aspirin). Our advice would be guided by the insight that any old pain killer would do. The corresponding explanation, say, of the relief of the headache in terms of the functional property of the pill as being an analgesic is what Frank Jackson and Philip Pettit call a *program* explanation,[13] to be contrasted with a *process* explanation, which would cite the actual chemical agent triggering the "pain relieving" program at a particular occasion. In other words, the functional property "programs" for a specific causal route without itself being the *triggering* cause of the corresponding causal process at any given time. Thus one or another member of a large disjunctive class of triggering causes may be actually operative on any given occasion, yet from the broader perspective of the functional explanation it is entirely immaterial which particular trigger causes the program to unfold.

By contrast, a *process* explanation of the same phenomenon would cite, as I have indicated, the actual chemical agent (the acetylsalicylic acid, as the case may be) operative in the given case. Similarly, in explaining why poor Jimmy's favorite toy car got crushed, one would rightly point to the stampeding herd that flattened it, even though in point of actual fact it was trampled upon by cow Clara. "Any old cow would have done!" one might rightly claim in justification, "Whether it was cow Clara or cow Ophelia that actually committed the brutal act is neither here nor there." Which is to say, the program explanation has *distinctive* explanatory merits, not matched by the corresponding process explanation, even though it may seem (but seem only!) that the latter (the process) and not the former (the program) cites the entity or process doing all the actual work. I hasten to add that this latter impression (of the process agent doing all the real causal work) is as yet entirely controversial and will be refuted, I hope, in the course of my essay. This point of the *distinctive* explanatory work accomplished by program- or macro-explanation may be borne out by the following two considerations.

First notice that just duplicating the *process* cause would not be duplicating the relevant causal facts. If we imagine cow Clara smashing the toy car just as she actually did, yet this time on her own rather than as a member of the stampeding herd, we would probably consider her to be alarmingly vicious, or just mad; she might just suffer from Mad Cow Disease! That is to say, the relevant causal roles implied by the process explanation are vastly different from the one suggested by the macro-explanation under consideration.

Second, notice in conclusion that even an explanation in terms of Clara's actual behavior *enriched with relevant environmental detail* (such as there being some

---

[13] Frank Jackson and P. Pettit, "Functionalism and Broad Content," *Mind* XCVII/387 (1988): 381–400; idem, "Program Explanation: A General Perspective," *Analysis* 50 (1990): 107–17.

neighboring cows running at the same speed) would still be a poor substitute for the macro-explanation couched in terms of the *stampeding herd*. For such an explanation in terms of two events accidentally operative at the same time (namely, the event of Clara running as well as the event of a bunch of bovine brothers and sisters running in synchrony in her immediate neighborhood) would not yield the relevant counterfactual insight that had the toy car not been crushed by Clara, it would still have been flattened by some other bovine member of the panicking herd.

All of this is of vital importance to the methodology and the philosophy of the cognitive sciences. In a functionalist theory of mind, at any rate, mental states are, as I have indicated, essentially characterized by the fact that a certain causal role gets filled, regardless of what it is that fills the role. So what a functionalist theory of mind requires for psychology is *program* explanation, not process explanation.

The general point here is that process explanations, in spite of appearing to give *more* detailed information may, on the contrary, be *less* informative than the corresponding program explanation couched in terms of more general functional properties. Process explanations are often lacking in what Robert Wilson has called "causal depth."[14] As a result they are not resilient across slight counterfactual changes as program explanations are. Thus the toy would have been crushed even if Clara had not stepped upon it. And the headache would have subsided even if a different chemical agent with the same analgesic powers had been administered. The weakness of process explanations is precisely that they are what Alan Garfinkel, in the context of explanations in the social sciences, has called *hyperconcrete*.[15] These considerations demonstrate the *distinctive* explanatory role of functional explanations and thus their indispensability over and above what process explanations may have to offer. They thus count against Kim's theory of supervenient causation with the attendant charge of epiphenomenalism in the case of mental causation.

This point is important enough to bear repetition. Given that program explanation is a species of *macro*-explanation, then if macro-explanation were to reflect nothing over and above mere *supervenient causation*, reducible, as Jaegwon Kim has argued, to the "real" causal relations holding between relevant subvenient events, mental causation would be merely epiphenomenal, a species of *noblesse* with nothing to do, in Samuel Alexander's words.[16] Thus psychology would be stripped of any independent explanatory force—such is the challenge presented by Kim's argument. However, what I hope my argument from functional explanation has added to Kim's overly simple model of supervenient causation is the vital insight into the *distinctive* causal role played by supervenient entities or processes. As a result, explanations in terms of supervenient causal relata have an explanatory force unmatched by, *and irreducible to*, any possible explanation in terms of the processes (or realizer states) involved.

3.4 Functional Explanation in View of Multiple Supervenience

The foregoing considerations in one way or another were based on the *multiple realizability* of the macro-properties by relevant lower-level properties, either

---

[14] Robert A. Wilson, "Causal Depth, Theoretical Appropriateness, and Individualism in Psychology," *Philosophy of Science* 61.1 (1994): 55–75.

[15] Alan Garfinkel, *Forms of Explanation: Rethinking the Questions in Social Theory* (New Haven, Conn.: Yale University Press, 1981).

[16] Samuel Alexander, *Space, Time, and Deity*, vol. 2, 2nd ed. (London: Macmillan, 1927).

directly, as when I pointed to the unmanageability of the heterogeneous realizer states (section 3.2), or indirectly as when I emphasized the explanatory irrelevance of the (indefinitely varying, multiple) *realizer* states *vis-à-vis* the *role* states (section 3.3). However, even more powerful positive arguments in favor of the indispensability of macro-explanations advert to *multiple supervenience* rather than to multiple realizability. An analogy with dispositional explanation may illustrate the point.

Dispositions, just like macro-properties, fail to produce causal effects on their own independently of their categorical base. And yet their explanatory power clearly differs from, or exceeds, that of their bases. This becomes intelligible when we recognize that one and the same categorical base realizes more than one disposition. Even so, only one of those is usually relevant for a given event. Thus Mary's death is related to the electrical conductivity of her aluminum ladder. But the categorical base thereof (the cloud of free electrons permeating the metal) also realizes such diverse dispositions as the thermal conductivity or the opacity of the metal.[17] Clearly, a proffered explanation for the tragic accident that cited only the underlying categorical base would be incomplete or even misleading but in any case, *objectively* less informative than one that cited just the relevant disposition. Moreover, the accident would have occurred even if another categorical base had realized the same disposition of electrical conductivity. Thus, this disposition is apparently relevant and its actual categorical base is causally efficacious. Together this constitutes good reason for calling the disposition in question the relevant causal factor.

This analogy may also cast light on why functional explanations cannot be reduced to causal explanations. The reason constitutes the mirror image of the phenomenon of multiple realizability, often cited in such contexts, because the irreducibility of dispositional explanations clearly does not consist in the fact that a given disposition can be realized by otherwise unrelated categorical bases, but precisely in the opposite fact that a given categorical base does not identify which of the dispositions realized is explanatorily relevant in a given case. In other words, it is not multiple realizability but rather *multiple supervenience* of dispositions onto one and the same subvenient categorical base that necessitates citing *that* disposition, which is responsible for the given effect. Thus the special character of higher levels of organization in nature can be vindicated in principle, and more effectively than by the conventional appeal to multiple realizability. They necessitate the need for macro-explanations with the causal depth and the theoretical appropriateness corresponding to the grain of the explanatory level in question.

It is extremely important to appreciate the distinctive force the *argument from multiple supervenience* marshals with respect to the explanatory power of macro-explanations as compared to that conferred by the previous argument based on considerations having to do with *multiple realizability*. Even though both arguments point in the same direction of the distinctive explanatory power furnished by macro-explanations, they put their fingers so to speak on quite different aspects of the use and the significance of functional explanation. And of the two arguments, I submit, the argument from multiple supervenience holds the greater promise of cogency and effectiveness in at least two important respects. Not only will it supply a philosophically satisfactory account of how a relationalist psychology may yield *bona fide*

---

[17] The example is due to David Lewis; cf. P. Menzies, "Against Causal Reductionism," *Mind* XCVII (1998): 551–74; also cf. Frank Jackson and P. Pettit, "Causation in the Philosophy of Mind," *Philosophy and Phenomenological Research* L, Supplement (1990): 195–214.

scientific explanations without falling victim to action-at-a-distance obscurantism, but it will also prepare the transition to a *systems account of downward causation* without falling victim to the mystifications of traditional emergentism. Let us therefore compare the two arguments each in their own turn.

According to the argument from multiple realizability the reason why we have to resort to macro-explanation is that even though on any given occasion the realizer state "fixes" the functional state involved in causing a particular effect, it is not *essential* that the base state have taken place for the effect to occur; any other role-realizing state would have accomplished just the same job. So, according to this argument, it has got to be the role-state, not the realizer-state, that is causally salient. If *it* had not occurred, the effect would not have taken place either.

However, the argument from *multiple supervenience* goes on to point out that generally speaking it just is not true that the realizer state as such suffices to "fix" the causal sequence of events, *not even in a particular case*. Very often, most certainly in the case of mental causation (which is essentially a relationalist affair), the realizer state fixes a given causal path *only in conjunction with certain "broader" contextual facts* pertaining to the effect in question. These contextual facts prompt a given realizer state to play a *distinct* causal role different from the role that very same state might be playing in relevantly different contexts. So in the case of multiple supervenience, when we single out the macro-event as the true causal agent, we are in fact taking into account the causal nexus of the explanandum in a much richer systemic way, one that includes relevant initial conditions, such as the ladder being in contact with an electrical wire rather than its having been exposed to the heat of the sun for a while. These initial conditions in their turn selectively activate one particular disposition rather than another. As a result, whatever thus gets activated constitutes the next causal component of the causal trajectory which, in the case of Mary's death, terminates in the unfortunate accident. Here we have macro-causation with a vengeance, for *the very same realizer state* (a particular cloud of free electrons) might have been involved, under different circumstances, in causing Mary to burn her feet, as the exposure of the ladder to the heat of the sun might have activated the *thermal* conductivity, not the electrical conductivity, of the aluminum it is made of. Thus the actual realizer state is not merely inessential because a different state might have realized the very same causal role; rather, it is inessential because *the very same realizer state* may yield a wide range of very different causal trajectories. Just citing the occurrence of the realizer state would thus leave the actual causal sequence of events entirely indeterminate. Conversely, the electrical conductivity *is* essential for the effect to occur. If *it* had not happened, the initial conditions would not have led to the fatal accident.

3.5 System Explanations and Downward Causation

The foregoing considerations naturally lead us to a fifth and final category of cases, for the example we have used to illustrate multiple supervenience may profitably be regarded as a micro-system (albeit not one with a particularly intricate degree of organization). It traces a salient causal path from a relevant initial condition, through the dispositional state selectively activated by it, to the eventual effect. This, I submit, can be used as an exemplar for what is the common object of research in the so-called special sciences. What gets studied in the special sciences is in fact huge systems of concatenated micro-systems which are systematically organized in such a way that their typical causal antecedents prompt typical patterns of causal processing to eventuate in typical effects, which in their turn serve as typical inputs for yet other causal sequences of events to take place. Regimented in this way the system produces emergent effects that have no salience at the level of physics and yet constitute the preconditions for the recurrence of the sequence in question or for the emergence of related processes that are significant at that same level of special science description. Thus, for example, the process of meiosis (whereby paired entities are split up by a force in such a way that each member of each pair gets assigned to a descendant entity) is a pair-splitting process involving, at each instance, myriad different molecules and molecular processes. However, the bewildering molecular details of the various mechanisms and processes involved obfuscate rather than illuminate the relevant cytological factors at work. That is to say, the pair-splitting process is a natural kind only at the level of cytology, not at the level of molecular biology. To be sure, each particular instantiation of the meiotic process is far from miraculous from the point of view of molecular biology, yet molecular biology as such lacks the conceptual resources to illuminate, let alone explain, the pattern of events involved. This is because the causal regimentation of the physical processes involved includes broad contextual factors exercising downward control on whether a potential causal pathway gets activated or not. The result is a virtually synergistic orchestration of forces and processes yielding characteristic effects which are significant only at the level of description proprietary to cytology. Thus it was *meiosis* that caused a particular cell to reproduce, not an unorganized random aggregate of molecular events. And the causality expressed in that statement constitutes genuine causality as betokened by the fact that it is supported by true counterfactuals corresponding to it. For if genes had not been redistributed in the way prescribed by meiosis, no reproduction would have taken place.

Indeed we can take this one step further. Even events much closer to the level of physics, such as earthquakes, should be regarded as patterned sequences of events instantiating a natural kind not of physics, but rather of geology or seismology. Combining the innumerable physical events taking place during a large-scale catastrophic accident into the single category of an *earthquake* is characterizing a sequence of events far exceeding the actual physical events taking place at the time of the calamity. Among other things it is taking into account such antecedent structuring causes having to do with continental drift, with the emergence of faults, and with the gradual build-up of pressure-imbalances in the crust of the earth along the lines of friction. A mere duplication of the physical events actually occurring at the time of the earthquake, yet this time, as the case may be, as part of the gravitational implosion of a celestial body, say, *without duplicating the antecedent structuring causes characteristic of earthquakes* would not amount to an *earthquake* inasmuch as it would fail to exhibit the relevant causal patterns typically operative in earthquakes in general. (Compare: merely duplicating the cloud of free electrons and Mary toppling off the metal ladder would not amount to a tragic case of electrocution unless the causal pattern characteristic of *electrical conductivity* had been selectively activated by some causally relevant antecedent condition. With no further specification of the dispositional state actually operative the duplicate case would remain woefully indeterminate regarding which causal work had actually been accomplished).

4 Conclusion

I believe the model of multiple supervenience I have presented goes a long way toward showing that macro-explanations are not only indispensable for practical and epistemological purposes, but also truly grounded in the causal structure of their respective ontological domains. Thus we are not reduced to assigning merely instrumental value to macro-explanations. Nor, conversely, do we set our philosophical standard too high if we demand of macro-explanation what we are intuitively inclined to demand of explanation *tout court*, namely, that it is justified only to the extent that it derives its explanatory force from a true representation of the causal powers actually operative in the relevant domain. Thus macro-explanations are justified only insofar as they reflect genuine macro-*causal* relations truly operative in the causal patterns exhibited by nature at a multitude of distinct organizational levels. In fact, physics in this respect is hardly less "special" or more innocent of macro-explanation than the special sciences so-called. However, this fact in itself would have spelled little philosophical comfort if it had not been possible to defend macro-explanation *independently*—along lines such as I have tried to set forth in this essay—for this makes it possible to defend macro-explanatory practices in psychology and elsewhere without having to plead not-guilty merely by association (namely, with that unimpeachable paradigm of science: physics), or to plead guilty, but then guilty in company too good to feel overly worried.

But there is more. Not only has this analysis shown that macro-causation is a real force in nature at multiple levels of existence, it also has lent us insight into the nature of *downward causation*. To believe in downward causation is not tantamount to the belief in brutely emergent fundamental laws (proper to a certain level of intricately organized systems of physical events and processes such as organisms or minds), with concomitant causal interference at lower levels of organization *in violation of the laws of micro-physics*. This in effect was what the British emergentists such as C.D. Broad and Samuel Alexander believed and defended in the first part of this century. Thus our analysis has made progress: *pace* Kim,[18] downward causation may be assigned a stable place in our picture of how the world is organized without upsetting our conception of the various domains of physics as constituting a closed and complete system of physical events at the physical level of description. Reflecting upon the phenomenon of multiple supervenience we have come to realize that organizing principles are themselves genuine causal factors actually operative in channeling and orchestrating the quantum flux of irreducibly statistical events to yield stable recurrent patterns that are, as a result of the systemic organization of their parts, self-sustaining or self-reproducing.

---

[18] Jaegwon Kim, "The Nonreductivists' Troubles with Mental Causation," in *Mental Causation*, J. Heil and A. Mele, eds. (Oxford: Oxford University Press, 1993).

III. FROM SCIENCE AND PHILOSOPHY
TO CHRISTIAN ANTHROPOLOGY

Anthropological & Theological Issues:

Philip Clayton

Arthur Peacocke

Ian G. Barbour

Stephen Happel

Ted Peters

Neuroscience & Religious Experience:

Fraser Watts

Wesley J. Wildman & Leslie A. Brothers

NEUROSCIENCE, THE PERSON, AND GOD: AN EMERGENTIST ACCOUNT[1]

Philip Clayton

1 Introduction

1.1 What Theologians Seem to Want

Much could be (and has been) said about neuroscience from the perspective of theology; theologians (in this volume and elsewhere) have not been shy about offering their own interpretations of brains, thoughts, and spirits. Of course, turnabout is fair play: neuroscientists have also not been shy in commenting on theology, God, free will, and other concepts both human and divine. After all, it is in part an empirical question why a brain with a structure like ours—one that shares our evolutionary history, stores and retrieves information as ours does, and responds like ours to electrochemical stimulation and viruses—would produce religious ideas and religious experience as ours does.

As fascinating as such disputes might be, they are not the subject of this essay, at least not directly. I am convinced that direct battles between neuroscience and theology (or, for that matter, direct concordances) will not even become conceivable until a new, deeper mediation between the two fields has been achieved. The neurosciences raise a question much closer to home than disputes about God: the question of who *we* are. Progress in neuroscience challenges, or at least is often taken to challenge, cherished notions of what it is to be a human person: self-consciousness, soul, "thinking being," free will. Unless and until we manage to defend a notion of the person that preserves concepts such as these *in light of* what we now know about the human brain, language about God, and any work such language is supposed to do within the human psyche, will appear gratuitous.

This is not to say that theological doctrines cannot be of any assistance.[2] Some essays in this volume describe theological doctrines that impinge, directly or indirectly, on work in the neurosciences and the social sciences. Imagine, for example, how one's notion of personhood might be affected by acceptance of the Christian doctrine of creation, the belief that we are "dust" and yet nonetheless indwelt by the Spirit of God. Divine creation would introduce purpose, for example, and purpose means an arrow of time—the belief that, besides the brute-fact given of the natural world, there is the directing influence of an unconditioned will who is both source and *telos* of this world. Christian theologians then supplement creation

[1] The brilliance of the Vatican Observatory-CTNS research model lies in the sustained collaboration and criticism that it spawns among the authors. I owe more debts of this sort than I can list here. Arthur Peacocke and Tom Tracy must nonetheless be singled out for particularly supportive comments and helpful criticism (respectively). The volume editors generously provided suggestions for improvements; Theo Meyering and Nancey Murphy in particular wrote detailed commentaries which resulted in valuable additions and crucial changes during the final rewriting.

[2] I am grateful to Mark Richardson for discussions of the following theological constraints on the theory of personhood.

beliefs with content from the biblical texts (and their tradition(s) of interpretation), which are taken to provide an important record of divine communicative intent. When the theologian has this much, she already has huge constraints on her theory of personhood. She has, implicitly, a Christian ethic of "willing the will of God." Moreover, the biblical texts speak of *covenant*, which implies a set of divine commandments for living—but also mutual agreement and responsibility (hence ethics again). Covenant gives rise to the notion of sin, which Thomas Tracy defines elsewhere as "the bias of the will toward an orientation of alienation from God."[3] The idea of sin then gives rise to the idea, and thus the possibility, of reorientation. Possibility is taken to have become actuality through divine initiative or grace.

I include this well-known list to show how detailed is the anthropology (theory of human nature) that theologians bring to the discussion table. (Note that "detailed" does not entail "nonnegotiable.") It includes, at least for traditional theologians, not only the existence of at least one purely spiritual being—hence the possibility of disembodied agency—but also the notions of will and of freedom, which come in both finite and infinite flavors. With will, so understood, comes consciousness: Christians conceive of God as conscious agent, an agent enough like human agents that the predicate "person" can also be attributed, if only in an analogous fashion, to the divine. On this view, humans and God are also *moral* agents; persons exercise their agency in light of real obligations to other persons (indeed, to the world as a whole) and to God. Finally, these agents are *social* agents. Religious notions of *community* emphasize a union among humans in light of the divine presence and the covenant which makes of us "one." The Christological expression of community is *kerygma*, the particular story of Jesus Christ, culminating in the belief in an eschaton or second coming. In short, the Christian parameters for talking about persons stretch from the moment of creation to the consummation of history, and from individual birth through life and death and on to the hope of a final reconciliation.

1.2 What Neuroscientists Know

How does theological anthropology contrast with recent results in the neurosciences? To get a sense of the sort of data and emphases, consider just a few of the recent findings on the physiological bases of human cognition, which offer a representative sample of the sorts of connections that have now been established:

- Recent advances in genetic research have led to a better understanding of how sensory receptors function to convey sensory inputs from the environment to the brain. By comparing the genetic constitution of the receptors (for example, receptors for smell and taste) from many different species, we are beginning to understand how sensory receptors first evolved and subsequently became more highly specialized in higher primates and eventually in *Homo sapiens*.[4]
- Studies involving the prefrontal cortex of humans and monkeys have led to a deeper understanding of short-term memory. Using implanted electrodes and PET scanning, specific areas of the prefrontal cortex have

---

[3] See Thomas Tracy, "Evil, Human Freedom, and Divine Grace," in *Human and Divine Agency: Anglican, Catholic, and Lutheran Persepctives*, F. Michael McLain and Mark Richardson, eds. (Lanham, Md.: University Press of America, 1999).

[4] William Hodos and Ann B. Butler, "Evolution of Sensory Pathways in Vertebrates," *Brain, Behavior, and Evolution* 50 (1997): 189–97.

been shown to be involved in spatial working memory, performance-ordered tasks, verbal working memory, object working memory, and analytic reasoning. These measurement techniques allow neuroscientists to study subjects as they perform a variety of tasks and to determine which specific areas of the brain are being stimulated during which activities.[5]

- Studies of patients with frontotemporal degeneration (FD) and patients with Alzheimer's disease (AD) have led to further insight into how the brain functions in the use of language. Persons suffering from FD show damage to the left frontal and anterior temporal brain regions, which leads to an accompanying decline in grammatical abilities. AD patients have defects from cerebral profusions in the inferior parietal and superior temporal regions of the left hemisphere, which correlates with a decline in semantic ability. A recent study by Murray Grossman and his research team demonstrated that, although language ability is centered in the left hemisphere, no single brain region is responsible. Instead, linguistic ability is the product of "a neural network distributed throughout the left hemisphere subserving different aspects of language comprehension."[6]

- People suffering from Huntington's Disease often show frontal lobe atrophy (and eventually total brain atrophy). Early manifestations of frontal lobe atrophy include specific cognitive losses: problems with attention, concentration, planning, and memory. Similar cognitive losses are characteristic of persons suffering from lesions in the frontal lobe.[7]

- Other studies have demonstrated that William's Syndrome (WS) is a result of the loss of the end of one of the copies of chromosome 7, involving perhaps fifteen or more genes. People who suffer from this condition are typically diagnosed mildly to moderately retarded, scoring in the 50s to 60s on IQ tests (equivalent to people with Down's Syndrome.) However, while people suffering from WS have limited writing and arithmetic abilities, their verbal communication skills and their ability to recognize faces and facial features often surpass those of their non-WS peers. In addition, some WS patients have an almost uncanny musical ability, and close to perfect pitch, even though they are unable to read written music. Although their overall brain mass is reduced, the frontal lobes are preserved (including the temporal lobes, which are associated with visual memory); one also finds an enlarged primary auditory complex and a comparable limbic area, which is associated with emotion.[8]

- Corticospinal excitability has been studied using focal, single-pulse transcranial magnetic stimulation (TMS) applied to the scalp. Using fourteen right-handed subjects, the researchers studied the effects of TMS on motor ability on each side of the body. When the subjects induced

---

[5] Tim Beardsley, "The Machinery of Thought," *Scientific American* (August 1997): 78–83.

[6] Murray Grossman, F. Payer, K. Onishi, M. D'Esposito, D. Morrison, A. Sadek, and A. Alavi, "Language Comprehension and Regional Cerebral Defects in Frontotemporal Degeneration and Alzheimer's Disease," *Neurology* 50 (January 1998): 157–63.

[7] E.H. Aylward, N.B. Anderson, F.W. Bylsma, M.V. Wagster, P.E. Barta, M. Sherr, J. Feeney, A. David, A. Rosenblatt, G.D. Pearlson, and C.A. Ross, "Frontal Lobe Volume in Patients with Huntington's Disease," *Neurology* 50 (January 1998): 252–58.

[8] Howard M. Lenhoff, P. Wang, F. Greenberg and U. Bellugi, "William's Syndrome and the Brain," *Scientific American* (December 1997): 68–73.

themselves to think sad thoughts, this facilitated greater motor potentials in the left hemisphere of the brain, while self-induced happy thoughts evoked greater motor potential in the right hemisphere. J.M. Tormos and his colleagues take these results as a clear sign that the brain evidences some form of lateralized control of moods.[9]

Results such as these present a clear challenge to those who would rend thought and affect from its physical substratum. The influences are both deep and bidirectional; they involve the deepest areas of mental functioning. Whether humanists and religionists are ready to acknowledge it or not, the neurosciences are now producing alternative explanatory candidates in the study of the human person.

1.3 Putting the Pieces Together

In this essay I will argue that the perceived tensions between theology and the neurosciences call for *renewed reflection on the nature of the person*. Formulating a philosophically adequate account of what it is to be a human person provides crucial guidance on how to relate these two diverse fields. Conversely, attempts at a direct connection (or reduction) of the one area to the other threaten to mislead unless one keeps the question of human personhood at the center of attention.

One feature of this discussion should be clearly acknowledged from the outset: in the nature of the case, physical sciences such as the neurosciences do tend to push one in the direction of *physicalism*, the view that all things that exist are physical. It is a basic assumption of good neuroscience, as with the other natural sciences, that only traceable physical causes be employed and that only physical mechanisms be introduced in explanations. "Physical" here has an interesting double meaning: it is a methodological standard ("the sorts of things that physicists can study") as well as an ontological thesis ("the sorts of things built up out of the fundamental particles and energies that we have discovered in the natural world").[10] So the question cannot be, Should we do a different kind of neuroscience, say, one that talks more about souls and their actions? Instead, the guiding question in the dialogue between theology and the neurosciences is, *How far* can a position go toward the physicalist assumptions that are basic to the empirical study of the brain without denying (or implicitly rejecting) factors necessary for the viability of religious belief?

To start with the obvious: holding a physicalist view of the world that denies the very existence of God—perhaps on the grounds that God is not a physical being, hence (given physicalism) God cannot exist—would leave little or nothing over of traditional Christian metaphysics. Short of ruling out the existence of God, however,

---

[9] J.M. Tormos, C. Canete, F. Tarazona, M.D. Catala, A.P. Pascual, and A. Pascual-Leone, "Lateralized Effects of Self-Induced Sadness and Happiness on Corticospinal Excitability," *Neurology* 49 (August 1997): 487–91.

[10] To say that physics is committed to *materialism*, the view that all things that exist are composed out of matter (or basic material particles such as atoms), would be to assert that physicists are committed to a particular metaphysical thesis; but this is to place metaphysics prior to the actual conclusions and methods of science, a move we should reject. What is interesting about *physicalism* is that, if it is to have any special warrant, it must appeal to the actual results of physics: the physicalist is committed to the ultimacy of just those entities and explanations that our best physics commits us to (or: the sorts of entities that physics could eventually come to know). Given this definition, to resist physicalism, as I do, is to resist the claim that all adequate explanations will ultimately be given in terms of physical laws and entities (or in terms of some idealized version of present-day physics).

one might find oneself offering interpretations of neuroscience that fundamentally conflict with belief in God or in divine action—which would represent virtually as complete a rejection of Christian theology as an outright denial of the existence of God. Or (somewhat less obvious but still crucial) one might be tempted to adopt a theory of knowledge that makes all theological accounts extremely unlikely (or meaningless) as explanations.[11] Finally, there are views and approaches that make the concept of God explanatorily unnecessary, a sort of appendix to the overall body of explanation. Although the last three views are not immediately fatal to Christian belief, they cast this belief deeply into question, opening the door for a well-argued case that we have "no need of that hypothesis."

I suggest that *these* sorts of tensions, rather than direct disputes over theological claims, actually represent the core of the discussion with the neurosciences. One needs to be fully aware of the stakes of the debate before turning to concrete theological proposals. If one holds that theological assertions are mere constructs or "evocative metaphors," then one has already ceded the case; one might as well admit that theology does no explanatory work whatsoever. If theology is to *do* anything in this discussion, then it must first be shown that there is something about the human person which *cannot* be captured, directly or derivatively, in neuroscientific terms.

2 Methodological Parameters for the Neuroscience-Theology Debate

2.1 Views on the (Actual or Potential) Impact of Neuroscience on Theology

Before entering the actual debate, let us consider a few of the now-dominant views on the probable impact of neuroscience on theology. In what follows I will challenge the first five and will defend a version of the sixth.

1. One position, which we might call *the Arbib Credo*,[12] states that all data about the human person will eventually be best explained in neuroscientific terms. On this view, theology has no explanatory power of its own; its terms pick out no theological entities or properties in the world but are rather constructs of individuals or societies. Moreover, says the Credo, the neurosciences (among other sources) give us sufficient reason to abandon the traditional explanatory claims of theology.

2. A variant of *1*, which one might call *"watch-out-ism,"* admits that nothing in current neuroscience falsifies theology... yet. But it warns that the results of neuroscience will eventually be disastrous for theology. For example, a proponent argues, we will eventually be able to predict an individual's actions based on the inputs from his environment and the succession of brain states, at which time free will will be in trouble. Or, she argues, we'll eventually understand the neural factors that incline people to have what they call "religious experience," at which time religious beliefs will cease to serve as credible explanations of religious experience.

3. The opposite position relies on *soul-based explanations*. On this view, souls exist. They are the kind of thing which can be explained only with theological terms such as "the image of God," salvation, or immortality. The nature of the soul makes it inaccessible, even in principle, to neuroscientific investigation.[13]

---

[11] See Philip Clayton, "Inference to the Best Explanation," *Zygon* 32 (1997): 377–91.

[12] I defer to Michael Arbib, who has presented one of the most sophisticated alternatives to theism (see his essays in this volume and the literature cited there).

[13] Of course, as Stephen Happel points out, "soul" could be, and has been, used in less dualist senses, e.g. to emphasize the religious dimension in psychological and even biological

4. One might also hold a view of *instrumentalism and agnosticism*. On this view, the neurosciences help us to understand human behavior and cognitive functioning; they are indispensable because of their usefulness in prediction and scientific understanding. But this view is agnostic about theological questions; it views them as falling into another category altogether, one not addressed by neuroscience.

5. The *no conflict* view believes it can overcome the potential conflicts presupposed in *1* and *3*. It does so, however, by making all the changes on the theological side: theology means something different than one used to think. One way to avoid conflict is to espouse a purely naturalist theology, one which does not make claims for the existence of any super-natural beings or properties. Theologians might still speak of the spiritual significance of the universe, using labels such as "sacred" or "ecstatic" naturalism, but on this view they do not (should not) mean these terms in a way that conflicts with physical explanations of all phenomena in the world. A related way to avoid conflict is to view all theological statements as metaphors in such a way that any actual or possible conflict is ruled out *ab initio*.

6. Let us call the view *compatibilism* which holds that the concrete results of neuroscience neither prove theology nor disprove it. Rather, on this view, the data are at least consistent with what one would expect to find if, say, a Christian theological anthropology is true. For example, theology also might advocate a theory of human nature in which conscious life is biologically based. Of course, consonances of this type do not prove the truth of Christianity, but they do tell against "conflict" views of the neuroscience-theology relationship. In this essay I defend *emergentist monism* as one viable compatibilist answer.[14]

2.2 Four Models of the Rationality of Religious Belief

I have argued elsewhere that scientific results are rarely the direct building blocks of theology: seldom are theological explanations constructed, directly or indirectly, using the language of scientific results and explanations.[15] Instead, the results of science must first be metaphysically interpreted and their underlying assumptions brought to the surface before they can tell for or against theological assertions.[16] Such meta-physical or meta-scientific assumptions have to be made explicit—and sometimes fought over—before one can get to the really interesting discussions. (In fact, most of the underlying assumptions of science are *not* directly theological in nature, even though we focus here on those metaphysical assumptions that can help

---

phenomena. This is surely correct. Because soul-language has traditionally served as the bulwark of dualist metaphysics, however, I use it here as the antithesis to the Arbib Credo and avoid it in my own constructive proposal below.

[14] "Compatibilism" in this sense must not be confused with compatibilism in the free will debate, which is the view that moral responsibility is compatible with all of one's actions being genetically or environmentally determined. Note also that compatibilism in this sense is weaker than an entailment relationship; it does not imply that results from neuroscience will someday be able to decide between (say) a Christian and a Muslim anthropology.

[15] See Clayton, *Explanation from Physics to Philosophy: An Essay in Rationality and Religion* (New Haven: Yale University Press, 1989); idem, *God and Contemporary Science* (Edinburgh: Edinburgh University Press; Grand Rapids: Eerdmans, 1998).

[16] Theo Meyering correctly points out (personal communication) another type of science-theology connection: "meta-inductive or speculative *extrapolations* that may be more or less plausibly drawn from the respective theories and thus are at best associated with them rather than being implied by them."

to structure the debate between the empirical sciences and theology.) If the theology-neuroscience discussion turns in large part on clarifying the framework assumptions (such as a theory of personhood) that determine how the two fields are to be connected, then we must come to agreement on what standards of evidence to apply to meta-empirical theories of this sort. I suggest four possibilities:

1. One might hold to the standard of *proof*, requiring that a compelling (or even logical) demonstration be given of any meta-empirical or religious belief. This approach was favored by natural theologians earlier in the modern period. The difficulty with such standards is that rigorous deductive proofs are not available even in most of the natural sciences (for example, in most of the biological sciences). Further, linear proofs are an inappropriate standard for any hermeneutical discipline, that is, for any discipline whose structure involves a two-way movement or interdependence between data and theories.[17] Since religious questions are clearly hermeneutical in this sense, the standard of proof seems unavailable from the start.

2. One might then only require that religious beliefs be *empirically probable*. Beliefs are empirically probable when they are based or grounded on empirical evidence. This standard is represented by the natural sciences, although philosophers of science have lately raised serious objections to the claim that one can move directly from empirical evidence to theoretical explanations. Of course, there are obvious problems with demonstrating the empirical probability of religious belief, since (unlike science) the beliefs in question involve truths and being(s) that are claimed to transcend the empirical universe. The claim that there can be knowledge only of empirical matters, never of metaphysical ones, traces back to Immanuel Kant, although there are now reasons to be suspicious of Kant's sharp dichotomy.[18]

3. A weaker standard would be to say that religious beliefs need only *not be counter-indicated* by the empirical evidence. According to this view direct empirical probability is not required; still, one's beliefs should be in no worse shape in the face of the available evidence than are the competing beliefs. If one is aware of evidence against one's position, for example, one has the obligation to reject that position.[19] I agree that theologians ought to submit themselves to this standard.

4. A final position holds that empirical evidence is simply *irrelevant* to religious belief. This view, often called *fideism*, holds in its most extreme form that faith itself is sufficient to ground belief. But I shall use the term in a looser sense to stand for any position which does not think that countervening empirical evidence is sufficient reason to question a religious belief or set of beliefs.

---

[17] See Clayton, *Explanation*, chap. 3. Clearly the natural sciences are hermeneutical in that they raise questions of interpretation (and are pursued by interpreting subjects). But there is an added interpretive dimension, which Anthony Giddens calls a "double hermeneutic," in the human (and theological) sciences: interpreting Shakespeare (or salvation history) involves intentional agents both in the act of interpretation and in creating the object to be interpreted.

[18] See Clayton, *Das Gottesproblem: Gott und Unendlichkeit in der neuzeitlichen Philosophie* (Ferdinand Schöningh Verlag, 1996); trans. forthcoming as *Infinite and Perfect? The Problem of God in Modern Thought* (Grand Rapids: Eerdmans, 1999), chap. 5.

[19] This epistemic standard draws significantly from the theory of falsification in the philosophy of science, as developed by Karl Popper and modified by Imre Lakatos. For further references, see my *Explanation*; and Nancey Murphy, *Theology in the Age of Scientific Reasoning* (Ithaca, N.Y.: Cornell University, 1990). There are also parallels to the theory of defeaters and "defeater defeaters" developed by Alvin Plantinga, e.g. in Plantinga and N. Wolterstorff, eds., *Faith and Rationality* (Notre Dame: Notre Dame Univ. Press, 1983).

My own preference is for 3. This is not special pleading, for there is good reason, on the one hand, to hold that many of our deeply held beliefs, religious and otherwise—beliefs about which it seems extremely counterintuitive to say that one *ought not* to hold them—do not meet the standards of proof or empirical-scientific probability. On the other hand, agents are justified in rejecting a position when they have reasons to think that it is false, and this pushes one toward 3 over 4.[20]

What is the significance of siding with 3? It implies that one's metaphysical beliefs are not and do not need to be direct inductive inferences from the empirical world as it is known through the natural sciences. Hence 3 means siding with fallibilism, in the sense advocated by philosophers of science such as Karl Popper and Imre Lakatos, rather than with positivism or other induction-based theories of knowledge. Each person holds a variety of meta-empirical beliefs; she may well be justified in holding these beliefs, even if they are not directly grounded in experience, as long as she holds them in a nondogmatic manner. A fallibilist epistemology means, first, that one will look for *conflicts* with experience—say, possible conflicts with the results of science—and that one will change one's higher-order beliefs appropriately. It means, second, that one will observe carefully whether one's various beliefs about the human person (religious, empirical, experiential) fit together into a theory that is both internally coherent and scientifically viable.

2.3 The Insufficiency Thesis

The debate about neuroscience, psychology, and mind presents one with a confusing clutter of possibilities. And yet in one sense one finds oneself returning again and again to one basic choice. Many neuroscientists, but not all, maintain the *Sufficiency Thesis*. It is the view that in the future neuroscience will be sufficient to explain all that we know about the human person. This view contrasts with the *Insufficiency Thesis*, which predicts that neuroscience will *not* be sufficient to explain all we come to know about the human person. I defend the *Insufficiency Thesis* in what follows not because of blindness to the power of the neurosciences (far from it!), but because there are parts of what it is to be a person that lie in principle beyond their reach. This "something more" has been called variously *consciousness* (David Chalmers, Thomas Nagel, Frank Jackson, Colin McGinn), *original intentionality* (John Searle), or perhaps *caring* (John Haugeland).[21] What it is and why it should play this role will concern us further below.

---

[20] See Clayton and Steve Knapp, "Ethics and Rationality," *American Philosophical Quarterly* (1994): 97–107. See also the methodological essays in Mark Richardson and Wesley Wildman, eds., *Religion and Science: History, Method, Dialogue* (London: Routledge, 1996), part II.

[21] David John Chalmers, *The Conscious Mind: In Search of a Fundamental Theory* (New York: Oxford Univ, Press, 1996); Thomas Nagel, *The View from Nowhere* (New York: Oxford Univ. Press, 1986); Frank Jackson, *Mind, Method, and Conditionals: Selected Essays* (New York: Routledge, 1998); Frank Jackson, ed., *Consciousness* (Brookfield, Vermont: Ashgate, 1998); Frank Jackson and David Braddon-Mitchell, *The Philosophy of Mind and Cognition* (Oxford: Blackwell, 1996); Colin McGinn, *The Problem of Consciousness: Essays towards a Resolution* (Oxford: Blackwell, 1991); idem, *Minds and Bodies: Philosophers and their Ideas* (New York: Oxford Univ. Press, 1997); idem, *The Mysterious Flame: Conscious Minds in a Material World* (New York: Basic Books, 1999); John R. Searle, *The Rediscovery of the Mind* (Cambridge: MIT, 1992); idem, *The Mystery of Consciousness* (New York: New York Review of Books, 1997); John Haugeland, *Having Thought: Essays in the Metaphysics of Mind* (Cambridge: Harvard Univ. Press, 1998).

Note some of the major features of the debate between the Sufficiency and Insufficiency Theses:

1. It is not settled by any current empirical data.

2. It is future-oriented. Indeed, its status is closest to that of a *wager.* Current scientific results and scientific progress to date are relevant to which side I wager on and how much I am willing to wager (that is, how strong is my commitment to the one view or the other). But other assumptions—metaphysical assumptions—also play a role in the different predictions of the Sufficiency and Insufficiency theorists. Think, for example, of the stock market: those who invest are willing to wager money on a future state which no one knows for sure. Even when one is a specialist in the vast amount of data that indicate whether one should invest in one or another firm, every investment remains a speculation.

3. *Sufficiency vs. insufficiency* is, in this sense, a classic philosophical debate, not a scientific debate. In this sense it is more like the debate about universals or free will than like the question of the explanation of thermodynamic phenomena. For example, among the Greeks, Democritus might have been on the side of the Arbib Credo and Aristotle on mine.

4. The debate bypasses the debate about dualism. Like "positivism," the word "dualism" seems today to be used only as a term of derision, at least in debates with or written for neuroscientists. Dualism is, strictly speaking, a species within the genus *substance ontology*, that is, it is a theory of being in which the world is divided into two basic types of existing things called substances. As such, it presupposes that a theory of being can and ought to be developed, an assumption not made in this essay. But the real differences and interesting questions raised by the neurosciences today are not adequately grasped within the framework of (traditional) substance ontology. So defending the Insufficiency Thesis is not the same as advocating dualism.

5. The Insufficiency Thesis is compatible with believing in the great explanatory power of neuroscience. It need not be an *anti*-scientific position, and I do not advance it as such. Nothing in the present essay blocks or diminishes the importance of neuroscientific research. It denies only one thing: the final sufficiency of the neurosciences for explaining the human person.

3 Toward a More Productive Debate on Neuroscience and Personhood

3.1 Progress in Neuroscience

From the outset one should be honest about how strong the tug is in the direction of the Arbib Credo; there is no point in hiding one's head in the sands of a pre-scientific age that denied the dependence of the mental on the physical. As we saw above, specific types of cognition—perceptions, memories, emotions—do correlate with specific state changes in specific brain regions. In some cases we can predict with a high degree of accuracy what neural processes will accompany which sorts of subjective experiences. Note that knowledge of the connections is increasing in both directions: neuroscience can predict more of the subjective experiences that will follow specific types of brain stimulation, as well as more about the sorts of neural activity (and in what regions) that will underlie particular psychological experiences.

Specific types of brain damage also lead to specific changes in subjective experience. In one famous case, Lawrence Weiskrantz reports on a subject whom he calls "D.B." who had had part of his visual cortex removed so that he was unable to see things located in a certain part of his visual field.[22] But if D.B. was asked to *guess* what the supposedly invisible object was, he could do so with nearly 100 percent accuracy. This phenomenon has come to be known as "blindsight." Weiskrantz suggests that D.B. may have been drawing on information located in the lower temporal lobe, suggesting that there are regions that are necessary for *conscious* awareness of what one perceives, even though one may draw information from other regions of whose processing one has no conscious awareness. (Interestingly, D.B. was able to learn with some practice to have at least limited awareness of what he "knew" in the lower temporal lobe.)

The brain sciences have thus established that, and how, specific types of mental experience correlate with specific brain functions. It is the *brain* that does the processing when you calculate 73×37 or when you feel fear after hearing a threat. In coming years we will learn massively more about what neurological states are *necessary* for certain mental experiences; we will find more and more such necessary conditions; and we will be able to specify the underlying brain states and processes with greater and greater precision. Gradually, we will be able to cause more and more specific mental or emotional responses by means of carefully controlled stimulation to the brain, and we will be able to model more and more of them on computer-based systems.[23]

As the neurosciences develop, we will be able to give increasingly complete accounts of how perceptions are represented, how they are recalled, and what is happening in the brain when a subject reports that one thought gives rise to another. We will understand the functions of emotions and why brains that have emotions like ours would confer survival value on an individual. We will also learn precisely which brain regions or distributed systems are active when a person reports having certain emotional, aesthetic, or religious experiences. We will know why brains such as ours would be prone to aesthetic and religious experiences of these sorts and what kinds of neural stimulations (or lesional damages) tend to increase or suppress such experiences. Some argue—though others dispute—that in the limit case we could learn the precise brain states that would have to occur if a human subject is to enjoy particular kinds of mental (phenomenal) experience.

Now some readers may find the prospect of such successes in neuroscience greatly exciting; others may find it greatly threatening. Whether a massively successful neuroscience would be a good or bad thing is not the topic of the present essay. Instead, I want to ask: If neuroscience is successful in this way, will we have *proven* that all things that exist are physical, or that the conscious self is an illusion? By no means! To move from successes in neuroscience to the doctrine that only physical things exist—whether one then advocates the falseness of belief in God or interprets the self as "merely metaphorical or constructed"—is, as I shall attempt to

---

[22] Lawrence Weiskrantz, "Neuropsychology and the Nature of Consciousness," in *Mindwaves*, C. Blakemore and S. Greenfield, eds. (Oxford: Blackwell, 1987).

[23] I recall reading recently in the popular press that Air Traffic Control already has an "awareness meter" that allows supervisors to monitor when an air traffic controller is losing conscious attention ("dozing off"). Perhaps it would be useful for professors to employ awareness meters for those students who tend to doze off during lectures!

show, a category mistake. First let us look at some of the options, and then I shall make a case for a mediating framework that I think is preferable to the two extremes.

3.2 Getting Rid of the Extremes

The sorts of neuroscientific results that I have just summarized tend to pull people in one of two directions. Some find here strong evidence that human cognitive behavior will ultimately be fully explained in terms of brain activity (the Sufficiency Thesis). Others find here no threat to their dualist intuitions: thought and emotion are still properties of the "spiritual self," they insist, and spirits or souls are just not the *kind* of thing that brain science can really tell us anything about. I will argue that *neither* of these more extreme views does justice to the data we have about the human self. There are serious issues in neuroscience and religion, but they depend on drawing careful distinctions closer to the middle and not on battles fought at the edges.

Let us call the two more extreme positions I have just mentioned *strong reductionism* and *metaphysical dualism*. Strong reductionists argue that human thought and mentality are in principle fully explainable by, because wholly caused by, neural firings. To understand why regions of the brain react in the ways they do would be to understand human thought, human emotions, human religious experience. According to so-called identity theorists, thoughts *just are* the neurological events studied by brain scientists.[24] Metaphysical dualists on the other end of the spectrum argue that there is an ontological entity such as the soul or mind (*Geist*) that is forever inaccessible to natural scientific study, even in principle, and that is the basis for and possessor of all mental events: ideas, wishes, emotions, intentions, and the like.

Other authors in this volume have given good summaries of why dualism is no longer a tenable position; I shall not repeat their arguments here. Is there an equally clear and compelling argument against at least the strongly reductionist programs which dominate much of the literature in the neurosciences? Yes. Bracketing for the moment the causal question, I would argue that there is a *difference in kind* between physical explanations of thoughts, feelings, and emotions on the one hand, and explanations of those ideas in their own terms on the other. Thoughts have a quality which philosophers (following Edmund Husserl) call *intentionality*. The simple definition of intentionality is *aboutness*; it is the characteristic of referring to something else. The referring relationship is intrinsically different from the causal relationship, where A causes B to occur. Causal relationships are clearly physical; they are the bread and butter of the physical sciences, whereas the reference relationship—which we all employ whenever we speak *about* something—works according to a vastly different "logic." Brian Cantwell Smith states the difference graphically:

> Reference—plain, ordinary, vanilla reference, of the sort out of which even the most trivial conversation is made—is manifestly able to leap amazing gaps in space, time and possibility: backwards to the first 10^{-23} seconds of the universe, forward to the death of the solar system, sideways into other possible worlds (such as to a world where Apple responded positively to Bill Gates's 1985 offer to license the Mac OS).... This non-

---

[24] See David M. Armstrong, *A Materialist Theory of the Mind*, rev. ed. (New York: Routledge, 1993); Patricia S. Churchland and Terrence J. Sejnowski, *The Computational Brain* (Cambridge: MIT Press, 1992); Patricia S. Churchland, *Neurophilosophy: Toward a Unified Science of the Mind-Brain* (Cambridge: MIT Press, 1986).

effectiveness [of reference] is in direct and exact contrast to physical causality, which is famously... proscribed from performing any such fancy long-distance or counter-factual maneuvers.... You can refer to the sun, I take it, *right now*; it doesn't take 8 minutes for your reference to reach its destination![25]

In reference there is no limitation to the speed of light. The particular nature of intentions and conceptual references helps to explain why identity theories (the alleged identity of mental experience and brain states) are inadequate. Imagine that you could (in principle) know exactly what neurological events occur when Michael is asked to define "justice" and makes a verbal response. Still, these events would never be identical to his definition of justice. Consider the analogy with what goes on in your computer's processor: knowing all the facts about the on and off states of some sixteen million registers is not the same as knowing that your computer is currently solving a differential equation.[26]

3.3 *Continuing Differences Between the More Moderate Positions*

It is still the case that the two more extreme positions garner most of the popular (and media) attention in debates about mind. Let us take it as shown, however, that these views—strong reductionism, or "the identity thesis," and metaphysical dualism—are not tenable. Do we then find a natural middle position emerging? As nice as that would be, it does not seem to be happening.[27] One still finds deep divisions between even the more moderate positions. Thinkers such as Jerry Fodor and Hilary Putnam argue that psychology is (more or less) independent of neural considerations; the neurosciences do not play a major constraining role in doing cognitive psychology.[28] (Similar responses are often given by humanistic psychologists.) On the other side, Patricia and Paul Churchland, Andy Clark, William Lycan and others argue that psychology and neuroscience must coevolve: any genuine

---

[25] Brian Cantwell Smith, "God, Approximately," unpublished paper, p. 8. This paper provides a brief summary of the broader argument in Smith's *On the Origin of Objects* (Cambridge: MIT Press, 1996). Michael Arbib argued correctly in criticism of an earlier draft of this essay that leaping gaps in space and time and other "long-distance maneuvers" is not in itself sufficient to show that reference is nonphysical. But if one considers the full range of what reference involves, I think it is clear that it is more closely associated with the logic of the mental than with the logic of physical-causal explanations; on this issue, see William Stoeger, Nancey Murphy, and Theo Meyering in this volume.

[26] It is true, as Theo Meyering has pointed out, that this is a special case of the more general divergence between structural and functional explanations. My point is that, although the physical or structural facts may determine the emergence of the mental, the mental is something more than its conditions of origination. (Admittedly, the computer analogy may raise further problems of its own because of differences between human and machine intelligence.)

[27] Those who have listened in on such debates in the past know that many of the arguments involve members of one moderate and viable research program accusing their opponents of actually holding one of the extreme positions just cited. Clearly, such comments generate more heat than light.

[28] See Jerry Fodor, *The Elm and the Expert: Mentalese and its Semantics* (Cambridge: MIT Press, 1994); idem, *A Theory of Content and Other Essays* (Cambridge: MIT Press, 1990); idem, "Special Sciences," in *Readings in the Philosophy of Psychology*, Ned Block, ed. (Cambridge: MIT Press, 1981); Hilary Putnam, "Philosophy and our Mental Life," in Putnam, *Mind, Language and Reality; Philosophical Papers*, vol. 2 (Cambridge: Cambridge Univ. Press, 1975), 291–303.

progress in psychology will give rise to resultant progress in neuroscience.[29] (Note that "coevolution" is not a mediating position, since it amounts to the denial of autonomous psychology, or folk psychology, in the sense advocated by Fodor and others.)

Still, we have now at least arrived at the playing field on which any fruitful debate of the deeper questions in neuroscience and religion (for example, religious experience) must take place. Also, note that we have now managed to formulate a clear disagreement. The one side argues that functionalist neuroscience can eventually provide as much reliable knowledge of human cognition as humans will ever get, whereas the other side denies this premise, maintaining that there are other types of knowledge of human cognition. What arguments can the two sides develop on this topic?

Let us start with the latter position. In recent years, a more moderate version of dualism—at least more moderate than Cartesian dualism—has been developed in the work of neuroscientists like Sir John Eccles and in several publications by Roger Penrose.[30] Penrose does believe that there is something like conscious substance, which is ontologically a different sort of thing than physical phenomena. He also maintains that there is "an essential *non*-algorithmic ingredient to (conscious) thought processes."[31] But even so he is not primarily interested in developing a sharply defined dualist metaphysics à la Descartes. Instead he asks, "What *selective advantage* does a consciousness confer on those who actually possess it?"[32] In my view, Eccles and Penrose are saddled with dualist dilemmas that admit of no easy solution. Yet clearly they represent research programs that are more scientifically respectable than classical Cartesian dualism.

On the other side one finds a more moderate version of the functionalist/neuro-scientific position, the "schema theory" as it has been developed by Michael Arbib.[33] Arbib is not a reductionist in the strong sense of the word. Schemas are "the basic functional unit of action, thought, and perception, a unit whose functionality is distributed—in the first instance—across the networks of the individual human brain."[34] He has also defined them as a "crystallization of some body of experience within a local situation" or simply as "parallel distributed adaptive computation." A schema can be "an internal structure or process (whether it is a computer program,

---

[29] See Paul Churchland, *The Engine of Reason, the Seat of the Soul: A Philosophical Journey into the Brain* (Cambridge: MIT Press, 1995); Andy Clark, *Associative Engines: Connectionism, Concepts, and Representational Change* (Cambridge: MIT Press, 1993); idem, *Being There: Putting Brain, Body, and World Together Again* (Cambridge: MIT Press, 1997); William Lycan, *Consciousness and Experience* (Cambridge: MIT Press, 1996).

[30] For example, see Roger Penrose, *The Emperor's New Mind: Concerning Computers, Minds, and the Laws of Physics* (New York: Penguin Books, 1989); Sir John Eccles, *Evolution of the Brain: Creation of the Self* (New York: Routledge, 1989); idem, *How the Self Controls its Brain* (New York: Springer-Verlag, 1994).

[31] Penrose, *The Emperor's New Mind*, 404.

[32] Ibid., chap. 10.

[33] For example, see Michael Arbib, "Schema Theory," in *The Encyclopedia of Artificial Intelligence*, S. Shapiro, ed. (New York: Wiley, 1992), 1427–43; Arbib, E. Jeffrey Conklin and Jane C. Hill, *From Schema Theory to Language* (New York : Oxford University Press, 1987).

[34] Michael Arbib, "Computing the Self and the Horrors of Humanity," unpublished paper, p. 6.

a neural network, or a set of information-processing relationships within the head of the animal, robot, or human)," or it can be "an external pattern of overt behavior." These two basic types of schemas can give rise to a "social schema, a schema which is held by the society *en masse*."[35]

What is interesting about making the schema concept basic for neuroscience is that it is a logical structure which *could* be given either a causal/functionalist or an emergentist interpretation. If one looks at schemas solely in a causal fashion, however, as summaries of causal mechanisms, then eventually they must be reduced down to the basic causal units of neural activity, neuronal firings; and this is precisely what Arbib does. Thus he writes that they are "functional units," that is, "composable units of brain function/neural activity."[36]

But note that schema theory is a logical device which could in principle be used in a more holistic fashion. One could speak of schemas as phenomena which emerge only at higher levels, when one abstracts from many of the composite parts. For example, cells are schemas—complex wholes—but they are also existing things in their own right. Likewise, my awareness of an orange, or of a situation of injustice, is a highly abstract phenomenon which includes, but goes beyond, countless observations, neural traces, composites, and other influences. Imagine that we gain massive understanding of the workings of your dog's brain; imagine that our efforts at predicting your canine's behavior succeed beyond our wildest expectations. Would this prove that your dog does not have subjective experiences (*qualia*) such as fear, concern, or affection, or that these qualia do not play any causal role in her actions? Even vastly successful neuroscience thus leaves open the key questions about mental life. It is *these* questions, not progress in neuroscience per se, that are of life-and-death concern for those interested in the claims of religion.

So I advocate a kinder, gentler schema theory. We need a study of mental phenomena which allows us to focus on higher-order units as (sometimes) genuine existents, not just composites of the parts of which they are composed. In particular, it is necessary to think of persons as distinctive units of activity, as agents capable of forming intentions, making references, and having subjective experiences in the fashion described above. *We therefore need a "science" of the person of which neuroscience is one, but only one, contributing part.* Such a study of the emergent person is genuinely holistic, however, only if it retains a place for speaking of one higher-order event (for example, a thought or *quale*) causing another without insisting that the whole story can be told in terms of neuronal firings. Arbib, in his well-known work in neural modeling, does not give adequate place to this possibility, though I think schema theory leaves room for it. Still, the crucial fact is that Arbib and others employ a logical framework which could in principle be read *either* in a holistic *or* in an atomistic fashion.

3.4 Closer to a Mediation: Information Biology and Virtual Reality

In the last section I argued that moderate dualist theories of mind on the one hand, and theories of the mental as "composable units of brain function/neural activity" on the other, represent positions that are still too far out along the spectrum of positions on the human person. This is not a straw-man dismissal; both are sophisticated positions, and one can see what features in the contemporary study of psychology

---

[35] See Arbib, "Crusoe's Brain: Of Solitude and Society," in this volume, p. 429.

[36] Ibid., 422.

would drive the authors to their diverse positions. Nonetheless, I believe the strongest theory of the human person lies in between these views. The question then becomes: What view of the self starts neither with theological claims as "obviously given" nor with the "obvious ultimacy" of neuroscientific explanations? What would such a mediating position look like?

First, the successful theory will have to grant what we know already from our experience in the world: that our thoughts are not found apart from the functioning of brains, and that damage to the brain can modify or eliminate subjective experience. At the same time, I have argued, the answer must allow for the emergence of mental phenomena and for mental causation. The resulting view will therefore begin the line of causation at the physical level, in a manner similar to Arbib's schema theory, but at the same time it must insist that a line of causal influence can also be traced (in the appropriate way) *among* the highly complex and abstract "schemas" that we call mental phenomena.[37] Any adequate theory of the human person will have to understand the effect of interactions with the surrounding environment upon mentality, while at the same time doing justice to the irreducible subjectivity of experience.

With regard to the former requirement, the field of information biology has begun to comprehend the way in which all organisms exchange information with the environment around them. In *The Tree of Knowledge: The Biological Roots of Human Understanding*, for example, the biophysicist Humberto Maturana and the cyberneticist Francisco Varela describe the "structural couplings" that arise between an organism and its surroundings.[38] The organism cannot be decoupled from its environment without dying. The feedback loop that exists between environment and organism is more than just an incidental connection between it and its world. Instead, the way in which those links are set up are physically *and ontologically* constitutive of the organism itself. This is certainly a far cry from a dualist position, where the soul is essentially different from (or uncoupled from) its physical world. At the same time, the information that arises out of these links, and the influence that one's understanding subsequently exercises via the body on the world, require *a new level* of explanatory concepts. Information-processing agents bring the dimension of subjectivity as one element in this biologically mediated two-way interaction with the environment.

The subjective element in this two-way semantic connection is expressed in some recent experiments in 360° virtual reality. For example, in one well-known artwork in this genre, Char Davies's "Osmose," the participant dons a head helmet and a special suit and enters into what seems to be a clearing that surrounds him. He is able to rise in the clearing by breathing in and to lower himself by breathing out; movements are made by gentle leanings from side to side. In the world through

---

[37] I admit that one can speak of causal influences among ideas, and of ideas on the brain, only in a sense that diverges from the standard use of the term "causality" in science. Here (as in the perplexing "nonlocality" results in quantum physics) we need nothing less than a new theory of causality. This theory must supplement the so-called efficient causality on which modern science has been based with a way of speaking of the "causal" influence of form or structure, of function, of information, or of the whole on its parts—yet without falling back into the four-fold causality of medieval metaphysics (formal, final, efficient and material).

[38] Humberto Maturana and Francisco Varela, *The Tree of Knowledge: The Biological Roots of Human Understanding*, rev. ed., trans. by Robert Paolucci (New York: Random House, 1992).

which he now "floats" the edges of objects are unclear, and he is able to move through, above, and below the trees and plants at will. This new set of structural couplings between "mind" and "world" can have a profound effect on participants.[39]

What occurs when one is in a 360° surround-sound virtual world? It seems clear that the experience gives to subjects a new sense of being embodied—they actually *are* embodied in a different way, thanks to the computer interface. This is why some describe the experience, even months later, as being "within me." New mental experiences *arise out of* one's being given new structural couplings with the world, altering one's mental experience as a result. (Similar mental transformation can arise out of the physical changes associated with drug experimentation, brain disease, or amputation—certainly more brutal forms of cognitive alternation!)

But the lesson to be drawn from these transformations is not the functionalist-reductive one that some neuroscientists would have us accept, for the transformation, though physically dependent, is a *mental* transformation, one whose explanation involves (among other things) psychological concepts. One's particular experience will rely on one's particular set of structural couplings with the world—and, in the case of brain damage, it may be altered by damage to receptors and processing regions—but it is also (irreducibly) about the new mental state that is caused. Equally importantly, these states in turn give rise to a new manner of being embodied in the world and to a new manner of acting causally upon the physical world.

This is a key insight; let me generalize it. The causal line seems to move "up" from the physical inputs and the environment to the mental level, then *along* the line of mental causation—the influence of one thought on another—and then "down" again to influence other physical actions, to make new records and synaptic connections within the brain, to produce new verbal behaviors, and so forth. This view is monist, not dualist: there is only one physical system, and no energy is introduced into that system by some spiritual substance external to it. At the same time, it seems, subsequent states of the entire system cannot be specified without reference to the causal influence exercised by the higher-level phenomena.

In a famous thought experiment by the Harvard philosopher Hilary Putnam, the reader is asked to imagine a "brain in a vat," a brain that has been removed from its body by a team of scientists and kept alive in a vat.[40] The imaginary scientists have re-established all of the myriad links that the brain had to its body, replacing them with computer inputs which exactly simulate the body and the environment that the person had experienced before the removal of his brain.

The thought experiment may suggest more than Putnam intended, however. Its practical difficulty, verging on inconceivability, underscores how the mental life is highly, even extremely, dependent on our structural couplings with the physical world. The brain possesses an incredibly large number of receptors for information from other parts of the body, and interrelates them with an amazing 10^{14} synaptic connections. Our mental experience is conditioned beyond what we can imagine by the body's vast input to the brain and by the complex way in which the brain

---

[39] Some persons emerge from fifteen minutes in this virtual world deeply touched, and (by their reports) sometimes profoundly transformed. Some report that they later experience being embodied in the nonvirtual world in a different way than in the real world, and a few have said, "I am no longer afraid of dying."

[40] See Hilary Putnam, *Reason, Truth, and History* (New York: Cambridge University Press, 1981).

processes it. And yet *there is a subjective experience of that world which is different from those physical inputs* and which in turn helps cause the variety of the outputs which constitute our action in the world. (Obviously the brain in the vat would have to be given not only massive inputs, but also the impression that it is acting within the world if it is not to "know" that it has been so rudely imprisoned.) The language of mental impressions, intended references, and mental causes is an irreducible part of the full story, just as in a virtual reality chamber the full story includes not only the new physical inputs to the brain, but also the irreducibly mental dimension of the experience—the transformed mental "place" that arises out of this new virtual-physical surrounding.

4 Emergentist Nonreductionism

4.1 Toward a Theory of the Person

The study of the human person, therefore, involves not only all the knowledge we can glean about the brain and its workings, but also study of the emergent level of thought, *described and explained not only in terms of its physical inputs and nature, but also in terms intrinsic to itself.* My first task has been to argue for the existence of both levels, and to understand the way in which the mental emerges out of the physical. The second task is to begin to integrate these two levels. What is the best framework for doing this? I suggest beginning with the notion of the human person as *psychosomatic unity.* Humans are both *body and mind,* and both in an interconnected manner. How does this work?

It is not difficult to describe what is normally connoted by the word "person." A person is one who is able to enter into human social interaction: praising your tennis partner, planning your dinner party for next Friday, carrying out your intention to graduate from college by next May—and being aware of (at least some) other humans as moral agents who have value and rights equal to your own. These are concepts of personhood that are basic to research in the social sciences (psychology, sociology, and cultural anthropology); they are reflected in the literature of various cultures around the world, as well as in multiple religious traditions. Of course, there are many questions that still leave us unsure: When does personhood start? Does it demand a metaphysical basis, such as the introduction of the soul or person-substance? Does it develop and end gradually? Can it be effaced within a human being? Is it a legal or social fiction, or a metaphysical reality? Such broader philosophical questions are crucial to the complete definition of personhood and hence are part of the discussion that neuroscientists and theologians must have if they are to find any common ground at all.

Personhood is therefore a level of analysis that has no complete translation into a state of the body or brain—no matter how complete our neuroscience might be. Of course, it presupposes such states; yet personhood represents an explanatory level that is distinct from explanations at the level of our "hardware." As Brian Cantwell Smith writes:

First, you and I do not exist in [physical explanations]—*qua people.* We may be material, divine, social, embodied, whatever—but we don't figure *as people* in any physicist's equation. What we are—or rather what our lives are, in this picture—is a group of roughly aligned not-terribly-well delineated very slightly wiggling four-dimensional worms or noodles: massively longer temporally than spatially. We care tremendously about these noodles. But physics does not: it does nothing to identify them, either as personal, or as unitary, or as distinct from the boundless number of other worms that could be inscribed on the physical plenum...[41]

The languages of physics and of personhood only partly overlap; one cannot do justice to the one using only the tools of the other. To give a purely physics-based account of the person is like saying that, because a club or church cannot survive without being financially viable (for example, receiving income from some source), it *just is* the economic unit which economists describe in terms of income and expenditures. The confusion, one might say, is a confusion of necessary and sufficient conditions. A living body and a functioning brain are *necessary* conditions for personhood, yet the wide discrepancy in the "logic" of the vocabularies suggests that they are not *sufficient* conditions. Personhood is not fully translatable into "lower-level" terms; persons experience causal and phenomenological properties (qualia) that are uniquely personal.

4.2 Separating the Questions of Science and Ontology

But is this answer permanent or temporary? What if, some time in the future, neuroscience succeeds beyond our wildest imaginings? What if we are someday able to model human behaviors precisely in complex computational machines? Will we not have shown that personhood is best understood as (something like) a sufficiently complex software system running on the right sort of hardware?

I do not think so. The debate between physicalist and nonphysicalist views of the person, after all, is not only about science; it is also about what actually or really or finally exists. We must ask: Are the properties measured by natural scientists—and recall that we have defined physicalism in terms of the methods of physics—the only sorts of properties that this particular object in the world has? In debating the issue it is important to distinguish the ontology of the phenomena (that is, of the world as we experience it) from the ontology of the *best explanation* of the phenomena. A cultural anthropologist, for example, might note that the subjects of her study report discussions with the spirits of animals and give explanations of her arrival in their village which conflict with the world as she experiences it (for example, she is the embodied spirit of one of their ancestors). In *describing* their beliefs, she suspends judgment on their truth, attempting to be as accurate as possible in re-presenting the world as they see it. In her explanations, however, she will feel free—indeed, it is required of her—to offer explanations which use an ontology (an account of what really exists in the world) that may diverge widely from their own.

The key question under debate, then, is the question of how much of subjective experience or "folk psychology" is irreducible, that is, how much of it actually belongs in a correct explanation of human experience. Some theorists defend an explanatory ontology that consists of brains and other physical organs and their states alone. At the opposite end, others argue that only minds exist, or that both minds and bodies represent primitive substances, defined as radically different sorts of things. Still other thinkers (for example, social behaviorists) hold that both brains and their social contexts exist, that is, both brains and whatever things we are committed to by an account of social contexts. The view to be defended here, *emergentist supervenience*, holds that brains, social context, and mental properties exist; which means (if I am right) that the correct explanatory ontology has to introduce at least three levels of "really existing properties." Yet more extensive ontologies are of

---

[41] Brian Cantwell Smith, "God, Approximately," unpublished paper, p. 3.

course available, such as those involving the real existence of ethical predicates, religious predicates, and various religious beings or dimensions. But nothing in emergentist supervenience immediately commits one to other types of properties than the mental.

4.3 Emergentist Supervenience

I agree with several of the other authors in this volume that the philosophical notion of supervenience is especially attractive as a bridge framework when discussing neuroscience and the person. Simply put, supervenience is intended to grant the dependence of mental phenomena on physical phenomena while at the same time denying the reducibility of the mental to the physical. Note that supervenience is about properties or groups of phenomena, and not about relations between substances (and the ontology that supports them).

Supervenience might be defined as follows:

> B-properties *supervene* on A-properties if no two possible situations are identical with respect to their A-properties while differing in their B-properties.[42]

The early uses of the concept of supervenience described the way in which ethical judgments are dependent upon certain physical states and yet not reducible to them. The notion made its major entrance into the mind-body debate in the article "Mental Causation" by Donald Davidson. Davidson writes,

> Although the position I describe denies there are psychophysical laws, it is consistent with the view that mental characteristics are in some sense dependent, or supervenient, on physical characteristics. Such supervenience might be taken to mean that there cannot be two events alike in all physical respects but differing in some mental respects, or that an object cannot alter in some mental respects without altering in some physical respects. Dependence or supervenience of this kind does not entail reducibility through law or definition: if it did, we could reduce moral properties to descriptive [ones], and this there is good reason to *believe* cannot be done.[43]

In the accounts examined so far, there is a direct and full relationship of dependence between the mental and the physical. I will call those views "*strong supervenience*" in which the physical determines the mental in its emergence and in all its subsequent behavior. Godehard Bruntrup writes of the strong supervenience relation, "Micro-properties determine completely the macro-properties (micro-determinism).... If mental properties are macro-properties in this sense, they are causally inefficacious qua mental properties."[44] In this construal of the physical-mental relationship, for example, one might hold that there are general physical laws such that, if they were known, the occurrence of any given mental event could be predicted from a thorough enough knowledge of the brain, its structure, and its past and present inputs.

---

[42] Chalmers, *The Conscious Mind*, 33. Contrast this definition with Chalmers's definition of *logical supervenience*: "B-properties supervene *logically* on A-properties if no two *logically possible* situations are identical with respect to their A-properties but distinct with respect to their B-properties" (p. 35).

[43] Donald Davidson, "Mental Events," reprinted in Davidson, *Essays on Actions and Events* (Oxford: Clarendon, 1980), chap. 11; 214.

[44] Godehard Bruntrup, "The Causal Efficacy of Emergent Mental Properties," *Erkenntnis* 48 (1998): 133–45.

There is a certain inherent tension in strong supervenience, however. As Jaegwon Kim, one of its best known (former) advocates, admits, "nonreductive materialism is not a stable position. There are pressures of various sorts that push it either in the direction of an outright eliminativism or in the direction of an explicit form of dualism."[45] One of the reasons for this instability is that such a position appears to leave no room for genuine mental causes; all the determination of outcomes seems to flow from the bottom (the physical substratum), leaving no "room for play" for the mental actually to do anything. At worst, mental phenomena become mere epiphenomena; their reality is bought at the cost of causal impotence.

So the question becomes: Can any framework that is consistent with what we know today about the brain, and with what we may reasonably be expected to *come* to know, also be consistent with a real causal influence of mental phenomena? Not only does folk psychology, the common-sense way of speaking of human persons, depend on a successful theory of mental causation, but the viability of (at least traditional) theological claims does as well. Strong supervenience theories might suggest how religious beliefs and experiences could arise. But however much the *function* of religious beliefs might be incorporated into such accounts, their *truth* could not be. There would be no place for religious insights *as correct* to alter behavior, and definitely no role for any influence of a disembodied divine force on the world. The supervenience concept seemed to offer the sort of framework required for drawing the links between the brain sciences and the mental life that we experience. But strong supervenience conflicts both with folk psychology and with theology.

Is it possible, then, to formulate a "weaker" version of the dependence relationship? I will define *weak supervenience* as the view that, although physical structures and causes may determine the initial emergence of the mental, they do not fully or solely determine the outcome of the mental life subsequent to its emergence.[46] This view amounts to a dependence of genesis, since it grants that the origins of mentality can be traced to the physical conditions without which there would be no mental phenomena. But it does not grant a full, bottom-up determination of the mental by the physical—hence the "Insufficiency Thesis" defended above—even though the degree of bottom-up influence will certainly far exceed our present knowledge. Weak supervenience thus retains the central tenet of supervenience theory: the mental is dependent on, yet not reducible to, the physical. One reason for choosing weak over strong supervenience is the belief in mental causation: there are genuine mental causes that are not themselves the product of physical causes. The causal history of the mental cannot be told in physical terms, and the outcome of mental events is not determined by phenomena at the physical level alone.

Weak supervenience is the stepping-off point for an emergentist theory of supervenience, and thus an emergentist theory of human personhood. The background for emergentist supervenience comes from the British Emergentists in

---

[45] See Jaegwon Kim, "The Myth of Nonreductive Materialism," in *The Mind-Body Problem*, Richard Warner and Tadeusz Szubka, eds. (Oxford: Blackwell, 1994), 242–60. More recently see Kim, *Mind in a Physical World: An Essay on the Mind-Body Problem and Mental Causation* (Cambridge: MIT Press, 1998).

[46] Note that "strong" and "weak" supervenience have been used in other (not always consistent) ways in the literature. I choose to run the risk of terminological confusion because of the particular appropriateness of these two terms to the position defended here.

the 1920s and '30s. As Jaegwon Kim notes, the early emergentists held "that the supervenient, or emergent, qualities necessarily manifest themselves when, and only when, appropriate conditions obtain at the more basic level; and some emergentists took great pains to emphasize that the phenomenon of emergence is consistent with determinism. But in spite of that, the emergents are not reducible, *or reductively explainable,* in terms of their 'basal' conditions."[47] Lloyd Morgan thus appeared to use "supervenient" as an occasional stylistic variant of "emergent."[48]

In a recent article, Timothy O'Connor has defined property emergence in a more careful manner:

Property P is an emergent property of a (mereologically-complex) object O
if and only if:
1. P supervenes on properties of the parts of O;
2. P is not had by any of the object's parts;
3. P is distinct from any structural property of O.

But after those three conditions we come to the big break in the philosophy of mind, the question that Kim calls "arguably the central issue in the metaphysics of mind"[49]: the question of mental causation. O'Connor formulates it this way in his final premise:

4. P has direct ("downward") determinative influence on the pattern of behavior involving O's parts.[50]

It is not difficult to provide a formal definition of emergence in this sense: "F is an emergent property of S if and only if (a) there is a law to the effect that all systems with this micro-structure have F; but (b) F cannot, even in theory, be deduced from the most complete knowledge of the basic properties of the components $C_1, ..., C_n$" of the system.[51] Note that emergent properties of this sort are genuinely novel. As Bruntrup writes, "Even if all the physical facts have been fixed, the emergence of consciousness is not implied with nomological necessity. . . . The existence of emergent properties could not be predicted by even a perfect knowledge of the underlying physical facts alone."[52]

A property is thus emergent only if laws cannot be formulated at the lower level that predict its occurrence *and* subsequent behavior, say, as a boundary condition of other well-established laws at that level. If, for example, we can relate the levels with the same bottom-up precision with which we can formulate the necessary physical conditions for the existence of conductivity or elasticity, then we do not have emergentist supervenience. A set of phenomena is designated as emergentist only when an exhaustive description of the underlying physical state of affairs, although necessary, is not sufficient for explaining the emergent properties. Thus an emergent

---

[47] Kim, *Supervenience and Mind*, 138.

[48] So Kim, ibid., 134.

[49] Ibid., xv.

[50] See Timothy O'Connor, "Emergent Properties," *American Philosophical Quarterly* (1994): 97f.

[51] Ansgar Beckermann, "Supervenience, Emergence, and Reduction," in A. Beckermann, H. Flohr, J. Kim, eds., *Emergence or Reduction? Essays on the Prospects of Nonreductive Physicalism* (New York: W. de Gruyter, 1992), 94–118; 104.

[52] Bruntrup, "The Causal Efficacy of Emergent Mental Properties," 140. Note that in this article Bruntrup does not accept the position that I am defending.

condition seems to be implied in Leslie Brothers' explanation of human social behavior in terms of "the representation of the generalized other" and the irreducible nature of first-person language—assuming that she means these terms to refer to a genuinely psychological reality that is something more than, and not just a different manifestation of, the underlying physical realities. One would also need to use the language of emergence if qualia (human subjective experiences, such as seeing red or being in love) are, at least in part, self-explaining.

I believe that emergentist supervenience offers the philosophically most adequate framework for conceptualizing mental properties in human persons. Does emergentist supervenience also offer a view of the person that is more compatible with theology than does strong supervenience as defined above? If true, would it represent, from the standpoint of theology, a better bridge principle? Clearly the answer is yes. Presumably theologians would have many more things to say about emergent properties and their source and ultimate purpose. They might also attempt to offer theologically based explanations of why the biological world could or would give rise to such emergent properties. Two caveats, however: when speaking in this way, theologians do not speak as scientists, and the status of such language vis-à-vis any presently available empirical verification should be made fully clear. Also, there is nothing in emergentist supervenience that *requires* a theological interpretation; it is not a form of natural theology. Emergentism is, in my view, a necessary condition for a theological interpretation of the human person, but it is emphatically not a sufficient condition for a theological anthropology.

Coming from the viewpoint of science, one might worry that such a position closes off research and hence progress in neuroscience. Does it introduce a constraint on the work of empirical scientists? I would argue not. Emergentists may have an equally vivid interest in knowing more about actual brain functions and in seeing neural explanations extended as far as possible. It is just that they wager that the "as far as possible" does not extend as far as an exhaustive explanation of the mental—unless part of that explanation is given in irreducibly mental terms! Talk about the subjective experience of being in love or the sense of self-awareness is irreducibly mental; such phenomena exercise a type of causal influence of their own.[53]

5 Persons and Explanatory Levels

By exploring the family of positions that eschew both dualism and strong reductionism, this essay has focused on that range of positions that seek to do justice both to neuroscience and to the human experience of personhood. Following contemporary usage, I have characterized these as the family of supervenience theories. We discovered that the same tension arises within this family as was present in the old *dualism versus reductionism* debate. One either does or does not accept the Sufficiency Thesis, the view that the causal explanations of human behavior will ultimately be given in neuroscientific terms.

Since both sides of this new debate accept the supervenience label, one might suppose that the ambiguity lies in the term itself. And indeed this is exactly right: one finds in the literature at least three different ways of characterizing the relation of mental to physical. All three presuppose that mental phenomena represent levels of

---

[53] Indeed, wouldn't it be a strange thing for a neuroscientist to find herself in the position of denying with passionate subjective conviction that there is any such thing as a *force* of subjective conviction?

complexification that depend on lower, simpler levels, yet that are in some sense not fully reducible to those lower levels:

1. The more complex level could be related to the lower level by a clear set of laws (call it *nomological supervenience*). In the present volume this appears to be the position of Theo Meyering, for whom the paradigm supervenience relationship is expressed by phenomena such as elasticity and conductivity. These are phenomena that are well understood scientifically in terms of the behavior of the particles making up the physical system in question, although the supervenient properties cannot be fully expressed except at the level of the set of particles as a whole. Nomological supervenience is also visible in the work of Richard M. Hare, who says explicitly that "supervenience brings with it the claim that there is some 'law' which binds what supervenes to what it supervenes upon." For Hare such laws are necessary conditions for supervenience: "what supervenience requires is that what supervenes is seen as an instance of some universal proposition linking it with what it supervenes upon."[54]

2. The higher level could have all of the attributes listed in *1*, yet without the condition just expressed by Hare, which we might call the "nomological condition." This second position is best known in the guise of what Donald Davidson calls "anomalous monism." Davidson holds that "mental entities (particular time- and space-bound objects and events) are physical entities, but...mental concepts are not reducible by definition or natural law to physical concepts."[55] Davidson disputes the law-likeness of mental events: mental events are of a different type than physical events, although there is a token identity of every mental event with a physical event. Still, in other respects his view stands fairly close to *1* above. Certainly he does not speak of mental phenomena as genuinely emergent. He insists only that at least one portion of the physical world does not admit of the kinds of causal explanation by means of natural laws that science has been successful in formulating in so many other areas. Yet no emergentist conclusions should be derived from this particular failure of law-like explanation, Davidson seems to say; the mental simply obeys different constraints than physical laws, such as the unique constraint of rationality.

3. The final type of supervenience is the one that I have been defending. It finds in mental phenomena and their dependence on the physical a supervenient relationship not unlike that accepted by the other positions. Yet it also finds grounds in the nature of this relationship for the ontological hypothesis that mentality represents an emergent level. That is, without questioning the dependence on the physical, it understands mental properties to be different in kind from those observed at lower levels and to exercise a type of causal influence unique to this new emergent level.

5.1 Minimalist Emergence

It is important to note that a majority of philosophers writing in the field still advocate either *1* or *2*. I believe that this reveals a shared sense of what is at stake in the present debate. If one wishes to avoid talk of self-consciousness (say, in the causal sense used by the German Idealists), or God-talk, or an opening for any other such religiously-tinged predicates, then one must insist that the mental be understood

---

[54] See Richard M. Hare, "Supervenience," *Aristotelian Society Supplementary Volume* 58 (1984): 1–16; 3.

[55] Donald Davidson, "Thinking Causes," in *Mental Causation*, John Heil and Alfred Mele, eds. (Oxford: Clarendon Press, 1995), 3–17; 3. See also idem, "Mental Events."

fully in terms of the physical world. By contrast, if one finds in the mental some sign of a new type of phenomenon within the world, then one has thereby introduced at least the *possibility* that there is something inherently mental or spiritual within the one world that we find around us. Clearly this possibility would represent an opening to theology that is of great significance to both sides. If one wishes to avoid such openings, then one must be sure at every cost that the mental is not interpreted in an emergentist sense. Conversely, it seems that those with theological interests—and with some motivation to integrate these interests with their understanding of science—will need to develop a theory of humans and their mental life that is either emergentist or establishes the same sort of minimal opening that emergentism defends. These are the stakes that make the present discussion and the present volume of such overwhelming importance. It is perhaps not too much to say that this debate about the human person expresses the crux of the battle between physicalist naturalism and its opponents today.

It is also a debate with no easy resolution, as we have seen. "Opening" means possibility, not proof; no one is talking of conclusive demonstrations here. One could easily accept emergentist supervenience and deny the truth of theism or religious belief in all its forms.[56] Still, even in this incredibly circumscribed form, emergentist supervenience presents one with what philosophers call a "forced choice." If one holds that all mental phenomena are only expressions of physical causes or are themselves, at root, physical events, then one has (at least tacitly) advanced a theory of the human person that is pervasively physical. It then becomes extremely unclear (to put it gently) why, *from the perspective of one's own theory of the human person*, a God would have to be introduced at all (except perhaps as a useful fiction). If a theologian espouses physicalism, she may be forging an alliance with the majority worldview within the neurosciences, but she may also be giving up the most interesting rapprochement between theology and the sciences of the person just as she approaches that debate's most decisive issue. By contrast, to introduce a soul-substance at this crucial juncture would be to abandon the debate altogether, for that move, almost by definition, leaves no common ground with natural science. Here I have argued not that supernatural souls exist but rather that human action reflects a type of mental causation that is something more than physical. This claim, minimalist as it is, may just be the necessary condition for a theology that is anything more than metaphorical. Theologians stand before their Rubicon and must either cross or not cross.

My strategy has been to map out a crossing where the river is most narrow (why add any unnecessary distance when the crossing itself is already difficult enough?). This helps explain why I have broken with dualist thinking and moved as far as possible in the direction of the natural sciences by arguing that:

- mental predicates represent a type of property, not a new form of substance;
- mental causation does not involve the addition of new energy into physical systems;

---

[56] Indeed, as I show in the final chapter of *God and Contemporary Science*, emergentist supervenience stands in a certain tension with traditional theological belief, which asserts a dependence of the physical on the spiritual. Either the dependence of the mental on the physical that I have defended must be corrected from another source, or it will require significant revisions in traditional Christian belief.

- mental processing does not occur without concurrent physical activity. Indeed, changes in brain structure and function (brain disease, lesions) have important and predictable effects on mental functioning;
- one's overall ontology should be monist. There is only one natural order, although it includes many different types of things. Mental causation is not supernatural; it is natural. It is thus amenable to explanation in this-worldly terms, although at least part of the explanation will need to employ irreducibly psychological concepts. I have not pleaded for supernatural interventions, nor have I construed mental functioning in any way analogous to the classic supernaturalist notion of intervention from outside. To put it bluntly: though there may be divine action on analogy with the action of embodied persons with the world, I have left no place for miracles in the sense of a countervening of natural law.[57]

I imagine, one final time, the objection, "Well, you have wagered against neuroscience, have you not?" The critic might object that I have introduced, if not a "God of the gaps," then at least a "mental causation of the gaps." Is not the more scientific response to expect that law-like explanations will eventually be possible "all the way down"—until all phenomena in the natural world have been explained from the bottom up? Doesn't "wagering" in this way amount to betting against science, and thus blocking the road for scientists? Indeed, is not the success of science heretofore good reason to conclude that bets on my side are backward-looking, obscurantist, and in general inhibitors to further scientific progress?

No. These well-worn objections tell against dualist positions, but they beg the question in the dispute between supervenience theorists. The reason it is absurd to postulate occult forces in the physical world (or "vitalist" forces in the biological world) is that we have learned that these realms operate in a fully law-like manner *based on explanatory successes in the relevant sciences.* What is really at stake in the present volume is the question of whether human persons are analogous— whether they can be exhaustively predicted and explained in a "bottom-up" manner. I have argued that we have good evidence to think not. Indeed, the hierarchy of the sciences itself offers evidence of principles which are increasingly divergent from "bottom-up" physicalist explanation.[58] Functionalist explanations play a role in the biological sciences (from cell structures through neural systems to ecosystem studies) that is different from the structure of explanation in fundamental physics, just as intention-based explanations play a role in explaining human behavior that is without analogy at lower levels. These emerging orders of explanation may also involve an increasing role for top-down explanations. Thus, for example, DNA embodies in its very structure the top-down action of the environment on the molecular biology of the human body. In intentional explanations it is even more clear that the goal for which the agent acts, or the broader context within which she understands her actions, influences the particular behaviors or thoughts.[59] An

---

[57] The details of what divine action would look like in this context are spelled out in my *God and Contemporary Science.*

[58] I cannot review the entire argument here. It is powerfully laid out in Arthur Peacocke, *Theology for a Scientific Age: Being and Becoming—Natural, Divine, and Human,* enlarged ed. (Minneapolis: Fortress, 1993) and is the topic of my forthcoming monograph on emergence.

[59] I thank George Ellis for pointing this out to me (in private communication).

emergentist view of the person is thus not an argument against science but rather one that is consistent with the pattern that we find emerging in the natural hierarchy of the sciences.

5.2 Emergence, AI, and the Social Sciences

This last point is important enough to bear restating: the case for emergent mental causation is not by itself a case for the existence of God, divine action, an eternal soul, or life after death; it is not directly a theological conclusion at all. Indeed, in some ways it might seem to be an *anti*-theological conclusion, because it understands mental phenomena to be "of a piece" with physical phenomena, and because the supervenience relationship asserts a dependence of the mental life on its physical basis—indeed, a high correlation between physical causes and mental effects— which is on the surface inconsistent with many parts of Christian teaching. To accuse this view of being a cheap theological concordism is to neglect all of these sharp differences. I have argued only that human mentality is an emergent feature of a very complex biological structure, the human brain, in its interaction with its environment.[60] This conclusion is not a No to science and a Yes to faith. Rather, it suggests that one will have to supplement the neurosciences with another set of sciences, say the human sciences, before one can provide the full explanation of that particular part of the natural world which is us.

Consider the parallels with the Artificial Intelligence (AI) program in computational theory and computer science. The challenge of the Turing test was to build a computational device whose outputs could not be distinguished by human agents from human outputs. In seeking to meet the test, computer scientists first worked at what is now called (by its critics) "brute force" AI. The hope was to achieve a functional similarity to human outputs by means of sheer computational power. However, already by the time of Deep Blue, the chess program that beat Gary Kasparov, a variety of other techniques had been introduced, effectively supplementing brute-force computation by a combination of heuristics and higher-level criteria. In the process of leaving behind brute-force AI, the very understanding of computational theory has been transformed. It has now been stretched— rightly, in my view—to include fundamental questions in semantics and the theory of meaning, so that theories of computation can now include holistic considerations such as the impact of broader semantic systems, contexts, and applications that go well beyond the actual computations in question. In very recent years, in what appears as a natural next step, some thinkers have even migrated from computational theory to fundamental ontology, the debate over how worlds are first constructed by means of information.[61]

One notes a certain irony here. The battle began over what is the most scientific approach to take in the study of humans. Those who offer a purely physicalist account of human mental predicates claim for themselves the laurel of scientificness and accuse their opponents of obscurantism. Yet according to the opposing position it will actually be more scientific to *deny* that human intentional actions can be

---

[60] Leslie Brothers emphasizes this dimension of sociality; on her account, the mental world grows (for instance) out of the sort of brain-brain interaction that we call conversation. See Brothers, *Friday's Footprint: How Society Shapes the Human Mind* (New York: Oxford University Press, 1997).

[61] See the detailed argument in Smith, *On the Origin of Objects*.

explained as law-like phenomena—*if*, as emergentists and anomalous monists believe, the phenomena in question are actually more than physical in nature.

The ongoing debate about the nature and methodology of the social sciences recapitulates (and sheds some helpful new light on) the discussion to this point. The two opposing camps appeal to the two warring fathers of modern social science, August Comte and Wilhelm Dilthey. Comteans argue for a predominantly natural-scientific approach to the social sciences, allowing no in-principle gap between them and the natural-scientific study of the human organism.[62] Present-day Diltheyans maintain that the object of study to which the human sciences are devoted is significantly different from the natural world. The natural world can be grasped using *causal* patterns of explanation, because such events really are the product of a series of causes. But human actions require the method of *Verstehen* or *empathetic understanding*, for human beings are subjects who are engaged in the project of making sense of their own world. Intentional actions can be understood only in terms of the logic of intentionality: wishing, judging, believing, hoping.[63]

The battle continues. A new round was launched by the successes of behaviorist social science, by Theodore Abel's oft-cited Comtean manifesto for positivism in the social sciences,[64] and more recently by the rapid advance of the neurosciences; shots were then returned by humanist psychologists and by more hermeneutically inclined theorists.[65] At the same time, analytic thinkers have carefully stressed the difference between explanations of human intentional actions and causal explanations of occurrences in the world, as in Georg Henrik von Wright's detailed defense of the logic of intentional explanations.[66] Whereas Carl Hempel tried to subsume the explanation of human actions under his general model of deductive-nomological explanation,[67] other leading philosophers of science such as Ernest Nagel underscored the unique nature of explanations of social action.[68] The net result is a clearer

---

[62] See Auguste Comte, *Cours de philosophie positive*, trans. as *Introduction to Positive Philosophy*, ed. Frederick Ferré (Indianapolis: Hackett Pub. Co., 1988).

[63] See Wilhelm Dilthey, *Hermeneutics and the Study of History*, Rudolf Makkreel and Frithjof Rodi, eds. (Princeton: Princeton University Press, 1996); idem, *Introduction to the Human Sciences*, Rudolf Makkreel and Frithjof Rodi, eds. (Princeton: Princeton University Press, 1989). Dilthey used this argument as the basis for his broader theory of the social sciences. The debate was repeated in the work of Wilhelm Windelband and others; see Windelband, *Encyclopädie der philosophischen Wissenschaften, in Verbindung mit Wilhelm Windelband*, Arnold Ruge, ed. (Tübingen: J.C.B. Mohr, 1912ff.).

[64] See Theodore F. Abel on explanation versus understanding in *The Foundation of Sociological Theory* (New York: Random House, 1970).

[65] For example, see Hans-Georg Gadamer, *Truth and Method*, Garrett Barden and John Cumming, eds. (New York: Seabury Press, 1975).

[66] Georg Henrik von Wright, *Explanation and Understanding* (Ithaca: Cornell Univ. Press, 1971).

[67] See Carl Hempel, "Typological Methods in the Natural and the Social Sciences," in Hempel, *Aspects of Scientific Explanation and Other Essays in the Philosophy of Science* (New York: Free Press, 1965).

[68] See Ernest Nagel, *The Structure of Science: Problems in the Logic of Scientific Explanation* (London: Routledge and Kegan Paul, 1961). In the end, however, Nagel's position comes out rather close to Hempel's.

sense of what it is that sets person-based explanations of individual and social action apart from causally based explanations.[69]

6 Separating Theology and the Theory of the Person

What relevance do such general reflections on the philosophy of social science have for the debate between the neurosciences and theology? The latter discussion is often set up as a special case of the more general debate over the independence of science. Anyone who suggests that there might be a limit to the scope of neurologically based accounts of human thought is understood to be doing something to scientific inquiry analogous to what supernaturalists do to physics when they insist that God can break into the natural order at any time, bringing about any result he pleases with no attention to natural causal influences or the requirements of the energy conservation principle.

Yet a moment's reflection shows that the two instances are precisely *not* analogous. The appeal to divine in-breaking and miracles is the appeal to actions carried out by an agent whose existence is contested and who is by definition unique, unlike any other agent. By contrast, explanations in terms of human intentions appeal to experiences that are (presumably) shared by every agent who can read and comprehend the words on this page. Cultural anthropology studies intentions shared by large groups of actors, as does sociology; and even psychology, with its focus on what is uniquely individual, still speaks in terms of shared personality types, motivations, complexes, structures, and pathologies. In all these cases, the explanatory (scientific) goal is to understand and explain the behavior of a large class of agents called human subjects—a type of natural entity with which each of us is deeply familiar, in part through direct introspective awareness. The question at issue, then, is not (in the first place) a theological question at all, but a basic question about human agency: Should we expect in the long run that neuroscientific explanations will be adequate to explain human behavior, or do social scientific explanations pick out a type of action in the world which demands an explanatory level of its own? One's motivation need not be in any way theological in order to defend an account of the human self that includes self-conscious intentions as basic building blocks of human behavior.

This distinction, albeit crucial, is more difficult than it may at first appear. A neuroscientist such as Michael Arbib, for example, might well insist that he too preserves a place for social scientific explanations. He might well (indeed, does!) insist that schema theory as discussed above supplements base-level neurological explanations of human behavior with a higher level of analysis, one which includes those schemas that make up the social world. The difference remains, however. Arbibian schemas are constructs composed out of neurological events and physical events in the world, which are their real foundation. Individual actors may *believe* that things such as societies and their institutions—and ideas such as freedom and responsibility, not to mention divine beings—really exist. But they are mistaken. Eventually, when we understand the physical nature of the human being well enough, we will be able to give a complete account of how ideas such as these came to be constructed. At that time we will leave behind the fictions of their independent reality

---

[69] Anthony Giddens explains the difference in terms of the "double hermeneutics" that characterizes social explanations in *New Rules of Sociological Method: A Positive Critique of Interpretive Sociologies* (London: Hutchinson, 1976).

and return to a purely physicalist account of ourselves and our computational products.

It is a very different thing to argue, as I have, that the social sciences pick out phenomena which, even though they have emerged from the physical world, are causally irreducible. This is an ontological claim, but it is also a claim about what the study of human persons entails, a claim about the form that (at least some) explanations of persons must take. It is the question that is ultimately at stake in the debate to which this volume is devoted.

Interestingly, we have found that the debate about neuroscience and personhood has a sort of fractal structure. When we left behind dualism and strong reductionism, it looked like we would find common ground; but the ground quickly fissured into (for example) schema theory and agent-centered theories. When we introduced supervenience theory as common ground between those views, we discovered multiple meanings of supervenience, which mirrored the tensions already encountered at the first two levels. Presumably, if there were space to explore the concept of emergence in greater detail, we would find the same disagreements occurring again at this level. And yet the iterations have helped to give sharp profile to the recurring dispute: Are neurological explanations finally sufficient or insufficient?

7 Emergentist Monism

One might ask, "What does it all mean? What kind of ontological position do these emergent properties entail? Is it monism, property dualism, or panpsychism? And where does this all leave theology?"

The ontological view that I defend might be called *emergentist monism*.[70] Monism asserts that only one kind of thing exists. There are not two substances in the world with essentially different natures, such as the *res cogitans* and *res extensa* (thinking and extended substance) propounded by René Descartes and the Cartesians. But unlike dual-aspect monism, which argues that the mental and the physical are two different ways to characterize the one "stuff," emergentist monism conceives the relationship between them as temporal and hierarchical.

In one sense, monism is a necessary assumption for those who wish to do science. For instance, we can (and must) assume that the total physical energy of the universe as a whole is conserved. No action that you perform, no thought that you think, can add to the total energy of the system without invalidating calculations based on physical laws. Incidentally, this is the problem with dualism, and with direct interventions into the world by a God who breaks natural laws: if a spiritual agent gives rise to a physical effect, it has brought about physical change without a physical cause or the expenditure of physical energy, and this fractures the natural order in a way that would make science impossible. There could be no scientific study of a world where cups spontaneously fly across the room and objects released from your hand could go either up or down according to spiritual forces. Science does not need full determinism (see the next paragraph). But it does need the world to reflect at least patterns of probability over time.

(Note that monism is not only in the interests of science; one can *also* give theological arguments in defense of monism. Monism makes the assertion that the world is one, that it constitutes a distinct order. Theologians speak of the universe

---

[70] The term was developed in ongoing conversations with Arthur Peacocke; see his essay in this volume.

as a whole as *finite*, in order to specify its single ontological status and to contrast it with a Creator whose nature is essentially infinite. Herein lies the theological importance of the phrase, "the unity of nature": in comparison to the Creator, all things in the universe share a common nature. Theologians have also argued that creatures can only exercise freedom within an ordered world that has an integrity and law-like structure of its own.)

I do not care if you want to think of this monism as a *sort* of materialism, but only if you mean by that that the "things" in the world—rocks and computers and persons—are all made out of *some material or other*. What is crucial is that you develop theories which do justice to the specific qualities that we actually find associated with the various "things" in the world. For example, after Isaac Newton we thought that physics presupposed at least the possibility of a fully determinate, and determined, account of the world. But when we found out that microphysical or quantum events simply do not work this way, we developed an essentially stochastic or probability-based science to deal with them. Likewise, when scientists began to research chaotic "systems," or systems far from thermodynamic equilibrium, they discovered that they were *essentially* unpredictable (for finite agents). But science did not end; instead, a fascinating new science of chaotic systems has been developed. An equally complex story would have to be told about the convertibility of matter and energy.

Now we come to a *very* complex object in the world: humans. With 10^{14} neural connections, the brain is the most complex interconnected system we are aware of in the universe. This object has some *very* strange properties that we call "mental" properties—properties such as being afraid of a stock market crash, or wishing for universal peace, or believing in divine revelation. On the one hand, to suppose that these features will be fully understood in terms of physics as we now know it is precisely that: a supposition, an assumption, a wager on a future outcome. A deep commitment to the study and understanding of the natural world (which I share with most, and probably all, the contributors to this volume) does not necessitate taking a *physicalist* approach to the human person—if by that one means that the actions of persons must be explained through a series of explanatory sciences reaching down (finally) to physics, or, more simply, that all causes are ultimately physical causes. (Note that under this definition there could be both reductionist and nonreductionist versions of physicalism.) On the other hand, *for both scientific and theological reasons*, I do not therefore advocate introducing an occult entity, such as Descartes's soul substance, in order to explain the person. To say that the human person is a *psychosomatic unity* is to resist both positions. It is instead to say that the person is a complexly patterned entity within the world, one with diverse sets of naturally occurring properties, each of which needs to be understood *by a science appropriate to its own level of complexity*. We need multiple layers of explanatory accounts *because* the human person is a physical, biological, psychological, and (I believe also) spiritual reality, and because these aspects of its reality, though interdependent, are not mutually reducible. Call the existence of these multiple layers *ontological pluralism*, and call the need for multiple layers of explanation *explanatory pluralism*, and my thesis becomes clear: ontological pluralism begets explanatory pluralism. (Or, to put it differently: the best explanation for explanatory pluralism is ontological pluralism.)

Elsewhere in this volume, Nancey Murphy draws on the work of Francisco Ayala in distinguishing the multiple meanings of "reductionism." Given her definitions, note that an emergentist position rejects causal reductionism, since it accepts mental causes. It therefore rejects explanatory (theoretical, epistemological) reductionism, insofar as mental properties need to be explained using a theoretical structure appropriate to them. At first blush, emergentist monism may *seem* like a version of ontological or metaphysical reductionism, since it breaks with dualism and refuses to postulate nonphysical entities such as souls. But emergentists must finally declare themselves opposed to reductionism even with respect to ontological (metaphysical) questions, for their central assertion is that the history of the universe is one of development and process. The one order exists at each stage in its history, but *what it is* that exists is not identical through time. Genuinely new properties emerge which are not reducible to what came before, although they are continuous with it.

What *emerges* in the human case is a particular psychosomatic unity, an organism that can do things both mentally and physically. Although mental functions supervene upon a physiological basis, the two sets of attributes are interconnected and exhibit causal influences in both directions. We therefore need a science or mode of study that begins (as a science should) with a theoretical structure adequate to this level of complexity. To defend an emergentist account of the self is not to turn science into metaphysics. Instead, it is to acknowledge that the one natural world is vastly more complicated and more subtle than physicalism can ever grasp. You can *wager* that the *real* things that exist in the world are physical processes within organisms, and that everything else—intentions, free will, ideas like justice or the divine—are "constructs," complicated manifestations of neural processes. But I am wagering on the other side. I wager that no level of explanation short of irreducibly psychological explanations will finally do an adequate job of accounting for the human person. And this means, I have argued, the real existence and causal efficacy of the conscious or mental dimension of human personhood.

8 Some Potential Objections

I conclude with some objections that might be (or have been) raised against this position.[71]

8.1 Why Not Use the Label "Physicalism"?

As long as one is concerned with the physical sciences and the study of objects in the physical world, why not use the label "physicalism" as the overarching position? Perhaps (my critic might grant) it is important that scientists countenance causal influences at various levels, and thus also explanations at each of these levels as well. Put differently, perhaps it is necessary to resist causal and explanatory reductionism. But why resist ontological reductionism? Indeed, what is the point of arguing about ontology anyway?

Presumably it is not necessary for working scientists to take a strong position on ontological questions like "what really exists" (though this fact does not prevent some from being vociferous advocates of ontological positions like physicalism and materialism). Moreover, one can work in some areas of the philosophy of science without raising ontological questions. But surely *theologians* hold some important ontological commitments, for they are committed to the existence of a spiritual being or dimension which, while it may include the world, transcends it as well. The

---

[71] Although I have not cited persons by name in what follows, I am again grateful to the various members of the Vatican Observatory-CTNS working group for raising these (or related) criticisms over the two years of the project.

biblical doctrine of the image of God (*imago dei*) suggests that something of the spiritual nature of this God is reflected in the nature of human beings (and perhaps in other parts of the natural world as well). Surely this fact commits theologians to an ontological thesis: the thesis that human persons, correctly and fully understood, include a spiritual dimension which, whatever else it is, is more than physical. It is for this reason that theologians, at least, cannot eschew the ontological question and cannot, at the end of the day, be satisfied with the label "physicalist."

8.2 Is this Theory Crypto-Dualist?

Clearly the position defended here is not a version of substance dualism; there has been no suggestion of mental substances intervening in the physical order. But is it a variant of *property* dualism, the view that, even if there is only one kind of substance, it has two fundamentally different kinds of properties?

Such a criticism rests on a misunderstanding. I have not portrayed a world divided into two distinct types of qualities, but rather a world with a vast array of different types of properties. Though there is no justification for the "dualism" label, the theory could fairly be called *property pluralism*, since it countenances a wide range of properties depending on their position in the complexity hierarchy. In a similar vein, Roger Sperry writes of his own position, "Because it is neither traditionally dualistic nor physicalistic, the new mentalist paradigm [in the study of consciousness] is taken to represent a distinct third philosophical position. It is emergentist, functionalist, interactionist [in the sense that it sees mental phenomena 'as primarily supervening on rather than intervening in the physiological processes'], and monistic."[72] Once one has grasped the hierarchical structure of the physical world, one can leave the old opposition between physicalism and dualism behind.

8.3 Why Not Think That Full Neuroscientific Explanations of Consciousness Will Eventually Be Available? Doesn't the View Taken Here Block Progress in Neuroscience?

"One should be cautious about wagering against scientific progress!" this critic complains. "Who would have imagined what we would learn through the new scanning technologies, or microsurgery techniques, or computer modeling? For that matter, who could have imagined fifty years ago what computers would be capable of today, and who can guess what they will be doing fifty years from now? Never bet against science!"

But my wager against the Sufficiency Thesis does not stem from underestimating the likely advances in neuroscientific theory or from dreading the coming advances in this field; far from it. Instead, it stems from *limitations in principle* which neuroscientists, philosophers, and even some theologians have neglected. Consider a parallel case: when a quantum physicist tells you that she has reason to think that even major advances in her field will not overcome physicists' inability to know both the location and the momentum of a subatomic particle with full precision, she is not being a pessimistic reactionary, for there are compelling scientific reasons to think that this limitation on our knowledge is intrinsic to quantum phenomena. Likewise, there are strong reasons to think that qualia are *intrinsically* subjective experiences. As the immunologist Gerald Edelman (who is certainly no dualist!) writes,

---

[72] Roger W. Sperry, *Science and Moral Priority: Merging Mind, Brain, and Human Values* (New York: Columbia University Press, 1983), 165.

We cannot construct a phenomenal psychology that can be shared in the same way as a physics can be shared.... What is directly experienced as qualia cannot be fully shared by another individual as an observer. [And later:] There is something peculiar about consciousness as a subject of science, for consciousness itself is the individual, personal process each of us must possess in working order to proceed with *any* scientific explanation.[73]

As in the quantum case, there is something in the case of qualia—an essentially first-person aspect—that makes them irreducible to the third-person scientific perspective. This aspect, which philosophers knew (and all human subjects know?) as consciousness or self-awareness, represents perhaps the single strongest argument on behalf of mental qualities as genuinely emergent in the sense defended in this essay. If Edelman is right, qualia cannot be exhaustively explained by neuroscience because they are the precondition for there being any scientific explanations in the first place.

8.4 Can You Have Emergence Without Falling into Vitalism, Idealism or Other Scientific Heresies?

Analogous to Arthur Peacocke's essay in this volume, I have sought to make the case that emergentist monism is not a quirky or anti-scientific metaphysical position. The strength of Peacocke's *Theology for a Scientific Age* is that it introduces emergent properties not just at the spiritual or mental level, or at the origin of life, but as a pervasive principle running through the hierarchy of the sciences. Emergent phenomena might be seen to occur even at the level of physical chemistry; as the chemist Joseph Earley has recently written, "Chemical combination generates properties and relations that are not simply related to the properties and relations of the components."[74] (Of course, one would have to appeal to a broader theory of emergence of the sort defended here to show that chemical phenomena are not merely physical phenomena under a different description.)

What is especially intriguing about the emergentist position is that it makes mental phenomena not an *exception* to the patterns in other sciences but rather yet another instance of them—albeit "higher," more complex, and in some respects stranger than any other properties of the natural world known to us. The vitalists, neo-idealists, and (to a lesser extent) the British Emergentists of the 1920s were all committed to a strongly metaphysical position which they brought *to* the biological sciences. This is why critics are justified in dismissing especially the first two positions as in tension with scientific results and methods. By contrast, I am advocating a theory of emergent properties no stronger than is required to interrelate results up and down the hierarchy of the sciences. I claim that this theory does a *better* job of interpreting the connections (and discontinuities) between various scientific disciplines than do any of its competitors.

Of course, one can speculate further about emergent properties at still higher levels (spiritual properties, say), or about orders of reality beyond the natural order as a whole; and surely systematic and philosophical theologians will find it necessary

---

[73] Gerald M. Edelman, *Bright Air, Brilliant Fire: On the Matter of the Mind* (New York: Basic Books, 1992), 114, 138.

[74] Joseph E. Earley, Sr., "How Constrained is the Origin of Coherence in Far-from-Equilibrium Chemical Systems?" delivered at the Second Conference on the Philosophy of Chemistry, Cambridge University, August 3–7, 1998, manuscript, p. 3.

to pursue some of these lines of reflection, as I have done elsewhere.[75] But to defend an emergentist theory of mental properties in dialogue with the neurosciences, as has been done in these pages, does not immediately commit one to a full-bodied (or: a fully *dis*embodied) theology or theory of the supernatural. In one sense I have sought nothing more here than to resist an unnecessarily physicalist interpretation of recent neuroscientific results that would bring them into conflict with those other disciplines (and experiences) that must also play a role, finally, in a full theory of the human person.

---

[75] For example, I have suggested some possible threads to pursue in the final chapter of *God and Contemporary Science*, esp. 257ff. See also Clayton, "The Case for Christian Panentheism," *Dialog* 37 (Summer 1998): 201–8; and the four critiques of this view, together with my response to them, in *Dialog* 38 (Fall 1999).

THE SOUND OF SHEER SILENCE:
HOW DOES GOD COMMUNICATE WITH HUMANITY?[1]

Arthur Peacocke

[Elijah] got up, and ate and drank; then he went in the strength of that food forty days and forty nights to Horeb the mount of God. At that place he came to a cave, and spent the night there.

Then the word of the Lord came to him, saying, "What are you doing here, Elijah?" He answered, "I have been very zealous for the Lord, the God of hosts; for the Israelites have forsaken your covenant, thrown down your altars, and killed your prophets with the sword. I alone am left, and they are seeking my life, to take it away."

He said, "Go out and stand on the mountain before the Lord, for the Lord is about to pass by." Now there was a great wind, so strong that it was splitting mountains and breaking rocks in pieces before the Lord, but the Lord was not in the wind; and after the wind an earthquake, but the Lord was not in the earthquake; and after the earthquake a fire, but the Lord was not in the fire; and after the fire a sound of sheer silence. When Elijah heard it, he wrapped his face in his mantle and went out and stood at the entrance of the cave.

I Kings 19: 8–13 (NRSV)

1 Introduction

When Elijah *in extremis* and in flight from the wrath of Jezebel sought a message from God and stood expectantly on the "mount of God," Horeb, what brought him to the mouth of his sheltering cave was not the great wind, the earthquake, or the fire, but—we are told—"a sound of sheer silence," from the depths of which Elijah is addressed by God.[2] The story encapsulates the directness and immediacy of such experiences and at the same time exemplifies their baffling character. For it is not only these archetypal figures and events in the tradition which have this character,

---

[1] This essay amplifies and extends a train of thought concerning the significance of "whole-part constraint" in relation to divine action which has engaged me since 1987; cf. fn. 1, p. 263, of my "God's Interaction with the World," in *Chaos and Complexity: Scientific Perspectives on Divine Action*, Robert J. Russell, Nancey Murphy and Arthur Peacocke, eds. (Vatican City State: Vatican Observatory; Berkeley, Calif.: Center for Theology and the Natural Sciences, 1995), henceforth *CAC*. I have used the term "whole-part constraint" to avoid any possible Humean implications of "downward/top-down causation" previously employed in this context. Perhaps this was unnecessarily cautious (cf. my guarded language in fn. 22, p. 272, in *CAC*!), since I continued to envisage a causative influence of the "whole" on the parts in complex systems (i.e., of the system on its constituents), as my essay in *CAC*, 272–76, 282–87, shows. Here I take the opportunity to emphasize this and to take account of other concepts that have been used to describe the whole-part and the mind-brain-body relation so that the inclusive notion of "whole-part influence" (as I here denote it) can be applied as an analogy for divine action, especially in relation to God's communication with humanity, that is, with possible divine effects on human consciousness (an approach which I developed earlier in my *Theology for a Scientific Age*, 2nd enlarged edition, [London: SCM Press; Minneapolis: Fortress Press, 1993], esp. in chap. 11—henceforth *TSA*).

[2] I Kings 19: 12 (NRSV). The implicit paradox is well illustrated by the alternative translations: "a low murmuring sound" (NEB); "a faint murmuring sound" (REB); "a sound of gentle stillness" (RV, footnote); and, of course, the familiar "a still small voice" of AV and RV.

but also the widespread "religious" experiences of humanity—both those inside and those outside of religious tradition.[3] The content of such experiences will be the concern of the last section of this essay, but their very existence raises questions about the general nature of God's interaction with the world and with humanity, especially when both are viewed in the contemporary perspectives of the natural and human sciences. The track of inquiries into "scientific perspectives on divine action" in the CTNS-Vatican Observatory series of research conferences has inevitably led to the question of how God possibly *can* communicate with a humanity that is part of the natural world and evolved in and from it. The natural and human sciences clearly provide a context entirely different from the cultural milieu of the legends concerning Elijah—and indeed from that of even a hundred years ago. The dominance of the essentially Greek, and unbiblical, notion in the Christian world that human beings consist of two distinct kinds of entity (or "substance")—a mortal, physical body and an immortal "spirit" (or "soul")—provided a deceptively obvious basis for envisaging how God and humanity might communicate. The divine "Spirit" was thought then to be in some way closely related to, and capable of communication with, the human "spirit"—both were capable of being, as it were, on the same wavelength for inter-communication. This ontology of "spirit" was not physicalist insofar as it was understood that "spirit" was not part of the causal nexus of the physical and biological world which the natural sciences continue to explicate.

The basis for such an ontology has been undermined by the general pressure of the relevant sciences towards a monistic nondualist view of humanity. In what follows we shall examine (2.1) the perspectives of science on the world[4] and advocate an "emergentist monism" as the epistemology and ontology most appropriate to these perspectives. The relation of wholes to parts in the systems of the world, which bears upon how effects and influences are transmitted in the world, is discussed (2.2) and the idea of "whole-part influence" is again utilized (2.2.1). Other terms used in this context are also surveyed and related to this notion (2.2.2). The idea of a "flow of information" between, and even in, systems proves to be illuminating (2.3), especially when the world is viewed (2.4) as a "System-of-systems." The mind-brain-body relation is considered (2.5) in the light of the foregoing and it transpires that the details of the relation between cerebral neurological activity and consciousness (the concern of many of the essays in this volume) cannot in principle detract from or particularly illuminate the causal efficacy of the content of the latter on the former. In other words, "folk psychology" and the holistic language of personhood are held to be justified and vindicated. The nature of communication between persons is then analyzed (2.6) and found to be mediated entirely through patterns within the physical constituents of the world, consistently with the monist feature of this approach and without eliminating the place for consciousness and intention in interpersonal communication.

With this as background, the inquiry can then move on to considering God's interaction with the world (3) and to distinguishing between various modes of this relation (3.1). In section 3.2, reasons for eschewing any attribution of "intervention" by God will be given, while recognizing that the key problem of the "ontological gap(s)" at the "causal joint" of divine interaction may, in principle, never be

---

[3] See also sec. 4.1 below, "Revelation and 'Religious Experience'," and fn. 80.

[4] Here, and elsewhere, the "world" = "all-that-is," including humanity—that is, everything other than God.

soluble—though its location can usefully be discussed and affirmed to be holistic and everywhere. How God may be best conceived as bringing about events, or patterns of events, in the world will be addressed in section 3.3, and an earlier hypothesis of the author—of divine holistic action on the world-as-a-whole by "whole-part influence"—will be further developed. This leads to a reinstatement of the traditional model of God as a personal agent in the world, albeit in a new perspective.

On this foundation it proves possible to move on to the question of how God could affect the content of human thinking instantiated in human-brains-in-human-bodies—that is, of how God could communicate with a humanity embodied in the natural world. This will entail consideration (4.1) of the status of what has traditionally been called "revelation" in human experience and, more particularly, religious experience. Finally, we can then examine (4.2) how God might be considered as communicating with humanity and whether such communication can be regarded as "personal."

As it happens, a perceptive—indeed magisterial—treatment earlier this century by Oliver Quick, in relation to sacramental theology, provides a useful conceptual framework for linking the steps in this inquiry.[5] His approach was based on a working distinction in human experience which can be extended to God's relation to the world. There are two ways, he suggested, in which "outward" things or realities—those which occupy space and time and are in principle, at least, perceptible to human senses (basically, the "physical")—may be related to our "inward" mental lives, which do not occupy space and time and are not perceptible to the senses. The "outward" things or realities may take their character either (1) from what is *done* with them in implementing "inward" mental states; or (2) from what is *known* by and through them of "inward" mental states. The first is an *instrumental* relation and the second a *symbolic* one.

This broad distinction in human experience has a parallel in God's relations to the world, to "all-that-is," which, in the Jewish and Christian monotheistic traditions, may be viewed (1) as the *instrument* whereby God is effecting some purpose by acting on and doing something with and through it; or (2) as the *symbol* in and through which God is signifying and expressing God's eternal nature to those who have the ability to discern it. We need to postulate ways in which God can effect instrumentally particular events and patterns of events in the world, in order to render intelligible *how* God might be known symbolically through particular events or patterns of events. These are what they are, and not something else, because of God's intention and purposes to communicate to humanity. Quick's analysis points to the need to clarify the *instrumental* mode of God's interaction with the world in crder to underpin the possibility of God's *symbolic*, communicating action on human-brains-in-human-bodies, that is, in our thinking. So what are the features of the world unveiled by the sciences that are relevant to such an inquiry?

2 The World

2.1 Scientific Perspectives on the World—Emergentist Monism

The underlying unity of the natural world is testified to by its universal, embedded rationality which the sciences assume and continue to verify successfully. In the realm of the very small (the subatomic) and of the very large (the cosmic), the

---

[5] Oliver C. Quick, *The Christian Sacraments* (London: Nisbet, 1927; repr.1955), chap. 1.

extraordinary applicability of mathematics—the free creation of human ratiocination in elucidating the structures, entities, and processes of the world—continues to reinforce that it is indeed *one* world. Yet, the diversity of the same world is apparent not only in the purely physical—molecules, the Earth's surface, the immensely variegated denizens of the astronomical heavens—but even more strikingly in the biological world. New species continue to be discovered, in spite of the depredations caused by human action.

This diversity has been rendered more intelligible in recent years by an increased awareness of the principles involved in the constitution of complex systems. There is even a corresponding "science of complexity" concerned with theories about such systems. It will be enough here to recognize that the natural (and also human) sciences increasingly give us a picture of the world as consisting of a complex hierarchy—or more accurately, hierarchies—a series of levels of organization of matter in which each successive member of the series is a whole constituted of parts preceding it in the series.[6] The wholes are organized systems of parts that are dynamically and spatially interrelated—a feature (sometimes called a "mereological" relation) which will concern us further below (section 2.2). This feature of the world is now widely recognized to be of significance in relating our knowledge of its various levels of complexity—that is, the sciences which correspond to these levels.[7] It also corresponds not only to the world in its present condition but also to the way complex systems have evolved in time out of earlier simpler ones.

What is significant about this process in time and about the relation of complex systems to their constituents now is that the concepts needed to describe and understand—as indeed also the methods needed to investigate—each level in the hierarchy of complexity are specific to and distinctive of those levels. It is very often the case (but not always) that the properties, concepts, and explanations used to describe the higher level wholes are not logically reducible to those used to describe their constituent parts, themselves often also constituted of yet smaller entities. This is an epistemological assertion of a nonreductionist kind, and its precise implications have been much discussed. With reference to a *particular* system whose constitutive parts (or "elements") are stable (see footnote 6), I think it is possible to affirm that there can be "theory" autonomy in the sense indicated above (that is, the logical and conceptual nonreducibility of predicates, concepts, laws, etc., of the theories applied to the higher level) without there being "process-autonomy" (defined to mean that

---

[6] Conventionally, the series is said to run from the "lower" less complex systems to the "higher" more complex systems—from parts to wholes—so that these wholes themselves constitute parts of more complex entities, rather like a series of Russian dolls. In the complex systems I have in mind here, the parts retain their identity and properties as isolated individual entities. So the systems referred to are those which, loosely speaking, were the concern of the first phase of general systems theory. In those systems the parts ("elements") of the complex wholes are physical entities (e.g., atoms, molecules, cells) which are either individually stable or which undergo processes of change (as, e.g., in chemical reactions), themselves analyzable as being the interchange of stable parts (atoms in that case). The *internal* relations of such elements are not regarded as affected by their incorporation into the system.

[7] See, e.g., *TSA*, 36–43, 214–18, and figure 3, based on a scheme of W. Bechtel and A. Abrahamson in their *Connectionism and the Mind*, (Oxford: Blackwell, 1991), figure 8.1; for a bold extension of the schema developed there, see Nancey Murphy and George F.R. Ellis, *On the Moral Nature of the Universe: Theology, Cosmology and Ethics* (Minneapolis: Fortress Press, 1996), chaps. 2, 4.

the processes occurring at the higher level are *more than* an interlocking, in new relations, of the processes in which the constituent parts participate).[8]

When the nonreducibility of properties, concepts, and explanations applicable to higher levels of complexity is well established, their employment in scientific discourse can often, *but not in all cases*, lead to a putative and then to an increasingly confident attribution of a *causal efficacy* to the complex wholes which does not apply to the separated, constituent parts, for "to be real, new, and irreducible... must be to have new, irreducible causal powers."[9] If this continues to be the case under a variety of independent procedures and in a variety of contexts, then new and distinctive kinds of realities at the higher levels of complexity may properly be said to have *emerged*.[10] This can occur with respect either to moving synchronically up the ladder of complexity, or diachronically through cosmic and biological evolutionary history. This understanding accords with the pragmatic attribution, both in ordinary life and scientific investigation, of the term "reality" to that which we cannot avoid taking account of in our diagnosis of the course of events, in experience or experiments. Real entities have effects and play irreducible roles in adequate explanations of the world.

We have been assuming, with the "physicalists," that all entities, all concrete particulars in the world, including human beings, are constituted of fundamental physical entities—whatever it is that current physics postulates as the basic constituents of the world (which, of course, includes energy as well as matter). This is a *monistic* view (a constitutively-ontologically reductionist one)—everything can be broken down into fundamental physical entities and no extra entities are to be inserted at higher levels of complexity to account for their properties. I shall denote this position as that of "emergentist monism," rather than as "nonreductive physicalism," for those who adopt this latter label for their view, particularly in their talk of the "physical realization" of the mental in the physical, often seem to me to hold a much less realistic view of higher level properties than I wish to affirm here—and also not to attribute causal powers to that to which higher-level concepts refer.[11]

---

[8] See the Appendix to this essay and my *God and the New Biology* (London: Dent, 1986, repr. Gloucester, Mass: Peter Smith, 1994), chaps. 1, 2, henceforth *GNB*. Whether or not this statement about theory- and process-autonomy applies to the relations *between* distinctive systems is a matter which will be examined further in sec. 2.4 and the Appendix.

[9] Samuel Alexander, as quoted by Jaegwon Kim, "Non-reductivism and Mental Causation," in *Mental Causation*, John Heil and Alfred Mele, eds. (Oxford: Clarendon Press, 1993), 204.

[10] William C. Wimsatt has elaborated criteria of "robustness" for such attributions of reality for emergent properties at the higher levels. These involve noting what is invariant under a variety of independent procedures; this is summarized in *GNB*, 27–28, from Wimsatt's paper "Robustness, Reliability and Multiple-Determination in Science," in *Knowing and Validating in the Social Sciences: A Tribute to Donald T. Campbell*, Marilynn Brewer and Barry Collins, eds. (San Francisco: Jossey-Bass, 1981).

[11] My view of emergent monism is in harmony with that of Philip Clayton, to whom I am much indebted for his shrewd and useful comments on this essay. Note that the term "monism" is emphatically *not* intended (as is apparent from the nonreductive approach adopted here) in the sense in which it is taken to mean that physics will eventually explain everything. Note also that this position is distinct from that of "dual-aspect monism" or "two-aspect monism," which could appear to be purely epistemological, being about how an entity is *viewed* from two different perspectives. Even when the "two" and "dual" refer to distinct properties of a single entity, there is not in these terms any implication of a *causal* relation between the "aspects" (any more than between the wave and particle aspects of the single

If we do make such an ontological commitment about the reality of the "emergent" whole of a given total system, the question then arises: How is one to explicate the relation between the state of the whole and the behavior of parts of that system at the micro-level? The simple concept of chains of causally related events $(A{\rightarrow}B{\rightarrow}C...)$ in constant conjunction (à la Hume) is inadequate for this purpose. Extending and enriching the notion of causality now becomes necessary because of new insights into the way complex systems in general and biological ones in particular behave. This subtler understanding of how higher levels influence the lower levels, and *vice versa*, still allows application in this context of the notion of a "causal" relation from whole to part (of system to constituent)—never ignoring, of course, the "bottom-up" effects of parts on wholes, for the properties of wholes depend on the properties of the parts being what they are.

2.2 The Relation of Wholes and Parts in Complex Systems

A number of related concepts have been developed in recent years to describe these relations in both synchronic and diachronic systems—that is, respectively, both those in some kind of steady state with stable, characteristic emergent features of the whole, and those which display an emergence of new features in the course of time.

2.2.1 Whole-Part Influence (or Constraint)

The term "downward-causation" or "top-down causation" was, as far as I can ascertain, first employed by Donald Campbell[12] to denote the way in which the network of an organism's relationships to its environment and its behavior patterns together determine in the course of time the actual DNA sequences at the molecular level present in an evolved organism—even though, from a "bottom-up" viewpoint, a molecular biologist would tend to describe the organism's form and behavior, once in existence, as a consequence of those same DNA sequences. Campbell cites as an example the evolutionary development of efficacious jaws made of suitable proteins in a worker termite. There are imprecisions and a lack of generalizability in Campbell's example and I prefer to use actual complex systems to clarify this suggestion. One could cite, for example, the Bénard phenomenon[13]—at a critical point a fluid heated uniformly from below in a containing vessel ceases to manifest the entirely random "Brownian" motion of its molecules, but displays up and down convective currents in columns of hexagonal cross-section. Moreover, certain auto-

---

entity of the electron). Talk of "two aspects" is not strong enough to include an affirmation that the higher level is real and has causal efficacy.

[12] Donald T. Campbell, "'Downward Causation' in Hierarchically Organized Systems," in *Studies in the Philosophy of Biology: Reduction and Related Problems*, Francisco J. Ayala and Theodosius Dobzhansky, eds. (London: Macmillan, 1974), 179–86. A valuable and perspicacious account (with which I entirely agree) of emergent order, top-down causation (fully illustrated by its operation in the hierarchical organization of the modern digital computer), and the physical mediation of top-down effects has been given in Murphy and Ellis, *On the Moral Nature of the Universe*, 22–32. For brevity here, I refer the reader to that recent excellent exposition. For earlier expositions of the hierarchies of complexity, of the relation of scientific concepts applicable to wholes to those applicable to the constituent parts, and of top-down/downward causation and whole-part influence (as discussed below), see *GNB*, chaps. 1, 2; *TSA*, 39–41, 50–55, 213–18 (esp. figure 3); and *CAC*, 272–76.

[13] For a survey with references, see Arthur Peacocke, *The Physical Chemistry of Biological Organization* (Oxford: Clarendon Press, 1983, 1989), henceforth *PCBO*.

catalytic reaction systems (for example, the famous Zhabotinsky reaction and glycolysis in yeast extracts) display spontaneously, often after a time interval from the point when first mixed, rhythmic temporal and spatial patterns the forms of which can even depend on the size of the containing vessel. Many examples are now known also of dissipative systems which, because they are open, a long way from equilibrium, and nonlinear in certain essential relationships between fluxes and forces, can display large-scale patterns in spite of random motions of the units— "order out of chaos," as Ilya Prigogine and Isabelle Stengers dubbed it.[14]

In these examples, the ordinary physico-chemical account of the interactions at the micro-level of description simply cannot account for these phenomena. It is clear that what the parts (molecules and ions, in the Bénard and Zhabotinsky cases) are doing and the patterns they form are what they are *because* of their incorporation into the system-as-a-whole—in fact these are patterns *within* the systems in question. This is even clearer in the much more complex, and only partly understood, systems of genes switching on and off and their interplay with cell metabolism and specific protein production in the processes of development of biological forms. The parts would not be behaving as observed if they were not parts of that particular system (the "whole"). The state of the system-as-a-whole is affecting (that is, acting like a cause on) what the parts, the constituents, actually do. Many other examples of this kind could be taken from the literature on, for example, self-organizing and dissipative systems[15] and also economic and social ones.[16]

We do not have available for such systems any account of events in terms of temporal, linear chains of causality as previously conceived (A→B→C→...). Hence, in my recent writings I adopted the term "whole-part constraint" to describe the effects on the constituent parts of their being incorporated into systems of this kind, because the term "causation" often has tended to denote simply a regular chain of events (sometimes, too, simply in terms of a Humean conjunction). A wider use of "causality" and "causation" is now needed to include the kind of whole-part, higher-to lower-level, relationships that the sciences have themselves recently been discovering in complex systems, especially the biological and neurological ones.

Here the term "whole-part influence," will be used to represent the net effect of all those ways in which the system-as-a-whole, operating from its "higher" level, is a causal factor in what happens to its constituent parts, the "lower" level. Such a "causal" relation within a particular system is one that relates entities which are, because of the mereological nature of the system, in some sense the same; so this "causal" relation might, adding confusion, entice some to regard the higher level as possessing a somewhat "metaphysical" character.

2.2.2 Other Analyses

Various interpretations have been deployed by other authors to represent this whole-part relation in different kinds of systems (and notably the mind-brain-body one—see section 2.5), though not usually with causal implications.

---

[14] Ilya Prigogine and Isabelle Stengers, *Order Out of Chaos* (London: Heinemann, 1984).

[15] *PCBO*; Prigogine and Stengers, *Order Out of Chaos*; Niels H. Gregersen, "The Idea of Creation and the Theory of Autopoietic Processes," *Zygon* 33 (1998): 333–67.

[16] Prigogine and Stengers, *Order Out of Chaos*.

1. *Structuring causes*. The notion of whole-part influence is germane to one that Niels Gregersen has recently employed[17] in his valuable discussion of autopoietic (self-making) systems—namely that of *structuring causes*, as developed by Fred Dretske[18] for understanding mental causation. Gregersen and Dretske refer to the event(s) that produced the hardware conditions (actual electrical connections in the computer) and the word-processing program (software) as the "structuring causes" of the cursor movement on the screen connected with the computer; whereas the "triggering cause" is usually pressure on a key on the keyboard. The two kinds of causes exhibit a different relationship to their effects. A triggering one falls into the familiar (Humean) pattern of constant conjunction. However, a structuring cause is never sufficient to produce the particular effect (the key still has to be pressed); there is no constant relationship between structuring cause and effect. In the case of complex systems, such as those already mentioned, the system-as-a-whole often has the role, I suggest, of a structuring cause in Dretske's sense.

This idea helps in responding to two features that Thomas Tracy[19] has found to be problematic in my own earlier use of "top-down" explanations.[20] Tracy was, firstly, concerned with the supposition that "top-down explanations cannot be analyzed in terms of structures of bottom-up explanation."[21] The particular examples of systems already given and the considerations which lead to distinguishing structuring from triggering causes serve to explain why such "top-down" explanations could not, by their very nature, be analyzed in "bottom-up" terms. That is the whole point of identifying them as such. For example, in the Bénard case, it is not the properties, as such, of the *individual* molecules of the water in a heated beaker which explains why they suddenly abandon random collisions and move in serried ranks with the same velocity in one direction at the critical point—or why suddenly, in the Zhabotinsky reaction, in a particular spatially defined band at certain (periodic) positions vertically in the reaction test tube, *all* the cerous irons should become ceric. In both examples, it is a distinctive structuring property of the whole, and of the new relations among the constituents involved, that is the operative factor.

Tracy also finds problematic "the move from whole-part explanation to treating the whole (or the nature of the system) as a cause."[22] I have shared this concern, for that is why I moved away from Campbell's terminology of "causation."[23] However, provided "causation" is given a wider than chain-sequence (Humean) sense consistent with the holistic behavior of complexes, as already discussed, it can still be applied to the whole-part relation.

---

[17] Gregersen, "The Idea of Creation."

[18] Fred Dretske, "Mental Events as Structuring Causes of Behavior," in *Mental Causation*, 121–36. Another example of his is as follows. A terrorist plants a bomb in the general's car. The bomb stays there until the general gets into the car and turns the ignition key and then is killed by the detonation of the bomb. The "triggering cause" of his death is his turning on the engine, but the "structuring cause" is the terrorist's action.

[19] Thomas F. Tracy, "Particular Providence and the God of the Gaps," in *CAC*, 306–7, fn. 39.

[20] In the first 1990 edition of *TSA* and before my espousing the of "whole-part constraint" in the 1993 enlarged edition of *TSA* and, more especially, in *CAC*, 263–87.

[21] Tracy, "Particular Providence and the God of the Gaps."

[22] Ibid.

[23] In *CAC*, 272, fn. 22.

2. *Propensities*. The category of "structuring cause" is closely related to that of *propensities* developed by Karl Popper, who pointed out that "there exist weighted possibilities which are *more than mere possibilities*, but tendencies or propensities to become real"[24] and that these "propensities in physics are properties *of the whole situation* and sometimes even of the particular way in which a situation changes. And the same holds of the propensities in chemistry, biochemistry, and in biology."[25] Hence Popper's "propensities"[26] are the effects of Dretske's structuring causes in the case that triggering causes are random in their operation (that is, *genuinely* random, no "loading of the dice").

3. *Boundary (limiting) conditions*. In the discussion of the relations between properties of a system-as-a-whole and the behavior of its constituent parts, some authors refer to the *boundary conditions* that are operating.[27] It can be a somewhat misleading term—"*limiting condition*" would be better but I will continue to use it only in this wider, Polanyian, sense.

A more recent, sophisticated development of these ideas has been proffered by Bernd-Olaf Küppers:

> [T]he [living] organism is subservient to the manner in which it is constructed...Its principle of construction represents a boundary condition under which the laws of physics and chemistry become operational in such a way that the organism is reproductively self-sustaining.... [T]he phenomenon of emergence as well as that of downward causation can be observed in the living organism and can be coupled to the existence of specific boundary conditions posed in the living matter.[28]

Thus a richer notion of the concept of boundary conditions is operative in systems as complex as living ones. The simpler forms of the idea of "boundary condition" as applied, for example, by Polanyi to machines are not adequate to express the causal features basic to biological phenomena. Indeed the "boundary conditions" of a system will have to include not only purely physical factors on a global scale, but also complex inter-systemic interactions between type-different systems (see section 2.4 below and the Appendix).

Willem Drees has also emphasized, with respect to the Bénard phenomenon, the role of the conditions at the actual, *physical* boundary of the fluid in its physical environment in determining the behavior of the billions of constituent molecules. He asserts that in this case one could replace the term "top-down" causation by "environment-system" interaction. The environment determining the temperature is simply a physical system so, he argues,

> ... There is no sense in which the system-as-a-whole has a specific, "emergent" causal influence. All the causal influences can be traced locally as physical influences within the

---

[24] Karl Popper, *A World of Propensities* (Bristol: Thoemmes, 1990), 12.

[25] Ibid., 17.

[26] Cf. my urging in *TSA* that there are propensities in biological evolution, favored by natural selection, to complexity, self-organization, information-processing and -storage, and so to consciousness.

[27] For example, Michael Polanyi, "Life Transcending Physics and Chemistry," *Chemistry and Engineering News* (August 21, 1967): 54–66; and idem, "Life's Irreducible Structure," *Science* 160 (1968): 1308–12. In his discussion, and mine in this essay, the term "boundary condition" is *not* being used, as it often is, to refer either to the initial (and in that sense "boundary") conditions of, say, a partial differential equation as applied in theoretical physics, or to the physical, geometrical boundary of a system.

[28] Bernd-Olaf Küppers, "Understanding Complexity," in *CAC*, 100.

system or between the system and its immediate environment. Boundaries are local phenomena, rather than global states of the system-as-a-whole.[29]

But the system is what has a boundary—and only the system can have it. It is *because* the system-as-a-whole is an entity, immersed in a conditioning environment with which it has a boundary, that it undergoes holistic reorganization of its constituent units. Indeed the Bénard phenomenon is independent of the shape of the container provided its dimensions are large with respect to convection cell size (the very condition that makes physical boundary effects negligible). The theory that has to give an intelligible account of all this has to deal with properties of the system-as-a whole—the temperature dependence of the viscosity and density of aggregates of molecules, their thermal conductivity and the mutual interplay of all these factors together in the behavior of the whole assembly. It is not enough, therefore, to pinpoint only the "environment-system" interaction as uniquely determinative. For it is only because of the nature of the entire system-as-a-whole that under such boundary conditions the constituent molecules manifest their unexpected, bizarre behavior. It is a case of "whole-part influence" in the sense defined above.

There is a sense in which the system-as-a-whole, because of its distinctive configuration, can constrain and influence the behavior of the parts to be otherwise than if they were isolated from this particular system. Yet the system-as-a-whole would not be describable by the concepts and laws of that level and still have the properties it does have, if the parts (in the Zhabotinsky case, the ceric and cerous ions) were not of the particular kind they are. What is distinctive in the system-as-a-whole is the new kind of interrelations and interactions, spatially and temporally, of the parts.

4. *Supervenience*. Another, much debated term which has been used in this connection, especially in describing the relation of mental events to neurophysiological ones in the brain, is that of "supervenience." The term, which does not usually imply any "whole-part" causative relation, goes back to Donald Davidson's employment of it in expounding his view of the mind-brain-body relation as "anomalous monism."[30] The various meanings and scope of the term in this context had been formulated and classified by Jaegwon Kim as involving: the *covariance* of the supervenient properties with, the *dependency* of the supervenient properties on, and the *nonreducibility* of the supervenient properties to, their base properties.[31] Another definition has been proposed in this volume by Nancey Murphy.[32] In the wider context of hierarchical systems (prescinding from the mind-brain-body problem, for the moment—see section 2.5 below) the term "supervenience" may be taken

to refer to the relation between *properties* of the same system that pertain to different levels of analysis... higher-level properties supervene on lower-level properties if they are partially constituted by the lower-level properties but are not directly reducible to them.[33]

---

[29] Willem B. Drees, *Religion, Science and Naturalism* (Cambridge: Cambridge Univ. Press, 1996), 102.

[30] Donald Davidson, "Mental Events," in *Essays on Actions and Events* (Oxford: Clarendon Press, 1980).

[31] Jaegwon Kim, "Epiphenomenal and Supervenient Causation," *Midwest Studies in Philosophy* 9 (1984): 257–70; repr. in *Supervenience and Mind: Selected Philosophical Essays* (Cambridge: Cambridge University Press, 1993).

[32] Nancey Murphy, "Supervenience, and the Downward Efficacy of the Mental: A Nonreductive Physicalist Account of Human Action," in this volume.

[33] Murphy and Ellis, *On the Moral Nature of the Universe*, 23.

One can ask the question

> [H]ow are the properties characteristic of entities at a given level related to those that characterize entities of adjacent levels? Given that entities at distinct levels are ordered by the part-whole relation, is it the case that properties associated with different levels are also ordered by some distinctive and significant relationship?[34]

The attribution of "supervenience" asserts primarily that there is a necessary covariance between the properties of the higher level and those of the lower level. When the term "supervenience" was first introduced its attribution did not imply a causal influence of the supervenient level on the subvenient one.[35] Its appropriateness is questionable for analyzing whole-part relations, which by their very nature relate, with respect to complex systems, entities that are in some sense the same.

Yet, in the context of the physical and biological (and, it must also be said, ecological and social) worlds, the mutual interrelations between whole and parts in any internally hierarchically organized system often, we have seen, appear to involve causal effects of the whole on the parts. We shall continue, therefore, to use the term "whole-part influence,"[36] rather than the terms *1–4* above, to refer to the subtle interlocking influences of the whole of any particular hierarchically organized system on its constituent parts.

2.3 Flow of Information

A general concept which has often been found to be applicable to understanding the relation between higher and lower levels in a single, hierarchically stratified complex system is that of there being a *flow of information* from the higher to the lower level. The higher level is seen as constraining and shaping the patterns of events occurring among the constituent units of the lower one. Although "information" is a concept distinct from those of matter and energy, in actual systems no information flows without some exchange of energy and/or matter. Nevertheless, as an interpretative concept it is useful not only in the more obvious context of the mind-brain-body relation but also in considering the relation of environment to biological processes, including that of evolution.[37] Thus, the case of the worker termite cited by Donald Campbell could well be interpreted as manifesting a temporal flow of information: information about the environment is, over a long period of time, impressed indirectly (via the effect of the environment on the viability of organisms possessing mutated DNA) on the DNA. This DNA then shapes the functioning of the organism that is capable of producing viable progeny. The concept of information is indeed apt for situations in which a form at one level influences forms at lower levels. This process can at least be conceived as a process of transfer of information, as distinct from energy or matter. John Puddefoot has usefully distinguished between:

---

[34] Kim, "Non-reductivism and Mental Causation," in *Mental Causation*, 191.

[35] However, utilizing her definition of supervenience, Nancey Murphy has suggested (personal communication, July, 1998) "that the supervenient level may involve additional circumstances that cannot be described at the subvenient level, and these additional circumstances can have a causal impact on the series of events. Thus, the causal connections will show up (be intelligible) only at the supervenient level of description."

[36] It must be stressed that the "whole-part" relation is *not* regarded here necessarily, or frequently, as a spatial one. "Whole-part" is synonymous with "system-constituent."

[37] Cf. Jeffrey S. Wicken, *Evolution, Information and Thermodynamics: Extending the Darwinian Paradigm* (Oxford: Oxford University Press, 1987).

1. "Information" in the physicists', communication engineers', and brain scientists' sense, that of C.E. Shannon—the sense in which "information" is related to the probability of one outcome or case selected out of many, probable outcomes or cases. In this sense it is, in certain circumstances, the negative of entropy.

2. "Information" in the sense of the Latin *informare*, meaning "to give shape or form to." Thus, "information" is "the action of informing with some active or essential quality," as the noun corresponding to the transitive verb "to inform," in the sense of "To give 'form' or formative principle to; hence to stamp, impress, or imbue with some specific quality or attribute" (quotation from the *Shorter Oxford English Dictionary on Historical Principles*, sense II).

3. "Information" in the ordinary sense of "that of which one is appraised or told" (*Shorter Oxford English Dictionary*, sense I.3).[38]

Puddefoot points out that information *1* is necessary to shape or give form, as information *2*, to a receptor. If that receptor is the brain of a human being, then *inter alia* information *3* is conveyed. In this essay the term "information" (as well as its associates) is being broadly used to represent this whole process of *1* becoming *3*—and only modulating to *3* when there is a specific reference to human brain processes in which *1* acquires meaning for human beings. Briefly, the mathematical (often digital) information *1* is the necessary basis of *2* (often "syntax") which can in human mental experience become *3*, with semantic content. I am not intending here in any way to imply that *3* is *reducible* to *1*—that semantics is reducible to syntax—only that *1* is the necessary pre-condition for the manifestation and emergence of *3*.[39]

Information *1* and *2* are often applicable to the higher- to lower-level interactions in hierarchically stratified physical and biological systems. The transition from information *1* and *2* to information *3* is, of course, ambivalently related to the opaque mind-brain-body relation, though it has been widely employed in that context. The concept of information *1*, or its flow, has been used to attempt to define living entities[40] but biologists have often been skeptical as to its general usefulness in, for example, understanding development,[41] though it has an obvious application—one which was of historical significance—in interpreting the relation between nucleotide sequences in DNA and amino acid sequences in proteins and so in relation to heredity, that is, to "genetic information." The notion of "flow of information" is therefore a conceptual tool ready to hand, as it were, to interpret the relation of higher to lower levels in a particular hierarchically stratified complex, but it must be used warily.

---

[38] John C. Puddefoot, "Information and Creation," in *The Science and Theology of Information*, C. Wassermann, R. Kirby, and B. Rordoff, eds. (Geneva: Labor et Fides, 1992), 15 (my numbering). For further discussion, especially in relation to biological complexity, see *PCBO*, 259–63, and in relation to evolution, 263–68.

[39] The transition from *1* to *3* is also closely akin to that from semiotics to semantics and coheres well with the emergentist-monist position.

[40] See *PCBO*, 259–63; Frank J. Tipler, *The Physics of Immortality* (London: Macmillan, 1995), 124–27.

[41] Michael J. Apter and L. Wolpert, "Cybernetics and Development. I. Information Theory," *Journal of Theoretical Biology* 8 (1965): 244–57.

But what of the relation *between* distinct systems? To this latter issue we must now turn.

2.4 The World-as-a-Whole[42]—An Interconnected and Interdependent "System-of-Systems"

The world consists of myriads of individual systems which are themselves very often hierarchically stratified complex systems of stable parts. We have been exploring their internal ("whole-part") relationships. But these individual systems themselves can interact in a highly ramified manner across space and time. For distant events in space (for example, flaring spots on the Sun shower cosmic rays on the Earth which affect its climate and the evolution of its living organisms); and in time (for example, the elliptical orbits of the planets about the Sun, hence the seasons of terrestrial life; and the relation of the Earth's axis to the plane of its motion to the north-south seasonal patterns). The individual systems of the world are increasingly demonstrated by the sciences to be interconnected and interdependent in multiple ways with, of course, great variations in the strengths of mutual coupling. Thus all wave functions of all sub-atomic particles (indeed of all matter) only go asymptotically to zero at an infinite distance from their maximal value, so that there is a finite, if small, chance of finding that particular particle anywhere.[43] On the Earth's surface, the ecological interconnectedness of all forms of life (including human), as well as their matter and energy cycles, themselves related to atmospheric and geological ones, has in recent years become increasingly apparent in all its baffling intricacies. These interactions between individual systems over space and time are as real in their mutual influencing as anything else described by the natural sciences, and their existence cannot be ignored in our reflections on the nature of the world and of God's relation to it, simply because we can never have one comprehensive theory of them. This character of the world-as-a-whole suggests that it is metaphysically plausible to perceive it as a System-of-systems (using the word "system" with the weight already attached to it in the light of complexity theory of individual systems). Such an epistemological assertion would have, as always, a putative ontological significance. In that case, the "world-as-a-whole" is not "simply a concept"[44] nor "an abstract description,"[45] but could at least provisionally be regarded as an holistic reality at its own level—even if the coupling *between* systems is much looser and more diffuse, and therefore less classifiable, than it is *within* a particular individual hierarchically stratified system clearly demarcated from its environment. The apprehension of all-that-is in its holistic unity as a System-of-systems is, of course, scarcely vouchsafed to the limited horizons and capacities of humanity, though every

---

[42] By the "world-as-a-whole," I here mean all-that-is, or ever has been; all that is created, i.e., all that is not God. (The outer dashed circle in figure 1 on p. 238 is meant to denote this).

[43] Recall also the notorious gravitational effect of the motion of an electron at the edge of, say, our galaxy on the collisions of macroscopic billiard balls; Michael Berry, "Breaking the Paradigm of Classical Physics from Within," Cercy Symposium on *Logique et Théorie des Catastrophes*, 1983.

[44] Niels Hendrik Gregersen, "Providence in an Indeterministic World," *CTNS Bulletin*, 14.1 (Winter, 1994): 26.

[45] Idem. "Three Types of Indeterminacy," in *The Concept of Nature in Science and Theology*, part I, vol. 3 of *Studies in Science and Theology* of the European Society for the Study of Science and Theology (Geneva: Labor et Fides, 1995), 175.

advance in the sciences serves to reveal further cross-connections between its component systems. Such interconnectedness would be transparent to the omniscient Creator, who continuously gives its constituents and its processes existence and in Whom all-that-is exists, from a sacramental, panentheistic perspective.

The relation between higher and lower levels *within* an individual hierarchically stratified system I have been designating by the pantechnicon term "whole-part influence."[46] This influence, I suggested, can often be regarded as a flow of information. We now have to ask: Can these notions be applied to the relations *between* systems in the world-as-a-whole? In order to respond to this question, it turns out to be necessary to clarify the relation between theory- and process-autonomy and this issue is discussed in the Appendix. There I conclude that, although theory-autonomy can occur without process-autonomy with respect to the internal relations of a particular system of stable parts, in the relation between two mutually interacting type-different systems both the theories applicable to and processes of each can be autonomous with respect to the other.

From that discussion it transpires that we shall have to recognize that the interactions and relations *between* distinctive systems are unlikely to be describable in the same way as those *within* hierarchically stratified systems of stable parts. We are regarding the world as a "System-of-systems," but not as a hierarchically stratified one, so that the principles of "weak" nonreducibility do not have to apply to the relation *between* the component systems of the world.[47] Indeed, if we could have a cosmic-global science of the world-as-a-whole as a System-of-systems, the theories (and predicates, concepts, laws, etc.) of that science would be expected to manifest not only theory-autonomy but also *ex hypothesi* process-autonomy since the processes going on in that whole System consist of the changing relations among type-different component systems (often containing type-different component parts).

Earlier we noted that when a particular hierarchical system was considered, the idea that there can be envisaged a "flow of information" from the higher level to the lower one could sometimes be usefully employed. Is this notion of the "flow of information" any help in thinking of the multiple interactions between individual systems in the world "System"? Such interactions are obviously highly variegated, multiple, and overlapping, as Gregersen says, for "we face a *criss-cross interpenetration* of different kinds of operational systems... a world of naturally polycentric systems..., a nexus of realities,[48] or "a network of influences."[49]

The world may be conceived of as an interconnected web of type-different systems interacting in specific ways and mutually influencing each other.[50] A

---

[46] See sec. 2.2 above.

[47] For "weak" nonreducibility, see the Appendix. If the systems in question are themselves part of an actual hierarchy of organization and are themselves stable, then the analysis may well revert to that applicable to the internal relationships *within* a larger hierarchical system of stable parts, each of which is then itself a system. For the world-as-a-whole, it is the interaction between systems not so described that is chiefly under consideration—"...the reality of the 'world as a whole' is itself a result of the interpenetrations between the type- and code-different systems observed..." (Gregersen, "The Idea of Creation and the Theory of Autopoietic Processes," 337).

[48] Ibid.

[49] Gregersen (personal communication, 12 November, 1996), describing my own view.

[50] Gregersen (personal communication, March, 1998) has expressed this point to me thus: "[P]erhaps the most curious feature about our universe is that it starts out as a unity and ends

common factor then discernible in the multiple interactions between such systems (in the whole cosmic System) is the transfer of information whereby *patterns* of events in one system affect *patterns* of events in another—and the interchange between the myriad systems of energy and/or matter are *ex hypothesi* variegated beyond the possibility of generalization. The use of the concept of information is thus particularly apt for elucidating these interactions since it is, conceptually at least, independent of those of matter and energy—though in nature it never occurs without their exchange.

2.5 The Mind-Brain-Body Relation and Personhood

Much of the discussion of the relation of higher levels to lower ones in hierarchically stratified systems has centered on the mind-brain-body relation, on how mental events are related to neurophysiological ones in the human-brain-in-the-human-body—in effect the whole question of human agency and what we mean by it. A hierarchy of levels can be delineated,[51] each of which is the focus of a corresponding scientific study, from neuroanatomy and neurophysiology to psychology. Those involved in studying "how the brain works" have come to recognize that

> properties not found in components of a lower level can emerge from the organization and interaction of these components at a higher level. For example, rhythmic pattern generation in some neural circuits is a property of the circuit, not of isolated pacemaker neurons. Higher brain functions (e.g., perception, attention) may depend on temporally coherent functional units distributed through different maps and nuclei.[52]

So that even an in-principle physicalist, such as Patricia Churchland, can express (with T.J. Sejnowski) the aim of research in cognitive neuroscience thus:

> The ultimate goal of a unified account does not require that it be a single model that spans all the levels of organization. Instead the integration will probably consist of a chain of models linking adjacent levels. When one level is explained in terms of a lower level this does not mean that the higher level theory is useless or that the high-level phenomena no longer exist. On the contrary, explanations will coexist at all levels, as they do in chemistry and physics, genetics and embryology.[53]

The still intense philosophical discussion of the mind-brain-body relation has been broadly concerned with attempting to elucidate the relation between the "top" level of human mental experience and the lowest, bodily physical levels. In recent decades it has often involved considering the applicability and precise definition of some of the terms used above in section 2.2 to relate higher levels to lower ones in hierarchically stratified systems. The question of what kind of "causation," if any,

---

up in a plurality of systems .... forever *based* on the same uniform matter, always *interacting with one another* in ever-new constellations of mutual influences (thus certainly interlocked) but nonetheless *appearing* in type-different forms, thus also *operating* by virtue of type-different causalities" (emphasis original).

[51] As indicated in the legend to fig. 1 on p. 239, where the schema of Patricia S. Churchland and T.J Sejnowski is depicted ("Perspectives in Cognitive Neuroscience" *Science* 242 [1988]: 741–45). The physical scales of these levels are, according to these authors, as follows: molecules, 10^{-10}m; synapses, 10^{-6}m; neurons, 10^{-4}m; networks, 10^{-3}m; maps, 10^{-2}m; systems, 10^{-1}m; central nervous system, 1m, in human beings.

[52] Terrence J. Sejnowski, C. Koch, and P. Churchland, "Computational Neuroscience," *Science* 241 (1988): 1299–1306, see p. 1300.

[53] Churchland and Sejnowski, "Perspectives in Cognitive Neuroscience," 744.

may be said to be operating from a "top-down," as well as the obvious and generally accepted "bottom-up," direction is still much debated in this context.[54]

Earlier (section 2.2), when discussing the general relation of wholes to constituent parts in a hierarchically stratified complex system of stable parts, I used "whole-part influence" and other terms and maintained that a nonreductionist view of the predicates, concepts, laws, etc., applicable to the higher level could be coherent. Reality could, it was argued, putatively be attributed to that to which these nonreducible, higher-level predicates, concepts, laws, etc., applied; and these new realities, with their distinctive properties, could properly be called "emergent." When this emergentist monist approach is applied to the mental activity of the human-brain-in-the-human-body then, "we must look to vernacular ["folk"] psychology and its characteristic intentional idioms of belief, desire, and the rest, and their intentional analogues in systematic psychology" in order to elucidate its nature.[55] Mental properties are now widely regarded by philosophers as epistemolog-ically irreducible to physical ones, indeed as "emergent" from them, but also dependent on them[56]—similar terms have been used to describe the relation of "higher" to "lower" levels as in the context of nonconscious, complex systems (see section 2.2.2). In the mind-brain-body case the idea that mental properties can be "physically realized" has also been much deployed.[57] Jaegwon Kim has argued that, if this latter concept (which overlaps that of supervenience in many treatments) is taken to mean that a microstructure physically realizes a mental property by being a *sufficient* cause for that property, and if for mental properties to be real is for them to have new, irreducible causal powers, then the nonreductive physicalist is thereby committed to downward causation (in a strong nomological sense) from the mental to the physical levels.[58] Kim then argues that, because there is complete causal closure at the physical level alone, mental properties cannot, in fact, have real causal

---

[54] See, for example, the collection of papers in *Mental Causation*, Heil and Mele, eds.

[55] Kim, "Non-reductivism and Mental Causation," 193.

[56] Broadly, this is the "nonreductive physicalist" view of the mental-physical relation, which has been summarized (ibid., 198) as follows:
1. *Physical Monism.* All concrete particulars are physical.
2. *Anti-reductionism.* Mental properties are not reducible to physical properties.
3. *The Physical Realization Thesis.* All mental properties are physically realized; that is, whenever an organism, or system, instantiates a mental property M, it has some physical property P such that P realizes M in organisms of its kind.
4. *Mental Realism.* Mental properties are real properties of objects and events; they are not merely useful aids in making predictions or fictitious manners of speech.

[57] The idea of mental states being "physically realized" in neurons was expanded as follows by John Searle, *Minds, Brain and Science*, (Cambridge: Harvard University Press, 1984), 26 (emphasis added):

> Consciousness...is a real property of the brain that can cause things to happen. My conscious attempt to perform an action such as raising my arm causes the movement of the arm. At the higher level of description, the intention to raise my arm causes the movement of the arm. At the lower level of description, a series of neuron firings starts a chain of events that results in the contraction of the muscles...[T]he same sequence of events has two levels of description. *Both of them are causally real,* and the higher-level causal features are both caused by and realized in the structure of the lower level elements.

What follows in the main text here shows that I am not satisfied with Searle's parallelism between the causality of the mental and physical; it is not enough. I argue later on in this essay for a *joint* causality whereby the mental influences the physical level in the brain.

[58] Kim, "Non-reductivism and Mental Causation," 202–5.

powers irreducible to physical ones. Hence there is a conflict between the postulate of downward causation (derived from the nonreducibility, and the need for causal efficacy, of the mental) and the physicalist's assumption that a complete physical theory can in principle account for all phenomena (causal closure). Steven Cain has succinctly summarized these conclusions of Kim: "…the *nonreductive* physicalist cannot live without downward causation, and the nonreductive *physicalist* cannot live with it."[59] Crain argues (and I agree) that it is Kim's assumption that a physical microstructure in "physically realizing" a mental property is its *sufficient* cause, which leads to the exclusion of any causative role for mental properties, for in the wider range of physical, biological, and other systems discussed in section 2.2, the causative effects of the higher levels on the lower ones were real but different in kind from the effects the parts had on each other operating at the lower level. Thus, what happens in these systems at the lower level is the result of the *joint* operation of both higher- and lower-level influences—the higher and lower levels could be said to be jointly sufficient, type-different[60] causes of the lower-level events. When the higher-lower relation is that of mind/brain to body, it seems to me that similar considerations should apply.

Up to this point, I have been taking the term "mind," and its cognate "mental," to refer to that which is the emergent reality distinctive especially of human beings. But in many wider contexts, not least that of philosophical theology, a more appropriate term for this emergent reality would be "person," and its cognate "personal," to represent the total psychosomatic, holistic experience of the human being in all its modalities—conscious and unconscious, rational and emotional, active and passive, individual and social, etc. The concept of personhood recognizes that, as Philip Clayton puts it,

> We have thoughts, wishes and desires that together constitute our character. We express these mental states through our bodies, which are simultaneously our organs of perception and our means of affecting other things and persons in the world…[The massive literature on theories of personhood] clearly points to the indispensability of embodiedness as the precondition for perception and action, moral agency, community and freedom—all aspects that philosophers take as indispensable to human personhood and that theologians have viewed as part of the *imago dei*.[61]

---

[59] Steven D. Crain, in an unpublished paper, kindly made this available to me.

[60] See the illuminating discussion of type-different causalities by Gregersen in his "Three Types of Indeterminacy," 173–74. He remarks:

> The Humean concept of causality that still prevails in the philosophical debate …. thinks of causality in terms of general laws applicable on systems of events and processes…Non-Humean concepts of causality normally think of causality in terms of influencing conditions and events that in their totality make up the effects…. My suggestion is that there exist quite different types of causality that can neither be subsumed under general laws nor be measured through additions and subtractions" (173).

In line with this, he espouses an "holistic" supervenience theory as against Kim's "physicalist" one, as in his "Divine Action in a Universe of Minds," paper presented at the ESSSAT Conference, Durham, March 31–April 4, 1998.

[61] Philip Clayton, "The Case for Christian Panentheism," *Dialog* 37.3 (Summer 1998): 201–8 (quotation on 205); see also his "Rethinking the Relation of God to the World: Panentheism and the Contribution of Philosophy," chap. 4 in *God and Contemporary Science* (Edinburgh: Edinburgh University Press, 1997), in which the nuances of panentheism are well developed. Broadly, they amount to a stronger form of immanence in which God is seen as in, with, and under the very processes of the world almost in a sacramental modality. See also *TSA, passim*.

There is, therefore, a strong case for designating the highest level, the whole, in that unique system which is the human-brain-in-the-human-body-in-social-relations as that of the "person." Persons are *inter alia* causal agents with respect to their own bodies and to the surrounding world (including other persons). They can, moreover, report on aspects of their internal states concomitant with their actions with varying degrees of accuracy. Hence the exercise of personal *agency* by individuals transpires to be a paradigm case and supreme exemplar of whole-part influence—in this case exerted on their own bodies and on the world of their surroundings (including other persons). Thus, the details of the relation between cerebral neurological activity and consciousness cannot in principle detract from the causal efficacy of the content of the latter on the former and thus on behavior. In other words, "folk psychology" and the real reference of the language of "personhood" are both justified and necessary.

2.6 Communication Between Persons

We are aiming at understanding better, in the light of what we now know through the sciences about human nature, how God might be conceived of as communicating with humanity. Let us remind ourselves first how human persons communicate *with each other*. How do we get to know each other, not only by description, but also by acquaintance—that is, get to know what is, as we say, "in each other's mind"?[62]

All communication at its most basic level is mediated through the senses—hearing, sight, touch, taste, and smell. The physical intermediaries are vibrations in pressure in the air, electromagnetic waves, physical pressure, changes of temperature, molecules, etc. Our genes, culture, nurture, and education have enabled human beings to decode patterns of these physical intermediaries so as to convey information about the content of the consciousness of the one attempting to communicate. These patterns can be immensely complex, associated with long histories, for example, in language and in the objective carriers of a cultural heritage such as books, tapes, paintings, sculptures, CDs, etc. They can be woven in time, as in music, drama, and language; and they can be more bodily based, as we now know from research into "body language" and communication through "eye-to-eye" contact. In all these ways individual persons communicate with each other and also with the wider human community—past, present, and future.

The receptor of this "information"[63] in the individual person is the individual human brain which stores this variegated "information" that constitutes knowledge of an other's state of consciousness (which is, under a different description, of the state of an other's brain). This occurs at different levels and is integrated into a perception of the other person. Such knowledge of the other person can be recalled, with varying degrees of rapidity and accuracy, into consciousness. On a nondualist view, this process can be regarded as a re-activation of the brain to reproduce the original patterns that previously constituted this conscious awareness of the other person[64]—as long as it continues to be recognized that these conscious "mental"

---

[62] See the articles in this volume by Leslie Brothers and Marc Jeannerod.

[63] The scare quotes around "information" are meant to indicate that I in no way wish to pretend that the mind-brain-body relation will be eventually subsumed entirely into information theory, useful as that is in delineating key aspects of the relation; see sec. 2.3 above.

[64] Presumably it is therefore at some point in brain development and function that autism, in which interpersonal communication is impaired, is to be located, as was suggested to me by John Marshall in the conference discussions.

events are a nonreducible reality that is distinctive of the human-brain-in-the-human-body.

It seems that all the processes involved in inter-communication between human persons can be investigated and described at different levels by the methods and concepts appropriate to the level in question without invoking any ontologically distinct, special, "psychic" medium, unknown to the natural sciences, as the means of communication. This is not to say that the meaning of what is communicated can be reduced simply to physical patterns in the media in question, for the interpretation of these necessitates a recognition of their distinctive kind of reality. But it is to stress that all communication between human beings, even at the most intimate and personal level, is mediated by the entities, structures, and processes—that is, by the constituents—of the world. The subtly integrated patterns of these means of communication do in fact allow mutual comprehension between two human individuals of each other's distinctive personhood. This knowledge of two persons of each other, this knowledge by acquaintance, is notoriously not fully expressible in any of the frameworks of interpretation appropriate to the various modalities of the interaction process. There remains an inalienable uniqueness, and indeed mystery, concerning the nature of the individual person and of the interaction between two persons. Both the sense of personhood, of being a person, and also awareness of interpersonal relations are unique, irreducible emergents in humanity.

Recognition of the rootedness of the means of interpersonal communication in the constituents of the world does not diminish or derogate from the special kind of reality that constitutes persons and their mutual interactions. For in such communication between persons there occurs a subtle and complex integration of the received sense-data with previous memories of that person, under the shaping influence of a long-learned cultural framework of interpretation that provides the language and imagery with which to articulate the relation in consciousness. So recognition of the physical nature of the means of communication between persons in no way diminishes the uniqueness and "in depth" character that can pertain to personal relationships at their most profound level for the individuals concerned, which are, indeed, often the most real and significant experiences of people's lives.

3 God's Interaction with the World

3.1 Modes of Interaction

This interaction has been variously classified in the history of Christian thought[65]: (1) the creative activity of God; (2) the sustaining activity of God; (3) God's action as final cause; (4) general providence; (5) special providence; (6) miracles. Since we human beings are individuals, the question of how God can communicate with us is an instance of 5, God's special providence—namely, how God can affect our thinking and so events, or patterns of events, in our brains. That God might be able to do so at all is, of course, an aspect of 4 that results from the character of 1 and 2, into which 4 is often subsumed.[66] For the purposes of the ensuing discussion, I shall be taking a broadly panentheistic view of the relation of the being of God to that of the world.[67] When more distinctly Christian theological matters are under consider

---

[65] For example, by Michael J. Langford, *Providence* (London: SCM Press, 1981), 6.

[66] As discussed in *TSA*, chap. 9, where references are given.

[67] For my understanding of panentheism, see *TSA*, 158–59, 370–72. Briefly, it is "the belief

ation, I have a broadly "modalist" understanding of the Trinity insofar as such a view is apophatic (that is, reticent) concerning the ontology of God, but recognizes the threefold character of the Christian experience of the personal God as transcendent, incarnate, and immanent (the "economic" Trinity).[68]

3.2 Intervention?

The successes of the sciences in unraveling the intricate, often complex, yet rationally beautifully articulated, web of relationships among structures, processes, and entities in the world have made it increasingly problematic to regard God as "intervening" in the world to bring about events that are not in accordance with these divinely created patterns and regularities that the sciences increasingly unravel. Indeed for most scientifically educated Christians, their very belief in the existence and nature of the Creator God depends on this character of the world. The transcendence of God, God's essential otherness and distinct ontology from everything else, always allows in principle the theoretical possibility that God *could* act to overrule the very regularities to which God has given existence. But, setting aside the immense moral issues about why God does not intervene to prevent rampant evil, more fundamentally this gives rise to an incoherence in our understanding of God's nature, for intervention suggests an arbitrary and magic-making agent far removed from the concept of One who created and is creating the world science reveals. That world appears increasingly convincingly as closed to causal interventions from outside of the kind that classical philosophical theism postulated (for example, in the idea of a "miracle" as a breaking of the laws of nature).

So the problem is: How can one conceive of the God who is the Creator of this world affecting events in it without abrogating the very laws and regularities to which God has given existence and all the time sustaining it in existence? It has been intensified by the general skepticism among philosophers, theologians, and scientists (if not in the general public) about the existence of a "supernatural" world which, by postulating an ontological category of immaterial "spirit," provided a route or channel, as it were, along which divine influences could operate to manipulate matter and human beings. Such dualism is not intellectually defensible today, and has few supporters, not least with respect to human nature. Theists find themselves asserting that the only ontological dualism to which they are committed is that between God and the world—that is, to the absolute difference between an infinite and necessary

---

that the Being of God includes and penetrates the whole universe, so that every part of it exists in Him but (as against pantheism) that His Being is more than, and is not exhausted by, the universe," *Oxford Dictionary of the Christian Church*, 2nd rev. edit., Frank L. Cross and Elizabeth A. Livingstone, eds. (Oxford: Oxford University Press, 1983), 1027. In contrast to classical philosophical theism with its reliance on the concept of necessary substance, panentheism takes embodied personhood for its model of God—cf. Clayton, "The Case for Christian Panentheism"—and so has a much stronger stress on the immanence of God "in, with, and under" the events of the world. This was the thrust of my essay, "Biological Evolution—a positive theological appraisal," in *Evolutionary and Molecular Biology* (Vatican City State: Vatican Observatory; Berkeley, Calif.: Center for Theology and the Natural Sciences, 1998), namely, that "God is the Immanent Creator *creating in and through the processes of the natural order*," and that the very processes of biological evolution, as revealed by the biological sciences, "*are* God-acting-as-Creator, God *qua* Creator.... The processes are not themselves God, but the *action* of God-as-Creator." This, of course, is why I also do not wish to resort to any micro-interventionist action of God to steer evolution.

[68] See *TSA*, 347–49.

being and the contingency of the entire created order. This is also a premise of this essay. But, as Austin Farrer, long since noted,[69] this inevitably leads, in all hypotheses concerning how God might bring about particular events (5 in section 3.1) to the problem of the "ontological gap(s) at the causal joint," for if God in God's own Being is distinct from anything we can possibly know in the world, then God's nature is ineffable and will always be inaccessible to us so that we have only the resources of analogy to depict *how* God might influence events.[70]

From a panentheist point of view, the problem of God's interaction with the world is mitigated—though the intractable problem of evil remains—because the total web of natural events, in this perspective, is viewed as in itself the creative and sustaining *action* of God but, of course, not identical *with* God.[71] This points us in the direction of postulating that the "ontological gap(s)" between the world and God is (are) located simply everywhere[72]—or, more precisely, because the world is "in God," God can influence the world in its totality, as a System-of-systems.

3.3 Whole-Part Influence as a Model for God's (Special, Providential) Interaction with the World

I have elsewhere expounded this model (using the term "constraint," now replaced by "influence") in an attempt to render intelligible how God might be conceived of as influencing particular events, or patterns of events, in the world without interrupting the regularities observed at the levels the sciences study; the reader is referred to those texts for a fuller account.[73] Only with a plausible account of how

---

[69] Austin Farrer, *Faith and Speculation* (London: A. & C. Black, 1967).

[70] For a fuller discussion see *TSA*, 148–52.

[71] For panentheism, see *TSA*, chap. 9 and fn. 67, above. The metaphor of natural events as, in some sense, *God's* actions should not, in my view, be stretched to include a metaphor employed by some authors of the world as God's *body*. The first has, like all metaphors, an "is/is-not" aspect—namely, in this case, my emphasis on the ontological distinction between God and the world. The second might tempt us unwarrantedly to seek for a divine analogy for the human brains and nerves whereby human decisions effect events in their bodies!

[72] Cf. my remarks in *CAC* (p. 287, first para.) which apply here too: "[T]he present exercise could be regarded essentially as an attempt, as it were, to ascertain where this ontological gap, across which God transmits "information" (i.e., communicates), is most coherently "located," consistently with God's interaction with everything else having *particular* effects and without abrogating those regular relationships to which God's own self continues to give an existence which the sciences increasingly discover." This concurs with Gregersen in his article, "Three Types of Indeterminacy" (fn. 14, p. 184), in which he says: "We cannot expect to find the causal 'routes' of divine action and their subsequent 'joints' with natural causes. The most we can do, is to suggest meaningful localizations of possible divine actions."

[73] *TSA, passim*, especially 157–60; and *CAC*, 272–76, 282–87, where I proposed:

If God interacts with the "world" at a supervenient level of totality, then God, by affecting the state of the world-as-a-whole, could, on the model of whole-part constraint relationships in complex systems, be envisaged as able to exercise constraints upon events in the myriad sub-levels of existence that constitute that "world" without abrogating the laws and regularities that specifically pertain to them—and this without "intervening" within the unpredictabilities we have noted [I had in mind here the in-principle, inherent kinds, i.e., quantum events, though the remarks would also apply to the practical unpredictabilities of chaotic systems]. *Particular* events might occur in the world and be what they are because God intends them to be so, without at any point any contravention of the laws of physics, biology, psychology, or whatever is the pertinent science for the level of description in question (283).

Ernan McMullin has raised the question of how this proposal of mine relates to quantum indeterminism in his, "Cosmic Purpose and the Contingency of Human Evolution," *Theology Today* 55 (1998): 407 and fn. 50. As he points out, in my view God does not definitively

God can affect the world "instrumentally" can we proceed to address the question of how God might communicate "symbolically" with humanity (see section 1).

Initially, I will prescind from any analogy with the mind-brain-body relation or with personal agency. The model is based on the recognition that the omniscient God uniquely knows, across all frameworks of reference of time and space, everything that it is possible to know about the state(s) of all-that-is, including the interconnectedness and interdependence of the world's entities, structures, and processes. By analogy with the operation of whole-part influence in real systems (see section 2.2), the suggestion is that, because the "ontological gap(s)" between the world and God is/are located simply *everywhere* in space and time, God could affect holistically the state of the world (the whole in this context). Thence, mediated by the whole-part influences of the world-as-a-whole (as a *System*-of-systems) on its constituents, God could cause particular events and patterns of events to occur which express God's intentions. These latter would not otherwise have happened had God not so intended.

This unitive, holistic effect of God on the world could occur without abrogating any of the laws (regularities) which apply to the levels of the world's constituents[74]— by analogy with the exercise of whole-part influence in the systems discussed in section 2.2. Moreover, this action of God on the world may be distinguished from God's universal creative action in that particular intentions of God for particular patterns of events to occur are effected thereby—and the patterns could be intended by God in response *inter alia* to human actions or prayers.

The ontological "interface" at which God must be deemed to be influencing the world is, on this model, that which occurs between God and the totality of the world (all-that-is), and this, from a panentheistic perception, is within God's own self. What passes across this "interface," I have also suggested,[75] may perhaps be conceived of as a flow of information, but one has to admit that, because of the "ontological gap(s)" between God and the world, which must always exist in any theistic model, this is only an attempt at making intelligible that which we can postulate as being the initial effect of God seen from our side of the boundary, as it were.[76] Whether or not this use of the notion of information flow proves helpful in this context, we do need some way of indicating that the effect of God at this, and so at all, levels is that of pattern-shaping in its most general sense. I am encouraged in this kind of exploration by the recognition that the concept of the *Logos*, the Word,

---

know the future, but has a maximally conceivable capacity to predict it based on total knowledge of present events and of the laws and regularities of natural processes (*TSA*, 128–33). In the case of quantum events, this would, to respond to his query, have to refer to God's prediction of the *statistical* outcome of multiple quantum events and not individual ones—if the standard "Copenhagen" interpretation of quantum mechanics is assumed. In his article, McMullin's other query about the proposal concerns how the interaction between an ontologically distinct God and the world might be conceived of without being the forbidden sort of intervention. This is met by the suggestion of the interaction being analogous to a flow of information, as described later in this section.

[74] Note that the same may be said of *human* agency. Also, this proposal recognizes explicitly that the "laws" and regularities which constitute the sciences usually apply only to certain perceived, if ill-defined, levels within the complex hierarchies of nature.

[75] *TSA*, 161,164; *CAC*, 274–75, 285. John Polkinghorne has made a similar proposal in terms of the divine input of "active information" in his *Scientists as Theologians* (London: SPCK, 1996), 36–37.

[76] Morever, I would not wish to tie the proposed model too tightly to a "flow of information" interpretation of the mind-brain-body problem (see also fn. 63 above).

of God is usually taken to emphasize God's creative patterning of the world and so God's self-expression *in* the world.

The panentheistic inter-relations of God and the world and the interaction of God with the world, including humanity, I have attempted to represent in figure 1 (overleaf).[77] This is a kind of Venn diagram and represents ontological relationships. It has the limitation of being in two planes so that the "God" label appears dualistically to be (ontologically) outside the world; although this conveys the truth that God is "more and other" than the world, it cannot represent God's omnipresence in and to the world. This limitation may be surmounted by noting that "God," in the figure, is denoted by the (imagined) infinite planar surface of the page *on* which the circle representing the world is printed. For, it is assumed, God is "more than" the world, which is nevertheless "in" God. The page underlies and supports the circle and its contents, just as God sustains everything in existence and is present to all. So the larger dashed circle, representing the ontological location of God's interaction with all-that-is, really needs a many-dimensional convoluted surface[78] not available on a two-dimensional surface. The point and tail of a double-shafted arrow have been placed at the centre of this circle to signal God's immanent influence and activity *within* the world. The present form of this figure is meant to stress particularly the many-leveled nature of the human recipients of divine communication.

3.4 God as "Personal" Agent in the World

I hope the model as described so far has a degree of plausibility in depending on an analogy only with complex natural systems in general and on the way whole-part influence operates in them. It is, however, clearly too impersonal to do justice to the *personal* character of many (but not all) of the most profound human experiences of God. So there is little doubt that it needs to be rendered more cogent by the recognition that, among natural systems, the instance *par excellence* of whole-part influence in a complex system is that of personal agency. Indeed in the previous section, I could not avoid referring to God's "intentions" and implying that, like human persons, God had purposes to be implemented in the world. For if God is going to affect events and patterns of events in the world, then one cannot avoid attributing the personal predicates of intentions and purposes to God—inadequate and easily misunderstood as they are. So we have to say that though God is ineffable and ultimately unknowable in essence, yet God "is at least personal" and personal language attributed to God is less misleading than saying nothing!

That being so, we can now legitimately turn to the exemplification of whole-part influence in the mind-brain-body relation (section 2.5) as a resource for modeling God's interaction with the world. When we do so the ascendancy of the "personal" as a category for explicating the wholeness of human agency asserts itself and the traditional, indeed biblical, model of God as in some sense a "personal" agent in the world is rehabilitated—but now in a quite different metaphysical, nondualist

---

[77] This is an elaboration of fig. 1 of *TSA* to include a depiction of the multi-leveled nature of human beings. While it hardly needs to be said, the infinity sign represents not infinite space or time, but the infinitely "more" that God's being encompasses in comparison with that of everything else.

[78] Recall Augustine's representation of "the whole creation" as if it were "some sponge, huge , but bounded" floating in the "boundless sea" of God, "environing and penetrating it... everywhere and on every side" (*Confessions*, VII.7).

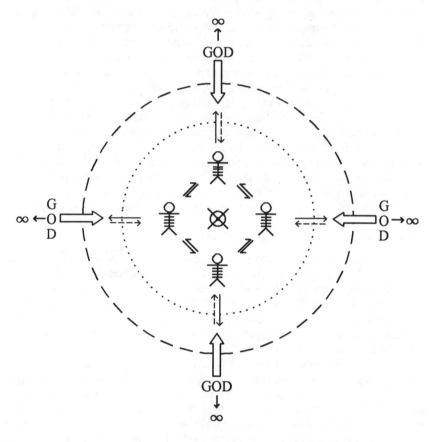

Figure 1.
Diagram representing spatially the ontological
relation of, and the interactions between, God
and the world (including humanity)

Legend for figure 1:

GOD   is represented by the whole surface of the page, imagined
             to extend to infinity (∞) in all directions

 the WORLD, all-that-is: created and other than God, and
             including both humanity and systems of non-human
             entities, structures, and processes

 the human WORLD: excluding systems of non-human
             entities, structures, and processes

 God's interaction with and influence on the world and its
             events

 tip and shaft of a similar double-shafted arrow perpendic-
             ular to the page; God's influence and activity *within* the
             world

---→   effects of the non-human world on humanity
———→   human agency in the non-human world

⇌   personal interactions, both individual and social, between
             human beings, including cultural and historical influences

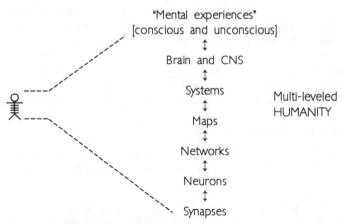

"Mental experiences"
[conscious and unconscious]
↕
Brain and CNS
↕
Systems   Multi-leveled
↕   HUMANITY
Maps
↕
Networks
↕
Neurons
↕
Synapses

Apart from the top one, these are the levels of organization
of the human nervous system depicted in fig. 1 of Patricia S.
Churchland and T.J. Sejnowski, "Perspectives on Cognitive
Neuroscience," *Science* 242 (1988): 741–45.

framework and coherently with the worldview (cf. section 3.2, above) which the sciences engender.[79]

When I was using nonhuman systems in their whole-part relationships as a model for God's relation to the world in "special providence," I resorted to the idea of a "flow of information" as being a helpful pointer to what might be conceived as crossing the "ontological gap(s)" between God and the world-as-a-whole. But now as I turn to more personal categories to explicate this relation and interchange, it is natural to interpret a "flow of information" between God and the world, including humanity, in terms of the "communication" that occurs between persons—not unlike the way in which a flow of Shannon-type information metamorphoses in the human context into information in the ordinary sense of the word.[80] Thus whatever else may be involved in God's personal interaction with the world, communication must be involved, and this raises the question: To whom might God be communicating? We would not be deliberating here on "scientific perspectives on divine action" if it had not been the case that humanity distinctively and, it appears, uniquely has regarded itself as the recipient of communication from an Ultimate Reality, named in English as "God." But in what ways has the reception of communication from God been understood and thought to have been experienced?

4 God and Humanity [81]

My account so far of how God interacts with the world has been chiefly concerned with devising a model for (1) the "instrumental" kind of relation. Now we have to think through the implications of this model for explicating (2) God's "symbolic" relation, that is, God's communicating relation to the world.

It is clear that all mutual interactions between human beings and the world (the solid and dashed single-shafted, double-headed arrows of fig. 1) are through the mediation of the constituents in the physical world of which human beings are part and in which human actions occur. Furthermore, all interactions between human beings (the pairs of solid single-headed arrows in fig. 1) also occur through the mediation of the constituents of the physical world, including the cultural heritage coded on to material substrates.[82] Such interactions include, of course, *communication* between human beings, that is, between their states of consciousness, which are also, under one description, patterns of activity within human brains. This raises the question: How, within such a framework of understanding, can one conceive of God's self-communication with humanity? This in turn raises the traditional question: How might God reveal Godself to humanity? How (in what way) can we conceive of God communicating with and to humanity in the light of the foregoing?

4.1 Revelation and Human Experience

In communication between human beings some of our actions, gestures, and responses are more characteristic and revelatory of our distinctive selves, of our

---

[79] See *TSA*, 160–66, and, more recently, *CAC*, 284–87, for an elaboration of this move and a discussion of the extent to which it is appropriate, if at all, to think of the world as the "body" of the ultimately transcendent God, who has a panentheistic relation to that same world.

[80] That is, Puddefoot's *1 → 2 → 3*; see sec. 2.3 above.

[81] The sequence of thought in this section is more fully amplified in *TSA*, chap. 11.

[82] Cf. sec. 2.6 above.

intentions, purposes and meanings, than are others. "It's not what you say but the way you're saying it." This prompts us to seek in the world those events and entities, or patterns of them, which unveil God's meaning(s) most overtly, effectively, and distinctively—constituting what is usually called "revelation," for in revelation God is presupposed to be *active*.

The ways in which such a revealing activity of God have been thought to occur in the different ranges and contexts of human experience can be graded according to the increasing extent to which God is said to be experienced as taking the initiative in making Godself explicitly known.

1. *General Revelation*. If the world is created by God then it cannot but reflect God's creative intentions and thus, however ambiguously, God's character and purposes[83]; and it must go on doing so if God continuously interacts with the world in the way we have proposed. Hence there can be a knowledge of God (and by inference, of God's purposes), however diffuse, which is available to all humanity through reflection on the character of the created world, its entities, structures, and processes, and in personal and social experience.

2. *Revelation to members of a religious tradition*. Belonging to a religious tradition provides one with the language and symbols to articulate one's awareness of God at any instant and as a continuing experience. The tradition provides the resources that help the individual both to enrich and to have the means of identifying his or her own experience of God. Thus there is a general experience of the ordinary members of a continuing religious community which may properly be regarded as a mode of revelation that is an enhancement of, and is more explicit than, the general revelation to humanity.

This kind of what one might call "religiously general" revelation arises when there is a confluence between, on the one hand, the streams of general human experience and general revelation and, on the other hand, those of the recollected and re-lived particular and special revelations of God that a tradition keeps alive by its intellectual, aesthetic, liturgical, symbolic, and devotional resources. These all nurture the unconscious of the adherents to that tradition and so shape their conscious awareness of God.

3. *Special revelation—revelation regarded as authoritative, and so as "special" in a particular tradition*. Some experiences of God by individuals, or groups of individuals, are so intense and subsequently so influential that they constitute initiating, "dubbing" experiences which serve in the community to anchor later references to God and God's relation to humanity, even through changes in the metaphorical language used to depict that ultimately ineffable Reality. The community then regards them as special, even if not basically different from those referred to in *2* above. So it is not improper to seek in history those events and entities, or patterns of them, which appear to have revealed God's meaning(s) most overtly, effectively, and distinctively.

That there should be such a knowledge is entirely coherent with the understanding of God's interaction with the world as represented in figure 1. The double arrows denote an input into the world from God that is influential in the whole-part constraining manner already discussed and thereby conceivable as an input of "information" in the sense of altering patterns of events in the world. The states of human brains can properly be considered to be such patterns so, in the model I am

---

[83] The *locus classicus* is, of course, Romans 1: 19–20.

deploying, there can be a general revelation to humanity of God's character and purposes in and through human knowledge and experience of the world.

The Jewish and Christian traditions have, more than most others, placed a particular emphasis on God's revelation in the experienced events of a history. Such "special" revelation, initiated (it is assumed) by God, has been regarded by Christians as recorded particularly in the Bible. How we are to receive this record today in the light of critical and historical study is a major issue in contemporary Christianity since it involves a subtle dialogue of each generation with its own past.

4. *Revelation and "religious experience."* My attempt to discriminate between modes of revelation according to the degree to which God is experienced as taking the initiative in making Godself explicitly known is helpful only up to a point, for there must be avoided the not uncommon tendency to press the distinctions too sharply and to ignore the smooth gradations between the different categories of revelation already distinguished. It is notable from a wide range of investigations how widespread such religious experience is, even in the secularized West, and that it is continuous in its distribution over those who are members of a religious community and those who are not.[84] The evidence suggests that the boundary between "general" revelation and revelation to members of a religious tradition is very blurred. But so also is the boundary between the latter and "special" revelation, for there are well-documented non-Scriptural accounts over the centuries of devotional and mystical experiences, regarded as revelations of God, among those who do belong to a religious tradition.

It is also widely recognized that the classical distinction between "natural" and "revealed" theology has proved difficult to maintain in modern times, for it can be held that the only significant difference between supposedly "natural" and supposedly "revealed" insights is that the former are derived from considering a broader (though still selected) range of situations than the latter. The same could also be said of the subsequently more widely favored distinction between "general" and "special" revelation, for the range of, and overlap between, the means whereby insights are gained into the divine reality has had to be recognized.[85]

There is therefore a gradation, but there are also differences in intensity and the degree of explicitness with which these "religious" experiences are received as revelations of *God* as their initiator—rather as a variegated and rough terrain may be accentuated to give rise to distinctive hills and even sharp peaks without loss of continuity. The questions now that follow are: How does our understanding of God's interaction with the world, including humanity, relate to human revelatory experiences of God? How can the notion of religious experience be accommodated

---

[84] See, for example, David Hay, *Religious Experience Today: Studying the Facts* (London: Mowbray, 1990). Typical questions concerning "religious experience" to which positive responses from between one third and one half of people in "Western" countries were obtained were: "Have you ever been aware of or influenced by a presence or power, whether you call it God or not, which is different from your everyday self?" or, "Have you ever felt as though you were very close to a powerful spiritual force that seemed to lift you out of yourself?"

[85] As David Pailin ("Revelation," in *A New Dictionary of Christian Theology*, Alan Richardson and John Bowden, eds. [London: SCM Press, 1983], 504–6) puts it, the ultimate justification of a supposed "revelation" is "by showing that the resulting understanding is a coherent, comprehensive, fruitful and convincing view of the fundamental character of reality."

by, be rendered intelligible in, and be coherent with, the understanding of God's interaction with the world that we have been developing?[86]

4.2 How Does God Communicate with Humanity?

If God interacts with the world in the way already proposed, through a whole-part constraining influence on the whole world system, how could God communicate with humanity in the various kinds of religious experience? It has been noted that the interpersonal relationships which we know of occur through the mediation of the constituents of the world. This suggests that religious experience that is *mediated* through sensory experience is intelligible in the same terms as that of the interpersonal experience of human beings. It is therefore plausible to think of God as communicating with human persons through the constituents of the world, through all that lies inside the dashed circle representing the world in figure 1—that is, *via* the nonhuman constituents represented by the inner dotted circle in the figure. God is seen as communicating "symbolically" through such mediated religious experiences by imparting meaning and significance to constituents of the world or, rather, to patterns of events among them.[87] Insights into God's character and purposes for individuals and communities can thereby be generated in a range of contexts from the most general to the special. The concepts, language, and means of investigating and appraising these experienced "signals" from God would operate at their own level and not be reducible to those of the natural and human sciences. The interpretation of mediated religious experience would have its own autonomy in human inquiry—"mystical" theology cannot be reduced without remainder to sociology or psychology, or *a fortiori* to the biological or physical sciences.

What about those forms of religious experience which are *unmediated* through sense experience? Brown subdivides them into the mystical, "where the primary import of the experience is a feeling of intimacy with the divine," and the numinous, "those experiences where awe of the divine is the central feature."[88] Swinburne divides them, on the one hand, into "the case where the subject has a religious experience in having certain sensations... not of a kind describable by normal vocabulary," and on the other hand, religious experiences in which "the subject... is aware of God or of a timeless reality.... [I]t just so seems to him, but not through his having sensations."[89] The experience of Elijah at the mouth of the cave on Mount Horeb was of all kinds: God communicated to him, not only through the natural phenomena of wind, earthquake, and fire, but eventually, apparently, and paradoxically, in an *un*mediated way—through "a sound of sheer silence," an image of absolute nonmediation.

In such instances, is it necessary to postulate some action of God whereby there is a direct communication from God to the human consciousness that is not mediated by any known natural means, that is, by any known constituents of the world? Is

---

[86] These questions are rendered more pertinent by the recognition of the important role played in recent years by "religious experience" as part of inductive and cumulative arguments which claim to warrant belief in God.

[87] This may properly be thought of as a "flow of information" from God to humanity, so long as the reductive associations of such terms are not deemed to exclude—as they need and should not—interpersonal communication.

[88] David Brown, *The Divine Trinity* (London: Duckworth, 1985), 37, 42–51.

[89] Richard Swinburne, *The Existence of God* (Oxford: Clarendon Press, 1979), 251.

there, as it were, a distinctive layer or level within the totality of human personhood that has a unique way of coming into direct contact with God? This was, as we saw in section 1, certainly the assumption when the human person was divided into ontologically distinct parts, one of which (often called the "spirit" or the "soul") had this particular capacity.

Now, we cannot but allow the possibility that God, being the *Creator* of the world, might be free to set aside any limitations by which God has allowed his interaction with that created order to be restricted. However, we also have to recognize that those very self-limitations which God is conceived of as having self-imposed are postulated precisely because they render coherent the whole notion of God as Creator with purposes that are being implemented in the natural and human world we actually have and which the sciences increasingly unveil. Such considerations also make one very reluctant to postulate God as communicating to humanity through what would have to be seen as arbitrary means, totally different in kind from any other means of communication to human consciousness. The latter would include the most intensely personal inter-communications, yet even these, as we saw above, are comprehensible as mediated subtly and entirely through the biological senses and the constituents of the world (section 2.6).

So, to be consistent, even this capacity for *un*mediated experience of God cannot but be regarded as a mode of functioning of the total integrated unity of whole persons—persons who communicate in the world through the world's own constituents. For human beings this communicating nexus of natural events *within* the world includes not only human sense data ("qualia") and knowledge stored in artefacts, but also all the states of the human brain that are concomitant (or whatever word best suits the relation of mind-brain-body) with the contents of consciousness and of the unconscious. The process of storage and accumulating both conscious and unconscious resources is mediated by the various ways in which communication to humanity can occur—and all these have been seen to be effected through the natural constituents of the world and the patterns of events which occur in them.

When human beings have an experience of God apparently *un*mediated by something obviously sensory—as when they are simply "waiting upon God" in silence—they can do so through God communicating *via* their recollected memories, the workings of their unconscious and everything that has gone into their *Bildung*, everything that has made them the persons they are. All of this can be mediated through patterns in the constituents of the world, including brain patterns. Experiences of God indeed often seem to be ineffable, incapable of description in terms of any other known experiences or by means of any accessible metaphors or analogies. This characteristic they share with other types of experience, such as aesthetic and interpersonal experience, which are unquestionably mediated through patterns in the events of this world. Experiences of God could take the variety of forms that we have already described and could all be mediated by the constituents of the world and through patterns of natural events, yet could nonetheless be definitive and normative as revelations to those experiencing them, for if God can influence patterns of events in the world to be other than they otherwise would have been but for the divine initiative—and still be consistent with scientific descriptions at the appropriate level—then it must be possible for God to influence those patterns of events in human brains which constitute human thoughts, including thoughts *of* God and a sense of personal interaction *with* God.

The involvement of the constituents of the world in the so-called unmediated experiences of God is less overt and obvious because in them God is communicating through subtle and less obvious patterns in the constituents of the world and the events in which they participate. The latter include the patterns of memory storage and the activities of the human brain, especially all those operative in communication at all levels between human persons (including *inter alia* sounds, symbols, and possibly Jungian archetypes), and the artefacts that facilitate this communication.

On the present model of special providential action—as the effects of divine whole-part influence—it is intelligible how God could also affect patterns of neuronal events in a particular brain, so that the subject could be aware of God's presence with and without the mediation of memory in the way just suggested. Such address from God, whether or not *via* stored (remembered) patterns of neuronal events, could come unexpectedly and uncontrivedly by the use of any apparently external means. Thus, either way, it would seem to the one having the experience as if it were *un*mediated. The revelation to Elijah at the mouth of the cave had both this immediacy and a basis in a long prior experience of God.

On examination, therefore, it transpires that the distinction between mediated and unmediated religious experiences refers not so much to the means of communication by God as to the nature of the content of the experience—just as the sense of harmony and communion with a person far transcends any description that can be given of it in terms of sense data, even though they are indeed the media of communication. We simply *know* we are at one with the other person. Similarly, in contemplation the mystic can simply be "aware of God...it just seems so to him" (as Swinburne puts it), and both experiences can be entirely mediated through the constituents of the world. So it is not surprising that those experiencing such communications from God experience them as intensely *personal*, for this is the kind of experience closest to them in ordinary life. What the treatment in this essay has therefore been pointing to is that an intelligible account can be given of how God can communicate *personally* to human beings within a world that is coherent and consistent with the descriptions of that world given at other levels by the natural and human sciences.

Certainly, for Elijah, "the sound of sheer silence" left no doubt about the *personal* nature of the command he had received and its meaning for him personally.

5 Appendix: *Concerning Theory- and Process-Autonomy*

The distinction between theory- and process-autonomy is important in relation to the possibility of regarding the world-as-a-whole as a "System-of-systems," to how the notion of reducibility (non-autonomy) applies to its processes and the theories applied to it; and to the usefulness, or otherwise, of the idea of a flow of information between systems (as distinct from that between levels). These considerations are relevant to the proposal that God has an effect on events in the world *via* a whole-part influence on it as a System-of-systems, possibly through an input analogous to a flow of information.

I myself have long argued,[90] following a seminal essay by Morton Beckner,[91] for the existence of a form of "weak" anti- or nonreducibility, in which:

---

[90] Arthur Peacocke, "Reductionism in Biology—A Review of Some of the Epistemological Issues and Their Relevance to Biology and the Problem of Consciousness," *Zygon* 11(1976): 307–24; *GNB*, chaps. 1, 2.

[91] Morton Beckner, "Reduction, Hierarchies and Organicism," in *Studies in the Philosophy of Biology*, Ayala and Dobzhansky, eds., 163–76.

1. There can be autonomy of the *concepts, theories, laws, etc.* (of the relevant science) applicable to the higher level(s)—in a *particular* system they can be logically not reducible[92] to those of the sciences applicable to lower levels of the system (theory autonomy).
2. The *processes* occurring at the higher level(s) in such systems are *not* autonomous with respect to those occurring at the lower one (process autonomy).

Statement *1* is the essence of *epistemological* anti- or nonreductionism. Statement *2* recognizes both that the whole system is composed of isolable, stable entities (the parts) subject to the concepts, theories, laws, etc., of the appropriate science applicable at that lower level, and also that the processes going on in the system regarded as a whole, to which the higher-level science applies, are simply the *same* processes as those going on in the lower level as perceived by the lower-level science, but now *in the complex inter-relations* that constitute the whole as the distinctive system it is.

Beckner showed conclusively that the form of epistemological anti-reductionism ("strong anti-reductionism" or "strong organicism"), which is a combination of both process- and theory-autonomy for the higher level, implies that some new entity (or entities) other than the separate, constituent parts is (are) *additionally* present in the whole system. For example, in the case of nonreduction of biology to physics and chemistry, this strong form of anti-reductionism leads to vitalism. In the present context, I am also not applying this strong form of epistemological anti-reductionist interpretation to the mind-body relation.

I agree with Beckner's analysis and have applied it in my discussion of *particular* hierarchically stratified systems; that is, I have espoused a "weak" form of epistemological anti-reductionism (theory- but not process-autonomy) with regard to, say, the relation of biology to physics and chemistry. In the case of the non-reducibility of the higher-level science, I have also argued for a, sometimes putative, attribution of reality to that to which the concepts, theories, laws, etc., of that science refer, so warranting the term "emergent," as already discussed section 2.1. When applied to *particular* systems of stable parts, any such attribution of an emergent reality to the higher level involves also attributing to it causal influence over the parts, the lower level. I have not taken this to mean that new processes are occurring at the higher level, but only that the basic lower-level processes are now so related at the higher level of the whole system that a new kind of reality has emerged at that level. When this is the case, note that I am not equating "process autonomy" with "causal efficacy" or saying there is a necessary conjunction between them.

This whole issue is a central concern of the essay by Niels Gregersen on autopoietic processes.[93] He wishes to retain the distinctiveness of *type-different*

---

[92] By the criteria of Ernest Nagel, *The Structure of Science* (New York: Harcourt Brace and Co., 1961), chap. 11.

[93] Gregersen, "The Idea of Creation and the Theory of Autopoietic Processes." This essay refers, in accord with a later phase in general systems theory, specifically to systems for which the parts themselves change in role and function on being incorporated into an autopoietic (self-organizing) system—as when a phonetic unit acquires new meaning in different kinds of systems (e.g., in a language, in a different cultural context). Gregersen and I have engaged in an extended correspondence over the years on these matters and on this recent essay of his in particular. To do justice to his insight and original approach requires a essay in itself. He has, with characteristic courtesy, taken great pains in that essay to compare his views with mine. In this current essay of mine I have taken very seriously his comments and criticisms,

systems and in that context argues for process-autonomy "because type-different systems operate on the basis of their own internal codes."[94] On reflection, I do not think Gregersen's point actually contradicts my position about *particular systems of stable* parts, as outlined above, but it does illuminate how we must regard interaction *between* distinct systems. Each hierarchically stratified system has its own internal set of higher-to-lower levels —distinctive nonreducible relations between its processes (its "code"[95])—and is therefore of a distinct "type" in Gregersen's (non-token) sense.[96]

In order to clarify the confusions arising from not clearly distinguishing discussions concerned with the internal relations *within* a hierarchical system and those concerned with relations *between* systems,[97] my proposal is as follows: when one is considering a particular system of stable parts and its internal relationships, "weak" nonreducibility can prevail and this then involves theory-autonomy of the higher level but not process-autonomy of that level, in accordance with the cogent arguments of Beckner. However, when one comes to consider the relation and interaction between *two type-different systems*,[98] each hierarchically stratified within itself, then even if such "weak" nonreducibility applies internally to each system regarded separately, yet in their interaction the processes going on in each of the respective systems at their corresponding higher levels are indeed autonomous with respect to each other. That is what is meant by taking the systems to be type-different at their higher levels, even if they have similar processes occurring at their lower ones.

---

and the main text here represents how I think our views might be brought into focus in a more coherent whole, with some adjustments on both our parts! However, within the scope of this present essay, I cannot give a detailed account with cross-references of my approach in relation to his, but only invite the reader to examine his essay and mine together, for I am increasingly convinced that the resolution of the problematic of "divine action" lies in the direction we are taking.

[94] Ibid., fn. 5.

[95] As specific, say, to a particular autopoietic system which operates, as Gregersen states (ibid.) on the basis of its own internal code.

[96] He writes: "Autopoietic theory…adopts the idea of *process-autonomy*, namely that different [the context implies "type-different"] systems are not causally reducible to one another…" (ibid.).

[97] Two particular sources of confusion which complicate this discussion of theory- and process-autonomy may be identified as follows. (1) As in the development of general systems theory, a distinction needs to be made among systems displaying nonreducible ("emergent") properties between those whose parts are stable and those whose parts change *because* they are incorporated into the system under consideration. This is the contrast between, say, the stability of water molecules as such in the Bénard convection phenomenon and the way a phonetic unit itself acquires new meanings in different systems. (2) The other confusion concerns what we mean by the "processes" occurring in a system. In yeast glycolysis, for example, the whole "process of glycolysis" (sometimes spatially and temporally self-organizing) may be regarded as a ramified inter-connection in space and time of the unit processes of individually identifiable chemical reactions, each of which is describable at its own level by its molecular transformations and associated chemical kinetics. Yet it is the complex *relations* of these unit processes that constitute "glycolysis." So it would be misleading to affirm that the "process of glycolysis" is autonomous with respect to the unit chemical processes, even though the concept, theories, laws, etc., applicable to the whole are now irreducible because of the new relationships between unit reactions.

[98] E.g., the system of the Sun and that of the Earth's atmosphere.

NEUROSCIENCE, ARTIFICIAL INTELLIGENCE, AND HUMAN NATURE: THEOLOGICAL AND PHILOSOPHICAL REFLECTIONS

Ian G. Barbour

1 Introduction

I hope to show that it is consistent with neuroscience, computer science, and a theological view of human nature to understand a person as a multilevel psychosomatic unity who is both a biological organism and a responsible self. We can avoid both materialism and body-soul dualism if we assume a holistic view of persons with a hierarchy of levels. The themes I will consider are embodiment, emotions, the social self, and consciousness. In the first three sections I look at these themes from the standpoints of theology, neuroscience, and research on artificial intelligence in computers. I then examine the concepts of information, dynamic system, hierarchical levels, emergence, and some philosophical interpretations of consciousness. Finally I suggest that process philosophy can provide a conceptual framework for integrating these varied perspectives on human nature.

We should note at the outset that theologians and philosophers bring their own conceptual frameworks to the interpretation of scientific theories. The theologian draws from the experiences, rituals, stories, and beliefs of an historical religious community. The philosopher seeks a coherent view of religious, aesthetic, moral, and cultural as well as scientific features of human life. We cannot expect neuroscience to provide a complete or adequate account of human nature because there are so many kinds of activity and levels of organization between neurons and persons in communities—including the relationships studied by evolutionary biology, developmental, cognitive, and social psychology, anthropology, linguistics, and even history, literature, and the arts.

But the concepts of the theologian or philosopher are not simply brought to the interpretation of science; they are also influenced by science. Scientific theories have implications for theology and philosophy, which may need reformulation or modification in the light of science. Conversely, philosophers and theologians can offer scientists wider intellectual and personal contexts for their work, suggestions of ways to relate it to other disciplines, and analysis of ethical issues arising from scientific theories and their applications. They can also encourage scientists to examine the philosophical assumptions underlying their judgments as to what features of phenomena are important to investigate and what types of concepts might be plausible—even though they recognize that scientific theories must be judged by their scope, consistency, compatibility with empirical data, and fruitfulness in suggesting further research.

2 The Self in Theology

I have chosen some themes to explore briefly from the history of Western theological reflection that are relevant to neuroscience and artificial intelligence.

2.1 Biblical Views

Four features of the biblical account of human nature are highlighted here.

2.1.1 An Embodied Self, Not a Body-Soul Dualism

The Bible regards body and soul as aspects of a personal unity, a unified activity of thinking, feeling, willing, and acting. Joel Green writes: "It is axiomatic in Old Testament scholarship today that human beings must be understood in their fully integrated embodied existence."[1] According to Oscar Cullmann, "the Jewish and Christian interpretation of creation excludes the whole Greek dualism of body and soul."[2] In particular, the body is not the source of evil or something to be disowned, escaped, or denied—though it may be misused. We find instead an affirmation of the body and a positive acceptance of the material order. Lynn de Silva writes:

> Biblical scholarship has established quite conclusively that there is no dichotomous concept of man in the Bible, such as is found in Greek and Hindu thought. The biblical view of man is holistic, not dualistic. The notion of the soul as an immortal entity which enters the body at birth and leaves it at death is quite foreign to the biblical view of man. The biblical view is that man is a unity; he is a unity of soul, body, flesh, mind, etc., all together constituting the whole man.[3]

According to the *Interpreter's Dictionary of the Bible,* the Hebrew word *nepheš* (usually translated as soul or self) "never means the immortal soul, but is essentially the life principle, or the self as the subject of appetites and emotion and occasionally of volition." The corresponding word in the New Testament is *psychē,* "which continues the old Greek usage by which it means *life.*"[4] When belief in a future life did develop in the New Testament period, it was expressed in terms of the *resurrection of the total person* by God's act, not the inherent immortality of the soul. Cullmann shows that the future life was seen as a gift from God, not an innate human attribute. Paul speaks of the dead as sleeping until the day of judgment when they will be restored—not as physical bodies nor as disembodied souls, but in what he calls "the spiritual body" (1 Corinthians 15:44). There were diverse strands in both Hebraic and Greek thought, and their influence on Paul's writing in the context of the Hellenistic world have been the subject of extensive discussion by biblical scholars.[5]

2.1.2 The Role of Emotion

"You shall love the Lord your God with all your heart, and with all your soul, and with all your mind" (Matthew 22:37). According to biblical scholars, these three terms—heart, soul, and mind—describe differing but overlapping human characteristics and activities rather than distinct faculties or components of the person. "The widely held distinction between mind as seat of thinking and heart as seat of feeling

---

[1] Joel B. Green, "'Bodies—That Is, Human Lives': A Re-examination of Human Nature in the Bible," in *Whatever Happened to the Soul? Scientific and Theological Portraits of Human Nature,* Warren S. Brown, Nancey Murphy, and H. Newton Malony, eds. (Minneapolis: Fortress Press, 1998), chap. 7, 158.

[2] Oscar Cullmann, *Immortality of the Soul or Resurrection of the Dead?* (New York: Macmillan, 1958), 30.

[3] Lynn de Silva, *The Problem of Self in Buddhism and Christianity* (London: Macmillan, 1979), 75.

[4] Norman W. Porteous, "Soul," in *Interpreter's Dictionary of the Bible* (Nashville: Abingdon, 1962), vol. 4, p. 428.

[5] For example, Brevard Childs, *Biblical Theology of the Old and New Testaments* (Minneapolis: Fortress Press, 1993), chap. 7.

is alien from the meaning these terms carry in the Bible.... the heart is the seat of the reason and will as well as of the emotions."[6] Paul writes: "If I understand all mysteries and all knowledge, and if I have all faith, so as to remove mountains, but have not love, I am nothing" (1 Corinthians 13:2). Love is of course not simply a matter of emotion, because it involves intention and action. But clearly it is not primarily a product of reason. Some portions of the Bible, such as the Wisdom literature, express the outlook of the wise person reflecting on human experience. But in most biblical texts we are called to be responsible agents rather than simply rational thinkers. Sin is understood as a defect of the will, not of reason. In much of Greek thought, the basic human problem is ignorance, for which the remedy is knowledge. But in biblical thought it is our attitudes and motives that lead us astray.

2.1.3 The Social Self

In the biblical tradition, we are inherently social beings. The covenant was with a people, not with a succession of individuals. Some of the psalms and later prophets focus on the individual (for example, Jeremiah speaks of a new covenant written in the heart of each person), but this was always within the context of *persons-in-community*. Judaism has preserved this emphasis on the community, whereas Protestant Christianity has sometimes been more individualistic. In the Bible, we are not self-contained individuals; we are constituted by our relationships. We are who we are as children, husbands and wives, parents, citizens, and members of a covenant people. God is concerned about the character of the life of the community as well as the motives and actions of each individual.[7] The religious community shares a common set of sacred stories and rituals. Even individual prayer and meditation take place within a framework of shared historical memories and assumptions.

2.1.4 In God's Image, but Fallen and Redeemed

The biblical assertion that humanity is created "in the image of God" (Genesis 1:27) is sometimes taken to refer to particular human traits (for example, rationality, free will, or moral responsibility) that distinguish us from other creatures. An alternative view in the history of Judaism and Christianity has been that the *Imago Dei* refers to the *relation* of human beings to God and indicates their potentiality for reflecting God's purposes. Human creativity can be seen as an expression of divine creativity.[8]

But the biblical tradition has also said that humans fall short of fulfilling these creative potentialities. In the light of evolutionary history, the fall of Adam cannot be taken literally. There was no Garden of Eden, no original state of innocence, free of death and suffering, from which humanity fell. But the story can be taken as a powerful symbolic expression of *human sinfulness,* where sin is understood as self-centeredness and estrangement from God and other people—and also, we might add, from the world of nature. Sin in all forms is a violation of relationships.[9] Original sin

---

[6] E.C. Blackman, "Mind," in *A Theological Word Book of the Bible*, Alan Richardson, ed. (New York: Macmillan, 1950), 145.

[7] Walter Eichrodt, *Man in the Old Testament* (London: SCM Press, 1951); Frederick C. Grant, *An Introduction to New Testament Thought* (Nashville: Abingdon, 1950), 160–70.

[8] Philip Hefner, "The Evolution of the Created Co-creator," in *Cosmos as Creation*, Ted Peters, ed. (Nashville: Abingdon, 1989).

[9] Marjorie Hewitt Suchocki, *The Fall to Violence: Original Sin in Relational Theology*

is not an inheritance from Adam but an acknowledgment that we are born into sinful social structures, such as those that perpetuate racism, oppression, and violence.

Redemption is the restoration of relationships—with God, with other people, and with other creatures—when brokenness and alienation are replaced by wholeness, healing, and reconciliation. The Christian tradition holds that this redemptive possibility is most clearly seen in the life of Christ and in our response to God's love made known in Christ. The doctrine of the Incarnation affirms Christ's full embodiment, underscoring again the importance of the body, while it also affirms his unique relationship to God and his total identification with God's will. *Imago Dei*, sin, redemption, and Incarnation can thus all be understood in relational terms rather than as attributes or state of individuals in themselves.

I will suggest that the first three themes above—embodiment, emotion, and the social self—are supported by neuroscience. The fourth theme—sin and redemption—is not supported by neuroscience, but when interpreted in the light of the other three themes it is not inconsistent with neuroscience.

2.2 Medieval and Modern Views

Greek thought included a diversity of views of human nature and of these the greatest influence on early Christian theology was Plato's view that a preexistent *immortal soul* enters a human body and survives after the death of the body. The Gnostic and Manichaean movements in the late Hellenistic world maintained that matter is evil and that death liberates the soul from its imprisonment in the body. The early church rejected Gnosticism, but it accepted the ontological dualism of soul and body in Neoplatonism and to a lesser extent the moral dualism of good and evil associated with it. Other forces in the declining Greco-Roman culture aided the growth of asceticism, monasticism, rejection of the world, and the search for individual salvation. Some of these negative attitudes toward the body are seen in Augustine's writing, but they represent a departure from the biblical affirmation of the goodness of the material world as God's creation.[10]

Aquinas accepted the Aristotelian view that *the soul is the form of the body*, which implied a more positive appraisal of the body. He said that the soul was created by God a few weeks after conception, rather than existing before the body. Aquinas gave a complex analysis of human nature and moral action that included an important role for emotions ("passions") in carrying out the good which is known by reason.[11] Medieval theologians expressed a sense of the organic unity of a world designed according to God's purposes. Nevertheless the concept of an immortal soul presupposed an absolute line between humans and other creatures and encouraged an anthropocentric view of our status in the world, even though the overall cosmic scheme was theocentric. With few exceptions, the nonhuman world was portrayed as playing only a supporting role in the medieval and Reformation drama of human redemption.

Descartes's dualism of *mind* and *matter* departed even further from the biblical view. The concept of soul had at least allowed a role for the emotions, as the biblical

---

(New York: Continuum, 1994).

[10] David Kelsey, "Human Being," in *Christian Theology*, 2nd ed., Peter Hodgson and Robert King, eds. (Philadelphia: Fortress, 1985).

[11] James Keenan, *Goodness and Rightness in St. Thomas Aquinas's Summa Theologiae* (Washington, D.C: Georgetown Univ. Press, 1992).

view had done. But mind, in the Cartesian understanding, was a nonspatial, nonmaterial "thinking substance," characterized by reason rather than emotion. Matter, on the other hand, was said to be spatial and controlled by physical forces alone. It was difficult to imagine how two such dissimilar substances could possibly interact. Descartes claimed that animals lack rationality and are machines without intelligence, feelings, or awareness.[12] The idea of the soul may have supported the dignity of the individual in Western history, but when understood individualistically it diverted attention from the constitutive role of the community in selfhood.

2.3 Contemporary Theology

An immaterial soul would be inaccessible to scientific investigation. Its existence could neither be proved nor disproved scientifically. But many *feminist theologians* today are critical of all forms of dualism for other reasons. They see in our culture a correlation of the dichotomies of mind/body, reason/emotion, objectivity/ subjectivity, domination/nurturance, and male/female. Male is associated with mind, reason, objectivity, and domination, which are given higher status than body, emotion, subjectivity and nurturance. Feminists decry the denigration of the body in much of Christian history; they seek a more positive evaluation of embodiment and a more integral view of the person.[13] *Environmentalist theologians* have criticized the soul-body dualism that postulated an absolute line between human and nonhuman life and thereby contributed to environmentally destructive attitudes toward other forms of life.

The theme of *the social self* is prominent among contemporary theologians. H. Richard Niebuhr defends "the fundamentally social character of selfhood," for "every aspect of every self's existence is conditioned by membership in the interpersonal group."[14] Niebuhr draws from George Herbert Mead and the social psychologists who say that selfhood arises only in dialogue with others. We are not impartial spectators but members of communities of interpreters. The social context is also evident in the idea of *the narrative self.* Alasdair MacIntyre and others maintain that our identities are established by the stories we tell about ourselves. These stories always involve other people.[15] Advocates of "narrative theology" insist that our personal stories are set in the context of the stories of our communities. Religious beliefs are transmitted not primarily through abstract theological doctrines but through the stories told by the religious community that provide the wider framework for our own life-stories.[16] We will see some parallels in the concept of the narrative self as it appears in recent writings by neuroscientists.

Theologian Keith Ward maintains that soul and body represent not two entities but *two languages* for talking about humans. In the tradition of British linguistic philosophy he considers the uses of differing types of language and their functions

---

[12] See Ian G. Barbour, *Religion and Science: Historical and Contemporary Issues* (San Francisco: HarperSanFrancisco, 1997), chap. 1.

[13] For example, Rosemary Radford Ruether, *Sexism and God-Talk* (Boston: Beacon Press, 1983).

[14] H. Richard Niebuhr, *The Responsible Self* (New York: Harper, 1963), 73.

[15] Alasdair MacIntyre, *After Virtue: A Study of Moral Theory*, 2nd ed. (Notre Dame, Ind.: Univ. of Notre Dame Press, 1984), chap. 15.

[16] James B. Wiggins, ed., *Religion as Story* (New York: Harper and Row, 1975); Michael Goldberg, *Theology as Narrative: A Critical Introduction* (Nashville: Abingdon, 1982).

in human life, concluding that soul-talk functions to assert the value and uniqueness of each individual and to defend human openness to God. Language about persons is used to interpret the lives of embodied agents capable of responsible actions.[17]

A *two-language approach* is also adopted by several psychologists with strong theological interests. Malcolm Jeeves holds that "mind" and "brain" are two ways of talking about the same events. He cites Donald MacKay's claim that the first-person agent's story of mental events is complementary to the third-person observer's story of neural events, and not in competition with it. For Jeeves, science and religion also present complementary perspectives or ways of perceiving the world. Elsewhere he suggests that there are different levels of activity in the brain to which differing concepts are applicable, and that activities at higher levels causally affect activities at lower levels.[18] Contemporary theologians have thus sought in various ways to recover the biblical themes of embodiment, emotion, and the social self.

3 Neuroscience and Selfhood

I suggest that current research in neuroscience shares with recent theological writings an emphasis on embodiment, emotions, and the social self. I will also examine neurological research on the role of consciousness and the implications of such research for our understanding of selfhood.

3.1 Embodiment

Perception is an evolutionary product of bodily action. Humberto Maturana and Francisco Varella maintain that historically the needs and actions of an organism affected the type of perceptual system it developed. In a frog's visual system, certain neurons respond only to small dark spots—undoubtedly an advantage in catching flies. So, too, human neurophysiology evolved in parallel with distinctive human goals and interests.[19] Michael Arbib argues that perception is not passive reception of data but an action-oriented restructuring of the world.[20] Mental representations ("schemas") provide information relevant to actions that could be carried out under the guidance of perceptions, expectations, and goals. Other studies have shown that the development of vision in newborn cats is dependent on bodily movement.

The influence of biochemical processes on mental events is evident in many types of research on the effects of hormones, "mind-altering" drugs, and therapeutic medications. For example, Peter Kramer examines the use of Prozac in the treatment of depression. He defends its value in correcting chemical imbalances (especially in the neurotransmitter serotonin), but he concludes that the most effective therapy combines medication with consideration of traumatic experiences and psychosocial factors in the patient's personal history.[21]

---

[17] Keith Ward, *Defending the Soul* (London: Hodder and Stoughton, 1992).

[18] Malcolm Jeeves, *Human Nature at the Millennium* (Grand Rapids, Mich.: Baker, 1997); idem, *Mind Fields: Reflections on the Science of Mind and Brain* (Grand Rapids, Mich.: Baker 1993). See also Donald M. MacKay, *Behind the Eye* (Oxford: Basil Blackwell, 1991).

[19] Humberto Maturana and Francisco Varela, *The Tree of Knowledge: The Biological Roots of Human Understanding* (Boston: Science Library, 1987).

[20] Michael Arbib, in this volume, and *The Metaphorical Brain 2: Neural Networks and Beyond* (New York: John Wiley, 1989), chap. 2.

[21] Peter Kramer, *Listening to Prozac* (New York: Viking Penguin, 1993).

The correlation of *brain sites* with particular cognitive functions can be studied by means of data on brain lesions or strokes occurring in human subjects or laboratory animals. PET scans can be used to monitor blood flow in small regions of the brain while the subject is carrying out an assigned cognitive task. Damage in a particular brain area has been found to prevent language acquisition without harming other skills. One patient with a brain lesion was able to write lucid prose but could not read it. Oliver Sacks describes patients who were unable to recognize faces but had no problem recognizing animals or objects.[22] Extensive research has been done on epileptic patients whose right and left brain hemispheres had been severed to control their seizures. Such patients might be able to follow instructions to pick up an object with the left hand, for example, but be unable to name it. But other mental functions seem to be widely distributed, and a given site may serve more than one function. Memory is distributed over many locations, with short-term memory differing markedly from long-term memory. Neural networks function globally and exhibit distributed properties. In all these cases, mental events are radically dependent on physiological processes at a variety of levels.

3.2 Emotions

Five approaches to the scientific study of emotions can be identified.

1. The Evolutionary Perspective: Darwin held that emotional behaviors are remnants of actions that were functional in evolutionary history. A dog's anger is evident in growling and baring the teeth, which embody a physiological readiness to act aggressively and signal such readiness to other creatures. A legacy of such behavior is seen when an angry person shouts and grimaces. Darwin claimed that a common evolutionary origin accounts for the universality of facial expressions of emotions in diverse cultures. Subsequent studies have found considerable cross-cultural consensus in the identification of photographs of faces expressing six basic emotions: anger, fear, happiness, sadness, disgust, and surprise. Proponents of sociobiology and evolutionary psychology have offered hypotheses concerning the adaptive value of many behaviors associated with emotions.[23]

2. The Body-Response Perspective: William James held that emotions are internal perceptions of physiological processes in our own bodies—tense facial muscles, sweaty palms, and especially the effects of the autonomic nervous system, such as a pounding heart, faster breathing, and higher blood pressure. He claimed that emotions are the result (and not the cause) of physiological changes that we perceive directly. More recent studies of patients with spinal cord injuries showed that the feedback from internal organs does affect the intensity of a person's experience of emotions.[24]

---

[22] Oliver Sacks, *The Man Who Mistook His Wife for a Hat* (New York: HarperCollins, 1985).

[23] Charles Darwin, *The Expression of the Emotions in Man and Animals*, (Chicago: Univ. of Chicago Press, 1965, orig. 1872); Carroll Izard, *Human Emotions* (New York: Plenum, 1977); John Tooby and Leda Cosmides, "The Past Explains the Present: Emotional Adaptations and the Structure of Ancestral Environments," *Ethology and Sociobiology* 11 (1990): 375–424.

[24] William James, *The Principles of Psychology*, (Cambridge: Harvard Univ. Press, 1983, orig. 1890); R.W. Levenson, P. Ekman, and W.V. Friesen, "Voluntary Facial Action Generates Emotion-Specific Autonomic Nervous System Activitiy," *Psychophysiology* 27 (1990): 363–84.

3. The Cognitive Perspective: Whether an animal (or a person) flees in fear or fights back in anger may be partly instinctive, but it also reflects a cognitive appraisal of the situation and a judgment about its potential danger. Authors in this tradition talk about the meaning of events and the expectations and goals that people bring to their appraisals. They insist that emotion cannot be separated from cognition. They also go beyond the six emotions studied by the physiologically-oriented authors above to consider complex human emotions such as guilt, shame, and embarrassment, or anxiety in the face of uncertainty.[25]

4. The Social Perspective: Here the role of culture in the social construction of emotions receives strong emphasis. Emotional feelings and their expressions are shaped by cultural meanings learned in infancy and throughout life. Anger at another person is often related to the belief that the other person is to blame for an offending action. Guilt is an acknowledgment that one has violated one's own norms, whereas shame is the feeling that one is not worthy in the eyes of others. Historians and social psychologists have described the role of emotions as a means of social control (by shame and guilt, for example, in Puritan New England). Other studies suggest that when children learn words for emotions and culturally approved actions to express them, their emotional experience is itself affected.[26]

5. The Neural Perspective: Research on the physiological structures of the brain can help us understand the functioning of emotions. The amygdala and hypothalamus in the limbic system have been shown to be crucial in several emotions. Some examples of such research are discussed below.

These five approaches are often viewed as competing theories. Research does sometimes yield data that support one approach rather than another. I suggest, however, that they should be viewed as alternative perspectives using *different levels of analysis* that are not necessarily incompatible with one another. Emotions are multifaceted: they are at the same time adaptive mechanisms, bodily feelings, cognitive appraisals, social constructions, and neural processes. Nevertheless, we must go on to ask how these levels are related to one another.

The work of Joseph LeDoux is conducted at *the neural level,* but it is entirely consistent with analysis at evolutionary, body-response, and cognitive levels.[27] He uses elevated blood pressure and heart rate as indicators of the emotion of fear in rats when they hear a sound to which they have previously been conditioned by association with an electric shock. He finds evidence of *direct* neural paths from the auditory system to the amygdala that allow a rapid (evolutionarily valuable) response. He also finds *indirect* paths to the amygdala by way of the cortex that are slower but provide for interpretation and discrimination among sounds (as proposed by the cognitivists). LeDoux distinguishes between emotions as objective body responses and brain systems, on the one hand, and the subjective feelings associated with them, on the other, which he says are inaccessible to scientific study.

---

[25] Magda Arnold, *Emotion and Personality,* 2 vols. (New York: Columbia Univ. Press, 1960); Richard Lazarus, "Progress on a Cognitive-Motivational-Relational Theory of Emotion," *American Psychologist* 46 (1991): 819–34.

[26] James Averill, "The Social Construction of Emotion: With Special Reference to Love," in *The Social Construction of the Person,* K.J. Gergen and K.E. Davis, eds. (New York: Springer-Verlag, 1985); Rom Harré, ed., *The Social Construction of Emotions* (Oxford: Basil Blackwell, 1986).

[27] Joseph LeDoux, in this volume, and *The Emotional Brain: The Mysterious Underpinnings of Emotional Life* (New York: Simon & Schuster, 1996).

Antonio Damasio has studied the relationships between *emotion* and *cognition* in people who have undergone damage in the prefrontal cortex. In a classic case, Phineas Gage recovered from a severe injury and retained his intellectual abilities but underwent a personality change in which he was unable to make decisions or observe social conventions. One patient with a prefrontal brain tumor was totally detached emotionally. When he viewed films depicting violence, he could describe appropriate emotional reactions but said he could not feel them, and he was unable to make decisions in daily life. Damasio argues that the cortex and limbic system work together in the construction of emotions. He suggests that both Descartes and modern cognitive scientists have neglected the role of emotion in cognition. Damasio also holds that consciousness and continuity of identity are provided by self-representation and the construction of a narrative that includes personal memories and intentions. He describes the self as a many-leveled unity. "The truly embodied mind does not relinquish its most refined levels of operation, those constituting soul and spirit."[28]

3.3 The Social Self

Neuroscience provides many types of evidence concerning the social character of cognition in animals and humans.

3.3.1 Social Interaction

The importance of social interaction in cognition has been defended by Leslie Brothers in her essay written with Wesley Wildman in this volume and in her recent book.[29] In her research she attached electrodes to the brain of a monkey watching videotapes of the face of another monkey. She found neurons selectively responsive to the other monkey's facial expression of emotions. She suggests that human infants are attentive to adult faces because they have been pre-wired by evolutionary history to respond to relevant facial signals. However, Brothers thinks emotions are so varied that it is misleading to group them together under the common term "emotion." Human emotions are expressed and recognized within a socially constructed communicative system. She insists that the mind is a social creation that cannot be understood by studying its neural basis alone. "I take the mind to be irreducibly transactional."[30] The person is part of a social-moral order, not something to be found in the neural account. Human actions are explained by reasons and historical narratives, not by physical and chemical causes. Through narratives we collaboratively create ourselves as persons as we enact our place in a social world.

Human language is of course a social product, even if the capacity for language is genetically based. Selfhood is intersubjective and relational, dependent on history and culture. The social world is internalized in the formation of one's self-image, which in turn affects one's interaction with other people. The whole field of social psychology is devoted to the study of phenomena that cannot be understood by analysis of individuals alone.

---

[28] Antonio Damasio, *Descartes' Error: Emotion, Reason, and the Human Brain* (New York: G.P. Putnam, 1994), 252.

[29] Leslie Brothers, *Friday's Footprint: How Society Shapes the Human Mind* (New York: Oxford Univ. Press, 1997).

[30] Ibid., 146.

3.3.2 Memory and Narrative Construction

The stories we tell about ourselves as agents and subjects of experience are part of our self-identity. Children learn mental predicates and self-referential language as their parents ascribe intentions, desires, and feelings to them. We have a continuing identity as subjects, but memory is always an active reconstruction rather than simply a retrieval of information. We seek coherence and plausibility in our stories; narratives are revised and related to future goals and plans. The tragedy of Alzheimer's disease is the loss of the long-term memory that is required for self-representation. Sacks describes the case of "the lost mariner" with a brain lesion and memory loss, for whom art and music aided the reconstitution of a new identity.[31] The stories we tell about ourselves are also influenced by the stories present in our culture, including those of our religious traditions.

3.3.3 Cultural Symbol Systems

Human beings form *symbolic representations* of the self and the world that are always partial and selective. We seek meaning and order by seeing our lives in a wider context that is ultimately cosmic in scope. We identify ourselves with purposes and goals that extend beyond our own lives, temporally and spatially. Religious traditions have provided many of the symbols through which individuals integrate conflicting desires and make sense of their lives in a more inclusive context. In story and ritual people participate in religious communities and share their historical memories and their experiences of personal transformation. These wider symbolic structures of order and meaning are indeed human creations, but I have argued that they are also responses to patterns in the world and in human experience, so they can be critically evaluated and revised.[32]

James Ashbrook and Carol Albright have proposed that models of ultimate reality can be found in neuroscience itself, particularly in Paul MacLean's idea of *the tripartite brain*.[33] The *upper brain stem,* which we share with creatures as far back as the reptiles, controls the basic life support systems, such as breathing and reproduction. It offers an analogy to the image of God as the sustainer of the conditions for life. The *limbic system,* which we share with mammals, is the center of emotions that mobilize action and make richer forms of relationship possible, including empathy and care of the young. These qualities lead us to recognize emotion and social relationships as part of reality and to envision a nurturing and interacting God. The *neocortex* as it developed in primates and humans is the center of interpretation, organization, symbolic representation, and rationality. Damage to the frontal lobes affects the ability to prioritize, make plans, and pursue long-term goals. The left hemisphere is associated with verbal and logical analytical thought, and the right with visual, spatial, and holistic synthesizing thought. The activities of the neocortex would parallel the idea of a purposeful God who rationally orders and pursues goals.[34] (We should note that critics of MacLean have argued that the

---

[31] Sacks, *The Man Who Mistook His Wife for a Hat*, 22–41.

[32] Ian G. Barbour, *Myths, Models and Paradigms: The Nature of Scientific and Religious Language* (New York: Harper and Row, 1974).

[33] Paul MacLean, *The Triune Brain in Evolution* (New York: Plenum, 1990).

[34] James Ashbrook and Carol Rausch Albright, *The Humanizing Brain: Where Religion*

relationships between the three regions of the brain are more complex than he recognized, but a distinction of three functions of the brain might still provide analogies for envisaging God, as these authors propose.)

Ashbrook and Albright say that human beings seek meaning by viewing their lives in a cosmic and religious framework that is itself a *human symbolic construct*. But they go on to say that such symbol systems are not just useful fictions if they seek to interpret coherently the data of human experience. Moreover, the brain is itself part of the cosmos and a product of the cosmos, so its structures reflect the nature of the cosmos and whatever ordering and meaning-giving forces are expressed in its history.

3.4 The Role of Consciousness

Finally, let us consider some recent research on consciousness in human life that may be relevant to our concept of selfhood.

3.4.1 Unconscious Information Processing

Many instinctive responses and changes in the hormonal and autonomic nervous system occur without our being aware of them; our attention would be drastically overloaded if we had to keep track of all these changes. It has long been known that under hypnosis, and in subliminal perception, events of which we are not aware influence subsequent behavior. A variety of more recent experiments show the presence of unconscious information processing. *Blindsight* occurs in patients with a lesion in area V1 of the visual cortex. They are unable to see an obstacle in their path, yet they will act as if they see it and will walk around it. In another type of experiment carried out by Benjamin Libet, subjects were told to record the exact moment when they voluntarily initiated a finger movement. Electric impulses were detected in the brain (the so-called *readiness potential)* up to one-third of a second before the subject's decision, suggesting that brain processes occur before the subject is aware of them.[35]

Daniel Dennett reports experiments on *metacontrast* in which the image of a disc is followed after a very short delay by the image of a ring. Subjects say they have seen only the ring, yet they report that there were two stimuli. Dennett offers three possible explanations: the first stimulus was overridden before it entered consciousness; it entered consciousness but memory of it was then obliterated; or information from the first stimulus was reinterpreted in the light of the second one.[36] The experiment provides one more example of information processing that occurs without our being aware of it.

3.4.2 The Evolution of Consciousness

Simple organisms have a minimal *sensitivity* and responsiveness to the environment. If a one-celled paramecium finds no food at one location it will use its coordinated oar-like hairs to move to another location. Perception of an elementary kind occurs

---

and *Neuroscience Meet* (Cleveland: Pilgrim Press, 1997).

[35] Benjamin Libet, "Unconscious Cerebral Initiative and the Role of Conscious Will in Voluntary Action," *Behavioral and Brain Sciences* 8 (1985): 529–66.

[36] Daniel Dennett, *Consciousness Explained* (Boston: Little, Brown & Company, 1991), 141–44.

when there is a selective response to information used to control actions. At somewhat higher levels *sentience* includes a capacity for pain and pleasure, which were presumably selected in evolutionary history for their contribution to survival. When a neural system is present, pain serves as an alarm system and an energizing force in avoiding harm. But continued pain may hinder action; even invertebrates under stress release endorphins and other pain-suppressant chemicals similar to those released in humans in response to pain, so it is reasonable to assume that they have at least some experience of pain.

Donald Griffin has studied the *mental abilities of insects and animals*. He associates consciousness with complex and novel behavior in changing or unfamiliar circumstances. Bees can communicate the direction and distance of food sources and can distinguish between water, nectar, and a possible hive site; they do their waggle dance only when other bees are around, but they have limited ability to modify their behavior in new circumstances. Griffin argues that the versatile and goal-directed behavior of animals is evidence of thought, feeling, and conscious awareness. Animals imaginatively compare possible courses of action and anticipate their consequences. Comparison of mental representations of alternative actions allows for more rapid, diverse, and adaptive responses to a changing environment. But Griffin holds that self-awareness is present only in certain species of primates. When looking in a mirror, a great ape will touch a mark previously placed on its forehead.[37] Terrence Deacon notes that primates have a limited capacity for symbolic communication. Teaching a few symbols to apes is a slow and arduous process requiring repeated rewards and punishments. The ability of primates to generalize and to follow logical rules (such as inclusion and exclusion) is impressive but far short of human capacities for language and abstract thought.[38] Such evidence would lead us to speak of *degrees of consciousness* rather than an all-or-nothing attribute.

3.4.3 The Construction of the Self

There are numerous versions of the thesis that mental activity is *modular*. Jerry Fodor's *Modularity of Mind* argues that the mind is a collection of relatively independent special-purpose modules.[39] Marvin Minsky's *Society of Mind,* making use of computational models, claims that the human mind is an aggregate of many small mindless components.[40] According to Arbib, "the you is constituted by the holistic net of schema interactions in your brain." The coherence of the schema is achieved by their interaction and not by a central organizer.[41] William Calvin compares mental activity to a choir that works together, coalescing into a harmonious chorus without a conductor. A higher-order model of the self and the narratives in which it is represented serves to coordinate diverse subsystems.[42] Dennett argues that "multiple drafts" (alternative interpretive narratives) momentarily compete for

---

[37] Donald Griffin, *Animal Minds* (Chicago: Univ. of Chicago Press, 1992).

[38] Terrence Deacon, *The Symbolic Species: The Coevolution of Language and the Brain* (New York: W. W. Norton, 1997).

[39] Jerry A. Fodor, *The Modularity of Mind* (Cambridge: Harvard Univ. Press, 1983).

[40] Marvin Minsky, *Society of Mind* (New York: Simon and Schuster, 1985).

[41] Arbib, *The Metaphorical Brain 2*.

[42] William Calvin, *The Cerebral Symphony* (New York: Bantam, 1989); idem, *The Cerebral Code* (Cambridge: MIT Press, 1996).

attention below the level of consciousness, and we are aware only of the winning versions.[43]

Michael Gazzaniga, on the other hand, introduces a more centralized *coordinating system*. He finds that split-brain subjects will carry out an action with one brain hemisphere that uses information from a visual input to the other hemisphere of which they are not aware; they will then try to explain their action by some other reason, unrelated to the visual input. He postulates an Interpreter (in the left hemisphere, the main site of linguistic abilities) that monitors and integrates the unconscious activities and tries to make sense of them in relation to belief systems.[44] Robert Ornstein's *Multimind* proposes many small modules with specialized skills, but also a governing self that links and coordinates these units.[45] A few brain researchers, including John Eccles, have continued to defend a dualism of mind and body in which the unity of conscious experience is an inherent property of mind, but this position has few adherents among scientists today.[46] There is thus considerable diversity among neuroscientists in their interpretations of modularity, but there is wide support for the idea that the unity of the self is achieved rather than given in the life of the individual.[47]

To summarize this section, recent work in neuroscience is consistent with the biblical emphasis on embodiment, emotions, and the social self. It offers some parallels with ideas found in recent theology concerning the narrative self. The findings of neuroscience on distributed mental activities and multiple levels of processing can be cited in support of holistic and multilevel ontologies, as we shall see. Current theories concerning consciousness are more speculative, and they are subject to a variety of philosophical interpretations, which will be examined later. The biblical view does indeed conflict with the determinist and materialist philosophical assumptions of many neuroscientists, but not, I suggest, with the data and theories of neuroscience itself.

4 Artificial Intelligence and Human Nature

We look now at recent work on computers and artificial intelligence (AI) and ask how it relates to neuroscience and our understanding of human nature.

4.1 Symbolic AI and the Computational Brain

AI research has a double goal: creating intelligent computers and understanding how the human brain functions. In an influential essay, Allan Newell and Herbert Simon maintained that a world of discrete facts can be represented by a corresponding set of well-defined symbols. They claimed that the relationships among symbols are abstract, formal, and rule-governed; symbols can therefore be processed by differing

---

[43] Dennett, *Consciousness Explained*.

[44] Michael Gazzaniga, "Brain Modularity: Towards a Philosophy of Consciousness," in *Consciousness in Contemporary Science*, A. J. Marcel and E. Besiach, eds. (Oxford: Oxford Univ. Press, 1988); see also Gazzaniga, *Mind Matters: How Mind and Brain Interact to Create our Conscious Lives* (Boston: Houghton Mifflin, 1988).

[45] Robert Ornstein, *Multimind* (Boston: Houghton Mifflin, 1986).

[46] John Eccles, *Evolution of the Brain: Creation of the Self* (London: Routledge, 1989).

[47] John Teske, "The Spiritual Limits of Neuropsychological Life," *Zygon* 31 (1995): 209–34.

physical systems (natural or artificial) with identical results. They asserted that the brain and the computer are two examples of devices that generate intelligent behavior by manipulating symbols.[48] Symbolic AI tries to explain all cognition in terms of information, but it is not necessarily physicalist or reductionist because information is not reducible to the laws of physics.

Proponents of symbolic AI have made four assertions:

- *The Formalist Thesis:* Intelligence consists in the manipulation of abstract symbols according to formal rules.
- *The Turing Test:* A computer is intelligent if in performing tasks it exhibits behavior that we would call intelligent if it were performed by a human being.
- *Substrate Neutrality* (or *Multiple Realizability):* Software programs can be run on differing physical systems (neuron-based or transistor-based) with identical results.
- *The Computational Brain:* The human brain functions like a computer. In popular parlance, *mind* is to *brain* as software *programs* are to computer *hardware.*

Critics of formalism have said that human language and perception are *context-dependent.* Hubert and Stuart Dreyfus have portrayed the importance of common-sense understanding, background knowledge, and nonlinguistic experience in the interpretation of human language. Linguistic and perceptual understanding, they insist, are active processes, strongly influenced by our expectations, purposes, and interests.[49] They have also emphasized *the role of the body* in human learning. Much of our knowledge is acquired actively through interaction with our physical environment and other people. We learn to ride bicycles not by studying physics or by acquiring a set of rules but by practice. We use the skills of "knowing how" rather than the propositions of "knowing that." Such "tacit knowledge" cannot be fully formalized. In a child's development, growth in perception is linked to action and bodily movement. These authors see in the formalist thesis the legacy of a rationalism that goes back to Plato: the assumption that knowledge consists of formal rational relationships that exist independently of the body and the material world. They claim that formalism is a new kind of dualism in which software and hardware, like mind and body, can be analyzed independently.

Terry Winograd, whose programs for robots that could manipulate blocks were hailed as early successes in AI, subsequently repudiated formalism and stressed the importance of *individual and social life* in human understanding. He now accepts Martin Heidegger's view that our access to the world is primarily through practical involvement rather than detached analysis. According to Heidegger, understanding is aimed not at abstract representation but at the achievement of our goals and interests. Our speech is communication for particular purposes, a form of action. Winograd also draws from Ludwig Wittgenstein, who insists that there is no private language or individual representation of the world, but only communication in contexts of social interaction. Language reflects our social practices, cultural

---

[48] Allan Newell and Herbert Simon, "Computer Science as Empirical Enquiry: Symbols and Search," originally published in 1976, reprinted in *Philosophy of Artificial Intelligence*, Margaret Boden, ed. (Oxford: Oxford Univ. Press, 1990).

[49] Hubert Dreyfus and Stuart Dreyfus, *What Computers Still Can't Do*, 3rd ed. (Cambridge: MIT Press, 1993).

assumptions, and "forms of life" in an interpersonal world. Winograd has redirected his own research and is working on the design and use of computers to facilitate human communication and social interaction, rather than to simulate individual human behavior in isolated domains.[50]

4.2 Learning, Robotics, and Embodiment

In most AI systems, discrete symbols that represent the world are processed serially. The development of *parallel distributed processing* (PDP), however, allowed many separate units to carry out operations simultaneously and to interact with each other without centralized control.[51] In task-oriented PDP networks, the system can be programed to modify itself in successive runs, so that it learns by trial and error. One such network can be trained by an instructor to pronounce a text, converting various combinations of letters into a recognizable sound output from a voice synthesizer. The information is stored throughout the network, rather than by a one-to-one correspondence between separate data items and separate memory locations. Patterns develop in the whole without prior specification of the parts. If the learning procedure is repeated, the network will not end up with an identical circuit configuration.[52]

A further step is taken by Rodney Brooks and others in the design of robots that are *embodied, situated agents*. They are embodied in the sense that they can interact with the world through perception (using visual, auditory, and tactile sensors) and through action, and they are situated in particular environments. They have a minimum of central control; their architecture is decentralized in relatively independent units that interface directly with features of the environment in the generation of actions. New modules are added as incremental layers without disrupting existing modules.[53] Such robots learn by doing, not by manipulating abstract symbols. Their mechanical bodies are of course very different from our biological bodies; what they learn from their actions will differ from what we learn from ours.

Anne Foerst, who has degrees in both theology and computer science, works at MIT with the group designing the humanoid robot, Cog. She describes four of its characteristics:

- *Embodiment:* The group holds that human intelligence cannot be separated from bodily action or reduced to computational abilities. Cog has a "head" and "hands" which can move and interact with its environment.
- *Distributed Functions:* Small independent processing units activate local motor controls. Modular units with loose connections between them, rather than large centralized programs, allow greater flexibility in

---

[50] Terry Winograd and Fernando Flores, *Understanding Computers and Cognition: A New Foundation for Design* (Norwood, N.J.: Ablex Publishing, 1986).

[51] David E. Rumelhart and James L. McClelland, eds., *Parallel Distributed Processing*, 2 vols. (Cambridge: MIT Press, 1986).

[52] Stan Franklin, *Artificial Minds* (Cambridge: MIT Press, 1995), chap. 12.

[53] Rodney A. Brooks and Luc Steels, eds., *The Artificial Life Route to Artificial Intelligence: Building Embodied, Situated Agents* (Hillsdale, Mich.: Laurence Erlbaum, 1995). See also Andy Clark, *Being There: Putting Brain, Body, and World Together Again* (Cambridge: MIT Press, 1997).

coordination, and facilitate the acquisition of new abilities without interfering with existing abilities.

- *Developmental Learning:* Like a newborn child, Cog learns visual-tactile (eye-hand) coordination from practice in grasping objects. Many of its capacities are developmentally acquired rather than preprogramed.
- *Social Interaction:* Cog practices the equivalent of eye contact and is programed to take into account some of the effects of its actions on people. These social features are at an elementary stage but are a goal of ongoing research.[54]

Foerst acknowledges that most of her colleagues think that *consciousness* is illusory and that they adopt a functionalist view of both human and robot capacities. Foerst herself says that there are "two stories" about human beings; computation provides models for only one of these stories. In our own lives we justifiably rely on our intuitive self-understanding. She calls for dialogue and mutual respect between theologians and computer scientists and recognition of the biases and limits of each discipline.

4.3 Socialization and Emotion in Computers

Recent work in robotics answers some of the objections raised against the symbolic AI program, but other questions remain in the comparison of artificial and human intelligence. The *process of socialization* in humans occurs over a span of many years. In computers, information processing is very rapid, but interaction with the environment takes considerable time. Robots might be socialized partly by being fed vast quantities of information, but if the critics of formalism are correct, participation in human culture and forms of life would require active interaction over a longer period of time. The Dreyfuses maintain that only computer systems nearly identical to the human brain and endowed with human motives, cultural goals, and bodily form could fully model human intelligence. That may be too strong a claim, but it points to the importance of culture as well as of body in human understanding and in any attempt to duplicate such understanding in machines.

The ability or inability of android artifacts to experience *emotions* has been a recurrent theme in science fiction, from Karel Capek's R.U.R. in 1923 to Commander Data in *Star Trek.* Most AI researchers claim only to simulate cognitive processes, and they hold that cognition is quite independent of emotions. Roger Schank writes:

> It would seem that questions such as "Can a computer feel love?" are not of much consequence. Certainly we do not understand less about human knowledge if the answer is one way or the other. And more importantly, the ability to feel love does not affect its ability to understand.[55]

Other authors hold that we can analyze the function of an emotion in evolutionary history and then try to construct an AI program that fulfills the same function. For example, the main behavioral function of fear is avoidance of danger, which could be programed directly. Aaron Sloman has developed a computational theory of

---

[54] Anne Foerst, "COG, a Humanoid Robot, and the Question of Imago Dei," *Zygon* 33 (1998): 91–111.

[55] Roger Schank, "Natural Language, Philosophy, and Artificial Intelligence," in *Philosophical Perspectives on Artificial Intelligence*, M. Ringle, ed. (Brighton, England: Harvester Press, 1979), 222.

emotions (understood as dispositions to behave in certain ways). He says that computers could not experience feelings but could represent the cognitive components of emotions—for example, the external causes of anger and its relation to one's beliefs and ensuing actions.[56]

Rosalind Piccard's research is directed toward building computers with the ability to *recognize and express emotions*. Her goal is to facilitate communication between computers and humans. For example, a computer instruction program could slow down or offer further explication when it perceived expressions of frustration or anger in the user's face or heartbeat. A computer voice synthesizer might deliver a message with an intonation conveying an appropriate emotional tone. Piccard cites Damasio's work on the positive role of emotions in human cognition and suggests that emotional abilities would also contribute to computer intelligence. She remains agnostic, however, as to whether future computers might actually experience emotions. If they did, she says, their experience would differ greatly from ours, which is linked to physiological and biochemical processes unlike anything in computers. Some emotions, such as shame and guilt, reflect distinctive experiences of selfhood. Piccard says that we do not know enough about human consciousness to speculate on whether it could be duplicated rather than imitated in a computer.

> Our feelings arise in a living and complex biological organism and this biology may be uniquely able to generate feelings as we know them. Biological processes may be simulated in a computer and we may construct computational mechanisms that function like human feelings, but this is not the same as duplicating them.[57]

4.4 Consciousness in Computers?

There are still enormous differences between *computers* and *brains*. A brain has 1,000 trillion neurons each connected to as many as 10,000 neighbors; the number of possible patterns in interconnecting them is far greater than the number of atoms in the universe. Signals between neurons are not digital but are encoded in continuously variable properties such as electrical potentials or neuron firing frequencies. Serial computers retrieve fixed information from local addresses; human memory is accessed through partial descriptive clues and is reconstructed in a more dynamic way. Gerald Edelman argues that parallel distributed processing in computer networks offers analogies to neural networks, but that neurons and brains have many properties unlike those of computer chips. During embryonic development, for example, nerve cells connect to particular types of cells, but there is no exact pre-wiring such as computers require.[58]

Human beings are hierarchically organized, with many levels between the atom and the self; computers can indeed be built with hierarchical architecture, but the levels are less diverse and lack the degree of integration found at higher levels in organisms. Most computers are designed to be reliable by following precise algorithmic rules. To be sure, the final states of distributed networks in computers that learn from experience are unpredictable, but their potential for creative novelty

---

[56] Aaron Sloman, "Motives, Mechanisms, and Emotions," in *Philosophy of Artificial Intelligence*, M. Boden, ed. (Oxford: Oxford Univ. Press, 1990); see also Keith Oatley, *Best Laid Schemes: The Psychology of Emotion* (Cambridge: Cambridge Univ. Press, 1992).

[57] Rosalind Piccard, *Affective Computing* (Cambridge: MIT Press, 1997), 136.

[58] Gerald Edelman, *Bright Air, Brilliant Fire: On the Matter of the Mind* (New York: Basic Books, 1992).

seems rather limited. New knowledge from neuroscience will undoubtedly affect future computer design, but we should not underestimate the differences or the difficulties.

Is it conceivable that a future computer or robot could be *conscious?* A human infant develops by participation in a social and linguistic community. Events in the human mind are dependent on cultural contexts that extend far beyond the individual. The prospects for the socialization of robots are rather uncertain. But once we recognize that there are gradations of consciousness at different stages of an infant's development from a fertilized egg, and differing forms of consciousness in diverse animal species, we will not have to assume that consciousness in computers, if it is possible, will be like adult human consciousness. I suspect that it will turn out that conscious awareness requires forms of organized complexity or properties of neural cells and networks that have no parallels in silicon-based systems. I do not think we can exclude the possibility of conscious computers on metaphysical grounds, but there may be empirical grounds for the impossibility of computer consciousness. Because we know so little about the physical basis of human consciousness or the directions of future research in computer science, I am willing to leave this question open.

The mathematician and theologian John Puddefoot emphasizes the gap between computers and humans today. "To be regarded as something approaching the human, a computer would need to grow, feel pain, experience and react to finitude, and generally enter into the same state of mixed joy and sorrow as a human being. In particular it would need to be finite, aware of its finitude, and condemned one day to die."[59] On the other hand, he does not think we can set limits as to what might be possible in future computers. He speculates that with structures closer to those of living organisms, and with processes of evolutionary change within computers themselves, an artifact might conceivably produce its own forms of mind. Puddefoot adds that it was through evolutionary processes, after all, that God created human minds.

Our view of computers and robots, like our view of animals, will influence *our own self-understanding.* In relation to both animals and robots, interpretations that abolish sharp lines between human and nonhuman forms seem at first to be *a threat to human dignity.* But human dignity is not threatened if we recognize that future robots would be more than information processors, and that they may share some of what we consider the higher human capacities. In the case of animals, the recognition of similarities with humans has led to calls for respect for animal rights and for the inclusion of other life forms in the sphere of *moral consideration.* In the case of robots a similar extension of moral status would be required. If they can suffer, as we believe animals suffer, we would have duties to minimize such suffering. Robots would also have moral responsibilities toward each other and toward humans. There are also important *psychological issues* concerning our fears of creatures different from us, whether aliens from space or human creations from the laboratory. Moreover, the dangers of human *hubris* and misuse of technological power (evident in myths from Prometheus and the Tower of Babel to Frankenstein) need exploration, but that would divert us from the topics of this essay.

---

[59] John Puddefoot, *God and the Mind Machine: Computers, Artificial Intelligence, and the Human Soul* (London: SPCK, 1996), 92.

5 Information, Systems, Levels, and Emergence

In attempting to relate studies of neurons and studies of persons, I find four concepts helpful: communication of information, dynamic systems, hierarchical levels, and emergence as an alternative to reduction.

5.1 The Communication of Information

The concept of information is applicable to biological, cultural, and computer systems, and can illuminate some similarities and differences among those systems. Information is *an ordered pattern* that is one among many possible sequences or states of a system. The pattern can be a sequence of DNA bases, alphabetical letters, auditory sounds, binary digits, or any other combinable elements. Information is *communicated* when another system (a living cell, a reader, a listener, a computer, etc.) responds selectively—that is, when information is coded, transmitted, and decoded. The meaning of the message is dependent on a wider *context of interpretation*. It must be viewed dynamically and relationally rather than in purely static terms as if the message were contained in the pattern itself.[60]

The *information in DNA sequences* in genes is significant precisely because of its context in a larger organic system. In the growth of an embryo, a system of time delays, spatial differentiation, and chemical feedback signals communicates the information needed so that the right proteins, cells, and organs are assembled at the right location and time. As Susan Oyama shows, complicated developmental pathways, with information flowing in both directions, connect genes with molecular activities and physiological structures.[61] Molecules of the immune system recognize an invading virus, which is like a key that fits a lock to release a specific antibody. The communication between molecules is dependent on properties of both the sender and the receiver. A receptor is part of an embodied action system that implements a response to signals.

In *sense perception*, transducers in the eye and ear convert physical inputs into neural impulses. As in all the cases above, the communication of information in the brain is a holistic property of a whole system. In itself, the frequency of firing of a neuron tells us very little about the information that is being communicated. Information is effective only in a context of interpretation and response. Once again, information flows in two directions. Information is constructed from sense data by active and action-oriented processes.[62]

Stored in the DNA is a wealth of *historically-acquired information* including programs for coping with the world. A bird or mammal uses specific visual or auditory clues to recognize and respond to a dangerous predator that it has not previously encountered. Individuals in some species are programed to communicate warning signals to alert other members of the species. Higher primates are capable of symbolic communication of information, and human beings can use words to express abstract concepts. Human information can be transmitted between

---

[60] Jeremy Cambell, *Grammatical Man: Information, Entropy, Language, and Life* (New York: Simon & Schuster, 1982).

[61] Susan Oyama, *The Ontogeny of Information: Developmental Systems and Evolution* (Cambridge: Cambridge Univ. Press, 1985).

[62] Maturana and Varela, *The Tree of Knowledge*; Francisco Varela, Evan Thompson, and Eleanor Rosch, *The Embodied Mind: Cognitive Science and Human Experience* (Cambridge: MIT Press, 1991).

generations not only by genes and by parental example but also in speech, literature, art, music, and other cultural forms.

It is tempting to use the concept of *information* to defend the possibility of immortality. In *Star Trek* the astronauts in a spaceship are instantly transported to a nearby planet by the transmission of information about all their molecular components, which are reassembled in a new location ("Beam me up, Scotty"). One might imagine that God has complete information about us and recreates us in another realm, which would have some similarities to the biblical idea of resurrection. However, the proposal appears reductionistic in its assumption that higher levels of selfhood are explainable by (and can be reconstituted from) information at the molecular level. I suggest that God knows us at all levels, including the highest level of our selfhood and subjectivity, and not just at the molecular level. God's relation to us is more personal than an inventory of molecules.

5.2 Dynamic Systems

Some authors have suggested that the study of *complex systems* provides principles applicable to phenomena as varied as neural, behavioral, and mental activities. They start from the thesis of Ilya Prigogine and Stuart Kauffman that complex systems exhibit global properties not predictable from their components. For example, a pattern of convection cells suddenly appears when a fluid is gradually heated from below; the cells all rotate in the same direction (left or right) but the direction is unpredictable. New forms of order appear when systems are near the border between order and chaos.[63] *Chaos theory* is the study of holistic temporal and geometric patterns without seeking reduction to detailed causal mechanisms. In nonlinear thermodynamic systems far from equilibrium, an infinitesimally small uncertainty (arising from quantum indeterminacy or from perturbations from outside any given system) can be greatly amplified, leading to large-scale consequences.[64]

Scott Kelso has used *dynamic systems theory* to compare the patterns found in neural, behavioral, and mental systems. Though the components at these various levels are very different, the trajectories mapped in the space of possible states show striking similarities. Nonlinear systems in metastable states make sudden transitions, often along bifurcating paths. Kelso studies situations in which an organism is coupled with its environment so that neither can be understood alone. He claims that underlying principles of self-organization operate at many levels and that global patterns of cooperative behavior can be most fruitfully analyzed by attention to collective variables.[65]

As one example, Kelso cites experiments in which subjects were told to clap their hands between the beats of a metronome. As the frequency of the beats was increased, the subjects suddenly switched from syncopated to synchronized claps. Using sensitive magnetic field detectors, he found that patterns in small groups of neurons became less coherent and then switched to new phase relations at the critical

---

[63] Ilya Prigogine and Isabelle Stengers, *Order out of Chaos* (New York: Bantam Books, 1984); Stuart A. Kauffman, *At Home in the Universe: The Search for Laws of Self-Organization and Complexity* (New York: Oxford Univ. Press, 1995).

[64] Stephen Kellert, *In the Wake of Chaos: Unpredictable Order in Dynamical Systems* (Chicago: Univ. of Chicago Press, 1993).

[65] J.A. Scott Kelso, *Dynamic Patterns: The Self-Organization of Brain and Behavior* (Cambridge: MIT Press, 1995).

frequency. As another example, he cites studies of recovery of functions after brain damage or loss of a limb or an eye; neural activities shifted to a nearby locus as the cortical map was reorganized, though the neurons themselves did not move. Distributed dynamic patterns, he claims, are more important than the physical structures themselves. As I see it, the results of such research are relatively modest so far, but they suggest the value of inquiry conducted on several levels at once. Systems analysis cannot replace microanalysis at the neural level, but it may provide principles for correlating events at multiple levels.

Alwyn Scott portrays *the emergence of consciousness* in a hierarchy of levels he calls "the stairway to the mind." He describes several examples of nonlinearity, including the diffusion equations for traveling waves in nerves, which cannot be derived from the equations of physics and chemistry. He calls his view "emergent dualism" or "hierarchical dualism."

> The idea that all can be reduced to the spare concepts of physics has been exposed as untenable because each level of the hierarchy is dynamically independent of its neighbors. Dynamic independence—in turn—arises from *nonlinearity*, which induces the *emergence* of new and qualitatively different atomistic entities at each level.[66]

5.3 A Hierarchy of Levels

Both neuroscience and computer networks support the idea that organized wholes exhibit systemic properties not evident in their parts. A *level* identifies a unit that is relatively integrated, stable, and self-regulating, even though it interacts with other units at the same level and at higher and lower levels. A living organism is a many-leveled hierarchy of systems and subsystems: particle, atom, molecule, macro-molecule, organelle, cell, organ, organism, and ecosystem. The brain is hierarchically organized: molecule, neuron, neural network, and brain, which is in turn part of the body and its wider environment. Human beings participate in the social and cultural interactions studied by the social sciences and humanities. A particular discipline or field of inquiry focuses attention on a particular level and its relation to adjacent levels.

Bottom-up causation occurs when many subsystems influence a system. *Top-down causation* is the influence of a system on its sub-systems. Higher-level events impose boundary conditions on chemical and physical processes at lower levels without violating lower-level laws.[67] Microproperties are not referred to in the specification of the macrostate by its global or collective properties. Network properties may be realized through a great variety of particular connections. Correlation of behaviors at one level does not require detailed knowledge of all its components. Just as the rules of chess limit the possible moves but leave open an immense number of moves that are consistent with but not determined by those rules, the laws of chemistry limit the combinations of molecules that are found in DNA, but do not determine the sequence of bases. The meaning of the message conveyed by DNA is not given by the laws of chemistry but by the operation of the

---

[66] Alwyn Scott, *Stairway to the Mind: The Controversial New Sciences of Consciousness* (New York: Copernicus, 1995), 187, italics in original.

[67] Donald Campbell, "'Downward Causation' in Hierarchially Organised Biological Systems," in *Studies in the Philosophy of Biology: Reduction and Related Problems*, Francisco J. Ayala and Theodosius Dobzhansky, eds. (Berkeley: Univ. of California Press, 1974), 179–86.

whole system. Communication of signals in neurons requires some expenditure of energy; what is communicated is not the energy, however, but the *form* of the signal in relation to input and output processes occurring at higher levels than the signal itself.

One way in which activities at higher levels influence lower-level activities is through *the feedback of information*. PET scans show that when people shut their eyes and think of mental images, the lowest level of the visual processing system (closest to the retina) is activated from higher levels without any input from the retina.[68] Learning programs in distributed processing networks result in patterns in the whole that have not been achieved by specifying the parts. The robot Cog is hierarchically organized with relatively autonomous distributed modules; information concerning the results of its activities is fed back so that it can learn by doing.

Holism is the claim that the whole influences the parts. The *whole-part* distinction is usually structural and spatial (for example, a *larger* whole). *Top-down causality* is a very similar concept, but it draws attention to a hierarchy of many levels characterized by qualitative differences in organization and activity (for example, a *higher* level). Levels are defined by functional and dynamic relationships and patterns in time, though of course they are inseparable from patterns in space.

John Polkinghorne has used the idea of *communication of information* as a model for conceiving of God's relation to the world. The opening verse of John's Gospel says: "In the beginning was the Word." The biblical concept of *logos* (word) combines the Hebrew idea of God's active communication and the Greek idea of rational structure, suggesting parallels with the concept of information. Arthur Peacocke has extended the idea of *whole-part relations* and suggests that God is the most inclusive whole. He also holds that the idea of *top-down causality* may be appropriate if God is imagined as acting from a level higher than any level in the realm of nature. I have discussed these proposals elsewhere.[69]

5.4 Reduction and Emergence

Epistemological reduction is the claim that theories at higher levels are derivable (in principle if not in practice) from theories at lower levels. Historically, higher-level theories have seldom been derived directly from previously existing low-level theories. Biological and psychological concepts, for example, are distinctive and cannot be defined in physical and chemical terms. But the theories of adjacent levels are not unconnected. The conceptual structure of theories at one level have typically been gradually altered in the light of theories at other levels. Moreover, inter-level theories may be proposed which are not derived from theories at either level alone. Lindley Darden and Nancy Maull discuss some historical examples and conclude that the unity of science is an important goal, but it is not achieved by theory reduction:

> An inter-field theory, in explaining relations between two fields, does not eliminate a theory or field or domain. The fields retain their separate identities, even though new lines of research closely coordinate the fields. . . . It becomes natural to view the unity of

---

[68] Erich Harth, *The Creative Loop: How the Brain Makes a Mind* (Reading, Mass.: Addison-Wesley, 1993), 64–83.

[69] Ian G. Barbour, "Five Models of God and Evolution," in *Evolutionary and Molecular Biology: Scientific Perspectives on Divine Action*, R.J. Russell, W.R. Stoeger, S.J., and F.J. Ayala, eds. (Vatican City State: Vatican Observatory; Berkeley, Calif.: Center for Theology and the Natural Sciences, 1998), 419–42.

science, not as a series of reductions between theories, but rather as the bridging of fields by inter-field theories.[70]

Churchland and Sejnowski describe *levels of organization and processing* in the brain, emphasizing network and system properties. Research at one level provides constraints and inspiration for research at other levels, both higher and lower. They portray the "coevolution of theories" as theories are revised and modified to take into account those at higher and lower levels.

> The ultimate goal of a unified account does not require that it be a single model that spans all the levels of organization. Instead, the integration will probably consist of a chain of models linking adjacent levels. When one level is explained in terms of a lower level this does not mean that the higher-level theory is useless or that the higher-level phenomena no longer exist. On the contrary, explanations will coexist at all levels, as they do in chemistry and physics, genetics and embryology.[71]

These authors believe that the integration of cognitive psychology and neuroscience will be the result of interdisciplinary interaction and not the replacement of one discipline by the other.

Ontological reduction is the claim that events at higher levels are determined by events at lower levels. It is a claim about reality and not just about theories. If events at a higher level have no causal efficacy, they are viewed as less real, or perhaps even as epiphenomena. I would defend ontological pluralism, a multi-leveled view of reality in which differing (epistemological) levels of analysis are taken to refer to differing (ontological) levels of events and processes in the world, as claimed by critical realism.[72] I take *emergence* to be the claim that in evolutionary history and in the development of the individual organism, there occur forms of order and levels of activity that are genuinely new and qualitatively different. A stronger version of emergence is the thesis that events at higher levels are not determined by events at lower levels and are themselves causally effective.

To sum up, the concepts of communication of information, dynamic systems, hierarchical levels, and emergence allow a more systematic elaboration of the view of the person as a multilevel unity—a view we have found consistent with biblical theology, neuroscience, and AI research. But they leave unresolved the problematic status of consciousness.

6 Philosophical Interpretations of Consciousness

Let us turn then to some philosophical interpretations of consciousness and its relationship to neuroscience and AI.

6.1 Eliminative Materialism

In *The Astonishing Hypothesis,* Francis Crick, codiscoverer of DNA, combines the presentation of data from the neurosciences with an explicitly materialist philosophy. He sees only two philosophical alternatives, a supernatural body-soul dualism or a materialistic reductionism. He equates dualism with religion, of which he is highly

---

[70] Lindley Darden and Nancy Maull, "Interfield Theories," *Philosophy of Science* 44 (1977): 60, 61.

[71] Patricia Churchland and Terrence Sejnowski, "Perspectives on Cognitive Neuroscience," *Science* 242 (1988): 741–45.

[72] Barbour, *Religion and Science*, 117–18.

critical, unaware that many contemporary theologians have rejected dualism. The volume opens with this statement:

> The Astonishing Hypothesis is that "you," your joys and your sorrows, your memories and your ambitions, your sense of personal identity and free will, are in fact no more than the behavior of a vast assembly of nerve cells and their associated molecules. As Lewis Carroll's Alice might have phrased it: "You're nothing but a pack of neurons."[73]

On the scientific side, Crick is critical of cognitive scientists for relying on computational models and neglecting neural research. His book is mainly devoted to research on visual processing and awareness. He proposes that consciousness is a product of the correlation of diverse neural systems through electrical oscillations of roughly forty cycles per second. He suggests that the activities of various brain regions are coordinated when these oscillations synchronize the local neuron firings. He does not totally dismiss the subjective character of consciousness, but he does not think that it can be studied by science. "What may prove difficult or impossible to establish is the details of the subjective nature of consciousness, since this may depend upon the exact symbolism employed by each conscious organism."[74]

Dennett holds that "consciousness is the last bastion of occult properties and immeasurable subjective states." Qualia (phenomena as experienced) are vague and ineffable. The self is *a linguistic fiction* generated by the brain to provide coherence retrospectively among its diverse narratives. Dennett holds that "multiple draft" scenarios of which we are not aware compete for dominance. The self is the "center of narrative gravity" of these scenarios. It is a useful fiction that we create to provide order in our lives. But the unity and the continuity of consciousness are illusions. There is no enduring Cartesian observer who unifies our diverse perceptions. Nor is there a continuous "stream of consciousness," as posited by William James or James Joyce. There are only unconscious processes unified intermittently by a representation of the self that the brain repeatedly recreates from memories of the past and new scenarios in the present.[75]

Dennett describes *the intentional stance* as the strategy of acting as if other people had intentions. The ascription of intentions is predictively useful, but we do not have to assume that intentional states are ever actually present. Dennett claims that he is an instrumentalist or functionalist who judges concepts only by their usefulness in describing behavior, without asking about their status in reality. But he seems to accept a metaphysics of materialism when he asserts that neuroscience will be able fully to explain intentional action. He says that he is not a "greedy reductionist" who expects to explain all higher levels directly in terms of the lowest level, but that he is a "good reductionist," expecting to explain any level in terms of the next lower one.[76]

6.2 The Irreducibility of Consciousness

In replying to eliminative materialism, several philosophers have maintained that consciousness and subjectivity are *irreducible and inaccessible to science*. Thomas

---

[73] Francis Crick, *The Astonishing Hypothesis: The Scientific Search for the Soul* (New York: Charles Scribner's Sons, 1994), 3.

[74] Ibid., 252.

[75] Dennett, *Consciousness Explained*.

[76] Daniel Dennett, *Darwin's Dangerous Idea* (New York: Simon & Schuster, 1995), 81–83.

Nagel holds that consciousness cannot be understood from the objective standpoint required by science (which he calls "the view from nowhere"). Conscious and intentional states presuppose a particular viewpoint. Scientific theories cannot explain phenomenal feelings or give an objective account of subjectivity. But science is not the only route to understanding, and in our practical life we inevitably attribute mental states to other people, and even to other species, though it is difficult to imagine what they are like. He cites evidence of the conscious inner life of animals but says that the experiential perspective can be understood only from within or by subjective imagination.[77]

Nagel does not defend a mind-body dualism but rather a *dual-aspect theory.* There is one set of events in the brain, of which mental concepts describe the subjective aspects and physical concepts describe the objective aspects. There is one substance with two sets of properties. Psychophysical laws connect the first- and third-person accounts which are both valid. Personal identity is unified and linked to memory and intention as represented in first-person accounts. Nagel holds that mental aspects are present only in relatively advanced organisms.

Colin McGinn holds that consciousness is beyond our comprehension because of *the limitations of human knowledge.* Evolution has endowed every species with limited powers of understanding developed for practical purposes. The senses are useful for representation of the spatial world in which we live, but consciousness is not spatial. The brain can be studied as a spatial object, and its parts have spatial coordinates and predicates such as size and shape. But the predicates of mental events are temporal rather than spatial. Knowledge of the correlations of phenomenal experience with physical data concerning the brain would not help us grasp the subjective character of consciousness, which cannot be described in the conceptual terms applicable to matter in space.[78]

McGinn believes that neural and mental events are correlated, but we cannot say how. Consciousness is a *causally emergent feature* of certain kinds of organized systems, but we cannot specify the necessary and sufficient conditions for consciousness to appear. Consciousness will remain an insoluble mystery, an intractable obscurity, because of our limited powers of comprehension. Both Nagel and McGinn seem to me correct in their critiques of reductionism, but I believe they underestimate the contribution of neuroscience to the study of patterns in mental events, even if science cannot capture the subjective feeling of such events.

6.3 Two-Aspect Theories

Owen Flanagan defends a *nonreductive naturalism* that draws from three sources: phenomenal first-person accounts, cognitive psychology, and neuroscience. He believes that the accounts can be correlated, though they have differing explanatory purposes. He takes seriously our conscious experience, our awareness of sensations, perceptions, emotions, beliefs, thoughts, and expectations. Flanagan describes neural correlates of visual experience, such as the neurons that respond to edges, shapes, colors, and motions, or the brain activities that are associated with fear and anger. But high-level concepts of the self are not expressible in neural terms. Human actions, for example, must be identified by the intentions that constitute them.[79]

---

[77] Thomas Nagel, *The View from Nowhere* (New York: Oxford Univ. Press, 1986).

[78] Colin McGinn, *The Problem of Consciousness* (Cambridge: Blackwell's, 1991).

[79] Owen Flanagan, *Consciousness Reconsidered* (Cambridge: MIT Press, 1992).

Flanagan acknowledges that *the self is constructed*. It is not given to us as a single entity or a transcendental ego. The newborn gradually builds an integrated self with the help of parents and other people. With maturation and socialization a distinct identity is formed, cast largely in narrative form in the stories we tell ourselves. The self is formed in active engagement with the environment and other persons. Our self-representations organize our memories of past events and our plans and aspirations for the future. Models of the self do not use concepts applicable to neurons, and they reflect our aims and values, which affect the choice of alternative patterns of action and human relationships.

In replying to Dennett, Flanagan agrees that the self is constructed but insists that it is not simply a useful fiction. Patterns of thought are real features of mental activity. The narrative self has *causal efficacy* as a complex and ever-changing self-representation. It causes people to say and do things; hence it has ontological and not merely linguistic status. Dennett had presented only two alternatives: either the self is an autonomous, enduring entity or else it is an illusion, a fiction that serves only instrumental functions. Flanagan offers a third: the self as a many-leveled reality that is constructed rather than given, in which activities at each level have some autonomy and yet are related to each other. This goes beyond Nagel's dual-aspect theory in arguing that there are causal relations between levels, rather than two perspectives on a single set of events. Flanagan does not share McGinn's pessimism about the contribution of neuroscience to our understanding of consciousness.

David Chalmers holds that *consciousness is irreducible* but argues that all other biological and psychological facts are determined by physical facts and are in principle explainable by physical theories. He holds that the cognitive sciences can provide reductive explanations for mental states considered as causes of behavior. Psychologists can even study awareness when it is viewed as access to information that is used to control behavior. They can give detailed functional accounts of memory, learning, and information processing, but they cannot say why these processes are accompanied by conscious experience, which is not defined by its causal roles. Phenomenal subjective experience is known firsthand in sensory perception, pain, emotions, mental images, and conscious thought.

Chalmers rejects materialism and functionalism and defends *a two-aspect theory,* which he also calls property dualism or a form of panpsychism. He proposes that *information states* are the fundamental constituents of reality, and are always realized both *phenomenally* and *physically.* "We might say that the internal aspects of these states are phenomenal and the external aspects are physical. Or as a slogan: experience is information from the inside; physics is information from the outside."[80] A dog has access to extensive perceptual information, so we can assume it has rich visual sense experiences. A fly has rather limited perceptual discrimination and also a lower level of experience with fewer phenomenal distinctions. Simple information states would be realized in simple physical structures and simple phenomenal experiences. "It is likely that a very restricted group of subjects of experience would have the psychological structure required to truly qualify as *agents* or *persons.*"[81]

Lynne Baker holds that neuroscience may provide the necessary and sufficient conditions for conscious events of a given modality, but not the conditions for

---

[80] David Chalmers, *The Conscious Mind: In Search of a Fundamental Theory* (New York: Oxford Univ. Press, 1996), 305.

[81] Ibid., 300.

particular reports of mental events. We cannot expect neuroscience to explain the specific content of consciousness. No study of neuronal activity could confirm or disconfirm the report "I realized that I believed Hal was trying to embarrass me." A person's belief that taxes are too high may be explained or predicted from other beliefs or events that psychologists and sociologists can study, but data at the level of neurons will not be illuminating. Beliefs are states of persons that help to explain their actions, not the interactions between neurons. Baker says that patterns of explanation at various levels indicate the reality of events at each level; she calls herself a metaphysical pluralist, not a dualist or a two-aspect monist.[82]

Of the three views in this section—eliminative materialism, the irreducibility of consciousness, two-aspect theories—it seems to me that the third is most consistent with human experience and with current theories in neuroscience. Process philosophy might be considered a form of two-aspect theory, but I suggest that it can better be described as dipolar monism.

7 Process Philosophy

The process philosophy of Alfred North Whitehead and thinkers influenced by him presents a coherent metaphysical framework within which many of the themes explored in previous sections can be brought together.

7.1 Dipolar Monism and Organizational Pluralism

Whitehead elaborated a set of philosophical concepts which emphasize becoming rather than being, change rather than persistence, creative novelty rather than mechanical repetition, and events and processes rather than substances. Whereas substances remain the same in different contexts, events are constituted by their relationships and their contexts in space and time. Whitehead and his followers hold that the basic components of reality are not one kind of enduring substance (matter) or two kinds of enduring substance (mind and matter), but *one kind of event with two phases.* In the objective phase a unitary event is receptive from the past; in the subjective phase it is creative toward the future. Every event is a subject for itself and becomes an object for other subjects. [83]

This philosophy is a form of *monism* because it insists on the common character of all unified events. "Dipolar" indicates an ontological claim, not merely an epistemological distinction, as some advocates of two-aspect monism propose. "Organizational pluralism" signals the recognition that events can be organized in processes in diverse ways, as emphasized by Charles Hartshorne, who reformulated and extended Whitehead's ideas. All integrated entities at any level have an inner reality and an outer reality, but these take very different forms at different levels. Both the interiority and the organizational complexity of psychophysical systems have evolved historically.[84]

---

[82] Lynne Rudder Baker, *Explaining Attitudes: A Practical Approach to Mind* (Cambridge: Cambridge Univ. Press, 1995).

[83] The classic source is Alfred North Whitehead, *Process and Reality* (New York: Macmillan, 1929); for an introductory account, see John B. Cobb, Jr. and David Ray Griffin, *Process Theology: An Introduction* (Philadelphia: Westminster, 1976).

[84] Charles Hartshorne, "The Compound Individual," in *Philosophical Essays for Alfred North Whitehead*, F.S.C. Northrup, ed. (New York: Russell & Russell, 1967).

Looking at diverse types of system, Whitehead attributes *experience* in progressively more attenuated forms to persons, animals, lower organisms, and cells (and even, in principle, to atoms, though at that level it is effectively negligible), but not to stones or plants or other unintegrated aggregates. David Griffin proposes that this should be called *panexperientialism* rather than *panpsychism,* because for Whitehead mind and consciousness are found only at higher levels.[85] Only in advanced life-forms are data from brain cells integrated in the high-level stream of experience we call mind. Experience at different levels varies greatly; consciousness and mind were radically new emergents in cosmic history.

An *atom* repeats the same pattern, with essentially no opportunity for novelty except for the indeterminacy of quantum events. Inanimate objects such as stones have no higher level of integration; the indeterminacy of individual atoms in an inanimate object averages out in the statistics of large numbers. A *cell,* by contrast, has considerable integration at a new level. It can act as a unit with at least a rudimentary kind of responsiveness. There is an opportunity for novelty, though it is minimal. If the cell is in a *plant,* little overall organization or integration is present; there is some coordination among plant cells, but plants have no higher level of experience. But *invertebrates* have an elementary sentience as centers of perception and action. The development of a nervous system made possible a higher level of unification of experience. New forms of memory, learning, anticipation, and purposiveness appeared in *vertebrates.*

In *human beings,* the self is the highest level in which all of the lower levels are integrated. Humans hold conscious aims and consider distant goals. Symbolic language, rational deliberation, creative imagination, and social interaction go beyond anything previously possible in evolutionary history. Humans enjoy a far greater intensity and richness of experience than occurred previously. The human psyche is the dominant occasion that integrates and harmonizes the diverse streams of experience it inherits. Its continuity is achieved as the route of inheritance of a temporally ordered society of momentary events.

Process thinkers thus agree with dualists that *interaction* takes place between the mind and the cells of the brain, but they reject the dualists' claim that this is an interaction between two totally dissimilar entities. Between the mind and a brain cell there are enormous differences in characteristics, but not the absolute dissimilarity that would make interaction difficult to imagine. The process view has much in common with two-language theories or a parallelism that takes mental and neural phenomena to be two aspects of the same events. But unlike many two-aspect theories, it defends interaction, downward causality, and the constraints that higher-level events exert on events at lower levels. At higher levels there are new events and entities and not just new relationships among lower-level events and entities.[86]

7.2 Embodiment, Emotions, Levels, and Consciousness

The themes in the neurosciences that were mentioned earlier are prominent in process philosophy.

---

[85] David Ray Griffin, "Some Whiteheadian Comments," in *Mind in Nature,* John B. Cobb, Jr. and David Ray Griffin, eds. (Washington D.C: University Press of America, 1977).

[86] David Ray Griffin, *Unsnarling the World Knot: Consciousness, Freedom, and the Mind-Body Problem* (Berkeley and Los Angeles: Univ. of California Press, 1998).

1. Embodiment: Every unified event is portrayed as a synthesis of past bodily events. There are no events that have a subjective phase without a prior objective phase. This can be called an ecological, relational, or contextual philosophy because it holds that every basic unit is constituted by its relationships. Moreover, we experience the causal efficacy of our own bodies. The senses such as sight always have a bodily reference rather than simply transmitting information about the world. The body is the vehicle of relationality with other persons. Process thought defends the idea of *the social self,* which is a product of the interaction of embodied persons and not of disembodied minds.

2. Emotions: Process thought recognizes the importance of nonsensory experience and the perception of feeling in our own bodies. Consciousness and cognitive thought occur against a background of feeling. Whitehead writes: "The basis of experience is emotional.... The basic fact is the rise of an affective tone originating from things whose relevance is given."[87] The technical Whiteheadian term "prehension" includes the communication of both conceptual and affective elements. The influence of one event on another is similar to the *communication of information*—including selective response by an interpretive system—but it includes an emotional component absent from most analyses of communication.

3. Consciousness: Whitehead says that consciousness first appeared in animals with a central nervous system as a radically new emergent. In human beings, most mental activity is unconscious. Consciousness occurs only in the last phase of the most complex occasions of experience, as a derivative byproduct of nonconscious experience. Self-identity consists in the continuity of processes most of which are below the threshold of awareness. Whitehead says that consciousness is "a late derivative phase of complex integration which primarily illuminates the higher phases in which it arises and only dimly illuminates the primitive elements in our experience."[88] It involves the unification of prehensions from the past and from the body with a new element: the contrast of past and future, the entertainment of possibilities, the comparison of alternatives.

4. A Hierarchy of Levels: Among process thinkers, Charles Hartshorne has developed most fully the idea of a series of levels intermediate between the atom and the self. He dwells on the differences between cells and mere aggregates such as stones.[89] His holistic outlook directs attention to system properties that are not evident in the parts alone. Process philosophy has always insisted on contextuality and relationality. But it recognizes that various levels may be integrated according to different principles of organization, so their characteristics may be very different. In a complex organism, downward causation from higher to lower levels can occur because, according to process philosophy, every entity is what it is by virtue of its relationships. The atoms in a cell behave differently from the atoms in a stone, as cells in a brain behave differently from those in a plant. Every entity is influenced by its participation in a larger whole. Emergence arises in the modification of lower-level constituents in a new context. But causal interaction between levels is not total determination; there is some self-determination by integrated entities at all levels.

---

[87] Alfred North Whitehead, *Adventures of Ideas* (New York: Macmillan, 1933), 226.

[88] Alfred North Whitehead, *Process and Reality*, corrected ed., David Ray Griffin and Donald W. Sherburne, eds. (New York: Free Press, 1978), 162.

[89] Charles Hartshorne, *Reality as Social Process* (Glencoe, Ill.: Free Press, 1953), chap. 1; idem, *The Logic of Perfection* (LaSalle, Ill.: Open Court, 1962), chap. 7.

5. *The Construction of the Self:* Whitehead was influenced by William James, who held that there is no enduring self but only the stream of experience. Thought goes on without a thinker, or even a succession of thinkers aware of the same past. Continuity of identity, James said, is guaranteed only by the persistence of memory. He held that we each use a constantly revised model of the self to impose order on the flux of experience. Whitehead also holds that the self is a momentary construction, but he asserts that it is a unified complex process. The unity of self is a unity of functioning, not the unity of a Cartesian thinker. We have seen that this view—that selfhood is constructed—is consistent with recent neuroscience.

However, I believe that Whitehead himself overemphasized the momentary and episodic character of the self. I have suggested that without accepting substantive categories we can modify Whitehead's ideas to allow for more continuity in the inheritance of the constructed self, which would provide for stability of character and persistence of personal identity.[90] Joseph Bracken agrees with my criticism of Whitehead and believes it can be remedied by emphasizing Whitehead's thesis that a temporal society maintains continuity among its momentary constituents ("actual occasions"). Bracken suggests:

> A much simpler way to preserve continuity among the discontinuity of successive actual occasions within human consciousness is to give greater importance to the Whiteheadian notion of a society as that which is created and sustained by a succession of actual occasions with a common element of form.[91]

Bracken proposes that a society that endures over time can be understood as a "structured field of activity" for successive generations of events. "When applied to the Whiteheadian notion of the human self as a personally ordered society of conscious actual occasions, this means that the self is an ongoing structured field of activity for successive actual occasions as momentary subjects of experience."[92] Such revisionist or neo-Whiteheadian proposals can remedy some of the problems in Whitehead's writings while supporting his fundamental vision of reality.[93]

7.3 The Status of Subjectivity

My own view is very similar to the *emergent monism* defended by Philip Clayton and Arthur Peacocke in the present volume. We share a commitment to explanatory pluralism and the diversity of levels of explanation, including the distinction between reasons for human actions and causes of physical effects. We share a commitment to organizational pluralism in a hierarchy of many levels rather than a mind-matter dualism. We join in advocating contextualism, in which every entity is constituted by its relationships. Emergent monists also have a strong sense of the temporality and historical character of reality, and it would not be inconsistent for them to accept the Whiteheadian emphasis on momentary events and dynamic processes and the process critique of enduring substances. We agree that consciousness and mind are

---

[90] Barbour, *Religion and Science*, 290.

[91] Joseph A. Bracken, S.J., "Revising Process Metaphysics in Response to Ian Barbour's Critique," *Zygon* 33 (1998): 407.

[92] Ibid, 408.

[93] See also Frank Kirkpatrick, "Process or Agent: Two Models for Self and God," in *Philosophy of Religion and Theology*, David Ray Griffin, ed. (Chambersburg, Penn.: American Academy of Religion, 1971); Paul Sponheim, *Faith and Process: The Significance of Process Thought for Christian Thought* (Minneapolis: Augsburg, 1979), 90–98.

emergent new properties found only at high levels of complexity, and that these potentialities were built into the lower-level components from the beginning.

However process thinkers diverge from emergent monism by holding that at least a rudimentary form of subjectivity is present actually, and not just as a potentiality, in integrated entities at all levels. What are the reasons for such attribution?[94]

1. The Generality of Metaphysical Categories: In Whitehead's view, a basic metaphysical category must be universally applicable to all entities. The diversity among the characteristics of entities must be accounted for by the diversity of the modes in which these basic categories are exemplified and by differences in their relative importance. The subjective aspect of cells may for all practical purposes be ignored, but it is postulated for the sake of metaphysical consistency and inclusiveness. Mechanical interactions can be viewed as very low-grade organismic events (because organisms always have mechanical features), whereas no extrapolation of mechanical concepts can yield the concepts needed to describe subjective experience. New phenomena and new properties can emerge historically, but not new basic categories. Wings and feathers may evolve from other objective physical structures, but subjectivity cannot be described in physical terms. The subjective character of events is also important in process theology, because it provides one of the routes of God's influence on the world. The Whiteheadian analysis of causality allows for formal and final as well as efficient causes in all events.[95]

2. Evolutionary and Ontological Continuity: There are no sharp lines between a cell and a human being in evolutionary history. Today, a single fertilized cell gradually develops into a human being with the capacity for thought. Process thinking is opposed to all forms of dualism: living and nonliving, human and nonhuman, mind and matter. Human experience is part of the order of nature. Mental events are a product of the evolutionary process and hence an important clue to the nature of reality. We cannot get consciousness from matter, either in evolutionary history or in embryological development, unless there are some intermediate stages or levels in between, and unless mind and matter share at least some characteristics in common.

3. Immediate Access to Human Experience: I know myself as an experiencing subject. Human experience, as an extreme case of an event in nature, is taken to exhibit the generic features of all events. We should then consider an organism as a center of experience, even though that interiority is not directly accessible to scientific investigation. In order to give a unified account of the world, Whitehead employs categories that in very attenuated forms can be said to characterize lower-level events, but that at the same time have at least some analogy to our awareness as experiencing subjects. Such a procedure might be defended on the ground that if we want to use a single set of categories, we should treat lower levels as simpler cases of complex experience, rather than trying to interpret our experience by concepts derived from the inanimate world or resorting to some form of dualism. It is of course difficult to imagine forms of feeling very different from our own, and we must avoid the anthropomorphism of assuming too great a similarity. Organizational pluralism allows for differences among levels and for the emergence of radically new phenomena, on which emergent monism rightly focuses attention.

---

[94] See Barbour, *Religion and Science*, 289–93.

[95] Barbour, "Five Models of God and Evolution," in *Evolutionary and Molecular Biology*, Russell, Stoeger, and Ayala, eds.

7.4 Immortality without an Immortal Soul

The process view of immortality, like its view of sin, redemption, and the Incarnation, is relational—that is, it is a relationship of persons to God and other beings, not a property of individuals in themselves. To articulate it adequately would require a longer discussion of the process view of God than we can undertake here. In process thought, God's attributes include distinctive forms of embodiment, emotion, consciousness, and social interaction. God is present in all time and space and knows all that can be known. God is eternal and unchanging in character and purpose but temporal in being affected by interaction with the world.

Process thinkers have defended two forms of immortality. *Objective immortality* is our effect on God and our participation in God's eternal life. Our lives are meaningful because they are preserved everlastingly in God's experience, in which evil is transmuted and the good is saved and woven into the harmony of the larger whole. God's goal is not the completed achievement of a static final realm but rather a continuing advance toward richer and more harmonious relationships.[96]

Other process writers defend *subjective immortality,* in which the human self continues as a center of experience in a radically different environment, amid continuing change rather than changeless eternity, with the potential for continued communion with God. John Cobb speculates that we might picture a future life as neither absorption into God nor the survival of separate individuals but as a new kind of community transcending individuality.[97] Marjorie Suchocki suggests that subjective and objective immortality can be combined, because God experiences each moment of our lives not merely externally as a completed event but also from within in its subjectivity. In that case our subjective immediacy would be preserved in God as it never is in our interaction with other persons in the world.[98]

In summary, process philosophy is supportive of the biblical view—which I suggested was consistent with the evidence from the neurosciences—that a human being is a multilevel unity, an embodied social self, and a responsible agent with capacities for reason and emotion. The dipolar monism and organizational pluralism proposed by process philosophy avoids the shortcomings of both dualism and materialism by postulating events and processes rather than enduring substances or entities. However, neither science nor philosophy—even when supplemented by data from the humanities and social sciences—can capture the full range of human experience or articulate the possibilities for the transformation of human life to which our religious traditions testify.

---

[96] Whitehead himself defended objective immortality but was open to the possibility of subjective immortality. See Cobb and Griffin, *Process Theology*, chap. 7.

[97] John B. Cobb, Jr., "What is the Future? A Process Perspective," in *Hope and the Future*, Ewart Cousins, ed. (Philadelphia: Fortress Press, 1972).

[98] Marjorie Hewitt Suchocki, *The End of Evil: Process Eschatology in Historical Context* (Albany: State Univ. of New York Press, 1988), chap. 5.

THE SOUL AND NEUROSCIENCE: THE EMPIRICAL CONDITIONS FOR HUMAN AGENCY AND DIVINE ACTION

Stephen Happel

Anima humana intelligit se ipsum per suum intelligere, quod est actus proprius eius, perfecte demonstrans virtutem eius et naturam.[1]

Thomas Aquinas

1 Introduction

Neural imaging, such as fMRI and PET-scans, has transformed contemporary research by providing "live" information about the interactive, functioning brain.[2] Instead of speculating about what happens physically when humans think, feel, or act, investigators can watch the dynamic neural patterns emerge, develop, and change through relatively noninvasive, nonviolent procedures. What was invisible or opaque is now visible and clear. With considerable assistance to therapeutic endeavors, such processes continue the intellectual war of the Enlightenment in favor of reason, visibility, and clarity against opaqueness, the unknown, and the superstitious. Just as Piranesi (1720–78) used his architectural etchings to excavate the antiquities of Rome and Dagoty (1648–1730) and Cowper (1660–1709) drew their dissected bodies, contemporary images of brain activity provide an entry into what is covered by skin, the physical operations of the mind wrapped in its sheathe.[3] The "Vesalian technique of opening a severed head, removed from the pathos of the total human subject, continues in present-day MRI scans."[4] Each involves a commitment to determine the physical conditions for memory, thinking, and action.

The significant role for visual analysis of neural activity means that images are not only models in inquiry (a heuristic), but also modes of communicating results to scientists and non-scientists (a rhetoric). The modeling and communicative dimensions of neural imaging involve the neurosciences in more extensive and controverted questions about the power of images in relationship to mathematical or verbal formulae.[5] They lift the veil on the larger cultural issue of the initiating agency of human beings. Are the images of the active brain simply the responses to external stimuli? Can the thinking or feeling brain initiate activity that exceeds or affects its neural base? Do the images tell us that "mind" is simply "brain" in action?

---

[1] "The human soul understands itself by its own understanding, which is its proper act, perfectly showing its power and nature." *Summa Theologiae*, I, Q. 88, a. 2, ad 3m (author's translation).

[2] For the role of neural imagining in contemporary science, see Barbara Maria Stafford, *Good Looking: Essays on the Virtue of Images* (Cambridge: MIT Press, 1996), 24–25.

[3] Barbara Maria Stafford, *Body Criticism: Imaging the Unseen in Enlightenment Art and Medicine* (Cambridge: MIT Press, 1994), esp. 47–129.

[4] Stafford, *Good Looking*, 133.

[5] See Gerald Holton, "Einstein's Influence on the Culture of our Time," in *Einstein, History and Other Passions: The Rebellion against Science at the End of the Twentieth Century* (Reading, Mass.: Addison-Wesley Publishing Co., 1995), 143.

1.1 Soul and Body

Classical philosophy and theology would have argued that mental activity, ethical intentionality, and feelings such as love or a passion for justice were to be located in the soul.[6] Whether this "soul," however, was to be understood by believers as a separate and distinct substance, immortal and invisible, "free" of the body so that human beings are defined in a dualist fashion as "body and soul" remains controverted. Despite recent popular interest in the soul, academic writers in science and religion, suspicious of any disembodied subjectivity, have tended to avoid soul-language.[7] Instead, they have focused upon mind, consciousness, and self-consciousness; but these terms in turn can be construed with or without bodily or physical content.

Medieval philosophers and theologians, while affirming the existence of the soul, did not reject corporeality. "Those who wrote about body in the thirteenth and fourteenth centuries were in fact concerned to bridge the gap between material and spiritual and to give to body positive significance. Nor should we be surprised to find this so in a religion whose central tenet was the incarnation—the enfleshing of its God."[8] Given the human experience of bodily birth and death, yet also a thirst for union with God, psychosomatic unity was a fact to be explained, not denied. How could the Christian tradition provide a theoretically differentiated, but practically applicable, interpretation of the human *persona* such that the scriptural beliefs in creation, sanctification, and eschatological transformation might be understood?

Language about the soul was later reified in the post-Cartesian philosophical environment.[9] Discussions about the material body and immaterial soul became embedded in the Enlightenment and Romantic themes of subjectivity, interiority, and individual autonomy. Debates about the value of these modern notions have themselves been controverted in the postmodern situation.[10] How is human subjectivity to be conceived? Is self-conscious life with its rational control of the planet a good or has it been destructive? Eighteenth-century philosophers initially provided epistemological and methodological foundations for Newtonian science and Cartesian or Leibnizian mathematics and freed human subjects from religious

---

[6] For a brief overview, see Warren S. Brown, Nancey Murphy, and H. Newton Malony, *Whatever Happened to the Soul? Scientific and Theological Portraits of Human Nature* (Minneapolis: Fortress Press, 1998), 2–8, 151–72.

[7] See Thomas Moore, *Care of the Soul: A Guide for Cultivating Depth and Sacredness in Everyday Life* (New York: HarperCollins, 1992).

[8] Caroline Walker Bynum, *Fragmentation and Redemption: Essays on Gender and the Human Body in Medieval Religion* (New York: Zone Books, 1991), 223.

[9] For a discussion of René Descartes and Nicolas Malebranche, see Michael J. Buckley, *At the Origins of Modern Atheism* (New Haven: Yale University Press, 1987), 146–47; see also in this context, Ian Barbour, *Religion in an Age of Science The Gifford Lectures, 1989–1991, Volume I* (New York: HarperSanfrancisco, 1990), 195–99, 207–9, and Arthur Peacocke, *Theology for a Scientific Age: Being and Becoming—Natural, Divine and Human* (Minneapolis: Fortress Press, 1993), esp. 223–54.

[10] See the conflict between Romanticism and the Enlightenment in Gerald Holton, "What Place for Science at the 'End of the Modern Era'?", in *Einstein, History, and Other Passions*, 3–39.

obfuscation and political repression.[11] By the twentieth century, these sciences, their supportive philosophies, and attendant technologies have been accused of destroying human freedom.[12] Instrumental or disengaged reason has evolved from being the savior of human culture to being the calculating perpetrator of fascist bureaucracy, gender discrimination, genocidal destruction, and emotional annihilation.[13]

Enlightenment figures understood religion as anti-humanist. But just as philosophy has developed since the eighteenth century, so too theology has migrated from its pre-Enlightenment marriage to folk medicine, ancient natural philosophy, and hierarchical politics, giving birth to policies of social liberation, recovery of human feelings, and the embrace of pluralist diversity in history.[14] Simultaneously, theologies after Friedrich Nietzsche and Martin Heidegger have proposed not only the death of God, but also the death of the subject, the closure of the book, and the end of history.[15] Hence, in the course of discussions between theology and science, it is crucial to know the terms of the debate and the identity of the dialogue partners.[16] In this essay, I am not claiming that science is a rationalizing, totalitarian conspiracy or that religion is a Romanticist extra-rational solution. Rather, I hope to put neuroscience in conversation with a thoughtful Christianity such that both may benefit by their location in contemporary culture.

1.2 The Agency of the Subject

The *power of the human subject* described by modernity is the context for the discussion of memory, the soul, and neuroscience. In effect, the anti-humanist and humanist interpretations of modernist subjectivity have claimed to know the meaning of human autonomy. On the one hand, autonomy understands freedom as determining oneself from within one's own resources without regard to any other; on the other hand, autonomous self-invention controls objects (and other subjects) in the world. The affirmative and negative attitudes toward these notions of individual autonomy can be found on both sides of the science-religion conversations.[17] How much agency is to be attributed to subjects? From what does this agency originate, if it exists? Is it self-generated; does it emerge as a command, rule, a task, and/or gift from others or an Other? Is it determined by nature and genetics, or history and

---

[11] For another reading, see David Ray Griffin, *Unsnarling the World-Knot: Consciousness, Freedom, and the Mind-Body Problem* (Berkeley: Univ. of California Press, 1998), 11–14.

[12] For what follows, see Alain Renaut, *The Era of the Individual: A Contribution to a History of Subjectivity*, trans. M.B. DeVevoise and Franklin Philip (Princeton, N.J.: Princeton University Press, 1997), esp. 3–57. See also Charles Taylor, *Sources of the Self: The Making of Modern Identity* (Cambridge: Harvard University Press, 1989), 305–54.

[13] The "dialectic of the Enlightenment" is aptly described in Martin Jay, *The Dialectical Imagination: A History of the Frankfurt School and the Institute of Social Research, 1923–50* (London: Heinemann, 1973), 253–99; see Holton, "The Public Image of Science," chap. in *Einstein, History and Other Passions*, 40–57.

[14] For a more lengthy description of these issues, see Stephen Happel and James J. Walter, *Conversion and Discipleship: A Christian Foundation for Ethics and Doctrine* (Philadelphia, Penn.: Fortress Press, 1986), 85–101.

[15] For a positive theological reading of philosophy after Jacques Derrida, see Mark C. Taylor, *Erring: A Postmodern A/theology* (Chicago, Ill.: Univ. of Chicago Press, 1984).

[16] See Brown, Murphy and Malony, *Whatever Happened to the Soul?*, 213.

[17] See Holton, *Einstein*, 3–57.

nurture? By both? The answers to these questions focus our understanding of the autonomy of human subjects in modernity and postmodernity. "[T]he overarching... principle is that the *modern* consists in a relation to the world according to which [humanity] posits itself as capable of providing the foundation for its own acts and representations, as well as for history, of truth, and the law."[18] The seeming liberation of the Enlightenment subject from past politics, history, and religion may be partially illusory, but in what does the "partially" consist?

Renaut claims that individualism is only one way of reading modernist subjectivity. He distinguishes autonomy from independence, with several differing forms of the subject in modernity: rationalist, empiricist, metaphysical, and criticist subjects. The one-dimensional reduction of subjectivity to individualism with its polar notion of autonomy prohibits a genuine understanding of what he considers to be *the* question in contemporary thought: the possibility of *transcendence in immanence*, of the reality of the other within and to the subject. His notion of autonomy would recover a *relative independence* in which subjects chose to limit their power through negotiated commitments to a common good.

Although I do not find Renaut's solution to this problem completely persuasive,[19] his contextualization is important for two reasons: (1) it is precisely the closure of the empirical, autonomous, self-determining agent to transcendence (or divine action) that has prohibited much dialogue between theology and the sciences; and (2) recent neuroscience is rife with underlying philosophical questions about human agency, self-originating autonomy, and the role of intersubjectivity and the social context in that agency. Here I will focus upon the philosophy, theology, and neuroscience of memory and time-consciousness. Why memory? Because it is often described as a passive repository for filed information, as though there were no available agent-subject accomplishing the tasks. It is also clear that in the pre-modern theological world, memory and the *anima* or *psychē* were intertwined.[20]

I therefore link together Edmund Husserl's contemporary philosophical interpretation of time-consciousness in section 2 with a re-reading of the Aristotelian and Thomist tradition on the soul and memory in section 3. In section 4 I present an understanding of the empiricist ego (or the subject as investigated by neuroscience) in a nonindividualist, nonmonadological fashion. With this opening to finite transcendence, it may be possible to illuminate how God characteristically operates through the interaction of finite subjects in our world to bring about divine ends.

2 Husserl and Time-Consciousness[21]

Husserl excludes from his analysis of time-consciousness "real Objective time," that is, the time computed by clocks and watches (23). It is not that he is uninterested in

---

[18] Renaut, *Era of the Individual*, 3.

[19] He creatively reads the contemporary solution through Emmanuel Levinas and Immanuel Kant.

[20] See Frances A. Yates, *The Art of Memory* (Chicago: University of Chicago Press, 1966), 27–49.

[21] Edmund Husserl, *The Phenomenology of Internal Time-Consciousness*, Martin Heidegger, ed., trans. by James S. Churchill (Bloomington, Ind.: Indiana University Press, 1964); for an overview, see Robert Sokolowski, *Husserlian Meditations: How Words Present Things* (Evanston, Ill.: Northwestern University Press, 1974), 138–67. Hereafter, page references to Husserl's *Phenomenology* appear parenthetically in the text.

physical, externally measured time, but that he wants to investigate the specific form of duration known in human consciousness and the ways in which the time of human consciousness knows temporal objects (94–96, 157–60).[22] In effect, he presumes the "external" time of nature, bracketing it for his study; and instead examines the experience of inner time-consciousness. "When we see, hear, or in general perceive something, it happens according to rule that what is perceived remains present for an interval although not without modification." But what remains, remains as "something more or less past, as something temporally shoved back [*Zurück-geschobenes*] (30)." In reflecting upon Franz Brentano's psychology, Husserl uses the example of a melody: individual notes do not disappear when a new note sounds, yet one does not hear all the notes simultaneously (for Husserl's differences with Brentano on intentionality, see 107–18).[23]

2.1 Intending Temporal Objects

What are we to understand about our interior ability to perceive a melodic line that requires not only knowing the present, but holding sounds from the past. "By *temporal Objects*, in this *particular sense*, we mean Objects which not only are unities in time but also include temporal extension in themselves.... That the expired part of the melody is objective to me is due—one is inclined to say—to memory, and it is due to expectation which looks ahead that, on encountering the tone actually sounding, I do not assume that that is all (43)." According to Husserl, all objects (except mathematics and some values such as beauty) are temporal objects, including the intending of human consciousness itself.[24] Human beings are a unity in time, yet have temporal extension. How does this kind of historical existence and its knowing succeed?

Husserl studies the phenomenology of a musical tone that retains its time, but "the sound vanishes into the remoteness of consciousness." In the way it appears to the hearer, it is continually different (45). The receding tone is replaced by the "actual now-point" that succeeds it, filling the time of the present. As it recedes, it loses clarity and finally "disappears" completely. Every temporal item appears in this "continually changing mode of running-off"; it is always something other, and yet in continuity with the same item in the present (47). Husserl sees these "running-off phenomena"(*Ablaufsphänomene*) as having their own continuities, receding into the past. In addition Experience will be filled with further objects; it is open-ended intending. The mode of running-off is constantly modified, every "now" being transformed into a "past (49–50)."

---

[22] Husserl's interest in "Objective time" is determined by his interest in how it arises, given the human temporal flux. Indeed, he sees the regularities of objective time, the time of natural objects, as constituted within the knowing process.

[23] For Brentano's understanding of intentionality and time consciousness, see Herbert Spiegelberg, *The Phenomenological Movement: A Historical Introduction* (The Hague: Martinus Nijhoff, 1969), vol. I, 38–44.

[24] By *intentionality* in this context, Husserl does not necessarily mean a "chosen, willed object." He means to describe the "neutral" arc, or a cognitive relation between knower and known; but the relation is reciprocal, i.e. the object can provoke the interpreter into "intending" its presence and the interpreter can "intend" an object without having a particular object in view. Intentionality can include the specific acts of feeling, sensing, conceptualizing, willing, and so forth. The term is meant to denote a neutral relational mutuality.

2.2 The Present as Retention and Anticipation

Once objects decay and recede into the past, no longer impressing themselves upon us in the present, they nonetheless remain in retention. Retention is not simply an after-image or a reverberation, according to Husserl; it is what he calls an "originary intuition (53)." "Retentional consciousness includes real consciousness of the past of sound, primary remembrance of sound, and is not to be resolved into sensed sound and apprehension as memory (54)." So a sound's simple reverberation is not the same thing as the retention of sounds that have just occurred. For Husserl, retention is based upon the "precedence of a perception or primal impression." The "now" is only thinkable within the horizon of a continuity of retentions, just as retentions are only thinkable as part of a continuum in which perceptual impressions continue into the future (55). So "primary remembrance or retention [is] a comet's tail which is joined to actual perception (57)."[25]

Recollection, or "secondary remembrance," is different from this primary retention. Remembering a melody from childhood "runs through" the melody in phantasy, but at the time of recall there still remains a now-point with primary remembrance blending in the continuum from the prior parts of the melody to its anticipated conclusion (58). We can still name this kind of recall "perception," but it is here an act that "*primordially constitutes* the Object (63)." This is different from the primary retentions, the shading off of a perception in the now in which the past continuum is presented to us precisely as perceived in the present. In the latter, the past is experienced as present to us as the past of what is even now being perceived.[26]

2.3 The Agent-Consciousness

For Husserl, the distinction between primary retention and secondary remembrance, between the "shading-off" of the now and reflective recall is crucial since it marks a bi-modal agent-consciousness from a purely passive receiver. His argument with Brentano's psychology is that by locating time-consciousness in phantasy, Brentano has left human beings with a passive consciousness. This misunderstands the arc of intentionality that exists between the knower and the known, between subject and object, given in the matrix of present experience. The primary intentionality or "originary consciousness" of knower and known has inscribed within it an intrinsic temporality that includes the presencing subject-object, primary retention (continual shading-off of present perceptions), and anticipations of a future (70). The originary time-consciousness is "precisely the transition from the actual now to the new now (141)." This temporal intention is *not* necessarily reflective, that is, either directly thought or willed; it constitutes the *agency* of human consciousness. "The originary appearing and passing away of the modes of running-off" is a process that we can

---

[25] Although my primary concern for the neuroscientific interpretations of time will be considered in section 4, Michael Arbib has pointed out that this is similar to the way schema assemblage combines current percepts with schema instances of continuing relevance that also project goals and expectations into the future. See Arbib, *The Metaphorical Brain 2: Neural Networks and Beyond* (New York: Wiley Interscience, 1989), sec. 2.1.

[26] For Husserl, perception is the "temporally constitutive consciousness with its phases of flowing retentions and protentions (176)." It includes the acts of sensation, but also judgment, even the act of observing an act. They are the unitary acts of time-consciousness.

only know indirectly (since all conceptual "capture" will be of the past "running-off"); in a particular sense, it is not freely chosen (61). On the other hand, time-consciousness is attentive to what emerges: "The wakeful [*wache*] consciousness, the wakeful life, is a living-in-face-of [*Entgegenleben*], living from one now toward the next (141)." This is the "flux," the point of the source from which "springs the now (100)."

Recollection, on the other hand, is a freely-willed activity, requiring selection and decision (71). In "real, re-productive, recapitulative memory," the temporal object is built up so that we seem to perceive it again (as in the remembered melody, 59). We can perceive it as a temporal unit with duration and succession (65). But in every case we perceive the item as a now with its retentional train (61). Recollecting a temporal object, like a melody, however, is not a "simple reproduction" or a repetition of a perception (69). The recollection or memory of a temporal object is accomplished within that primary retentiveness of the subject with its perceptual "running-off (71)." The reproduced flow of a temporal object coincides with or overlays the retentional flow (73). But since the recollected temporal object had its own anticipations or expectations and these terminated in the present temporal flow, "events which formerly were only foreshadowed are now quasi-present, seemingly in the mode of the embodied present (76)." In other words, the ability to recall an item/event from the past depends upon the perdurance of the event in the present, upon its continued effects (77). Everything new interacts with the old; every recollection has a definite coloring given to it by the now, because the now is in some way a product of what is being remembered.

This interconnection of retention, present, and expectation is central to Husserl's understanding of how temporal consciousness *is* human being. We do not have a "mere chain of 'associated' intentions, one after the other..."; we have an intentionality that is active, unifying past realizations and possible fulfillments. This intentionality, however, is an open-ended intention, without necessary contents (78). "Every perception has its retentional and protentional halo (139)." The intentionality has its contents in the history of retentions and expectations, its recollections and their anticipations into the present. In this sense, memory and expectation differ from each other. Not only does memory continue in the questions and fulfillments of the present (toward its future and therefore the future of a particular memorialized past), but also its truth can be discerned by its accuracy to what was perceived ("Have I really had this appearance with exactly this content? [80]"). Expectations, however, are fulfilled in perceptions themselves.[27] What is expected is about to be perceived; once in the now of perception, it becomes present and has gone into retention. Memory is the consciousness of "having-been-perceived"; anticipation, while rooted in the flow of the "running-off" present, is what is "about to be perceived (81)."

This does not mean for Husserl that all memories are secure sensations; memories can be fantastic, invented pasts.[28] Even when one is recalling a genuine

---

[27] This leaves aside the issue of deflated expectations in the present or unfulfilled anticipations. However, Husserl's account claims that the protentions that emerge from the past *must* continue into the present, even in an unfulfilled expectation in the present; in other words, there is no *completely* unanticipated event in the now. Without some intentional "overlap," the moment could not be recognized.

[28] Imagination constructs a present and future on the basis of possible conditions; it actively and practically mediates the past into the present. Fantasy has no possible conditions in the empirical world; its products can only be dream-like cartoons, a form of virtual reality. The

past experience (a "lighted theater" in childhood is Husserl's *exemplum*), it is not the re-presentation of the perception that is recollected. "What is meant and posited in the memory is the object of the perception together with its now, which last, moreover, is posited in relation to the actual now (82)." I bring the theater of memory to the present, but I also know it in its temporal flow as past, lying in continuity with the present perceptions that occupy intentionality. The object is constituted as past, within a flow of perceptions in the present that are not only its successors but its progeny. The "running-off" patterns of the present are the ancestors of current perception. What makes human consciousness a unity is this flow of temporal interactions. What defines the now of consciousness is the retention of prior nows as past; the past, present, and future exist only in a continuum of temporal consciousness.

This unity-in-difference of open, simultaneous intending and temporal succession constitutes consciousness. Intentionality and time are "inseparably constituted (104)." Can one "get behind" the temporal flow of conscious objects that fill consciousness to the flow of consciousness itself? Not exactly. The flow of sensation (even unnoticed), acts of willing, reproductions in memory can be noted, but primary consciousness (open simultaneous intending) can only be copresented within this flow (110). It is clear, however, that for Husserl that unity, indistinguishableness of the similar, underlies the "consciousness of otherness, of difference… (114)."

This philosophical model of intentional subjectivity (and objectivity) has occupied much of twentieth-century thought. Subject and object are linked mutually and are reciprocally defined within intentionality. The temporal flux through which consciousness continues to identify itself also correlates with the object as its aspects pass through time. The dialectic of presencing, retention, and anticipation provides the primordial subjectivity within which recollections of past continua or possible imagined (and fantastic) projections are inserted. It is precisely in intending objects that the passing of time is noted. Through this lens, let us re-examine the language about memory and the soul in classic medieval theology.[29] How, then, is Husserl's notion of consciousness and intentionality related to Aquinas and Aristotle?

3 Aquinas and the Soul

It is generally agreed that the early Greeks did not have a unitary notion of the body and the mind.[30] Only in Athens during the fifth century B.C.E. did *psychē* become the

---

boundaries between the products of the two intentional operations are, however, permeable. The term "phantasm" (the traditional medieval term) has migrated in usage from its neutral notion in Aristotle and Aquinas as the activated and activating product of the knowing process in sensing, understanding, or judging. Since the nineteenth century, it has tended to be identified with "fantastic," non-empirical, unverifiable. This is not its original meaning. In one sense, medieval phantasms can function not unlike schemas in Arbib's sense; see sec. 4 of Arbib's "Towards a Neuroscience of the Person," in this volume. See also "Religious Imagination," *The New Dictionary of Theology*, Joseph A. Komonchak, Mary Collins, Dermot A. Lane, eds. (Wilmington, Del.: Michael Glazier, 1987), esp. 502–5. In Ted Peters's arguments about the role of the human body in the Christian understanding of the afterlife (in this volume), he claims to make an imaginative, not a fantastic, assertion in theology.

[29] Although the medieval notions of the soul and memory apply in most areas to Jewish, Muslim, and Christian theologians (due to their derivations from Plato and Aristotle), I shall concentrate on Christian theology.

[30] See Bruno Snell, *The Discovery of the Mind: The Greek Origins of European Thought*,

center of human consciousness, a principle of discrimination in public life, influenced by the rise of literacy, politics, and law.[31] Aristotle's use of the term acknowledges that "to attain any assured knowledge about the soul is one of the most difficult things in the world."[32] Aristotle's *De Anima* had appeared in Europe by the thirteenth century, and Aquinas used it as the primary critical tool for interpreting the traditional Christian vocabulary. In Aquinas's treatise on the soul in the *Summa Theologiae* and the *Summa Contra Gentiles*, his dialogue with the Platonic thought of Augustine and the Muslim interpreters of Aristotle is evident.[33] Aquinas attempted a theological understanding that would both address the systematic needs of his treatises to provide a *basis for the practice* of the virtues (*ST*, I, Q.78; *SCG*, II, 5,1) and offer a *critique of popular images* of the soul as a small, inner, childlike body that was swept off to heaven, purgatory, or hell after death (see *SCG*, II, 49,11).[34]

In the interpretation of Aquinas that follows, I follow a reading of his thought in light of Husserl's study of consciousness and temporality as well as cognitional theory developed by his theological interpreters in this century.[35] I do not, however, repeat Aquinas in his own words or provide a complete presentation on cognition and the soul. As will be evident, Aquinas's understanding of the soul is not the contemporary notion of subject.[36] His language is metaphysical, like Aristotle's; his analysis is theoretic, a philosophical critique of medieval common sense (both biblical and cultural). In his interpretation, the soul is a metaphysical, theoretic reality, the first act of an organic body (see *SCG* II, 57, 14).[37] The soul is labeled a *substantia*, or more accurately, "body and soul are not two actually existing substances; rather, the two of them together constitute one actually existing

---

trans. T.G. Rosenmeyer, (New York: Harper & Row, 1960), 43–70, 226–63; and Jan N. Bremmer, *The Early Greek Concept of the Soul* (Princeton, N.J.: Princeton University Press, 1983), esp. 13–69.

[31] Bremmer, *Greek Concept of the Soul*, 68.

[32] Aristotle, *On the Soul*, in *The Basic Works of Aristotle*, Richard McKeon, ed. (New York: Random House, 1941), 535.

[33] *ST*, I, QQ. 75–89, and *Summa Contra Gentiles*, trans. James F. Anderson, (Garden City, N.Y.: Doubleday, 1962), II, QQ. 46–90. All further citations will be in parentheses in the text in the usual fashion.

[34] Colleen McDannell and Bernhard Lang, *Heaven: A History* (New Haven: Yale University Press, 1988), 17–110; on purgatory, see Jacques Le Goff, *The Medieval Imagination*, trans. Arthur Goldhammer, (Chicago: University of Chicago Press, 1988), 67–77.

[35] See Josef Maréchal, *Le Point du Départ de la Metaphysique* (Paris, Louvain, 1922–44); Karl Rahner, *Spirit in the World*, trans. William Dych, S.J., (New York: Herder and Herder, 1968; German text, 1957); and especially Bernard J.F. Lonergan, *Verbum: Word and Idea in Aquinas*, David B. Burrell, ed. (Notre Dame, Ind.: University of Notre Dame Press, 1967; orig. pub., 1946–49); idem, *Insight: A Study of Human Understanding* (New York: Longmans, 1957); and idem, *Method in Theology* (London: Darton, Longman & Todd, 1972). An historical survey can be found in Otto Muck, *The Transcendental Method*, trans. William D. Seidensticker, (New York: Herder and Herder, 1968).

[36] Lonergan, *Method in Theology*, 95–96; idem "Christ as Subject," in *Collection*, Frederick E. Crowe, ed. (New York: Herder and Herder), 174, n. 11. Aquinas does treat passion and will as dimensions of the soul, but post-Romantic understanding of the subject would require focusing the soul through feelings and history. The goal here is to focus upon the cognitive dimensions of the soul and conscious agency in the neurosciences.

[37] Lonergan, *Verbum: Word and Idea in Aquinas*, vii–xv.

substance" (*SCG*, II, 69, 2).[38] If one understands *substance* in a Lockean, post-Cartesian sense, one can misinterpret Aquinas's claim that soul is the "substantial form" of the body (*ST*, I, Q. 76, a. 4), thinking of it as a material thing, pre-dating or coterminous with an already constituted body. Aquinas is particularly clear that this is not what he means (*ST*, I, Q 75, a. 1; *SCG*, 49; 69,1). With some attention to detail, it is not difficult to think through Aquinas's understanding of the soul's operations and discover precisely the attention to the cognitive and practical intentions of which Husserl speaks. What Aquinas described in metaphysical terms, he based in an introspective psychology that noted human operations and developed an epistemology (in terms of a rational psychology) that supported the metaphysics.[39] He believed he had analyzed not only the empirical operations of an embodied soul, but had left room for God's interaction (Renaut's *transcendence-in-immanence*).

Lonergan argues that current inattention to the soul is actually a refusal to attend to the intelligibility of "what is."[40] According to William Barrett, the banishing of the soul is a rejection of consciousness in modernity.[41] By capitulating to an empiricist and reductionist view of the subject, modernity has denied that human understanding has an active role to play in apprehending the world as true; it neglects to analyze the kind of interaction that humans have with "what is." To discuss "soul" in Aquinas, therefore, is not to study an arcane "left-over" of pre-modern piety, but to think about the nature of human mental activity in our world. Is there anything unique about human cognitive interventions in the patterns of our world? Does the analysis of Aquinas tell us anything about what is necessary to approach the empirical ego of contemporary neuroscience?

3.1 The Soul and Consciousness

The soul, for Aquinas, is a concept that explains the data of consciousness in the human person.[42] Consciousness, however, should not be confused here with "reflective activity."[43] Consciousness is first of all the acts ("powers") of the intellect for sensing, asking questions, understanding concepts, weighing evidence, and judging the truth or falsity of things; it decides what is worthwhile or valuable. Consciousness is aware of itself. It knows itself in the performance of its tasks. As the epigraph states: "The human soul knows itself by its own act of understanding (*intelligere*)." This "knowing itself" is not an introspective intuitive grasping; consciousness knows itself in and through the objects it apprehends. The self-reflexive intellect "knows itself, and knows that it knows" (*SCG*, II, 66, 5). Aquinas distinguishes between our seeing an object and seeing the process of seeing, but points out that the second is not "extrinsic to the seeing faculty."[44] Sense, under-

---

[38] Lonergan, *Insight*, 436–37.

[39] See Lonergan, *Verbum*, 47, 72 n. 115, 74–85.

[40] Ibid., 20.

[41] William Barrett, *Death of the Soul: From Descartes to the Computer* (Garden City, N.Y.: Doubleday, 1986), 119–41.

[42] For the general approach, see Bernard Lonergan, "Isomorphism of Thomist and Scientific Thought," in *Collection*, F.E. Crowe, ed. (New York: Herder and Herder, 1967), 142–51.

[43] Lonergan, "Christ as Subject," 177 n. 14.

[44] Aquinas, *In III de Anima*, lect. 2 §591, quoted in Lonergan, "Christ as Subject," 181 n. 21.

standing, and reflection are conscious acts, even if only the latter can be self-reflexive and can therefore reflect back upon the activities of sensing (*SCG*, II, 58).

3.2 The Soul and Agency

Aquinas's analysis focuses upon "soul" as a term that labels the agency within the human person (*ST*, I, Q. 77, a. 1). If there were no agency, human beings would simply be a collage of exteriorly related pieces (*ST*, I, Q. 76, a. 3; *SCG* II, 56). Aquinas, in his dialogue with the Muslim commentators about the active intellect (*intellectus agens*), claims to understand the relatively autonomous agency of each individual person. "Soul" is the language that specifies the principle of activity in human life; it insures a voice by which subjects understand themselves, the world, and God.[45] Following Aristotle, he specifies the data for this agency as knowing (all cognitive activities from perceiving through judging), willing, and moving. Although these data include what contemporary thinkers would call intelligence, the body, and affective awareness, Aquinas discusses these issues only insofar as they relate to his theological concerns: the knowledge and love of God as well as the practice of virtue (*ST*, I, QQ. 75,78, introduction). Hence, in his treatises on the soul, he studies primarily the soul's essence, powers, and operations in desire and thinking. Since this is a normative, theological study, he examines the senses and the appetites only insofar as they impinge upon making decisions for the good and thinking about the world.

3.3 The Embodied Soul

Aquinas's reflection on the senses, however, is not minimal. Humans are for him "mixed" creatures, whose thinking can only take place in relation to images (phantasms) and whose decisions require a discrimination of desire in their choice and use of material objects. On the one hand, knowing is a passive, receptive activity in which the first level of awareness is sensing; yet this reception of images remains only a partial, potential knowing. On the other hand, the active intellect must interpret the data that have been abstracted, compose identities and definitions, and make judgments about their truth or falsity. However, in every case, the action of thinking must return to the data of sense, for evidence as well as for examples to illustrate its judgments (*SCG*, II, 73, 38).[46] Aquinas sees it as a fallacy to locate knowledge merely in a passive reception of data (*SCG*, II, 73, 26). In fact, he thinks this fallacy (empiricism *avant la lettre*) is rooted in the fact that some investigators simply have powerful memories and strong imaginations! The active and passive character of intelligence is a unity. Intelligence is potentially available for all objects within its horizon, but sometimes it is inattentive or asleep, and it can ask the wrong questions. At other times, it actively construes the world (*SCG*, II, 78, 4). At no time, however, can human knowledge take place without the senses. Even our knowledge of first principles (for example, noncontradiction), according to Aquinas, is derived from sensible things (*SCG*, II, 83, 26, 32).

---

[45] The "voice" is a coordinated/coordinating voice. As Arbib has pointed out to me, his neurological studies on schema theory (*Metaphorical Brain 2*, 211) and Leslie Brothers' on the social nature of brain activity (*Friday's Footprint: How Society Shapes the Human Mind* [New York: Oxford University Press, 1997]) show how the "unifying agency" of understanding is the interaction of distributed brain regions constructing patterns and overlapping foci of consciousness (see sec. 4).

[46] Lonergan, *Verbum*, 25–33.

The dependence of human beings upon objects that are external to the knowing subject is characteristic of a creature that is embodied. The soul is not an extrinsic, already-out-there spirit that rides herd on an already-constituted body; nor is it an intrinsic, already-in-there motor that moves the pre-formed machine (*ST*, I, Q. 76, a. 3). Actually, humans are a composite of *anima* and *materia prima*, of form and prime matter that appear as an ensouled body (see *ST*, I, Q. 77, a. 6).[47] The soul is known by its operations, by what it does in its embodiment; it is to be found in all levels of human knowing and desiring. In this way, human intelligence includes the principles of life that animate lower forms, such as plants and animals (*ST*, I, Q. 76, aa. 3–4; *SCG*, II, 58, 4–7). Touch, the primary sense for Aristotle and Aquinas, is an act of the sensitive soul, seemingly shared with animals; but it is an act of the entire conscious human being. The senses, hunger, sexual desire all operate within the unity of our intelligent consciousness, our one soul (*ST*, I, Q. 75 a. 3). To be human is to be a bodily soul or ensouled body (*ST*, I, Q. 76, a. 8). Every human act is both bodily and soul-full. There is a single soul in each individual that expresses itself in a bodily fashion, each part of the body accomplishing its animated purpose both individually (for example, seeing) and for the whole (interpreting and understanding in the mind) (*ST*, I, Q. 76, a. 8).[48] Aquinas even argues that it was necessary for the soul that it be endowed with intelligence *and* feeling (*ST*, I, Q. 76, a. 5); in the *Summa contra Gentiles*, he argues in addition that it is the "desire of the body" for its own perfection that acquires the kind of knowing (or soul) humans now have (*SCG*, II, 83, 14, 28).[49]

3.4 The Soul and Memory

Like Aristotle, Aquinas distinguishes *memory* and *reminiscence*.[50] In a culture in which memorized rhetorical commonplaces were crucial for political oratory, Aristotle concerns himself not so much with the sheer storage of sensations, but with the possibility of recovering issues, ideas, or examples that were once present. It is the ability to recollect, to make choices in one's investigation of the past that distinguishes humans from animals. Aquinas agrees: animals remember objects and places that are fearful or pleasurable (*ST*, I, Q. 78, a. 4). But what is called "instinct" in animals is actually cogitation or estimation in humans; in addition, humans have *reminiscence*, an intellectual application of the past to specific present intentions. But humans also have *memory* which is the free recall from the inventory of the past relevant information, values, and plans of action.

Intelligence is both passive and active; it receives sensations as data and inquires about their meanings (*ST*, I, Q. 79, aa. 2–3). Memory is part of the passive dimension of intelligence; it retains both universal ideas and specific particulars (*ST*, I, Q. 79, a. 7; *SCG*, II, 60, 3). But because it involves both universal and particular

---

[47] Karl Rahner, "The Theology of the Symbol," *Theological Investigations*, vol. IV, trans. Kevin Smyth, (Baltimore, Md.: Helicon, 1966), 247. For Rahner, the body *is* the material presence of the particular kind of *spirit* or transcendence that the finite soul is.

[48] Ibid., 245–49.

[49] Within an evolutionary biology, this appears as an intriguing premodern insight. For reflections on this topic, see Francisco J. Ayala, "Human Nature: One Evolutionist's View," *Whatever Happened to the Soul?*, Brown et al., eds., 31–48.

[50] Aristotle, *"De Memoria et Reminiscentia,"* in *Basic Works*, 607–17.

dimensions, it can be recovered by both sense and intelligence (*ST*, I, Q. 79, a. 6).[51] Since Aquinas thinks of historical particulars as "accidental" rather than essential, he locates human recollective memory within universalizing intellectual capacities. It holds the past as past individual events within sensation; it holds the ideas that integrate them in intellect. Willing, for Aquinas, integrates in its pursuit of the good, the senses, desire, intelligence, memory, and action. This claim for the integration of sense, intelligence, and will, however, shows how and why Aquinas can maintain that human beings do not know specific individual things in themselves, in their "inwardness" (*ST*, I, Q. 85, a. 1; Q. 86, a. 1) except through turning to the image "to perceive the universal nature existing in the individual" (*ST*, I, Q. 84, a. 7).[52] What we know, we know by our senses and by comparison to the senses (*ST*, I, Q. 84, a. 8). But this intimate relationship of knowing to the body means that when our bodies are tired or alert, our soul is likewise lethargic (*ST*, I, Q. 84, a. 7; *SCG*, II, 79, 11) or more attentive (*ST*, I, Q. 85, a. 7; *SCG* II, 90, 2). Thus, people with better imaginations, calculating powers, or stronger memories can think better.

3.5 The Soul as the Knowing Subject

Aquinas's language and his metaphysics may hide for the modern reader his careful attention to the knowing subject. What philosophical claims can he make that would guide current research and discussion? First, he states that human knowing is an active as well as a receptive process, both dependent upon the empirical world and independently critical in relationship to this world and to its own operations. Second, this knowing only takes place within the intimate cooperation of the individual's body. Third, intelligence is open-ended; it intends by wondering and inquiring about everything within its horizon. Fourth, this intelligence can reflect upon itself. Human beings genuinely understand themselves, "for we would not inquire into the nature of the intellect were it not for the fact that we understand ourselves" (*SCG*, II, 59, 10). Fifth, open-ended human intelligence can go beyond the senses, intending and estimating, even understanding the reality of God and of all things in God. Sixth, it is not this "higher agent" (that is, God) who makes us understand (*SCG*, II, 13). Human intelligence rightly apprehends reality ("what is"—*quid quod est*)[53] through its senses and makes correct judgments (even if revisable) on the basis of the evidence provided.

3.6 Aquinas and Husserl

Although an extensive correlation is not possible here, it is worth noting that Aquinas's metaphysical vocabulary about powers in intelligence coheres performatively with Husserl's notion of subjective time consciousness. How? The subtleties of temporal duration in Husserl are obviously not matched in Aquinas. However, in both, active and passive consciousness is present to itself prior to knowing objects, yet it is only available in and through the process of knowing itself. The pre-thematic

---

[51] This is related to current computational models in which schema instantiation may be data-driven (sense acquisition) and hypothesis-driven (top-down inquiry and questions). See Arbib, *The Metaphorical Brain 2*, sec. 5.3.

[52] This is not unlike Arbib's claim that schemas function somewhat differently in acquisition of data and re-use of the data at times of reflection or contemplation.

[53] Lonergan, *Verbum*, 16–25. Arbib has rightly pointed out that the neuroscientist *qua* scientist can find a resonance with the first four areas of Aquinas's concerns.

unity of consciousness escapes empirical detection except as a network of cognitive activities. It copresents itself in the process of knowing; at the same time, it also evades "complete" self-presentation in the sense that at no time is the temporal subject *totally* available, except in Aquinas's case, to God.[54] The images that emerge passively from the senses are converted into intelligibility by active intending, but they have become part of past memory. "Phantasms are temporal, new ones springing up in us every day from the senses" (*SCG*, II, 73, 31).

Intelligence for both is nonetheless (infinitely) open-ended, oriented toward the future. Aquinas's work does not constitute a sufficient analysis of subjectivity or of human temporality; but it does indicate that in a premodern environment of thought, a Christian thinker could analyze his own performance as a thoughtful believer and coherently thematize that performance through a systematic, theoretic vocabulary.[55] The world of interiority that Husserl examines turns from consciousness of the subject to a self-reflexive knowledge of that subject.[56] The metaphysical body and soul become a self-conscious subject, examining itself introspectively.

4 Neuroscience, Memory, and Human Subjectivity

The empirical subject of modern science has turned its investigative tools upon the human brain. What had first been studied by natural philosophy and metaphysics, and then by philosophical psychology, has now been analyzed by experiments, mathematics, and computers. At stake, however, are some of the same issues that have marked the earlier traditions: passive reception of data, the processes for understanding those data, the active nature of intelligence, and subsequent action. Investigators continue to ask "How do human beings know and learn?" "How is awareness of time encoded?" and "How is the brain an agent in the environment?" I would assert that these continue to be questions about the human soul as well as the human brain. One could modify Keith Ward's rhetoric slightly: the embodied "brain is the way the soul appears to others."[57]

Metaphors for Brain Activity. It would, of course, be possible to survey the multiple metaphors that neuroscientists use when speaking about the brain and brain functions. Indeed, as I (and others such as Michael Arbib and Mary Hesse) have argued elsewhere, metaphors function as guides and models, heuristically determining the paths of investigation and the rhetorical modes of communication.[58]

---

[54] The opening to a radically distinct form of Other is what makes Aquinas's cognitive agent *intrinsically* dialogical; part of the problem with Husserl's subject is its implied solipsism.

[55] For a popular account of Aquinas on the soul, see Stephen Happel and James R. Price, III, "Geography of the Soul: An Intellectual Map," in *Nourishing the Soul*, Anne Simpkinson, Charles Simpkinson and Rose Solari, eds. (New York: HarperCollins, 1995), 60–69.

[56] Lonergan, *Method in Theology*, 95–96, 259–60, 288.

[57] Keith Ward, *Defending the Soul* (Oxford: Oneworld, 1992), 142. The argument made by Philip Clayton in this volume is central here: the human person (and I would say the subject) is the mediating notion between neuroscience and religion.

[58] See my "Metaphors and Time Asymmetry: Cosmologies in Physics and Christian Meanings," in *Quantum Cosmology and the Laws of Nature: Scientific Perspectives on Divine Action*, Robert John Russell, Nancey Murphy, and Chris J. Isham, eds. (Vatican City State: Vatican Observatory Publications; Berkeley, Calif.: Center for Theology and the Natural Sciences, 1993), 103–34; and Michael Arbib and Mary Hesse, *The Construction of Reality* (Cambridge: Cambridge University Press, 1986).

Neuroscientists speak of landscapes and maps,[59] transmitters and receivers,[60] keyboards,[61] housing and rooms, mosaics,[62] networks,[63] machines, computing centers with inputs and outputs,[64] and even scripts, dramas, actors, and audiences![65] An incomplete inventory indicates that there are even more metaphors, without any clear priority at the investigative or communicative level. The nondominance, or perhaps better, occasional importance of one model over another, seems to be due to the relatively youthful nature of the sciences that study the brain and to the intellectual history that spawned the various perspectives on brain activity. But the diversity may also be required by the complex functions of the brain itself.

The presence of these metaphors in the scientific literature, however, makes this reader want to study the scientists as writers of literature, of a particularly precise, experimental literature, to be sure, but of prose nonetheless.[66] I shall not examine these metaphors extensively here. In the concluding third of this essay, I will focus briefly upon three themes of the neuroscientific literature: mental agency, time-awareness, and knowing as bodily and interpretive. Despite the diversity of these topics, I agree with Owen Flanagan that the unity of the empirical subject or the search for a theory of consciousness is not an "idle fantasy."[67] In many ways, these

---

[59] See Arbib, *The Metaphorical Brain 2*, 14; Jon H. Kaas, "The Reorganization of Sensory and Motor Maps in Adults Mammals," in *The Cognitive Neurosciences*, Michael S. Gazzaniga, ed. (Cambridge: MIT Press, 1995), 53ff; Antonio R. Damasio, *Descartes' Error: Emotion, Reason, and the Human Brain* (New York: Avon Books, 1994), xiv–xv.

[60] Eric R. Kandel, James H. Schwartz, and Thomas M. Jessell, *Essentials of Neural Science and Behavior* (Norwalk, Conn.: Appleton & Lange, 1995), 219–306; Damasio, *Descartes' Error*, 181.

[61] Charles S. Sherrington, cited in Marc Jeannerod, *The Cognitive Neuroscience of Action* (Cambridge: Blackwell, 1997), 2–3.

[62] A.R. Luria, quoted by Arbib, *The Metaphorical Brain 2*, 17; William H. Calvin, *The Cerebral Code: Thinking a Thought in the Mosaics of the Mind* (Cambridge: MIT Press, 1996), 2, 39–48.

[63] Ibid., 48; Trevor W. Robbins, "Refining the Taxonomy of Memory," *Science* 273 (6 Sept., 1996): 1354; Alan Prince and P. Smolensky, "Optimality: From Neural Networks to Universal Grammar," *Science* 275 (14 March, 1997): 1606–8. The complexities that figure in the use of "neural networks" can be seen in Daniel Gardner, ed., *The Neurobiology of Neural Networks* (Cambridge: MIT Press, 1993).

[64] Herbert Killackey, "Evolution of the Human Brain: A Neuroanatomical Perspective," in *The Cognitive Neurosciences*, 1247; Francis Crick, *The Astonishing Hypothesis: The Scientific Search for the Soul* (New York: Charles Scribners' Sons, 1994), 177–99.

[65] Bernard J. Baars, *In the Theater of Consciousness: The Workspace of the Mind* (New York: Oxford University Press, 1997), esp. 39–11; Joseph LeDoux, *The Emotional Brain: The Mysterious Underpinnings of Emotional Life* (New York: Simon & Schuster, 1996), 116–21; Brothers, *Friday's Footprint*, 80–99; Alan Searleman and Douglas Herrmann, *Memory from a Broader Perspective* (New York: McGraw-Hill, 1994), 125–27.

[66] For a recent useful, brief history of the role of imagination in the natural sciences, see Lorraine Daston, "Fear and Loathing of the Imagination in Science," *Daedalus* 127.1 (Winter, 1998): 73–95; for the role of "thematic imagination" in science, see Gerald Holton, *The Scientific Imagination: Case Studies* (Cambridge: Cambridge University Press, 1978), esp. 3–24.

[67] Owen Flanagan, *Consciousness Reconsidered* (Cambridge: MIT Press, 1992), 213. Flanagan's "naturalist" bias should be examined in the context of how far the notion of the brain is constituted by social experience and how far sociality should be extended (to include

themes or hypotheses could shape questions that an experimental program might answer; they could also establish some of the conditions for the particular kind of divine-human interaction that Christians and Jews claim is central to their religious commitments.

4.1 Mental Agents

The struggle of neuroscientists to understand the brain through experimental procedures is only an exacerbated example of nineteenth-century philosophical problems about how historical beings can understand their own history, without changing what they understand![68] By intervening in brain activity, other brains are changing the ways in which they themselves interpret the world. Even if there is no specific region of the brain given over to the "I," the subject, many would like to speak not just of a "powerful processing mechanism,"[69] but of some kind of executive, an editor. How the prefrontal cortex operates in working memory is central to this discussion.[70] It is possible, even in this context, to speak not so much of a single editor, but a group or "heterarchy, in which many different modules dominate the overall behavior as appropriate."[71] The problem of "initiation,"— whether humans react passively to the environment or whether they "script" the environment—is at the heart of these current studies. If human agency is not top-down hierarchy, is it a well-functioning committee?

The layering of brain activity converges in such a way that it acts as a unity, however conditioned by lower manifolds.[72] For perceptual and motor schemas, the model is "an active, information-seeking process composed of an assemblage of instances of perceptual schemas."[73] As Clayton, Murphy, and Peacocke argue in this

---

an infinite Other?); but his emphasis upon the lower neural pathways as the basis of consciousness must be taken seriously.

I will use the term "conscious" in the sense of some philosophical phenomenologists to include all forms of awareness by somaticized, neural pathways. That means that self-reflective activity is not the only form of conscious activity. There are pre-reflective forms of consciousness such as the immediacy of some sensations. These can be reflected upon; they may include a separate marker for such activity. There are also unconscious activities of the body (e.g. metabolism of one's cells) to which one attends only in the laboratory; but unconsciously, they are still occurring in the investigator. Much of human activity is pre-reflective, but that does not mean that it is not conscious; it has an intentional frame. See Lonergan, *Insight*, 319–47.

[68] This was examined in the evolution of psychology as a science and its relationship to biology; see Robert M. Young, *Mind, Brain and Adaptation in the Nineteenth Century: Cerebral Localization and its Biological Context from Gall to Ferrier* (Oxford: Clarendon Press, 1970).

[69] Mark S. Seidenberg, "Language Acquisition and Use: Learning and Applying Probabilistic Constraints," *Science* 275 (14 March, 1997): 1600.

[70] Ingrid Wickelgren, "Getting a Grasp on Working Memory," *Science* 275 (14 March, 1997): 1581–82.

[71] These are Michael Arbib's comments (personal communication) on William L. Kilmer, W.S. McCulloch, and J. Blum, "A Model of the Vertebrate Central Command System," *International Journal of Man-Machine Studies* 1(1969): 279–309.

[72] This is, of course, a "minimalist" statement; but the scientific constraints on the question require caution. James B. Ashbrook and Carol Rausch Albright, *The Humanizing Brain: Where Religion and Neuroscience Meet* (Cleveland, Ohio: Pilgrim Press, 1997), 113.

[73] Arbib, *Metaphorical Brain 2*, 18.

volume, there emerge at the upper layers characteristics that the network provides that are not available at lower levels of neuronal activity. On the one hand, the brain's electro-chemical impulses are produced by interaction with the environment; on the other, they are potential for agency.[74] The initiatory capabilities of the brain stir investigators to speak consistently of "higher level" operations and philosophers to speak of emerging mind or consciousness.[75]

4.1.1 Schemas and Consciousness

Arbib's *schemas* and Damasio's *somatic markers* perform this mediating role between sensation, evaluation, judgment, and action.[76] Like Kant's *schemata of the understanding*, they operate at a level differentiated from the neural input, grouping groups of sensations, selecting, and ordering for the sake of determining the composition of subjects and predicates, evaluating evidence, and prompting action. What is doing the "ordering?" Kant required an intermediate notion to function between sensibility and pure understanding.[77] On the one hand, he wanted to preserve the independent power of minds to calculate and reason (mathematics and theory) and at the same time to argue for the dependability of the senses (*contra* David Hume), thus grounding the two-fold inquiry of the new science. Although these schemata are products of the imagination, they provide a unity of "inner sense."[78] The schemata provide every concept with its sensible evidence. They are the process through which the mind emerges as an agent. They also determine the inner sense of time; they do not exhaust the "transcendental unity of apperception."[79] For Kant, even the rules that govern the normal internal processes of cognition are temporal, expressing a fundamentally unthematizable ongoing present.

Kant was, however, not able to foresee the scientific or technological future. He maintained that the schematism of the understanding "is an art concealed in the depths of the human soul, whose real modes of activity nature is hardly likely ever to allow us to discover, and to have open to our gaze."[80] With the PET-scan and

---

[74] Ashbrook and Albright, *Humanizing Brain*, 148.

[75] This does not necessarily mean that consciousness is uninvolved in neural circuitry. "As the brain evolves, it exhibits new patterns of neural activity which can provide new 'ecological niches' for the evolution of new neural circuitry to exploit those patterns. But once such new circuitry evolves, there is a new 'information environment' for the earlier circuitry—so it may evolve in turn to exploit these new patterns" (Arbib, *Metaphorical Brain 2*, sec. 7.2). It is precisely this "initiating" sequence (in which a new information environment "exploits" [i.e., actively perpetuates] an effect) that I think is at the heart of the philosophical, psychological, neuroscientific, *and* theological notions of consciousness. So Chalmers argues that "consciousness is not an explanatory construct, postulated to help explain behavior or events in the world. Rather, it is a brute explanandum, a phenomenon in its own right that is in need of explanation." See David J. Chalmers, *The Conscious Mind: In Search of a Fundamental Theory* (New York: Oxford University Press, 1996), 188.

[76] See Arbib, *Metaphorical Brain 2*, 3–84; Damasio, *Descartes' Error*, 165–201.

[77] Immanuel Kant, *Critique of Pure Reason*, trans. Norman Kemp Smith, (New York: St. Martin's Press, 1965), 180. See Mark Johnson's discussion of Kant on schemata of the imagination in his *The Body in the Mind: The Bodily Basis of Meaning, Imagination, and Reason* (Chicago: University of Chicago Press, 1987), 21–24, 152–57.

[78] Kant, *Critique of Pure Reason*, 182.

[79] Ibid., 183, 185.

[80] Ibid., 183.

fMRI work of neuroscientists, the governing patterns of neural input are indeed present to view. In light of these views, Arbib's schemas are processes, if I understand them correctly. They are internal models that guide "the organism's interaction with the world."[81] Arbib emphasizes that both visual and bodily maps are coded in our neural architecture, where intelligence is a "multiple, layered plexus of properties."[82] They exist not in a temporally linear string, receding into the past, but in a three-dimensional lattice that models the subject's way of being in the world. The network is an active information-seeking process.[83]

4.1.2 Schemas as Active

Schemas and subsequent schema assemblages are not passive associations.[84] They frame, script, and shape our perceptions and actions.[85] All knowing is hermeneutical; language itself for Arbib is an extension among humans of perception.[86] Long-term memory (LTM) is the network of schemas that not only stores our understanding of the world, but also of the subject who composes and interprets the world.[87] For Arbib, these schemas are an approximation of the way the mind thinks, an hypothesis that can be tested for probability and accuracy.[88]

4.1.3 Somatic Markers as Active

Where Arbib searches primarily for frames for cognition, Damasio studies the role of emotions in thinking and deciding. He distinguishes various nonintentional feelings: hunger; intentional, but pre-reflexive responses such as avoiding a reckless automobile; and reflexive choices such as befriending someone.[89] He notes that many would have argued that emotion does not have a role in thoughtful choice, because passion makes humans "subjectively" biased. He believes that the neurochemical tag that adheres to some images, questions, stories, or concepts is a *somatic marker* that functions as an "automated alarm signal" or a selector for pleasurable outcomes.[90] These markers mediate, sometimes preconsciously, among alternatives for decision-making. "Somatic markers do not deliberate for us," but they do focus certain options before others and permit rapid elimination of some possibilities.[91]

Somatic markers mediate among options, determining one's characteristic habits of action, of virtuous or vicious propensities.[92] Like Arbib's schemas, somatic

---

[81] Arbib, *Metaphorical Brain 2*, 9.

[82] Ibid., 6.

[83] Ibid., 18.

[84] Ibid., 37.

[85] Ibid., 47.

[86] Ibid., 19–23.

[87] Ibid., 31.

[88] Ibid., 48.

[89] Damasio, *Descartes' Error*, 167.

[90] Ibid., 173–75.

[91] Ibid., 174.

[92] Christian thinkers since Augustine have spoken of the *fomes peccati*, the physical, somatic marks of sinful actions in the actor's body, which, despite the believer's firm choice of amendment, require careful monitoring. See Karl Rahner, "The Theological Concept of

markers are interpretive, neither the "pure data" of sensations, nor conceptual abstractions. They have a history; they are the frameworks for our personal and social histories. Due to his own research on fear, Damasio sees these somatic markers as primarily related to survival. "Achieving survival coincides with the ultimate reduction of unpleasant body states and the attaining of homeostatic ones, i.e. functionally balanced biological states."[93]

The critical loci for this marking system are the prefrontal cortices.[94] Because somatic markers have to do with emotional history, they provide "categorized contingencies" that consider present options and project possible outcomes. The prefrontal cortices are "ideally suited" for this work because they are directly connected to all motor and chemical neural pathways.[95] Somatic markers can function both inside and outside reflection,[96] but in either case they partake of consciousness (in the sense of being aware), which is an intentional (but not necessarily deliberative) process. "Coherent mental activity" (knowing, in effect) requires "basic attention" and working memory, both of which function within the frameworks set by somatic markers.[97] The cumulative preferences that occur as a result of these interactions define the style of a person, her character, and anticipated courses of action. Because these markers are defined within a social environment, they also reflect and shape cultures.[98] As Brothers makes clear, "only brains in a social field can generate the kind of consciousness that includes 'I'."[99]

The study of schemas and somatic markers points to a series of important research questions. But in this context, the issue I wish to highlight is the following: Does this layered grouping of groups of sensitive data offer a neural, somatic way of asking about the agency within human experience? The empirical subject, described by early Enlightenment sciences, is often seen as either passive to its environment (sheer passive sensation as registration) or as manipulative of data outside itself. On the one hand, "inner" consciousness is seen as the victim of the "outside" environment; on the other, inwardness is an inner power controlling the other. Neither is accurate; neuroscientific studies struggle to determine the active, cognitive dimensions of "transcendence within immanence," the empirical, nonindividualist subject.

4.2. Memory and Time-Awareness

Just as the notions of somatic markers and schemas give some evidence for the search for agency in neurobiological research on human consciousness, so analyses

---

Concupiscentia," in *Theological Investigations,* vol. 1, trans. Cornelius Ernst, (Baltimore: Helicon, 1965), esp. 360–69.

[93] Damasio, *Descartes' Error,* 179.

[94] Ibid., 180.

[95] Ibid., 183.

[96] Ibid., 187.

[97] Ibid., 198.

[98] Damasio states: "*If we assume that the brain is normal and the culture in which it develops is healthy,* the device has been made rational relative to social conventions and ethics" (*Descartes' Error,* 200, my italics). The relationship between nature and nurture in this field could use a healthy dose of critical theory! For the social dimensions and a critique, see Brothers, *Friday's Footprint,* 117–18, 148, n. 10.

[99] Brothers, *Friday's Footprint,* 103.

of memory point to empirical interpretations of a perduring, self-constituting subject. The "convergence zones" in the prefrontal cortices record the "unique contingencies of our life experience."[100] When we lose our ability to recall our history, we lose our selves.[101]

In Aristotle's and Aquinas's terms, current research concerns not only reminiscence or estimation, which humans share with other animals, but also the non-deliberative and deliberative memory that structures human consciousness. There is the cogitation or estimation that our primordial, reptilian brains share with the animals; in addition, there is memory grounded not only there but in the prefrontal cortices. Experiments are conducted to study somatic markers at the level of embodied emotions, recollection by choice of explicitly memorized conceptual data, and the reminiscences of learned skills and habits. "Many current memory researchers still think that there is little or no evidence that genuine *forgetting...* occurs from LTM in healthy people."[102] But what is available is not always accessible.[103] Consolidation of memory occurs over time so that trivial, non-significant input does not get stored.[104]

4.2.1 Memory as Active

Research in memory recognizes that there is likely no single memory storage facility in the brain. Researchers distinguish LTM and short term or working memory (WM), each contributing in a three-dimensional way to the texture of the subject.[105] How and why impressions get from working memory to long term memory is an important question. Some postulate a "central executive" in the working memory that coordinates attentional resources that derive from speech-based information and visual-spatial data.[106] Nondeclarative learning (sometimes called implicit memory) is inaccessible to reflective mental activity; declarative knowledge (or explicit memory) can be learned in a single trial and stored for ready use.[107] Selective, filtered forgetting too is crucial. There are the basic limits to retrieval due to data that intervene and take the place of what has receded from immediate attention. Some diseases, certain types of head injuries, and toxic agents (such as alcohol) can affect our ability to recall.[108] But besides retrospective memory, the recovery of our past,

---

[100] Damasio, *Descartes' Error*, 182.

[101] Ibid., 196–97.

[102] For the survey that follows, see Alan Searleman and Douglas Herrmann, *Memory from a Broader Perspective* (New York: McGraw-Hill, 1994), 57–58. More technical surveys may be found in Elizabeth L. Bjork and Robert A. Bjork, eds., *Memory* (New York: Academic Press, 1996); and Norman E. Spear and David C. Riccio, *Memory: Phenomena and Principles* (Boston: Allyn and Bacon, 1994).

[103] Searleman and Hermann, *Memory*, 109.

[104] Ibid., 166–67.

[105] Ibid., 213–18.

[106] Ibid., 69–70.

[107] Ibid., 103. For extensive discussion, see Peter Graf and Michael E.J. Masson, eds., *Implicit Memory: New Directions in Cognition, Development, and Neuropsychology* (Hillsdale, N.J.: Erlbaum, 1993).

[108] Searleman and Hermann, *Memory*, 105. Note how this discussion conditions the 'idealized" time-consciousness in Husserl's philosophical account.

there is also prospective memory, the ability to plan for future tasks. In both cases, it should be noticed that it is the constituted and self-constituting self that is operating. We gain our personal reputations for effective dependability by remembering what we are to do in the future.

Researchers have developed accounts of the multiple biological and neural frameworks to account for the layered, temporally consolidating nature of human memory.[109] Explicit or declarative memory seems to require the temporal lobe system; implicit or nondeclarative memories involve the cerebellum, amygdala, and for simple sense storage, specific sensory or motor systems. The latter form of memory-learning may operate through stimulus and response in a nonreflective manner; the former requires reflective attention.[110] At the level of neurotransmission and synaptic interaction, LTM and WM may simply be two points of a graded process.[111] Sensitization, classical conditioning, and nondeclarative forms of memory/learning, indicate how both genetic and developmental elements affect the regulation of both long-term and short-term effectiveness of synapses.[112] These changes, extending to the somatic sense of self in the human subject, indicate the fundamental biological and biochemical processes that constitute memory.

4.3 Bodily Knowing

Researchers postulating schemas and investigating memory struggle to understand the levels of self-awareness that humans attribute to themselves.[113] Even though "consciousness appears to be a process, not a place,"[114] it is nonetheless completely embodied. Note that I did not suggest that consciousness is "enbrained" (even if such a barbarous neologism were possible). No thinking or feeling process in the brain takes place without connection among multiple regions of the brain; in addition, data registration, encoding in memory, etc., invariably involve complete bodily awareness. In the process of cognition, "the body is not passive." For Damasio, "mind derives from the entire organism as an ensemble...."[115]

---

[109] See Kandel, Schwartz, Jessell, *Essentials of Neural Science*, 651–94; for alternate views, see for example Raymond P. Kesner, "The Neurobiology of Memory: Implicit and Explicit Assumptions," in *Neurobiology of Learning and Memory*, Gary Lynch, James L. McGaugh, Norman M. Weinberger, eds. (New York: Guilford Press, 1984),111–18.

[110] See the research on two memory systems in Mortimer Mishkin, B. Malamut, and J. Bachevalier, "Memories and Habits: Two Neural Systems," in *Neurobiology of Learning and Memory*, 65–77.

[111] Searleman and Hermann, *Memory*, 671. Or as Michael Gazzaniga argues, they may be part of what he calls "superordinate organizing systems" (similar to Arbib's schemas?). See Michael Gazzaniga, "Advances in Cognitive Neurosciences: The Problem of Information Storage in the Human Brain," in *Neurobiology of Learning and Memory*, 78–88. See also Joaquín M. Fuster, *Memory in the Cerebral Cortex: An Empirical Approach to Neural Networks in the Human and Nonhuman Primate* (Cambridge: MIT Press, 1995).

[112] Searleman and Hermann, *Memory*, 692. For the social dimensions of memory-formation, see Brothers, *Friday's Footprint*, 41–42, 98.

[113] Thus, a philosopher like John Searle (whatever one thinks of his solutions) does not want to cede the territory of consciousness either to dualists or absolute materialists. He thinks the questions of consciousness are still worth asking for both scientists and philosophers; see *The Rediscovery of Mind* (Cambridge: MIT Press, 1994).

[114] Ashbrook and Albright, *Humanizing Brain*, 155.

[115] Damasio, *Descartes' Error*, 225.

4.3.1 Subject and Perception

Perception is hermeneutical; as an interpretive procedure, it operates in a feedback loop in which the history of one's perception informs future action and the attentions of the present affect how one understands the past.[116] Sensory processing is guided by what one is looking for; it is neither sprayed against a *tabula rasa* nor is it blind fumbling.[117] Perceptions initially define a body-space which humans inhabit with other human beings. They are intrinsically relational—to an environment of things and of other people. Neural networks require an "other" to exhibit their characteristic operations.[118] This builds a "dynamic map of the overall organism anchored in body schema and body boundary" which could not be constructed in one portion of the brain, but rather in coordinated areas.[119] Thus in seeing something, one also "knows" one is seeing something, although this kind of prereflective cognition (note: not pre-conscious, since it is aware) is in the background. It is a "sense" of spatio-temporality that provides the screen against which specific "experiences" are highlighted.[120]

4.3.2 Subject and Body

Subjectivity should be understood as the integrating process in which the body-state and all *reflexive* awareness (perception, image-formation, schema formation, etc.) are being integrated.[121] Mark Johnson uses "balance" in bodily experience to develop an understanding of how metaphor and imagination function in cognition and the natural sciences.[122] The "background" (or pre-reflective process) always exceeds our ability to reflect upon it, since reflection is necessarily about what has just passed temporally. Nonetheless, the "background," however unthematized, continues as long as one has not died.

---

[116] Marie Barinaga, "Visual System Provides Clues to How the Brain Perceives," *Science* 275 (14 March, 1997): 1583; for discussion of bodily-feedback on emotional responses, see LeDoux, *Emotional Brain*, 291–96. There has been considerable discussion on this claim among the contributors to this volume. To say that perception is hermeneutical is *not* to claim that the senses are nonveridical. It may locate the discussion of their empirical validity, but it neither denies nor affirms their truth-telling. In principle, I would argue that the senses provide relatively adequate data for truth-claims.

[117] Ingrid Wickelgren, "Getting the Brain's Attention," *Science* 278 (3 October, 1997): 35–37.

[118] Brothers, *Friday's Footprint*, 66–110. Ashbrook and Albright, *Humanizing Brain*, 5. As Brothers concludes, mind is "irreducibly transactional," (146).

[119] For a discussion of the molecular constitution of this body-space in relation to the environment, see Kandel, Schwartz, Jessell, *Essentials of Neural Science*, 335–46.

[120] Damasio, *Descartes' Error*, 234.

[121] This seems to be *contra* Damasio (*Descartes' Error*, 242–43) who argues that subjectivity emerges when there is a self-reflective knowledge of one's perceiving and image-forming activities. This view would create a problem for calling infants fully human, let alone autistic or schizophrenic individuals.

[122] Johnson, *Body in the Mind*, 74–100.

5 Conclusions and Questions

Ashbrook and Albright argue that the soul is our unique ability to integrate "the sensory and the symbolic processes of meaning making."[123] Far from being a non-material entity, the soul is stamped with the unique identity of each brain. At the same time, brains "affect and change" bodies.[124] This seemingly "naturalist" view of the soul, while it creates problems for traditional philosophical notions of immortality,[125] coheres at least with Aquinas's emphases upon the soul as "form of the body," with the body and images as always implicated in cognition, and with the role of various levels of memory in the classical theological tradition.

The soul as a theological concept explains the agency as well as the passivity of human consciousness; it exists, however, within the concrete neural schemas of the human brain in their interactions with the body and its location in an environment of things and people. The integration of vegetative, sensitive, and rational aspects of soul in Aquinas, while maintaining the priority of self-conscious life, nonetheless marks humans as an embodied species with all the lower-level information loops that mark plant and animal species. The "lower" levels of tropic and sensitive dimensions are not erased by the rational soul; they are integrated. Reminiscence as instinct or the implicit memory of skills is not subverted by intelligence; it is enveloped by deliberative forms of memory. In short, the more passive dimensions of cognition cooperate with the active ones; but the two aspects of human consciousness intertwine in what contemporary thinkers would call a feedback loop. Body and soul are terms used to interpret the unity of the human subject.

The distinctions that Husserl draws between ordinary temporal consciousness and the consciousness that is the active, unthematizable presencing of the agent-subject are more complex than Aquinas's psychology. However, I think that both strands of thought underlie the discussion in neuroscience about editors, complex processing mechanisms, schemata, etc. Is there a way to discuss or investigate the "immediate" presence of the subject to itself? If all investigation studies what has just passed into memory, then how or who is the active integrator (including the investigator)? Husserl postulates this deeper form of memory tracing. Could it be related to the somatic background? What sort of research program would seek such an *active presenting subject*?

A neuroscientific and philosophical anthropology that "leaves room" for human subjects is a crucial condition for Christian theology. These transactional subjects must be embodied, conscious, and intelligent. Otherwise, subjects could not see

---

[123] Ashbrook and Albright, *Humanizing Brain*, 15.

[124] Ibid., 42, 92.

[125] Although it does not necessarily create problems for Christian notions about the resurrection of the body, as Peters's argument in this volume attests. I have deliberately avoided discussion of the Thomist commitment to the immortality of the soul separated from the body (*ST*, I, Q. 75, aa. 5–7; Q. 89, a. 1). For Aquinas, the conscious human intelligence *naturally* desires always to exist because it intends universals (Q. 75, corp.); a natural desire cannot be in vain. Aquinas recognizes the problem of cognition without the senses after death (Q. 89, a. 1) and argues that human beings after death know through graded participations in the divine intellect (Q. 89, a. 1, ad 3). One must recall that Aquinas is here most evidently a theologian, attempting to make coherent a doctrine of personal continuity from death through eternal life. The residual Platonism, however, in Aquinas remains clear: "The separated soul... has a greater freedom of intelligence, since the weight and care of the body is a clog upon the clearness of its intelligence in the present life" (Q. 89, a. 2, ad primum).

themselves as either a *task* to be completed or as a *responsibility* to be weighed and evaluated. More importantly, however, the element that evades this discussion altogether is how human subjects experience themselves as *gift*, as contingent beings in the world, originating in some Other. Ashbrook and Albright see the role of humans to be making order and meaning—a constructive, investigative role.[126] But humans' role in the world, the way in which cognition occurs within human subjects, is interactive, cooperative. Christians and Jews claim that this interaction is so cooperative that the subjects' very ability to perceive another object or subject is itself a gift from an Other. This openness to the other characterizes empirical subjects in all their investigative, task-oriented strategies. Can it extend beyond the empirical world, as Aquinas would argue? What would be the conditions under which the empirical subject could, first, entertain such a possibility, and second, affirm its presence or absence within the world?

I have argued that there are links between the so-called transcendental subject of nineteenth- and twentieth-century philosophy (Husserl) and the empirical subject (neurosciences) since the Enlightenment, and I have placed these links in dialogue with the neo-scholastic tradition concerning the soul. By focusing upon memory and subjectivity in each, it may be possible to think through the forms of "transcendence within immanence" in contemporary culture.[127] Examining neurocognitive emergent transcendence may provide some of the empirical, "objective" conditions within human subjects for the possibility of divine action.

Presumably for Christians, God interacts with all levels of creation in distinct fashion. In this account of the soul, it is especially within and with human beings that God has chosen to cooperate with human agency.[128] In the world of early Christianity, when God's existence was less problematic, this strategy was described as the "divine condescension"—God's loving choice to work within the patterns of creation at any level that was necessary so that God could be experienced. Without that generosity and the neural network that is one of its conditions, we humans would not be able to appreciate the glory of God. It would stun us into silence.

---

[126] Ashbrook and Albright, *Humanizing Brain*, 33–35.

[127] Given the fact that the brain is a "far from equilibrium" subject, the topic of self-organizing systems is also relevant to the issues at hand. If this research group were to build on its prior studies, some examination and evaluation of these theories would be appropriate. For work in this area, see Karl H. Pribram, *Origins: Brain & Self Organization* (Hillsdale, N.J.: Erlbaum, 1994).

[128] For me, that is why the criteria with which Leslie Brothers and Wesley Wildman end their essay are important. Are there criteria for making a judgment that schizophrenics or those with temporal lobe epilepsy are ill and in need of medical and pharmacological assistance and those claiming religious experiences (without the other patterns) are not? To answer that there are differences does not invalidate schizoid individuals as possible saints; but it may be *despite* their illness, not *because* of it!

RESURRECTION OF THE VERY EMBODIED SOUL?

Ted Peters

1 Introduction

To be a human person is to be embodied. If the soul is essential to the self, the soul needs to be united with its body. Soul and body come together in a single package.

Embodiedness seems to be the assumption of the modern world, certainly modern naturalism as we find it in what many call the "scientific worldview." And, despite intellectual rumors to the contrary, this has been an enduring commitment within Christian theology. The creedal confession, "I believe in... the resurrection of the body," does not affirm that an immortal soul lives on in a disembodied state. Rather, it affirms that salvation requires bodily resurrection.

This consonance between modern assumptions and essential Christian commitments is worth pointing out, because contemporary discussions of the mind-brain problem fostered by advances in the neurosciences frequently make the false assumption that Cartesian dualism represents Christian belief. When Descartes's substance dualism is dubbed dismissible, frequently the presumption is that theological tenets regarding resurrection are dismissible right along with it. This is a mistake, however. Dismissing Christian belief on the grounds that it proffers salvation in terms of a disembodied soul commits the *straw theology fallacy*—that is, the alleged Christian view dismissed is not in fact the Christian view, or at least not the essential Christian view.

Now, to be sure, I am not going to suggest that advances in the neurosciences or cognitive theory will lead to a secular affirmation of resurrection. What I intend to show is more modest, namely, that some consonance exists between philosophical reflection on brain research and essential Christian commitments to understanding the human reality as embodied selfhood or personhood. So strong is this Christian commitment that visions of redeemed selfhood and eternal consciousness before God incorporate visions of a transformed physical nature. That we have a promise of future divine action to raise us from the dead comes solely from the revelation of a promising God, not from science or philosophy. That this resurrection needs to be envisioned in terms of embodied consciousness, however, provides an open gate to contemporary cognitive theory that may be worth passing through.

The primary task of this essay will be to correct an elementary though wide-spread error in current discussions of cognitive theory, namely, that if current brain theory repudiates Cartesian substance dualism then it simultaneously renders nonsensical the Christian commitment to a future resurrection. The assumption seems to be: If Cartesians can no longer live in their disembodied souls, then the souls of Christians can no longer be immortal. This is an error because the concept of a disembodied soul, mortal or immortal, is not fundamental to Christian eschatology. Resurrection of the body is.

In what follows I plan to provide examples of the straw theology fallacy in the science of Francis Crick and the philosophy of John Searle. I will then refine the concept of "Descartes's error" by showing that in contemporary neuroscience we frequently find a sustained attempt to demonstrate that the intellectual mind—what many rightly or wrongly identify with the soul—is embodied; and this embodiment includes brain activity sponsoring feeling, emotion, and social interaction. On this topic I will recognize briefly the work of Joseph LeDoux, Antonio Damasio, and Leslie Brothers. At this point I will introduce the theological agenda, showing historically how belief in bodily resurrection is essential to Christian thinking and showing constructively what elements necessarily go into the concept of a resurrected body. The historical discussion will rely on the work of Caroline Walker Bynum. Raising the question of how best to tie together cognitive theory with theology, I will turn initially down two roads that will prove to be blind alleys, or at least detours. One blind alley will be the notion of the "humanizing brain" in the work of James Ashbrook and Carol Albright. The other blind alley will be the Artificial Intelligence model of the human soul as a disembodied information processor proposed by Frank Tipler. Returning from these two blind allies, I will follow the road mapped out by Wolfhart Pannenberg in connecting the resurrected body to God's eschatological act wherein time is taken up into eternity, and where God provides the continuity of our identity even when our bodies and our souls disintegrate over time.

2 Working Vocabulary

Before we proceed, a word about words. Without stipulating exhaustive definitions of all the relevant vocabulary, I would like to assume the usability of certain terminology during the forthcoming discussion. Even though I regularly employ vocabulary such as "body and soul" or "brain and consciousness," I do not wish to presume a metaphysical commitment to substance dualism, according to which mind and body are separate substances.[1] Rather, it is to allow some carryover from discussions of the relationship between brain function and mind toward discussions of selfhood and salvation. Each of us has a soul in the sense of an essential personhood; but that soul is embodied both in this life and in the resurrection.[2]

My own position comes closest to that of emergent holism—wherein the soul is a distinguishable human property even if embodied. Yet the arguments I put forth here could also support a physicalism which posits that no distinguishable soul or mind exists. Both emergent holism and physicalism could be consonant with the Christian view that when the body is raised the whole person is raised. What these two positions share is divestment of commitment to the soul in disembodied form.

I find it quite understandable that critics might mistakenly conflate Cartesian dualism with Christian eschatology, because Christian vocabulary and some

---

[1] "The distinguishing claim [of substance dualism]...is that each mind is a distinct nonphysical thing, an individual package of nonphysical substance, a thing whose identity is independent of any physical body to which it may be temporarily attached." Paul M. Churchland, *Matter and Consciousness*, rev. ed. (Cambridge: MIT Press, A Bradford Book, 1996), 7.

[2] Reinhold Niebuhr associates the self with the body; but he avoids reducing the self to the body. Although the self is internally related to the body, it enjoys a quality that permits a partially objective view of this body. A person or a self can speak of "my body," analogous to though quite different from the idea of "my property." When Niebuhr then employs the word "soul," he refers to the organic unity of consciousness with its body. "The self is soul insofar as it has an experience of the unity. But it is more than soul insofar as it can think of its body as an object even while it is an inner experience of the bodily organic unity." *The Self and the Dramas of History* (New York: Scribner's, 1955), 26. What needs to be added to this is the self—body plus soul—over time, in history with a biography.

Christian ontologies have frequently worked with dualities such as body and soul. Until recently, theologians have not been forced to clarify the distinction between two overlapping ways of conceiving personal salvation. One, rooted primarily in the ancient Hebrew understanding, pictures the human person as entirely physical, as dying completely, and then undergoing a divinely appointed resurrection. The other, a later view influenced by Greek metaphysics, pictures the human person as a composite of body and soul; and when the body dies the soul survives independently until reunited with the body at the final resurrection. The first of these could be made consonant with the materialist implications of contemporary neuroscience—the second probably not. In both pictures, however, the resurrection of the body is decisive for eschatological salvation. Or, to say it another way, the essential Christian commitment is to embodied, not disembodied, salvation.

   I also distinguish between neuroscience, cognitive science, and philosophical speculation or cognitive theorizing. The neurosciences engage in the empirical study of brain functions, gathering data about perception, memory, and control of movement to be used in developing brain theory. The field of cognitive science includes researchers in artificial intelligence, cognitive psychology, linguistics, and such; but it does not for the most part focus on consciousness. I use the term "cognitive theory" in a generic way to identify attempts to construct a theory that elaborates on the connection between brain function and, among other things, human consciousness. Although it utilizes the work of neuroscientists and cognitive scientists, cognitive theory deals with questions bequeathed to it by the philosophical tradition, such as the mind-body problem.[3] The challenge to Christian belief in resurrection—to the extent that a challenge exists—comes from cognitive theorizing, not from science proper.

3 A Scientific Rejection of the Disembodied Soul

Nobel Prize winning geneticist Francis Crick provides us with an example of the straw theology fallacy. Crick is a scientist at war with religion. He identifies dualism with religion in general; and he identifies Christianity as one of the many religions holding the view that the soul can be extracted and exist beyond death independently from the body. Then he pictures modern science on the "attack" against such outmoded religious beliefs. The ammunition with which he arms himself for the attack is reductionism—that is, the assumption that a complex system such as human consciousness can be explained by the behavior of its parts and their interactions. The military objective he marches toward is this: "the main object of scientific research on the brain is... to grasp the true nature of the human soul."[4] Jingoistically, he throws down the gauntlet: "This is in head-on contradiction to the religious beliefs of billions of human beings alive today."[5] The weapon, he trusts, will bring him the victory he names the "Astonishing Hypothesis," which is that "'You', your joys and

---

   [3] Michael Arbib illustrates the mission of "cognitive science" in terms of a twofold aim: "to enhance our understanding of human thought and behavior and to provide new strategies for building intelligent machines.... [C]all us cyberneticians, model-builders, brain theorists, cognitive or computational neuroscientists..." *The Metaphorical Brain 2: Neural Networks and Beyond* (New York: John Wiley & Sons, 1989), 4.

   [4] Francis Crick, *The Astonishing Hypothesis: The Scientific Search for the Soul* (New York: Charles Scribner's Sons, 1994), 7.

   [5] Ibid., 261.

your sorrows, your memories and your ambitions, your sense of personal identity and free will, are in fact no more than the behavior of a vast assembly of nerve cells and their associated molecules."[6]

What we have here is the rejection of the notion of a disembodied consciousness under the name "soul" and the substitution of a materialist reductionism, according to which all soul functions are hypothetically explainable according to neuron activity. Regardless of whether the hypothesis will eventually be sustained through empirical research, Crick commits at the outset the straw theology fallacy by assuming that religious belief can be written off as holding a simplistic dualism.

Reflecting on the role that a view such as Crick's plays in modern secular thinking Mary Midgley comments: "According to the majority secularist view, there is obviously nothing *but* a body... If certain confusions do result from Descartes's having sliced human beings down the middle, many people feel that the best cure is just to drop the immaterial half altogether.... Amputation of the mind is to be performed by reductive techniques, translating whatever needs to be said about minds into statements about bodies."[7] One might get the impression from modern science in general and the neurosciences in particular that religious intuition regarding spiritual reality is so effete that it is about to pass into oblivion. Not so. Perennial philosopher Huston Smith argues that "we should keep on thinking of it [Spirit] as substantial and thing-like as matter is, while differing categorically from matter."[8] He goes on: "matter is that which is sensible, multiple, and subject to time. Spirit, by contrast, has none of these traits. It does not impact our senses. It is single, and ultimately beyond numerical considerations altogether. And, while it is present in time, it is itself timeless.... Ontologically, Spirit is more real than everything else. Causally, it occasions everything else. And axiologically, it excels everything else by being perfect."[9] Smith represents the billions of believers with whom reductionistic cognitive theorists are in head-on collision.

4 A Philosophical Rejection of the Disembodied Soul

Like Crick the scientist, Searle the philosopher commits the straw theology fallacy. However, in the process he offers a more careful analysis that aids in clarifying issues.

The contribution of John R. Searle is that he steadfastly maintains that consciousness has ontological status. He holds that human subjectivity is an unavoidable datum requiring scientific analysis. Even though consciousness is a first person phenomenon, its existence as such makes it a legitimate domain for third person scientific investigation. Even though it is experienced subjectively, it warrants objective examination.

Searle's thesis is that the brain causes consciousness. Consciousness is real, but it has a biological cause. Labeling his own view "biological naturalism," he argues that "the brain is an organ like any other; it is an organic machine. Consciousness is caused by lower-level neuronal processes in the brain and is itself a feature of the

---

[6] Ibid., 3.

[7] Mary Midgley, "The Soul's Successors: Philosophy and the Body," in *Religion and the Body*, Sarah Coakley, ed. (Cambridge: Cambridge University Press, 1997), 53–68; 53.

[8] Huston Smith, "The Ambiguity of Matter," *Cross Currents*, 48.1 (Spring 1998): 49–60; 53.

[9] Ibid., 55.

brain."[10] The link between the brain and consciousness is a causal one. "Brain processes cause consciousness in all its forms."[11]

Searle positions himself between Cartesian dualism, on one side, and reductionist materialism, on the other. He opposes dualism because, having drawn a strict distinction between two realities—what is mental and what is physical—dualism has been unable to make the relation of the two intelligible. This loss of intelligibility renders the view unavailable to scientific investigation or, more precisely, out of step with the modern scientific worldview. He similarly opposes materialism, a monism which posits only one reality—the physical—and thereby eliminates the mental from existence. Materialism may be scientific, but it is also unnecessarily reductionistic. "The way out is to reject both dualism and materialism," writes Searle, "and accept that consciousness is both a qualitative, subjective 'mental' phenomenon, and at the same time a natural part of the 'physical' world."[12]

As a part of the physical world, consciousness should in principle be available for scientific investigation. Searle believes he can show how this is the case by distinguishing between the epistemic objectivity of scientific *method* and the ontological subjectivity of the *subject matter*. At first, one might think that consciousness would be off limits to objective study, because we experience consciousness subjectively. Searle repeatedly uses the example of physical pain: if someone pinches me I recoil in pain. Pain is a qualitative state of my consciousness. Even though the pain belongs to my subjective experience, the existence of the pain is an objective datum.

> The epistemic objectivity of method does not preclude ontological subjectivity of subject matter. To state this in less fancy jargon: the fact that many people have back pain, for example, is an objective fact of medical science. The existence of these pains is not a matter of anyone's opinions or attitudes. But the mode of existence of the pains themselves is subjective. They exist only as felt by human subjects.[13]

Thus, Searle grants ontological status to qualitative states of consciousness, and renders them data for someone else's scientific investigation.

Might this lead to a research program? Searle answers in the affirmative. Here is what he proposes: "we need a neurobiological account of exactly how micro-level brain processes *cause* qualitative states of consciousness, and how exactly those states are *features* of neurobiological systems."[14] In short, Searle's contention that brains *cause* consciousness functions as an hypothesis. It is not a conclusion drawn from scientific experimentation, because that experimentation has not yet been done. Searle's view is rather a philosopher's proposal, a possibility, a projection. His rejection of Cartesian dualism and maybe even materialist reductionism is not based

---

[10] John R. Searle, *The Mystery of Consciousness* (New York: New York Review of Books, 1997), 17.

[11] John R. Searle, *The Rediscovery of the Mind* (Cambridge: MIT Press, 1992), 247.

[12] Searle, *Mystery of Consciousness*, xiv.

[13] Ibid., 122.

[14] Ibid., 129. Searle can say, "we know in fact that brain processes *cause* consciousness" (Ibid., 191); so his proposed research program must be for the purpose of confirming what Searle already claims we know as fact. With dualists and materialists disputing him from both sides, it is curious that he begins self-assuredly with what he declares to be a fact; but then he formulates it as a question and asks scientific researchers to answer it. "If we could answer the causal questions—what causes consciousness and what does it cause—I believe the answers to the other questions would be relatively easy." Ibid., 192.

upon scientific knowledge; it is rather a hypothetical rejection based upon a guess that the brain will in fact turn out to be causally responsible for consciousness.

When it comes to concepts we have inherited from our religious traditions, Searle likes to situate himself squarely within what he perceives to be the scientific worldview.[15] This applies to religious beliefs regarding the immortality of the soul and the existence of God.

> Nowadays… no one believes in the existence of immortal spiritual substances except on religious grounds. To my knowledge, there are no purely philosophical or scientific motivations for accepting the existence of immortal mental substances. [and later,] Our problem is not that somehow we have failed to come up with a convincing proof of the existence of God or that the hypothesis of an afterlife remains in serious doubt, it is rather that in our deepest reflections we cannot take such opinions seriously. When we encounter people who claim to believe such things, we may envy them the comfort and security they claim to derive from these beliefs, but at bottom we remain convinced that either they have not heard the news or they are in the grip of faith.[16]

From the perspective of one who has heard the news regarding the scientific world-view, those who still hold outdated religious beliefs are to be pitied because they are still in the "grip of faith." Searle's position is determined by his general self-situating within the modern scientific worldview; it is not due to science having successfully proved that his position is correct.

The task of a Searle-inspired research program, it seems to me, would be for research into brain functions to explain as much of human consciousness as possible. Searle may eventually be proven right, even if at present he is only guessing.

I should like to add that I see no theological investment in preserving any semblance of Cartesian dualism. I see no theological advantage in protecting what Daniel Dennett calls an "ineffable residue" of a soul disentangled from bodily connection.[17] What Searle has rightly acknowledged is that first-person subjectivity in the form of consciousness exists as an indubitable fact, and that if biological explanations prove adequate—or prove inadequate as well—this fact will remain.

5 Descartes's Error

As we move toward serious dialogue between science and theology, we need to distinguish issues internal to each domain from issues that connect each domain. One issue that, on the face of it, might look theological but is in fact internal to the neurosciences is the question of the tie or lack of tie between reason and emotion, between the rational and affective dimensions of being human.

If thought can be mapped onto the brain, can emotion as well? Joseph LeDoux throws down the gauntlet:

> Cognitive science is really a science of only a part of the mind, the part having to do with thinking, reasoning, and the intellect. It leaves emotions out. And minds without emotions are not really minds at all. They are souls on ice—cold, lifeless creatures devoid of any desires, fears, sorrows, pains, or pleasures.[18]

---

[15] "To accept dualism is to deny the scientific worldview that we have painfully achieved over the past centuries." Ibid., 194.

[16] Searle, *Rediscovery of the Mind*, 27, 90–91.

[17] Daniel C. Dennett, *Consciousness Explained* (Boston: Little, Brown and Co., 1991), 388.

[18] Joseph LeDoux, *The Emotional Brain* (New York: Simon and Schuster, 1996), 25.

LeDoux wants to expand the scope of cognitive science—perhaps even renaming it "mind science"—to include emotion along with intellect.[19] The one brain is responsible for both, and the one brain responsible for both is embodied. "Emotion and cognition are best thought of as separate but interacting mental functions mediated by separate but interacting brain systems."[20] An embodied brain that includes both intellectual and emotional functions seems to be the prevailing vision of neuroscience today.

"Emotion, feeling, and biological regulation all play a role in human reason," writes Iowa neurologist Antonio Damasio.[21] Reason cannot operate alone, insists Damasio, because it is bodily; and thereby reasoning is influenced by all that is bodily. This leads Damasio to repudiate Descartes's famous dictum, "I think, therefore, I am" (cogito ergo sum). Descartes presupposed that the mind is a "thinking thing" (res cogitans) separable from the nonthinking body extended in space with component parts (res extensa). What empirical brain research demonstrates, says Damasio, is that the mind's rationality functions in inextricable concert with—not in isolation from—bodily actions such as feeling and emotion. So, he sings the song sung by today's scientific and philosophical chorus: "This is Descartes's error: the abysmal separation between body and mind."[22]

Descartes's legacy lives on today in the form of a metaphor—one can find this metaphor in the Artificial Intelligence (AI) community—drawn from information theory: the mind is an information processor. The terms of the metaphor are these: the mind and brain are related as are software and hardware in a computer. Just as the software cannot function without the hardware, so the mind cannot function without the body. This bothers Damasio. Because the mind includes so much more than merely software-like information processing, the analogy is misleading. The mind is thoroughly biological, and the complementary influence of feeling and emotion demonstrate this. Descartes's AI disciples are in error too.

"Feelings form the base for what humans have described for millennia as the human soul or spirit," says Damasio as he develops his point that soul and body are inextricably destined to exist together.[23] "Love and hate and anguish, the qualities of kindness and cruelty, the planned solution of a scientific problem or the creation of a new artifact are all based on neural events within a brain, provided that brain has been and now is interacting with its body. The soul breathes through the body... "[24]

6 Feeling, Emotion, Soul, Society

Psychiatrist and neural researcher Leslie Brothers refines Damasio's position on emotions. What we who speak English identify as discrete emotions—fear, rage, disgust, happiness, surprise, sadness, etc.—do not map onto the brain. Experiments

---

[19] Ibid., 68.

[20] Ibid., 69. The central thrust of LeDoux's argument is that what the brain does regarding our emotions is not at the conscious level. "What the brain does during an emotion occurs outside of conscious awareness." Ibid., 267.

[21] Antonio R. Damasio, Descartes' Error: Emotion, Reason, and the Human Brain (New York: Avon, 1994), xiii.

[22] Ibid., 249.

[23] Ibid., xvi.

[24] Ibid., xvii.

have failed to show consistent correspondences between specific emotions and specific brain activity. This does not mean such emotions do not exist, but that they are not isomorphic with specific brain functions—therefore, they must be socially constructed. Emotion is a category; but it is a cultural and not a biological category.

Emotional constructs derive from the interaction of biological feelings with the social environment. Brothers distinguishes feeling from emotion, the latter being dependent on, but not reducible to, the former. "Feeling" does not refer to something separate from and additional to bodily actions—it is simply a way of talking about the bodily action itself.[25] Brain circuitry underlies emotion, but discrete emotional forms are shared with others in society. Mental states are brain dependent, to be sure; but this brain dependence is accompanied by dependence upon social interaction as well. "In contrast to contemporary cognitive neuroscience, which views the mind as a kind of closet with entities like emotion, linguistic rules, and memory arranged inside," she writes, "I take mind to be irreducibly transactional.... Just as gold's value derives not from its chemical composition but from public agreement, the essence of thought is not its isolated neural basis, but its social use."[26]

In sum, the "soul" includes rational and linguistic capacities, but also the affective dimensions we associate with emotion. All this has a basis in brain circuitry; but the biology does not operate in isolation. At least one important component of the soul, the emotional component, is the result of biological feeling and social interaction. Damasio says it forcefully: "The truly embodied mind... does not relinquish its most refined levels of operation, those constituting its soul and spirit."[27]

The issue internal to neuroscientific research and to reflection by cognitive theory is the interaction of emotion with intellect, not whether the intellect is embodied or disembodied. The soul, if inclusive of both emotion and intellection, is presumably embodied. And this embodiment is relational. It includes social interaction.

Theologians from the Enlightenment to the present have been insisting that the human person must be viewed as a psychosomatic unity—that is, as an integrated whole including the physical as well as mental and spiritual and even communal dimensions. Although the works of LeDoux, Damasio, and Brothers in neuroscience understandably have nothing to say about resurrection, they provide indirect support for an integrative anthropology. A modest yet noteworthy consonance exists here.

7 Is Descartes's Error Also a Christian Error?

If contemporary scientific research successfully establish that René Descartes was in error by positing the independence of the intellectual mind over against the emotional dimensions of the body, is Christian theology then guilty of the same error? No. The essential commitments of Christian theology do not depend upon the substance dualism presumed in Cartesian philosophy or its predecessor, Platonic philosophy. Such confusion in contemporary thought is understandable, because theology frequently formulates concerns in Platonic and Cartesian terms.

This leads Arthur Peacocke to ask rhetorically: "Surely Christian theism is committed to the doctrine of the immortality of the soul? Not so, though many Christians have presumed this doctrine in a Cartesian and Platonic form." So

---

[25] Leslie Brothers, *Friday's Footprint: How Society Shapes the Human Mind* (New York and Oxford: Oxford University Press, 1997), 117–18.

[26] Ibid., 146.

[27] Damasio, *Descartes' Error*, 252.

Peacocke endorses a version of "Christian materialism," a nonreductionistic or holistic view of the human person. "Let us agree to reject an ontological dualism, a two-tiered division of human beings into body and mind or body and soul, two distinct entities to which may be attached mutually exclusive predicates."[28]

Ian Barbour notes that against Cartesian dualism several objections have been raised. For one, it would seem that if mental events could influence physical events, then this would violate the conservation of energy. A second and more relevant objection is that the postulated mental and physical substances are so dissimilar that it is difficult to imagine how they would interact. Thirdly, the portrayal of mind as totally unlike matter does not seem to fit evolutionary theory, because no route is available for explaining how mind evolved from matter. Dualism does not allow for an intermediate stage or bridge between matter and mind.[29] These are nontheological arguments. Barbour also tenders a theological suggestion: "neuroscience is consistent with the biblical emphasis on embodiment, emotions and the social self."[30]

Recent feminist theology is similarly willing to distance itself from Cartesian dualism as well as from dualisms in general. Feminist thinking is pro-body and holistic. Roman Catholic theologian Susan Ross relies on the Incarnation and the use of physical elements in the sacraments as evidence that God is pro-body.

> Dualisms such as nature/history, body/spirit, female/male are oppositional... A Catholic feminist perspective bases its critique of these dualistic conceptions on a retrieval of the Incarnation, seeing God's taking on the condition of humanity as God's own self-expression. Sacramentality grows out of human embodiment and its connection to the natural world, not in contrast to it. Feminist theology thus argues for a closer connection between nature and history, body and soul.[31]

8 The Ancient Christian Commitment to Resurrection of the Body

What Peacocke, Barbour, and Ross offer as systematic observation can be buttressed historically. Historian Caroline Walker Bynum has argued persuasively that, contrary to conventional wisdom, Christian teaching regarding the resurrection has consistently insisted on embodied salvation. Whether in this life or the next, we cannot be our true self apart from our body, either our present earthly body or our transformed resurrected body. Here is her thesis:

> I argue that for most of Western history body was understood primarily as the locus of biological process. Christians clung to a very literal notion of resurrection despite repeated attempts by theologians and philosophers to spiritualize the idea. So important indeed was the literal, material body that by the fourteenth century not only were spiritualized interpretations firmly rejected, soul itself was depicted as embodied.[32]

Then she concludes by drawing implications for us in the modern world.

---

[28] Arthur Peacocke, "A Christian Materialism?" in *How We Know*, Michael Shafto, ed. (New York: Harper, 1985), 146–68; 148.

[29] Ian Barbour, *Religion in an Age of Science* (San Francisco: Harper, 1990), 195–96.

[30] Ian Barbour, in this volume, p. 261.

[31] Susan A. Ross, "God's Embodiment and Women," chap. 8 of *Freeing Theology: The Essentials of Theology in a Feminist Perspective*, Catherine Mowry LaCugna, ed. (San Francisco: Harper, 1993), 193.

[32] Caroline Walker Bynum, *The Resurrection of the Body in Western Christianity, 200–1336* (New York: Columbia University Press, 1995), xviii.

The materialism of this eschatology expressed not body-soul dualism but rather a sense of self as psychosomatic unity. The idea of person, bequeathed by the Middle Ages to the modern world, was not a concept of soul escaping body or soul using body; it was a concept of self in which physicality was integrally bound to sensation, emotion, reasoning, identity—and therefore finally to whatever one means by salvation.[33]

As evidence Bynum retrieves the controversies of the first centuries following the New Testament era to show how resurrection was envisioned, namely, as transformation of our present body into an immortal body that would maintain our identity but escape those physical factors that cause suffering or pain or death. She reminds us how Tertullian wrote that "our flesh shall remain even after the resurrection… but at the same time impassible, inasmuch as it has been liberated by the Lord for the very end and purpose of being no longer capable of enduring suffering."[34]

A millennium later, when much had changed in Christendom and the notion of the immortal soul had gained ground, physical resurrection remained decisive.[35] Thomas Aquinas combined the two views. The thirteenth century cultural context of Thomas, influenced as it was by the Platonic and neo-Platonic traditions, held that the soul of the deceased could continue to exist in a disembodied state such as purgatory. In the face of this, Thomas retrieved Aristotle's view and described the soul as the *form of the body*. "Body and soul are not two actually existing substances; rather, the two of them together constitute one actually existing substance."[36] This tendered a somewhat more materialist view of human nature.

This also led to a certain ambivalence. On the one hand, Thomas could say that when the soul of a just person is separated from the body, it sees God. On the other hand, it is unfitting and imperfect for the soul to remain forever without the body. So Thomas argued that finally the soul would be reunited with the body at the advent of God's kingdom. Almost repeating Tertullian, Thomas said, "So no one living in this mortal flesh can see God. But the flesh the soul resumes in resurrection will be the same in substance and will through divine gift have no corruption."[37] Despite such ambivalence in Thomas and elsewhere, the theology of the era leads Bynum to

---

[33] Ibid., 11.

[34] Tertullian, "On the Resurrection of the Flesh," *Ante-Nicene Fathers*, (Buffalo: Christian Literature Publishing Co., 1885), III:590, cited in Bynum, *Resurrection of the Body*, 43. Justin Martyr put it this way: "We expect to receive again our own bodies, though they be dead and cast into the earth, for we maintain that with God nothing is impossible." "First Apology of Justin," XVIII, in *Ante-Nicene Fathers*, I:169.

[35] In the fifth century Augustine spoke of two resurrections, the first a resurrection of the soul or spirit for the righteous plus a second bodily resurrection in which the unrighteous would join the righteous for the final judgment. "There are two regenerations… —the one according to faith, and which takes place in the present life by means of baptism; the other according to the flesh, and which shall be accomplished in its incorruption and immortality by means of the great and final judgment—the one the first and spiritual resurrection, which has place in this life, and preserves us from coming into the second death; the other the second, which does not occur now, but in the end of the world, and which is of the body, not of the soul, and which by the last judgment shall dismiss some into the second death, others into that life which has no death." *City of God*, XX:6.

[36] Thomas Aquinas, *Summa Contra Gentiles*, II:69:2. See the discussion of Aquinas by Stephen Happel in this volume.

[37] Thomas Aquinas, *Expositions in Job*, chap. 19, *lectio* 2; cited in Bynum, *Resurrection of the Body*, 258.

remark, "No mainstream theologian of the late Middle Ages denied the doctrine of bodily resurrection."[38]

At this point, I must make a concession to the cognitive theorists who criticize the Christian position. The concept of a disembodied soul akin to the way it was conceived in ancient Greece belongs to the tradition; and its presence in Christian thought on the eve of Descartes's entrance onto the philosophical scene was influential.[39] The question of just how essential the idea of a disembodied soul is to Christian belief is a matter for theologians to debate among themselves. It is my judgment, based on both historical and systematic criteria, that what is so essential as to be indispensable is the affirmation of bodily resurrection following the model of Jesus' Easter resurrection. This would be the case regardless of how the question of a temporarily disembodied soul is resolved.

9 Four Knotty Issues

The concept of resurrection into an incorruptible body raises four theological issues: the status of the flesh, the role of divine power, the problem of identity, and finally the question of chain consumption.

9.1 The Flesh

Over against the Gnostics and Docetists, for whom the physical body was too earthy and too degrading for divine inhabitation, many Christian apologists emphasized resurrection of the flesh. Rather than refer merely to the body (*sōma*), they emphasized the transformation of the source of human desire, the flesh (*sarx*). The desires of the flesh can themselves be cultivated as desires for God. Mystics such as Mechtild of Magdeburg lift up desire: "Your longing will live, for it cannot die, because it is eternal. Let it yearn on until the end of time, when soul and body will unite again."[40] This positive appropriation of flesh with its desiring sits at the extreme edge of the discussion; yet it serves to point up the seriousness of the central commitment to embodiment. Contrary to mistaken opinion, Christian anthropology is ready for a positive appropriation of the fleshly body.

9.2 Divine Power

The ancient theologians also emphasized that resurrection is not natural. Rising from the dead is not a phenomenon of nature. If there is to be a resurrection, it will come in the form of direct divine action, as an act of the God who created the world in the first place. Only God has the power to raise the dead. No principle or element in our

---

[38] Bynum, *Resurrection of the Body*, 276.

[39] Because "the description of death as the separation of body and soul is... used so much as a matter of course," writes Karl Rahner, "we must consider it the classical theological description of death." *On the Theology of Death* (New York: Seabury, Crossroad, 1973), 16. He adds, "The resurrection of the body, at the end of the world, is a dogma of faith." Ibid., 25. Rahner's unique contribution to the discussion is his notion of the soul becoming pancosmic at death. "The soul, by surrendering its limited bodily structure in death, becomes open towards the universe and, in some way, a codetermining factor of the universe precisely in the latter's character as the ground of personal life of other spiritual corporeal beings." Ibid., 22. For detailed discussion of the biblical texts, see the essay by Joel Green in this volume.

[40] Mechtild of Magdeburg, *Das Licht*, book 6, chap. 15; cited in Bynum, *Resurrection of the Body*, 340–41.

natural body (*sōma*) or even in our soul (*psychē*) survives death. Resurrection is a new act of God, an act of divine power tantamount to the original act of creation.

9.3 Identity

Next, there is the question of continuity of identity. As previously noted, some ancient Christian thinkers pictured a future resurrection of the very fleshly bodies we inhabited while on earth. Granting that it would be difficult yet still possible for an all-powerful God to reassemble and revive our various body parts, this view presumes that continuity of the flesh would guarantee our identity in the resurrection.[41] Yet, future resurrection is also thought by some to be a re-creation, a new creation. In this context, new creation would mean more than merely transformation; it would mean resurrection into a new body without physical continuity with the earthly body. Finally, theologically, we need to admit that we are not responsible for our own identity beyond the grave—God is. Whether we understand the resurrection crudely as a reconstitution of our previous body or as a change so radical that a new body is virtually created, the continuity as well as the transformation that marks our identity will be something God guarantees.

9.4 Chain Consumption

This leads an issue in which the previous three issues converge. It is the curious problem of chain consumption. If one's vision of resurrection consists in reassembling the physical elements that once constituted the fleshly body of an individual person, then God has a logical problem. The problem is this: given that life eats life in continuous cycles, which elements belong to which person? In the ancient world the question arose with regard to cannibalism. If one person eats another, does the flesh of the eaten one then belong to the eater or to the eaten? How will God discriminate on judgment day? Although cannibalism itself was rare, human beings eaten by wild beasts—such as Christian martyrs in the Roman arenas—was common, feared, and puzzling.

The more modern version of the chain consumption problem would incorporate a longer sense of evolutionary time, the cycling and recycling of nature's elements through death, fertilization and growth of plants, digesting the plants as food, death, and so on. It would also incorporate our notion of metabolism, the death of old cells and birth of new ones on seven year or similar cycles. If resurrection means reassembling previous physical elements belonging to an individual, then it is not clear which elements belong to which individual. In his *City of God*, Augustine addresses the issue.

> For all the flesh which hunger has consumed finds its way into the air by evaporation, whence...God Almighty can recall it. That flesh, therefore, shall be restored to the [person] in whom it first became human flesh. For it must be looked upon as borrowed by the other person, and, like a pecuniary loan, must be returned to the lender.[42]

---

[41] A particularly subtle version that avoids the reassemblage problem is developed by Origen, who posited a "germ" or miniature schema of the body—following the seed analogy of St. Paul in 1 Cor. 15:37—that dies, is sown in the ground, and then grows up again at the resurrection. Like grain, the germ is "implanted in them which contains the bodily substance." *De Principiis*, II:10:3.

[42] Augustine, *City of God*, XXII:20.

Augustine thinks he has solved the chain consumption problem by giving priority to the first human being to possess bodily elements. He assumes that each person begins life with a fresh set of hitherto unused physical elements. Given our modern understanding of the interrelatedness of elements in the ongoing life cycle, it would be difficult to arrive at such a proprietary understanding of physical elements. We all share, to greater or lesser degrees, the same physical elements.

This leads Arthur Peacocke to solve the problem of chain consumption differently from Augustine. He solves it by distinguishing slightly between transformation and re-creation. "It is only too clear," writes Peacocke, "that the constituents of human bodies are at death irreversibly dispersed about the globe, eventually contributing to the bodies of other, later persons (as well as other living organisms). Hence the actual transformation of individual human bodies could not itself secure the continuity of personal identity through death." Transformation understood as merely immortalizing our mortal bodies is conceptually inadequate, because over time we share the elements of our bodies with other creatures. Thus, an adequate conceptual account must include a component of new creation, a *"re-creation* into a new mode of existence."[43] Peacocke emphasizes that the resurrected body of Jesus and our resurrected bodies are not miracles within the existing natural order; rather they are eschatological realities belonging to God's re-creation. This implies that the continuity of our identity through the resurrection will be something achieved by the creative act of God.

Could we say more? Rather than requiring God to locate and piece together all the molecules of our previous body, might we say rather that who we are is found in the patterning of the molecules? Rather than the molecules as matter, might we say that the form is what counts? Does God remember and reincarnate our form or pattern? Peacocke, though emphasizing the new in new creation, partially reiterates the vision of Origen: "The previous form does not disappear, even if its transition to the more glorious occurs... although the form is saved, we are going to put away nearly [every] earthly quality in the resurrection... [for] 'flesh and blood cannot inherit the kingdom' (1 Cor. 15:50)."[44] Whereas for Origen the "form is saved," for Peacocke the resurrected body is tantamount to a new creation.

10 Two Roads Not Taken

Which way is forward? The map already shows two roads. Yet, I am not inclined to follow either of them. Although at first they appear to take us in the right direction, following them for a short distance shows that they are more likely to detour us.

10.1 The First Road Not Taken: The Humanizing Brain

The first of the two roads not taken is marked with the sign, "The Humanizing Brain." This path leads toward uniting the scientific understanding of the brain with religious thinking via analogy. It is a natural theology pursued by James B. Ashbrook and Carol Rausch Albright. They use a method of analogy or correlation whereby identifiable brain functions become associated with theological descriptions of God. This correlation method comes in two forms: the split brain form and the triune brain

---

[43] Arthur Peacocke, *Theology for a Scientific Age*, enlarged ed. (Minneapolis: Fortress Press, 1993), 285.

[44] Origen, "Fragment on Psalm 1:5," cited in Bynum, *Resurrection of the Body*, 64.

form. In both cases, this method uses brain analogies en route to the same goal, namely, the propagation of a holistic or integrative religious vision.

Split brain theories describing bimodal consciousness so popular in the 1970s distinguished separate but complementary tasks performed by the two brain hemispheres, the right and the left. Intuitive and affective operations became associated with the right brain, and the more analytical operations with the left. The two were genderized, associating the right with the feminine and the left with the masculine. New Age spirituality capitalized on the dualism apparently rooted in the double brain and began trumpeting the need for holistic integration.[45] James Ashbrook builds upon the analogy of right-left brain competition overcome by cooperative integration to draw a theological picture. He associates the left brain with "proclamation and saving the world" by "naming reality and analyzing experience." He associates the right brain with "savoring the world" by immersing us in sensory experience and imagining liberating possibilities. Then he adds: "However, a third style appears more prevalent than the other two: the whole brain integrating its experience cognitively and affectively, thereby caring for the world. This is most apparent in all forms of prophetic theology—liberation, feminist, Third World, political."[46]

This is the double brain analogy. There is also a triple brain analogy. Together Ashbrook and Albright extrapolate from Paul D. MacLean's view of the triune brain—reptilian brain, old mammalian brain, and new brain or neocortex—"three minds... suggestive of various ways of understanding God's ways of being God."[47] What are the three ways for God to be God signaled by MacLean's triune brain? First, the sensory based reptilian brain which we share with reptiles, birds, and other mammals is focused on survival, reminding us of God whose "eye is on the sparrow." Second, the old mammalian brain is concerned with personal attachments, emotional responsiveness, and meaningful memory; and this is consistent with views of God as nurturer, as love, as provider of meaning for our lives. Third, the neocortex seems to employ an ordering power that represents the order of the universe, suggesting God as the one who creates the universe through the divine Word or *logos*. In addition, the frontal lobes of the new brain support our ability to empathize with others, suggesting a God who knows our needs better than we do ourselves.[48]

Ashbrook and Albright sum up their method: "We regard the brain as a primary lens with which to study and understand the cosmos and its dynamic source, God."[49]

---

[45] New Age holism was popularly articulated in the 1970s and 1980s by *Brain-Mind Bulletin* editor, Marilyn Ferguson, especially in her widely read book, *The Aquarian Conspiracy* (Los Angeles: Tarcher, 1980). Ashbrook and Albright do not cite Ferguson. Rather, they reference resources such as Robert E. Ornstein, *The Psychology of Consciousness*, 2nd ed. (New York: Harcourt Brace Jovanovich, 1977); numerous articles by Roger W. Sperry in *Zygon*; and Julian Jaynes, *The Origin of Consciousness and the Breakdown of the Bicameral Mind* (Boston: Houghton Mifflin, 1976).

[46] James Ashbrook, "Interfacing Religion and Neuroscience: A Review of Twenty-Five Years of Exploration and Reflection," *Zygon* 31.4 (Dec., 1996): 545–82; 563.

[47] James B. Ashbrook and Carol Rausch Albright, *The Humanizing Brain: Where Religion and Neuroscience Meet* (Cleveland: Pilgrim, 1997), xxxii. They rely on Paul D. MacLean, *The Triune Brain in Evolution* (New York: Plenum, 1990) among other sources.

[48] Ashbrook and Albright, *Humanizing Brain*, xxxiii.

[49] Ibid., 163.

Whether in two parts or three parts, distinguishable brain parts become analogues for what Ashbrook and Albright would like to say about God and theology. And what they most want to say is this: "Our universe is in the hands of the whole-making, integrating, emerging God! May we, as cocreators, work also toward whole-making."[50]

With regard to our focal concern here, Ashbrook and Albright affirm that the human soul must be embodied in the brain. "The soul, far from being a nonmaterial entity, bears and expresses the unique stamp of each person's brain/mind…"[51] Soul is inextricably connected to memory. Consciousness is constituted by memory. Memory is a brain function. When this brain function ceases, the soul disappears. "When people lack memory—as in amnesia or in not taking time to remember—then they have lost their soul. And in losing their soul they lose contact with the sacred, the depth of meaning itself."[52] This marks a rather significant shift from classical Christian theology, wherein the threat to the soul was found in surrendering to the power of sin and rebellion against God, not in losing brain function. Here "the human brain embodies the *human* meaning of divine purpose."[53]

This is not the method I wish to pursue. Such an analogical method presupposes in advance just what the theologian wishes to say about God; then it simply hangs this message on whatever hooks seem to protrude from what appears to be scientific understanding. Such a method has no essential link to scientific research. It could be equally applied to any phenomenon of doubleness or tripleness one might find in nature or culture. The three-leaf clover provided a trinitarian analogy for St. Patrick in Ireland just as the tripartite brain does here. No ontological connection between the being of the clover and the being of God was ever intended by the Irish saint. Such a connection is not firmly established here either.

When it comes to understanding the human soul, I can accept the hypothesis that the soul as mind is inextricably linked to the brain and thereby embodied. But then to so identify the human meaning of divine purpose with this brain embodiment leaves no room for a transcendent God to raise the ensouled body.

10.2 The Second Road Not Taken: Immortal Information Processing

In his book, *The Physics of Immortality*, physicist Frank Tipler argues for immortality in the form of disembodied information processing that simulates or emulates human consciousness as we know it in embodied form. Starting with the astrophysical assumption that we live in a closed universe that will eventually double back on itself and collapse in a big crunch making all physical life as we know it impossible, he goes on to posit that the human race may still escape to a supra-physical dimension of reality in which conscious experiencing will go on forever. Tipler thinks we can rely on a physical mechanism that leads to resurrection of living beings. What is this mechanism? It is computer simulation. By defining a living being as essentially an information processor, a definition common to the Artificial Intelligence (AI) school, Tipler projects an eschatological computing era in which all previously living beings will be simulated—duplicated or replicated or

---

[50] Ibid., 165.

[51] Ibid., 42.

[52] Ibid., 93.

[53] Ibid., 105.

emulated—along with their respective environments to live in never ending subjective time. Objective time along with its physical cosmos may self-destruct; but the supra-physical society of resurrected emulations will live on eternally at what he calls "Point Omega." *"We shall be emulated in the computers of the far future....* the reality we as resurrected individuals shall inhabit in that far future is 'virtual reality' or 'cyberspace'."[54]

The New Testament language, as found in 1 Corinthians 15:44, connects resurrection to a "spiritual body." Tipler believes his emulated information processors provide an explanation for what St. Paul means by spiritual body.

> Borrowing the terminology of St. Paul, we can call the simulated, improved, and undying body a "spiritual body," for it will be of the same "stuff" as the human mind now is: a "thought inside a mind... The spiritual body is thus the present body (with improvements) at a higher level of implementation... an emulated person would observe herself to be as real, and as having a body as solid as, the body we currently observe ourselves to have. There would be nothing "ghostly" about the simulated body, and nothing insubstantial about the simulated world in which the simulated body found itself.[55]

We should congratulate Tipler for being sensitive to a number of theological concerns here. First, he is sensitive to the promise of perfection. Our present state of human existing is not adequate. We do not hunger simply for life beyond death as a continuance of what we have now. We hunger for salvation. So, without using the term "salvation," Tipler announces that the simulated body will transcend the previous model by eliminating bodily defects such as missing limbs. Youth will be substituted for old age; sight for blindness; etc. Second, Tipler is aware of the problem of continuity of identity. His view maintains it. Against objections that total death followed by total re-creation denies continuity, Tipler sensitively argues that continuity in conscious self-identity is both necessary and possible. To be resurrected as a replica of one's former self does not deny that it is the same self. The key here is the identity of the information patterns within which we are aware of our experience of the world and ourselves.

> An exact replica of ourselves is being simulated in the computer minds of the far future. This simulation of people who are long dead is "resurrection" only if we adopt what philosophers call the "pattern identity theory"; that is, the essence of identity of two entities which exist at different times lies in the (sufficiently close) identity of their patterns. Physical continuity is irrelevant.[56]

What is curious about the Artificial Intelligence model of the mind as an information processor is that, contrary to the direction cognitive theory is going, we end up here with a disembodied soul, a supra-physical consciousness.[57] Today's

---

[54] Frank J. Tipler, *The Physics of Immortality* (New York: Doubleday, 1994), 220, italics in original.

[55] Ibid., 242.

[56] Ibid., 227.

[57] The irony is that in an age of opposition to the disembodied mind of substance dualism, the disembodied mind returns with AI. "But *this* difference [i.e., the difference between an organic body and a machine simulation in a computer] is no more relevant to the question of conscious intelligence than is a difference in blood type, or skin color, or metabolic chemistry, claims the (functionalist) AI theorist. If machines do come to simulate all of our internal cognitive activities, to the last computational detail, to deny them the status of genuine persons would be nothing but a new form of racism." Churchland, *Matter and Consciousness*, 120. With this in mind, computer scientist and theologian ("cognoboticist" is the term she uses) Anne Foerst contends that Artificial Intelligence researchers operate with a different world-

philosophers first tell us we cannot have minds disconnected from bodies; then they proceed to disconnect minds from bodies and stick them into machines. Or, perhaps more precisely, they surgically remove the mind or soul from the organic body and transplant it into an inorganic body, a computer. With this irony as background and with a bow to embodiment, Tipler recognizes that the soul cannot enjoy itself fully without its body. So, he gives it a virtual body; and then he calls this "real."

The simulations which are sufficiently complex to contain observers—thinking, feeling beings—as subsimulations exist physically. And further, they exist physically by definition; for this is exactly what we mean by existence, namely, that thinking and feeling beings think and feel themselves to exist. Remember, the simulated thinking and feeling of simulated beings are real.[58]

The nature of a soul's experience is itself in relation to an environment, and this environment is experienced as physical. Yet a simulated environment is enough for Tipler. In order for it to be enough, he must follow Bishop Berkeley on this: to be is to be perceived.[59] If as a computer simulation we perceive physicality, then physicality exists thereby.

Although this replica theory may appear at first to be religiously unsatisfying, Tipler's road has a side street that leads back to our main highway. Tipler faces squarely the terrifying emphasis of St. Paul, namely, that we do undergo total death and total re-creation. Theologically speaking, neither our physical atoms nor our soul's content perdures beyond death on its own. Despite the apparent coldness of the image of computer processing, Tipler's image has the advantage of correlating squarely with Paul's image of the seed dying and then sprouting.

We note in passing that the Easter resurrection of Jesus Christ plays no role in Tipler's eschatology. Tipler actually rejects the claim that Jesus rose from the dead. In fact, on the basis of this rejected claim, he declares himself to be a non-Christian and an atheist.[60] Tipler's view of resurrection is strictly limited to the future, with no Christological prolepsis or anticipation.

Even as future, though, resurrection is more historical or evolutionary than it is eschatological. Resurrection is the result of developmental processes already begun; not the result of a divine act. This future evolutionary reality arises from life understood as information processing taking hold of its own destiny and creating a supra-physical environment for its existence just prior to the moment when the physical world self-destructs.[61] Yet, almost in contradiction to his rejection of the constitutive role of Easter in our resurrection, he curiously uses his view of our future resurrected forms to explain phenomena attached to Jesus' past resurrection appearances: how Jesus' disciples did not recognize him until he willed it; how a simulated person can be erased from one part of the universe and then instanta-

---

view from that of theologians, and the theologians are much more inclined to take the embodiment of the mind seriously. "At present the human body seems far beyond understanding; it is impossible, for instance, to correlate neuronal activity with intelligent tasks. Yet, AI researchers believe that their research supports the scientific approximation of a complete analysis of humans. They believe that there is a mechanistic and functionalistic explanation of everything that is going on in humans." "Artificial Intelligence: Walking the Boundary," *Zygon* 31.4 (Dec., 1996): 681–94; 685.

[58] Tipler, *Physics of Immortality*, 210, italics in original.

[59] Ibid., 211.

[60] Ibid., 305; 309–13.

[61] Ibid., 225.

neously reappear in another, making possible his walking through walls; and even how Jesus could eat or be touched even though he appeared like a ghost.[62] Tipler here is merely offering speculations about the Easter phenomenon. He sees no proleptic or redemptive power in Jesus' resurrection that has any impact on the rest of us. At best, for Tipler, there is but a loose connection between Jesus' Easter body and our future resurrected forms.

The connection is also too loose between individual survival beyond physical death and the no longer real environment replaced by a virtually real environment. Eschatological resurrection for the Christian is inextricably tied to the advent of the Kingdom of God, to the renewal of all creation. Janet Martin Soskice, critical of Tipler, connects human hope with nonhuman destiny.

> If a resurrection faith becomes too exclusively a hope for the next life for humans (an etiolated orthodoxy of which Tipler's would be a crude caricature), there is no hope of the triumph of God's justice on earth, no point in praying that God's kingdom will come and will be done *on earth* as it is in heaven, and no salvation for nonhuman creation (trees that 'clap their hands' and hills that 'rejoice').[63]

Wolfhart Pannenberg makes the same point *vis-à-vis* Tipler. Over against Tipler's view that life, once it has begun to evolve, must therefore continue to evolve forever, outliving the very physical universe that gave it birth, Pannenberg concedes that life could come to a total end. The end of life would not thwart God's eschatological salvation, however, because salvation includes resurrection as a divine act. In addition, resurrection does not apply strictly to the perdurability of individual information processors; rather, it applies to the fulfilled and renewed creation as a whole.

> The phase of contraction of the universe after the disappearance of organic life could even be considered a condition for the replacement of this world by a "new heaven and a new earth" in the eschaton or, more precisely, for its transformed (by participation in God) "simulation", as Tipler puts it…. Salvation cannot be conceived as occurring separately for humans at the end of history. What happens to human beings has to be related to the entire world process.[64]

11 The Road to Take: God's Resurrection of the Whole Person

The better road to follow, in my judgment, is the one partially mapped out by Pannenberg. Although he does not analyze the significance of the neurosciences or cognitive theory for resurrection, what Pannenberg contributes to our understanding of eschatology forcefully makes the point that the whole person is the subject of eternal salvation. More than merely a matter of souls and bodies, resurrection from time into eternity completes, fulfills, and consummates the identity we seek.

Theologically, the primary issue arising in ancient Christianity regarding the immortality of the soul was not that it would be embodied or disembodied. Rather, it was the question of the divinity of the soul in its natural state. Immortality was the

---

[62] Ibid., 244.

[63] Janet Martin Soskice, "Resurrection and the New Jerusalem," in *The Resurrection*, Stephen T. Davis, Daniel Kendall, and Gerald O'Collins, eds. (Oxford: Oxford University Press, 1997) 41–58; 57.

[64] Wolfhart Pannenberg, "Theological Appropriation of Scientific Understandings: Response to Hefner, Wicken, Eaves, and Tipler," chap. 18 in *Beginning with the End: God, Science, and Wolfhart Pannenberg*, Carol Rausch Albright and Joel Haugen, eds. (Chicago: Open Court, 1997), 438.

chief quality of divinity for the ancient Greeks. If, as in the Platonic tradition, the soul is by nature immortal, then its survival after the death of the physical body would be guaranteed apart from any action taken by God. In contrast, the Christian belief is that the soul, like the body, belongs to the created order of things; it is not immortal by nature. Any immortality we as finite creatures might have could come only from an action taken by God. Raising the dead to new life is the particular form taken by that divine action. Raising Jesus on the first Easter is the prototype that God promises to follow for our future resurrection.

What is raised is the whole person inclusive of our various aspects: body, soul, and everything else. This does not forbid positing the existence of a human soul. What it forbids is the notion of a permanent disembodied soul that bears the essence of a person's identity. Theologically, it would be a mistake to think of our soul as our true person, thereby rendering the body an unnecessary appendage or a prison from which to seek escape. It would be a mistake to think of the soul as having a built-in capacity for immortality that can bypass the need for a divine act of resurrection. Pannenberg makes the point this way: "In distinction from Plato's view of the deity of the soul Christian theology views us as creatures in both body and soul, destined indeed for immortality in fellowship with God, yet not possessing it of ourselves, nor able to secure it for ourselves, but receiving it only as a gift of grace from God."[65]

One of the limitations in formulating the issue strictly in physical or metaphysical terms is that we easily overlook who we are over time. The mere existence of our body or our soul or both does not in itself determine who we are. Our individual biographies do. The brute survival of a body or a soul into immortality is of little consequence unless immortality is for someone, for some person in particular. Who we are as a whole person includes our whole life story.

The whole person of whom we speak is the whole person stretched out over time—that is, his or her entire life story is what is precious to God. Theologically, we do not reduce a person to who he or she is in any given moment, not even the crisis moment at death. One's past and future adhere to each moment. Who we have been in the past is always copresent to what we are doing and feeling at any given moment. We experience this in part through the presence of memory and expectation, the latter being a projection based on what exists in the former. "Great is the power of memory, an awe-inspiring mystery, my God," exclaims Augustine; "So great is the power of memory, so great is the force of life in a human being whose life is mortal."[66] Such a respect for human memory leads Ashbrook and Albright to equate it with the soul. They report how two forms of memory—*declarative memory*, what can be declared or brought to mind as a proposition or image, and the *procedural memory*, which includes various skills such as catching a ball—can be mapped onto the brain.[67] So, when the brain ceases to support our memory, our soul dies. Yet, to my mind, this does not end the matter. I would add that God has a memory, so to speak; or better, God has the power to catch temporality up into eternity.

---

[65] Wolfhart Pannenberg, *Systematic Theology*, 3 vols. (Grand Rapids: William B. Eerdmans, 1991–98) III:571.

[66] Augustine, *Confessions*, tr. by Henry Chadwick (Oxford: Oxford University Press, 1991), X.17:194.

[67] Ashbrook and Albright, *Humanizing Brain*, 93.

Pannenberg approaches this significant item by observing, "In the awareness of our identity as we march through time we keep the past and future of our lives in some sense present to us by recollection and expectation. In this way our sense of time in our lives participates in eternity."[68] Each moment of our awareness is framed by our memory and our anticipation. Each moment of our being is framed by our past and our future. Our identity consists of continuity-in-difference over time. As finite beings, we experience the continuity-in-difference as a sequence of moments stretched out over time. These sequences may seem broken, discreet, or separated. Death looms as the biggest break of all, the ultimate separation. Our perduring identity becomes a bit of a mystery as we contemplate these temporal separations. Yet, God's eternity has a way of sweeping up all these moments, maintaining their distinctiveness, but overcoming their separation. In the resurrection "when this corruptible will have put on incorruption," writes Pannenberg, perfected finitude "will no longer have the form of a sequence of separated moments of time but will represent the *totality* of our earthly existence."[69]

Does this mean eternity is timeless? No. Life without time for finite creatures would be eternal death, not life. We must not think of eternity as the antithesis of time. Rather, it is the consummation of temporal history, the ontological copresence of all chapters in the history of God's creation. This line of thought will require refinement of Boethius' classical definition of eternity as the "total, simultaneous and complete possession of unlimited life."[70] The celebration of life is essential, because God's eternity is our salvation. Yet a question for us remains: Does simultaneity negate time?

Karl Barth offers refinement with his notion of duration. Eternity provides what time lacks, namely, duration from moment to moment. "Eternity has and is the duration which is lacking to time," writes Barth; "It has and is simultaneity."[71] Therefore, eternity is not merely the "negation of time." Eternity does not negate beginning, succession, and end. "The duration of God Himself is *the* beginning, succession, and end.... It is itself that which begins in all beginnings, continues in all successions and ends in all endings.... To that extent it is and has itself beginning, succession and end."[72] Here eternity is not simply another metaphysical category

---

[68] Pannenberg, *Systematic Theology*, III:562.

[69] Ibid., III:561. Joseph Ratzinger takes advantage of this intersection of time with eternity to defend the immortal soul while still affirming resurrection of the body. Ratzinger is impatient with contemporary theologians who reject the immortal soul in order to accommodate the materialism of modern science. "Acceptance of the unity of the human being may be well and good but who, on the basis of the current tenets of the natural sciences, could imagine a resurrection of the body?" *Eschatology*, vol. 9 of *Dogmatic Theology*, (Washington: Catholic University of America Press, 1988), 106. He then turns to theological resources to affirm that the soul carries us from time to timeless eternity. He connects time with a "form of bodily existence. Death signifies leaving time for eternity with its single today. Here the problem of the intermediate state between death and resurrection turns out to be only an apparent problem. The between exists only in our perspective. In reality, the end of time is timeless. The person who dies steps into the presence of the Last Day and of judgment, the Lord's resurrection and parousia." Ibid., 107–8.

[70] "*Aeternitas igitur est interminabilis vitae tota simul et perfecta possessio.*" Boethius, *The Consolation of Philosophy*, V:6:4.

[71] Karl Barth, *Church Dogmatics*, 4 vols. (Edinburgh: T. & T. Clark, 1936–62), II:1:608.

[72] Ibid., 610.

along with time. Eternity is the presence of God. God is the constant, the enduring one, the one who provides continuity over time for us while holding all things together eternally.

Pannenberg is critical of Barth here because this notion of simultaneity leaves out a description of the succession of moments in time and fails to proffer a present that embraces past and future.[73] What deserves attention is how we conceive of the present moment. Is the present moment a *temporal point*, a discreet segment that is differentiated from its past and future? Or is the present moment a *duration*, a moving now that integrates its past and future? Each of us as a finite creature has a limited sense of duration propped up by memory and expectation. For eternity to be life-giving, it must be a divine duration within which the whole of who we are over time is unified.[74] Eternity provides our identity, regardless of whether or not we reassemble the parts of the fleshly bodies we had at death.

> The future of consummation is the entry of eternity into time. For it has the content that characterizes eternity but that is lost in the disintegration of time, namely, the totality of life and therefore also its true and definitive identity. For this reason the eschatological future is the basis for the lasting essence of each creature that finds manifestation in the allotted duration of its life and yet will achieve its full manifestation only in the eschatological future.... All of us are still on the way to becoming ourselves, and yet all of us are in some sense already the persons we shall be in the light of our eschatological future.[75]

The concept of eternity as God's duration provides a path to follow toward solving the four knotty problems inherited from the ancients. First, the central commitment to embodiment is reaffirmed; in this case temporal embodiment. Second, the resurrection of the body is not a natural event; it results from divine action. Third, the continuity of identity is bequeathed to the resurrected one by the divine eternity which holds every chapter of one's temporal biography together. Fourth, the problem of chain consumption is overcome because eternity holds the actual biography of each person together—as our personal histories enter eternity— thereby avoiding any dependence upon reconstituting the elements of the fleshly body at the moment of death.

12 Conclusion

Where have we been? We began by observing that cognitive theory relying upon advances in brain research is asserting that no soul exists, or if it does exist it is totally dependent on brain function. Within the domain of the neurosciences this means that intellectual thinking is conditioned by emotion, that both reason and feeling derive from the same embodied source, the brain. Where the domain of neuroscience meets that of philosophy and theology, this means that science cannot underwrite the notion of a disembodied mind or soul. The substance dualism of

---

[73] Pannenberg, *Systematic Theology*, III:596, n.222.

[74] "Eternity is the undivided present of life in its totality. We are not to think of this as a present separated from the past on the one hand or as the future on the other. Unlike our human experience of time, it is a present that comprehends all time, that has no future outside itself." Ibid., II:92.

[75] Ibid., III:603–4. John Polkinghorne writes, "The resurrection of Jesus is a great act of God, but its singularity is its timing, not its nature, for it is a historical anticipation of the eschatological destiny of the whole of humankind." *The Faith of a Physicist* (Princeton: Princeton University Press, 1994), 121.

Cartesian philosophy is in error. And, if the dualistic vocabulary and conceptuality inherited by Christian theology from the Platonic tradition begins to look too much like Cartesian substance dualism, then theology is also in error. This error would be particularly grave if the concept of immortality would hang on such a dualism. Yet, as I have tried to show, the essential Christian commitment is not to a disembodied soul but rather to resurrection of the whole person. Even though in some epochs Christian theology has employed the vocabulary of the immortal soul, this in itself has not defined salvation. Salvation always consists of resurrection of the body.

This essay has not set out to certify by scientific means the Christian doctrine of the resurrection. Nor does it even hint at this. Rather, much more modestly, it has tried simply to show that even if neuroscientific research eventually demonstrates that all human intellection and emotion derive from the physical operations of the brain, the Christian understanding of salvation is not in jeopardy. Cartesian substance dualism might be in jeopardy, but the Christian promise of resurrection is not.

The Christian promise of resurrection does not depend upon the physical operations of the brain beyond death, nor upon the survival of a disembodied mind in the form of a metaphysical soul, nor upon any other natural process identifiable by science or philosophy. Rather, building upon the witnessed resurrection of Jesus Christ on the first Easter, the Christian promise points our gaze toward an eschatological transformation—a new creation—to be wrought by an act of God wherein, like Christ, we too will be raised from the dead.

COGNITIVE NEUROSCIENCE AND RELIGIOUS CONSCIOUSNESS

Fraser Watts

1 Introduction

Issues about divine action come into sharp focus when considered in relation to cognitive neuroscience. When divine action is considered in relation to the physical sciences, or to evolution, we are grappling with the rationality of faith, with whether it is credible in our contemporary scientific world to believe in a God who is active in creation. However, when we come to the cognitive neurosciences, we are also dealing with the credibility of daily religious life and practice, and in particular with the prayerful relationship of Christian people with the God in whom they believe.

It is a bedrock Christian assumption that there is a God with whom people can have some kind of communion. It is assumed that, through prayer and meditation, Christians can enter into some kind of relationship with God, that they can open themselves to God's influence, and that they can in some measure discern God's will. The question is how these assumptions square with the scientific study of mind and brain.

Another, more theological, way of putting the special issues that arise when divine action is considered in relation to the cognitive neurosciences is to say that divine action cannot here be considered in isolation from revelation. Divine action in relation to the natural world may not involve revelation. However, because human beings are capable of receiving revelation, it is to be expected that divine action in relation to people *will* normally involve revelation. The special questions about the scientific credibility of how God *acts* that arise in relation to human beings are therefore the same as those that arise in relation to how God *reveals* Godself to human beings.

It is not being suggested that divine action in relation to individual human beings is absolutely restricted to their conscious awareness of such action. God could presumably act in relation to people who, through incapacity, had no conscious awareness of anything. However, it seems that divine action in relation to people is particularly powerful and effective when they are conscious of that action and acknowledge it. The same point can be expressed differently by saying that God can act in relation to people more effectively when they believe in God, though disbelief does not take them altogether outside the orbit of divine providence.

The central place of revelation in God's action in relation to human beings means that "religious experience" also has an important place. However, the concept of religious experience is a problematic one (as Wesley Wildman and Leslie Brothers also note in this volume). With somewhat different emphases, though with much convergence, a sustained critique of many conventional assumptions about religious experience has been mounted by philosophers of religion such as Steven Katz and Wayne Proudfoot, and theologians such as Nicholas Lash, though with a riposte from Robert Forman.[1] Though it has often been assumed, in a tradition stemming from

---

[1] Steven T. Katz, *Mysticsm and Philosophical Analysis* (London: Sheldon, 1978); Wayne Proudfoot, *Religious Experience* (Berkeley, Calif.: University of California Press, 1986); Nicholas Lash, *Easter in Ordinary* (London: SCM Press, 1988); idem, *The Beginning and End of Religion* (Cambridge: Cambridge University Press, 1996); Robert K.C. Forman, ed.,

William James, that mystical experience is in some sense "pure" experience, these critics have emphasized the extent to which it arises from training, enculturation, and language. Also, though it has also often been claimed that mystical experience is invariant across cultures, the critics have drawn attention to the extent to which mystical experience is described in different ways in different cultures. Nevertheless, the idea of a common-core to such unitive experience across different cultures remains perfectly plausible, and compatible with an outer belt of culturally-variable descriptors. Though it is helpful to draw attention to the cultural background of unitive experience, there is no justification for going so far as to say that unitive experience is "nothing but" the manifestation of culturally-acquired habits of interpretation.

Even more important in the present context, these critics have emphasized that much religious experience is not mystical in character, and does not need to be. Lash, in particular, has rejected the idea that religious experience is experience of a qualitatively distinct kind, or experience of some special domain, and has proposed that it should, rather, be understood as the experience of anything whatsoever from the vantage point of religious belief. On this view, what makes an experience religious is how the experience is interpreted, rather than the special quality of the experience. There is obvious merit in this critical response, but sometimes critics of the concept of religious experience present their points in an over-strident way.

It is perhaps best to see the term "religious experience" as covering a variety of different kinds of experience, with family resemblances among them. That sidesteps the argument about whether religious experience is to be defined in terms of phenomenological quality *or* interpretative framework. I suggest that core examples of religious experience are distinct *both* in their phenomenological quality *and* their interpretative framework. However, there are also borderline cases of religious experience that lack one or other of these. Powerful unitive experiences are not always related to a conventional religious framework of interpretation; the empirical literature makes clear that atheists can have powerful unitive experiences.[2] Equally, many experiences that are interpreted religiously are not of the unitive kind. The key point in the present context is that a broad range of "religious" experiences can potentially be seen in terms of divine action.

2 Persons, Minds, and Brains

There are important issues to be considered regarding the neural basis of religious experience. Many have been tempted by the idea that there is some special affinity between the human mind and the divine mind, or between the human spirit and the Spirit of God.[3] In one sense, this is a correct assumption. I assume that, because people are self-conscious, they are capable of recognizing the presence and activity of God in a way in which other creatures cannot. (It is also true that the imperfections of human nature can mitigate against their actually doing so. The emphasis here is on potentiality rather than actuality.) Our special human capacity for relationships

---

The Problem of Pure Consciousness: Mysticism and Philosophy (New York: Oxford University Press, 1990); idem, "On Capsules and Carts: Mysticism, Language and the Via Negativa," *Journal of Consciousness Studies* 1 (1994): 38–49.

[2] David Hay, *Religious Experience Today* (London: Mowbray, 1990).

[3] Fraser Watts, "Towards a Theology of Consciousness," in *Consciousness and Human Identity*, J. Cornwell, ed. (New York: Oxford University Press, 1998), 178–96.

means that we can "relate" to God in a way that other creatures cannot, though equally we have a special capacity to *mis*understand God.

However, it would be a mistake to think that our special capacity for relating to God is *purely* mental or spiritual. To talk in this way would be to fall into the trap of thinking that the physical, mental, and spiritual can be separated from the physical in human beings, when in fact they can only be distinguished. (As Samuel Taylor Coleridge realized, many conceptual problems arise from treating mere distinctions as divisions.[4]) We relate to God as the composite physical-mental-spiritual creatures that we are. In short, it is as persons that we relate to God, not as mere minds, brains, souls, or spirits.

Just as God's relationship to us should not be seen as purely spiritual, we should not go to the other extreme of suggesting that God somehow "tweaks" our thought processes by controlling what goes on in our brains. This would be repugnant theologically and scientifically. Rather, when God acts in relation to us, or reveals Godself to us, this will be reflected at all levels of our personhood, brain processes, cognitive mechanisms, and phenomenal experience.

The mode of God's revelatory influence on human beings raises thorny issues, which have kept philosophers of mind occupied for at least the last half century. It is important to find a path between two extreme positions. One position to be rejected is the assumption that there is, or can be, some kind of separation of our mental (or spiritual) capacities from our physical nature. The spiritual can be distinguished from the physical, but not divorced from it. Along with this, we should reject the reificationist assumption that we have things called minds or souls that are separate from our bodies.[5] ("Thinghood" belongs to the physical world.)

Another position to be rejected, at the opposite extreme, is the idea that our higher capacities can be explained so fully in terms of our physical processes that it is redundant to talk about our mental capacities, or that in some sense they are not "real." There is as yet no scientific reason for thinking that mental processes and capacities can be completely explained in terms of physical processes. Certainly they are grounded in physical processes, but may not be fully determined by them. Also, even if they were fully determined by them, it would be a big jump to say that talk of mind (or soul) added nothing and was redundant, or that mind and soul were somehow not "real."

In rejecting these two extremes, I am aligning myself with most current philosophical discussion of "supervenience" and "nonreductive physicalism" (see the essays by Nancey Murphy and Philip Clayton in this volume.)

Implicit here is the broader question of the science and theology of the person. I take a broad view of the person, similar to that found, for example, in Rom Harré's trilogy *Physical Being*, *Social Being*, and *Personal Being*,[6] namely, that we are physical/personal/social beings. Despite the strong reductionism that still lurks in some quarters, found in writers such as Francis Crick and Richard Dawkins, the human sciences are taking discernible steps, as in Harré's work, towards seeing the human being in broad terms, including the biological, personal, and social. This kind

---

[4] See Owen Barfield, *What Coleridge Thought* (Middletown, Conn.: Wesleyan University Press, 1971), 18–21.

[5] Watts, "Towards a Theology of Consciousness."

[6] Rom Harré, *Social Being* (Oxford: Blackwell, 1979); idem, *Personal Being* (Oxford: Blackwell, 1983); idem, *Physical Being* (Oxford: Blackwell, 1991).

of view is congenial theologically for integrating the natural and the transcendent, as Karl Rahner, among others, has made clear.[7] We are transcendent creatures, but this transcendence arises out of the natural world that is God's creation; there is no opposition or disjunction between our transcendence and our naturalness. There thus seems to be a promising convergence between theology and the human sciences in taking a broad, integrative view of the person.

There are good theological reasons for including the physical brain in our concept of the person when we think about how God acts in relation to people. The physical brain is part of God's creation. Like everything else in creation, it would be seen, theologically, as existing within the life of God and being dependent on God. When God seeks to reveal Godself to people, it would be bizarre to suppose that God would wish, or need, to bypass this aspect of creation in order to do so. As cognitive neuroscience gains increasing understanding of the cognitive and neural processes subserving religious consciousness, it will probably become clear which brain processes are particularly involved in people coming to think in ways that are in tune with God. However, this need not lead us to ask whether religious experience is caused by the brain *or* by God. The two cannot properly be set up against each other in that way.

Unfortunately, there are various factors that lead people to think in this way. One of the chief difficulties at present is that neuroscience has not yet come to terms with "person-level" realities at all. It tends to operate with a simple dichotomy between brain process and everything else. Sometimes the reality of higher-level phenomena is simply denied. Alternately, even if their reality is accepted, there is little idea about how to integrate brain processes with more personal phenomena. I see this not as a necessary limitation of neuroscience but as a reflection of its present state of development. There are signs that some neuroscientists are attempting to move towards a broader person-level framework.[8]

Another unfortunate habit of thought is the tendency of the modern, scientific mind to ask, "What caused A? Was it X or Y?" This question has its uses in scientific investigation. However, in the human sciences the answer is usually that there are multiple causes, none of which is sufficient. Theories that "go for broke" on a single cause are usually wrong. For example, the question "Is depression caused by physical *or* social factors?" has not proved a fruitful one. The answer is that it is caused by both. There may be methodological value in pushing a single explanation as far as it will go and discovering its limitations; however, the lesson from the history of psychological research is that one generally does not get very far before needing to bring in a complex network of causal factors.

2.1 Complementarity

Complementary descriptions are paralleled by complementary explanations. Like different levels of description, different causal levels can be distinguished but should not be divorced from one another. Mental activities such as reviewing for an examination may provide the best explanation for why the examination is passed, but of course the physical brain is also involved in the process of reviewing. But which

---

[7] Karl Rahner, *Foundations of Christian Faith* (London: Darton, Longman and Todd, 1978).

[8] Leslie Brothers' work is one of the most interesting examples; see her *Friday's Footprint: How Society Shapes the Human Mind* (New York: Oxford University Press, 1997).

level is "in the driving seat"? Where can the most relevant explanation be offered? It is important to allow for compatibility and integration between different levels, but also to be free to prioritize the explanatory level that is most relevant in a particular context, rather than always prioritizing a particular level such as the neurological one.

The history of research in the human sciences ought to make us wary of asking whether an apparent revelation of God really is such *or* whether it has some other, natural explanation in terms of people's thought processes or brain processes. However, there are additional *theological* reasons for not asking whether something comes from God *or* has a natural cause. It would involve the mistake of seeing God as one cause in a series of possible causes, and would "naturalize" and limit God in a way that classical theism has always been careful to avoid. The mistake is to pit God as a cause against natural causes. Of course, it is always right to ask of a possible revelation whether it really is of God, and the tradition has worked out some helpful rules of discernment, looking for example at the effects of experiences, as Murphy has pointed out in relation to Catherine of Siena.[9] Such criteria will always be relevant in discerning whether particular experiences reflect the mind and will of God in a special way.

One interesting case where this issue has been focused is in terms of Ignatius of Loyola's concept of "uncaused consolation." William Meissner has provided an excellent analysis of this concept in relation to the human sciences, noting some of the available theories of religious experience in the human sciences, from the neurological to the psychoanalytic.[10] He concludes this exploration by rejecting the antithesis of natural and divine causes that underlies the concept of uncaused consolation. The antithesis between natural and divine causes, implicit in an uncritical use of Ignatius' concept of "uncaused consolation," has influenced how the church has examined miracles in relation to candidates for sanctification. It seems that to be established as a miracle, natural explanations have to be excluded. I suggest that this is theologically misconceived.

There are various ways in the literature of avoiding this kind of mistake. However, the basic solution to this problem that I favor is "perspectivalism": regarding theological and naturalistic perspectives as different in character but complementary to one another.[11] A theological perspective can illuminate how human processes and events can be seen as being in tune with the will and purpose of God. In contrast, a naturalistic perspective elaborates the cognitive and neurological processes by which this arises. Meissner has argued for this kind of relationship between the theology and psychology of grace; the theology of grace emphasized that it is a promise that comes from God, whereas the psychology of grace is concerned with how this works itself out at the human level.[12]

---

[9] Nancey Murphy, "What Has Theology to Learn from Scientific Methodology?" in *Science and Theology: Questions at the Interface*, M. Rae, H. Regan and J. Stenhouse, eds. (Edinburgh: T & T Clark, 1994), 101–47.

[10] William W. Meissner, *Ignatius of Loyola: The Psychology of a Saint* (New Haven: Yale University Press, 1992).

[11] Fraser Watts, "Science and Theology as Complementary Perspectives," in *Rethinking Theology and Science: Six Models for the Current Dialogue*, N. Gregersen and W. van Huyssteen, eds. (Grand Rapids, Mich.: Eerdmans, 1998), 157–90.

[12] William W. Meissner, *Life and Faith* (Washington, D.C.: Georgetown University Press, 1987).

The complementary perspectives with which we are dealing here are, up to a point, like the complementary relationships between mind and brain perspectives, illustrated, for example, by the relationship between the physical and mental aspects of going to sleep. The physical aspects can be measured in term of the changing electrical rhythms of the brain; the mental changes can be monitored in terms of changes in the content of thought processes and control over them. Both perspectives are needed to give a complete account of the single process of going to sleep—there is no question of eliminating one perspective or the other. It is important to distinguish between different levels of description, but not to divorce them so completely that integration is impossible. For example, different levels of description of going to sleep are not independent. Knowing what is happening at one level enables you to predict with reasonable (but probably not complete) accuracy what is happening at the other level.

Though it is helpful to recognize that complementary discourses arise in the cognitive neurosciences, it needs to be stressed here that the kind of complementarity that exists between mind and brain is not exactly like that which exists between theological and naturalistic approaches. Mind arises from brain, whereas the world is presumed to arise from God. Also, mind and brain discourses are more closely comparable to one another in their purpose and level of detail, whereas theology is a very different kind of discourse, more holistic than any scientific discourse.

Also, though it is sometimes assumed that talk of the complementarity of discourses implies that there is no contact between them, this only follows if both languages are taken in a nonrealist way. However, as do most contributors to the dialogue between theology and science, I espouse a kind of realism about both languages. If this is the case, then the implication that the discourses are separate and unconnected does not follow. Neither does it follow from espousing some kind of realism that talk of complementarity is merely epistemological.[13]

With a complementarity approach, it will be clear that there need be no incompatibility between (1) talking about God's influential presence in our thought processes, and (2) looking at the cognitive and neurological processes by which our thought processes are drawn into resonance with God's will and purpose. This needs to be spelled out because those who advance neurological theories of religious experience often see them as an alternative to traditional religious assumptions about the grounding of such experiences in God.

The mere existence of religious experience does not support belief in God, in the way that some philosophers of religion would like to suppose, because religious experience is too easily open to alternative explanations. But religious experience is, of course, perfectly compatible with the assumption of the existence of a real God. The issue here is similar to that which arises in connection with the explanation of new species in terms of evolution and natural selection. The array of species and their adaptedness does not constitute an argument for the existence of God, because there is an adequate naturalistic explanation that does not invoke God. However, it is perfectly possible to see evolution as reflecting the will and purpose of God; the facts of the relationship between species and their adaptedness to their environments are in no way incompatible with such an assumption.[14]

---

[13] Watts, "Science and Theology as Complementary Perspectives."

[14] See, for example, Robert J. Russell, William R. Stoeger, S.J., and Francisco J. Ayala, eds., *Evolutionary and Molecular Biology: Scientific Perspectives on Divine Action* (Vatican City State: Vatican Observatory; Berkeley: Center for Theology and the Natural Sciences, 1998).

My approach to religious experience is similar to that of Wildman and Brothers in this volume, who also emphasize the importance of holding together different approaches. They emphasize, as I do, that the phenomenal, neurological, and social-psychological approaches sit side-by-side. It is no doubt good practice to consider the relevance of all the various possible nontheological approaches, even though they may differ in importance in any particular case. For example, the social-psychological approach may at times be of great importance, and at other times of hardly any significance. Wildman and Brothers also introduce the theological perspective as yet another approach to religious experience. It is important to emphasize that the theological approach is not just one more approach in the list of approaches, but an approach of a rather different kind that focuses on ultimate meaning and significance.

3 Religion and Epilepsy

At this point, the best way of proceeding will be to review some of the main theories currently being advanced about how brain processes give rise to religious experience. I will have criticisms of these theories to advance, from the perspectives of both science and theology. In due course, I will try to show how a critique of current theories can point to a new generation of such theories that would at least be a step toward more adequate ones.

Probably the most widely canvassed theory is based on the idea that there is a relationship between religious experience and Temporal Lobe Epilepsy (TLE). Work on this link has attracted a good deal of popular interest and cannot be ignored even though, I will argue, a critical examination of the research indicates that the link is much more tenuous than is often assumed.

The claim has been made that people suffering from TLE have more religious experiences and preoccupations than others. This gives rise to the hope that the neural basis of TLE may provide a clue to the neural basis of religious experience. However, there is some doubt about the basic assumption of increased religiosity in TLE. A clear distinction needs to be made between religious experiences *while* people with TLE are undergoing seizures, and experiences and preoccupations *between* seizures.

There is quite often a religious component in seizure experiences, but again caution is needed, because the religious aspects of seizure experiences are in many ways *un*like religious experiences in other contexts. The emotional tone of seizure experiences is often disturbing, being associated with an element of anxiety,[15] whereas it is a fairly standard feature of powerful religious experiences occurring in other contexts that they result in a positive emotional tone, even when people have been severely stressed before the experience.[16] Also, patients undergoing seizure experiences generally know that what they are experiencing is no more real than, say, a dream,[17] whereas people who have powerful religious experience are generally

---

[15] See Peter Fenwick, "The Neuropsychology of Religious Experience," in *Psychiatry and Religion*, D. Bhugra, ed. (London: Routledge, 1996), 167–77.

[16] One good survey of religious experience is that of David Hay, *Exploring Inner Space* (Harmondsworth: Penguin, 1982). For a more general review see Ralph W. Hood, et al., *The Psychology of Religion: An Empirical Approach*, 2nd ed. (New York: Guilford, 1996).

[17] George W. Fenton, "Psychiatric Disorders of Epilepsy: Classification and Phenomenology," in *Epilepsy and Psychiatry*, E. Reynolds and M. Trimble, eds. (New York: Churchill Livingstone, 1981).

convinced of the reality of what they are experiencing. Finally, the experience of the presence of God that many religious people report is notably undramatic, and often does not have the rather weird quality of seizure experiences.

Though it has become the received wisdom—on the basis of accumulated clinical reports—that people with TLE are unusually religious at other times, this is not supported by more careful research. Careful studies by David Tucker and co-authors on a series of seventy-six people with TLE and two comparison groups found no evidence to support the supposed religiosity of people with TLE.[18] Other studies that have controlled for the brain damage, psychiatric illness, etc., have also not found a link between religious experience and TLE.[19]

In the early 1980s, Michael Persinger started the trail off again with studies claiming a correlation between (1) the frequency of epileptic signs and (2) the frequency of paranormal and mystical experiences, and a sense of presence of God.[20] However, the finding of a correlation between his two sets of signs is almost worthless scientifically because the two sets of questions on which it is based (about seizure experiences and religious experiences) overlapped so much that it was almost inevitable that a correlation would be found between the two.

Unfortunately, this rather unpromising line of research on the neural basis of religious experience seems reluctant to die, and it recently re-appeared in some newspapers; one carried the headline, 'God Spot' is Found in the Brain. The report arose from recent research by V.S. Ramachandran, comparing the responses of religious people and people with TLE to words invoking spiritual belief.[21] His only new evidence is that two patients who had both epilepsy *and* religious preoccupations showed strong physiological response to religious words. However, he himself is admirably restrained himself in the conclusions he draws, and remains open-minded about the validity of religious experience. At the present time, there seem to be no compelling scientific reasons for linking the neural basis of religious experience and TLE. A neurological theory of religious experience based entirely on the supposed link with TLE is unlikely to be an adequate theory. However, a multi-component neurological theory in which TLE-like mechanisms are one strand, such as that set out by Wildman and Brothers in section 3 of their essay in this volume, may still have promise.

4 D'Aquili's Theory of the Cognitive Operators Involved in Religion

Though the foregoing ideas about TLE have had the greatest exposure, I believe the approach to the neurological basis of religious experience taken by Eugene d'Aquili and his colleagues is in many ways much more sound. Though I will have criticisms

---

[18] David .M. Tucker, R.A Novelly, and P.J. Walker, "Hyperreligiosity in Temporal Lobe Epilepsy: Redefining the Relationship," *Journal of Nervous and Mental Diseases* 175 (1987): 181–84.

[19] See Fenwick, "The Neuropsychology of Religious Experience."

[20] Michael A. Persinger, *Neuropsychological Bases of God Beliefs* (New York: Praeger, 1987). There is a good, critical review of Persinger's research in Malcolm Jeeves, *Human Nature at the Millennium* (Grand Rapids, Mich.: Baker Books, 1997).

[21] V.S. Ramachandran and Sandra Blakeslee, *Phantoms in the Brain: Human Nature and the Architecture of Mind* (London: Fourth Estate, 1998).

to make of his approach too, I begin by paying tribute to the important ways in which it represents an advance on the idea of a link between religion and epilepsy.[22]

First, d'Aquili has a considered theory, of some sophistication, about the nature of religious experience; he knows what he is trying to explain. Second, his neuroscientific theory is grounded in a comprehensive analysis of the functions of the human brain, in terms of what he calls its "cognitive operators." Third, his theory is, I believe, on the right lines in trying to see how brain processes that subserve more general cognitive functions also have a particular application in the context of religious experience. Among other things, it is much more plausible from an evolutionary point of view, that the parts of the brain that subserve religion should have other functions as well. Fourth, it is attractive that he does not, as Persinger does, see religious consciousness as arising from a malfunctioning of the human brain. It would certainly sit uneasily with a classical theological perspective to suggest that God was dependent on neurological abnormalities in order to reveal Godself.

D'Aquili sees two different cognitive operators as being involved in different facets of religion: the "causal operator" and the "holistic operator." They subserve different aspects of religion. The causal operator is involved in the perception of the world as being controlled by God, and subserves that part of religion that seeks to control the world through relating to God. The holistic operator subserves the altered states of consciousness that are an important part of religious experience, especially the sense of unity that is central to mystical experience. These two will need to be considered separately.

In introducing the terms "causal operator" and "holistic operator" we already encounter one of the problems with d'Aquili's approach, namely, that these are not generally accepted terms in neuroscience. In a sense, that is not as worrying as it sounds. There is no generally accepted, comprehensive theory of the cognitive architecture and its neural substrate. (Schema theory is one contender, represented in this volume by Michael Arbib's work, but it has not won widespread acceptance.) Most researchers in cognitive neuroscience are not even interested in formulating a comprehensive model; their interest is only with the neural basis of highly specific cognitive functions. Also, because of the early stage of scientific work in this field, research thus far has focused chiefly on relatively low-level cognitive functions.

However, when we turn to the basis of religion in cognitive and neural processes, it is clearly not going to work to consider only a highly specific, localized system. The fact that researchers are at an early stage of work on a highly complex set of scientific questions puts a constraint on theological dialogue with this area of science. Religion, whatever else it may be, is clearly a high-level cognitive disposition that involves a broad array of cognitive processes; it seems clear that only a broad and fairly comprehensive theory of human cognitive architecture will suffice to explain the neural basis of religion. Sadly, in contemporary cognitive neuroscience, there is currently no such generally accepted theory. So, there is nothing unusual in those few researchers (like d'Aquili) who grapple with such high-level functions—and thus need a comprehensive theory of the cognitive architecture—having to make up their own theory. The unconstrained nature of such theorizing is unattractive scientifically, but currently there is no alternative.

---

[22] See Eugene G. d'Aquili and A.B. Newberg, "The Neuropsychology of Religion" in *Science Meets Faith*, F. Watts, ed. (London: SPCK, 1998), 73–91. For a fuller account, see idem, *Mystical Mind* (Minneapolis, Minn.: Fortress Press, 1999).

Moreover, I suggest that there is nothing particularly idiosyncratic about d'Aquili's theory of cognitive operators. There are seven of them, the holistic operator, the reductionist operator, the causal operator, the abstractive operator, the binary operator, the formal quantitative operator, and the emotional-value operator. It is a distinctive theory of cognition, but one that I believe is as defensible as any other at the present early stage of modeling the cognitive system as a whole.

Of the two parts of d'Aquili's theory relevant to religion, his treatment of the holistic operator is scientifically more sound. One of its more convincing aspects is the way it handles the ceremonial rituals that are an important part of religious practice in many different cultures. D'Aquili draws attention to the way in which rhythmic activity of some kind, be it visual, auditory, tactile, or proprioceptive, plays a key role in inducing a mystical sense of unity. He suggests that this leads to a high degree of activation of the sympathetic nervous system, with some spill-over into the parasympathetic system, leading to activation of the parietal-occipital region of the nondominant hemisphere; this is where d'Aquili locates the holistic operator. An alternative route by which the holistic operator can be activated is through the use of meditation. Here it is the parasympathetic system that becomes highly activated, with spill-over into the sympathetic system.

The holistic operator is only half of d'Aquili's theory. There is also the role of the causal operator in subserving the perception of the world as being controlled by God, which, he suggests, gives the religious practitioner a sense of control over the world. Whereas the holistic operator is located on the nondominant side of the brain, the causal operator is located on the dominant side, specifically in the anterior convexity of the frontal lobe, the inferior parietal lobule, and their interconnections. However, this half of d'Aquili's theory seems to me less well developed and less compelling than his theory of the role of the holistic operator in mystical experience.

Also, I have serious doubts about the plausibility of a single "causal operator." The perception and ascription of causality is extremely varied and heterogeneous in the level at which it proceeds. While I am willing to accept that the part of the brain labeled by d'Aquili as the causal operator is involved in the ascription of causality, it seems much more doubtful whether a single system is responsible for *all* ascriptions of causality. Someone ascribing a "cause" to a ball seen coming over the garden fence is doing something very different from parents worrying over where they went wrong in bringing up their children, though both are in different ways seeking to ascribe causality. Religious ascriptions of causality are more like the latter, yet I suspect that the more complex the causal ascriptions, the less satisfactorily they can be localized in the causal operator.

Perhaps the chief concern about d'Aquili's theory is the range of "religious experience" to which it is applicable. One problem is that religious experience is so diverse that it may indeed be inappropriate to look for a single neural theory of such diverse phenomena. The forms of unitive experience with which d'Aquili's theory is concerned are those deliberately induced by ceremonial ritual or meditation; it does not deal so explicitly with spontaneously occurring mystical experiences, though it has become clear from surveys that such experiences occur to a substantial proportion of the population.

However, an even more serious limitation is that mystical experience may not be central to religious belief or practice. It is clear that there is a great difference between unitive experience and what we might call more generally "the experience of God." Within d'Aquili's theory, one would look to the part of it concerned with the causal operator for help here, but there must be doubts about how much religious interpretations of the world are about ascribing causality, rather than ascribing religious significance to events in a more general way.

This is a point at which the umbrella term "religious experience" can mislead, because it includes such disparate cognitive and social phenomena. The ascription of causality to God, and the accompanying sense of control over the world, is perhaps most clearly a feature of the sort of primitive religions studied by social anthropologists. However, it is not clear that this is really what is going on in contemporary Christianity. Belief in divine providence may look and sound like a belief in God's control of the world, and intercessory prayer may look like an attempt to influence this "controller," but there are many features of how these "language games" are played that suggest that they are not quite what they seem. Most theologies of prayer generally do not see prayer as an exercise in controlling the world by influencing God.[23]

It should be noted that I have presented in very condensed form the detailed theory that d'Aquili and his colleagues have developed over twenty years. At present, it derives much of its scientific credibility from making sense of a wide range of circumstantial considerations surrounding the occurrence of mystical experience. More recently d'Aquili has undertaken more direct experimental investigation of the theory as it relates to religious practices. His theory seems to give a plausible account of the neural basis of mystical experience, and is probably the best such theory we have at present. However, these are early days in the investigation of a difficult and complex question, and no one would suggest that we already have the definitive scientific account.

5 Towards a Multi-Level Cognitive Theory

I now want to suggest how theories of the cognitive processes subserving religious consciousness might evolve. In particular, I suggest that there are some useful lessons to be learned from theories of the cognitive processes involved in emotion. Emotion is a "high-level" phenomenon of comparable cognitive breadth and generality to religion. Indeed, I have argued elsewhere that there is a useful scientific analogy to be drawn between religious and emotional experience.[24] By "cognitive" I mean here primarily the structures and processes which shape attention and memory, by which our responses to experience are mediated, and from which our understanding of the world arises. Cognition in this sense is not necessarily conscious, though some cognitive processing may be done consciously.

The first generation of cognitive theories of emotion were single-level theories.[25] For example, Gordon Bower developed a theory of the semantic networks subserving emotion. The network idea is essentially that each concept is a node in a semantic space, with links to semantically-related nodes. Bower suggested that when a particular emotion (for example, anxiety) is activated, the node concerned leads to a degree of activation of associated nodes. Tim Beck developed a similar but looser theory of emotional "schemata." A schema is a kind of template through

---

[23] Vincent Brümmer, *What Are We Doing When We Pray?* (London: SCM, 1984).

[24] Fraser Watts "Psychological and Religious Perspectives on Emotion," *Zygon* 32 (1997): 243–60.

[25] See J. Mark G. Williams, et al., *Cognitive Psychology and Emotional Disorders*, 2nd ed. (Chichester: John Wiley, 1997).

which experience is filtered and which may distort it, and from which memories and automatic thoughts may arise.

However, it became clear that such single-level theories were inadequate. For example, there was always some confusion over whether the emotion nodes in Bower's network theory were concerned with emotion concepts or with emotions themselves. A multi-level theory of emotion seemed necessary to handle this distinction. Also, single-level theories were quite unable to handle the clash between different levels, which is a common feature of emotions—for example, *knowing* that there is nothing to be frightened of, but nevertheless *feeling* frightened. It thus became clear that only multi-level theories would prove adequate, and I suspect that the same will prove to be true of cognitive science theories of religious consciousness.

There are now several multi-level cognitive science theories of emotion in the field,[26] and a consensus has not yet been reached. However, there is emerging agreement about the importance of the different levels of the cognitive system concerned with aspects of cognition relevant to emotion: I suggest that the distinction between them is also relevant to the scientific analysis of religious consciousness. In what I think was the first multi-level cognitive theory of emotion, Howard Leventhal distinguished three basic levels: sensory-motor, conceptual, and schematic.[27] Sensory-motor aspects of emotion are reflected, for example, in the way in which people with specific animal phobias develop remarkable perceptual tuning for the relevant stimulus and show a reflex-like response. We are coming to have a good grasp of the neuropsychological processes underlying this lowest cognitive level of emotion, and Joseph LeDoux's excellent work (summarized in this volume) represents a major breakthrough.

The distinction between Leventhal's other two levels (conceptual and schematic) has now been made more clearly in terms of John Teasdale and Philip Barnard's model of the cognitive architecture, "Interacting Cognitive Subsystems" (ICS).[28] This model is in turn a development of an earlier version that Barnard proposed to handle psycholinguistic data. It is one of the attractive features of the model that it has spawned a very diverse range of applications. ICS is a general model of the cognitive architecture, but it is not concerned with the neural substrate; I will not try to suggest here what that might be. However, it is often a good scientific strategy to get a fairly clear idea of the requirements of a cognitive model at a functional level before going into brain mechanisms.

ICS proposes a distinction between two different systems in the central engine of cognition, called the propositional (PROP) and implicational (IMP) systems. The IMP system is concerned with abstracting "meanings" or regularities from various lower-level subsystems, and it is through the meanings coded in the IMP system that emotion arises. Interestingly, these meanings are not coded in conceptual or propositional form, and so cannot be articulated directly. For this, they have to be translated into the different code of the PROP system. The IMP system in humans is presumed to be a development of a similar but less developed subsystem in other

[26] See Williams, et al., *Cognitive Psychology and Emotional Disorders*, chap. 11.

[27] Howard Leventhal, "A Perceptual-Motor Theory of Emotion," in *Advances in Experimental Social Psychology*, vol. 17, L. Berkowitz, ed. (New York: Academic Press, 1984), 118–73.

[28] John D. Teasdale and Philip J. Barnard, *Affect, Cognition and Change* (Hillsdale, N.J.: Lawrence Erlbaum, 1993).

species. However, the PROP system seems to be a more distinctively human one, associated with the special human capacity for language. There are many obvious benefits in this human capacity to translate meanings into a form in which they can be articulated, but there are difficulties too. For example, worry and insomnia seem to be distinctive human problems, reflecting our human PROP system.

Assuming the general validity of this distinction, I would want to argue for the importance in religious cognitive processing of the IMP system, a subsystem concerned with meanings, albeit at a nonpropositional level. Indeed, the discernment of such meanings seems to be at the heart of religion. However, this is probably not unique to religion. Another context in which similar cognitive processes seem to operate is the development of insights about oneself in psychotherapy. It is not unusual for important but difficult insights to be glimpsed initially at the level of inarticulate meanings before being re-coded in a propositional form capable of articulation.

The suggestion that religious consciousness comes into being initially at this level is consistent with much traditional theology, especially with the apophatic tradition of the unknowability of the transcendent God. However, in the light of what I am proposing here, the apophatic approach to theology takes on a new significance as reflecting not only the transcendence of God but also how the cognitive system works in arriving at an understanding of God. When the mystics have spoken of the ineffability of God, perhaps they have been talking not only about God's nature, but also at a human level about how there is a sense of loss as the felt meanings of the IMP system are re-coded in a form in which they can be articulated to oneself—and even more so to others. The IMP system is, in this sense, always somewhat "transcendent."

Of course, words and phrases may play an important role in revelatory experience; my point about the intrinsic ineffability of religious consciousness in no way denies this. Even when a phrase from the scriptures strikes home with revelatory power, there is often a further task of discerning exactly what its significance is for the person concerned in a particular situation. This leads to a kind of "cross-talk" between verbal scriptural material encoded in the PROP system and the more intuitive grasp of its significance that emerges in the inarticulate code of the IMP system. Far from calling the distinction between the two systems into question, the role of particular words and phrases in revelatory experiences helps to elucidate the contribution of the two systems and the inter-relationship between them.

The IMP system is essentially concerned with meanings and interpretations; it follows from this proposal about the central role of the IMP system in religious consciousness that cognitive processes of interpretation are also of central importance to religious consciousness. This fact relates to the debate referred to earlier about whether religious experience is to be defined in terms of qualitatively distinct experience or in terms of the religious style of interpretation of any experience. My own view is that this is a false dichotomy. While accepting the crucial role of interpretation—one that I would locate in the operation of the IMP system—I do not see this as ruling out the possibility of religious experience sometimes having a qualitatively distinct character. Above all, there is no need to make the reductionist move of seeing religious experience as being "nothing but" a particular mode of interpretation.

The distinction between the IMP system and the PROP system may be helpful in formulating discrepancies between different aspects of religious life, just as it is helpful in formulating discrepancies between different aspects of emotion. It seems to be quite common for people who reject religion at an intellectual (propositional) level to be drawn to it at a more intuitive (implicational) level. For example, people who do not "believe in God" may find themselves praying at times of severe stress. It is perhaps one of the current features of much of Christianity in the developed world that it manifests discrepancies between cognitive levels, with religious beliefs making sense intuitively but not intellectually.

These remarks on the role of the PROP and IMP systems in religious consciousness are rather preliminary, and clearly do not amount to a complete account of how the cognitive system is involved in religious awareness. Nevertheless, I hope that they point towards a new kind of theory that is hierarchically organized, and recognizes the to-and-fro between different parts of the overall cognitive system. Let me emphasize again that I see this kind of theory as being neutral as far as the reality of religious belief is concerned. On the one hand, a theory of the basis of religious consciousness in the meaning-generating processes of the IMP system could be coupled with a nonrealist view of God; religion could be seen as nothing more than a spin-off of the way that our IMP system functions. On the other hand, it can be taken, in conjunction with a realist view of God, as indicating how we are receptive to revelation of the nature and purpose of God and the significance of God for our lives.

6 God's Influential Presence in the World

I now want to consider how the kind of cognitive processes which I have suggested underpin the experience of God may also be involved in God's "action" in relation to people. I maintained at the beginning of this essay that it is to be expected that God's action in relation to people would be mediated through their experience of God.

First, let me set out some of the basic presuppositions that I believe should guide our thinking about divine action, especially about divine action in relation to human beings. Like all talk about God, talk of divine action is analogical, and the background metaphor is that human action provides a way of understanding God's influence within the world. When the analogy between divine action and human action is pressed too far it becomes misleading. This is another part of our inheritance from Enlightenment thinking, in which the analogy between God's relation to the world and the soul's relation to the body was drawn more tightly than before.[29]

Though every theologian would maintain the world's dependence on God, it remains the case that much theological talk about the "world" seems implicitly to assume a degree of autonomy for it that is foreign to what is assumed in the Bible (especially the Old Testament), or indeed by the early Fathers, such as Origen. There have been at least two particular points at which Christian theology has moved toward seeing the world as increasingly separate from God. One was in the thought of Augustine, driven largely by the problems of theodicy; the problem of evil in the world seemed less acute if the world had in some sense fallen away from God.[30] The other was during the seventeenth century, when the growing tendency to see the world in mechanical terms again emphasized its separateness from God. Of course,

---

[29] Edward Craig, *The Mind of God and the Works of Man* (Oxford: Clarendon Press, 1987).

[30] Crawford Knox, *Changing Christian Paradigms* (Leiden: E. J. Brill, 1993).

early-modern thinkers assumed that the merely mechanical nature of the universe showed its dependence on God all the more clearly. However, the fact remains that mechanical science increasingly led to the world being seen as separate from God.[31]

Once we stand back from religious thinking stemming from the Enlightenment period, we can readily see that it would be a theologically preposterous piece of idolatry to suggest that what we call "the world" could exist independently of the "maker of heaven and earth." There is no pantheism lurking here. I am not saying that God cannot be separated from the world, only that the world cannot be separated from God. I share the conventional theological assumption that the world is part of the life of God, but God is not just the life of the world. To say that the world cannot be *separated* from God is not to imply that it cannot be *distinguished* from God (recall Coleridge's point that distinctions should not be turned into divisions). Expressed in these terms, the classic theological assumption is that the world can be distinguished from God, but cannot be divorced from God (that is, it would have no existence apart from God.)

Those who have grappled with the problem of divine action in the modern dialogue between theology and science have generally seen the dangers of assuming that the world can be separated from God. However, the term divine "action" may in fact be a subtly misleading way of talking about God's influential presence in the world, and especially so when we consider the influence of God on people. One of the key problems is that it does not adequately suggest the *inter*active nature of God's involvement with people. To talk of divine "action" conjures up the notion of God acting unilaterally and independently. In contrast, much of Christian theology has seen God's influence on people as being interactive, or, in other words, as being dependent on God's initiative but also on people's response. The free will of human beings, which has long been a core assumption of Christian theology, means that God should be seen as leading and influencing people, but not as acting in a way that controls them.

In addition, the episodic nature of human action can mislead us into thinking that God's influence in the world is similarly episodic. In some ways divine "activity" might be preferable to divine "action" because it suggests a less discreet, episodic mode of God's influential presence in the world. Clearly, the metaphor of divine "action" has its uses, and I would not want to abandon it. However, alternative metaphors would allow us to become less dependent on that of "action." Given that every metaphor for God's involvement in the world is bound to be misleading in some way or other, we can reduce the extent to which we are misled by using more than one metaphor, thereby weakening the grip of any particular metaphor on our theological thought processes.

7 Resonance and God's Providence

One of the background issues to this discussion concerns the relationship between general and special providence. A central theological task is to find an appropriate way of balancing God's general upholding of the world with the particular events that we wish to attribute in a special way to God's influential presence. There are strategic choices here. There is no great difficulty in squaring at least a weak view of general providence that emphasizes the orderliness of the world with modern science. However, this leaves a great deal of work to be done in giving a scientifi-

---

[31] Mary Midgley, *Science as Salvation* (London: Routledge, 1992).

cally credible account of special providence. The general approach I take to this problem is to try to develop a strong view of general providence, and then to see special providence as arising as far as possible (albeit not entirely) out of general providence.

For this proposal to have any kind of plausibility, a richer account of general providence will be needed than is sometimes given. It would need to be one that allows more scope for the purposes of God than, for example, just seeing God's faithfulness reflected in the orderliness of nature. However, if we are prepared to postulate general propensities in nature that reflect God's creative, redeeming purposes, we can reasonably see many specific occasions of providence as significant manifestations of those general propensities.

Of course, there are problems in reconciling such an expanded view of general providence with modern science. However, there are good reasons—both theological and scientific—for challenging a view of the world as merely orderly, but purposeless. The fine-tuning of the universe and the underlying propensities of evolutionary biology would be key places to look for scientific support for a strong, rich view of general providence. My desire for a stronger but scientifically credible approach to general providence is consistent with my emphasis on the importance of retreating from an unhelpful implicit tendency to think of the world as separate from God.

Nevertheless, the problem of special providence cannot be sidelined in this way entirely. A conventional theology requires some scope for God to act in relation to specific events in the world. However, I suggest that a rich view of general providence makes it unnecessary to offer any distinct account of nonhuman special providence. It seems reasonable to suggest that all specific instances of providence that really *must* be handled as such arise in connection with human beings, rather than with the rest of the created world.[32] Seeing the task of providing an account of special providence in this more specific way changes the scientific and theological nature of the task.

A more explicitly trinitarian way of highlighting the distinctive issues that arise when considering divine action in relation to human beings is to say that we are thinking primarily of God the Spirit. In contrast, when divine action is considered in relation to the physical sciences or to evolution, we are focusing primarily on the activity of God the Father. We are led to look for the "footprints" of the Father-Creator in the created order. Enlightenment "natural theologians" were even tempted to say that one could observe evidence of God. It is now widely accepted that talk of "observation" and "evidence" in this context is not quite appropriate, because it discounts the crucial contribution of faith, reverence, and prayerful reflection in discerning the footprints of God in the world.[33] However, it is interesting that the

---

[32] A similar view has been taken by George F.R. Ellis in an essay in an earlier volume of this series, "Ordinary and Extraordinary Divine Action: The Nexus of Interaction," in *Chaos and Complexity: Scientific Perspectives on Divine Action*, R.J. Russell, N. Murphy and A.R. Peacocke, eds. (Vatican City State: Vatican Observatory; Berkeley, Calif.: Center for Theology and the Natural Sciences, 1995), 359–95.

[33] Nicholas Lash "Observation, Revelation and the Posterity of Noah" in *Physics, Philosophy and Theology: A Common Quest for Understanding*, R.J. Russell, William R. Stoeger, S.J., and George V. Coyne, S.J., eds. (Vatican City State: Vatican Observatory, 1988), 203–15.

work of the Father-Creator comes close enough to being observable for this mistake to be credible.

The work of the Spirit, in contrast, is known chiefly through human experience, both individual and collective. Indeed, the Spirit becomes known to us chiefly through our co-operation with the Spirit, and attempts to "observe" or "demonstrate" the work of the Spirit are even more misplaced here than they are with the Father-Creator. The Spirit is known primarily through obedient action. Acquaintance with the Spirit is primarily through what Francisco Varela has called "enactive" knowing.[34] The cognitive neurosciences are especially relevant to understanding how this enactive knowledge of the Spirit might proceed.

Because God's action in relation to human beings is especially interactive, it is with human beings that the metaphor of divine action can be especially misleading. I do not have a well worked-out proposal for an additional metaphor to set along side it, though I suggest that we might explore the contribution that ideas of "resonance" or "tuning" could make. What I have in mind is that people can be more or less "attuned" to God, rather as a receiver can be attuned to a transmitter, or in resonance with it. Of course, this does not exclude the notion of "action"; in physical resonance there is still a specific "input" from "outside." However, the metaphors of "resonance" or "tuning" seem to point us in helpful directions.

Above all, they suggest the interactive nature of God's action, something which is especially important in considering how the Spirit acts in relation to people. For the Spirit to act, there has to be receptivity. In the Johannine metaphor, the Spirit comes to "dwell" with believers; and this dwelling requires receptiveness. The metaphors of resonance and tuning also suggest action of a gentle and nondisruptive kind, which I suggest is how God normally acts in relation to people, allowing them the freedom to respond, or to go their own way. The notion of resonance is essentially an interactive notion, which is what is required when we consider God's action in relation to people.

Further, there is an implication of constancy in the way God seeks to bring us into attunement with Godself. The Christian tradition has generally emphasized that God does not seek to do this only at certain times, but constantly seeks to do so, even where there is no receptivity on the part of the people to whom he reaches out. This implies a helpful link between general providence and special providence that is to be welcomed, because the notion of special providence becomes particularly problematic when it is completely divorced from general providence. The implication is that the world, reflecting as it does God's creative purpose, is one in which our thoughts and actions are constantly being drawn into resonance or attunement with God's nature and purpose.

A final implication of the notion of resonance is that, when we become attuned to God, possibilities are opened up which might not otherwise arise. Resonance carries with it the possibility of potentiation when resonance is achieved. Thus there is a suggestion of empowerment when people act in resonance with God. In this way, the resonance metaphor is fully consistent with "mighty acts of God" occurring, once resonance is established.

---

[34] Francisco J. Varela, Evan Thompson, and Eleanor Rosch, *The Embodied Mind: Cognitive Science and Human Experience* (Cambridge: MIT Press, 1991). Varela and colleagues define the enactive approach as consisting essentially of two points: "[1] perception consists in perceptually guided action and [2] cognitive structures emerge from the recurrent sensorimotor patterns that enable action to be perceptually guided" (173).

It will be helpful to remember here the point made earlier in the essay about different domains of description and explanation. There are two pitfalls to be avoided. One is to see divine action simply as a particular way of describing things, but one that has no real explanatory power and reflects no real divine influence. The other is to see God's influence as real, but operating in a way that is totally divorced from all other causal influences in the world. The resonance metaphor is intended to point us toward a way of conceptualizing God's influential presence in the world in a way that avoids both pitfalls. It allows for the real and effective influential presence of God in the world, but it is intended to preserve an interactive compatibility between divine action and other levels of explanation.

The metaphor of resonance is, of course, fairly close to talk about the input of "information" from God (discussed by Arthur Peacocke in this volume). The concept of information has the possible advantage of not implying physically detectable energy. However, in the form in which it has been proposed so far, it seems to be a less interactive notion, and to lack a theologically helpful emphasis on human receptivity. However, there may not ultimately be much at stake here. If the concept of information-input was deemed preferable to that of resonance, it would be important to develop it in a way that was more explicitly interactive. On the other hand, the notion of resonance might be developed in a way that did not explicitly assume physically detectable energy, if that was thought desirable.

I put forward the concepts of "resonance" or "tuning" as metaphors; it remains to be seen how successful they will be as such, and whether they eventually become more than metaphors in guiding us to a scientifically congenial understanding of God's activity in relation to people. New metaphors take time to prove their worth (or to fail to do so). Too much should not be expected of them at the outset. I am more committed to the desiderata for a metaphor for God's activity in relation to people set out in the previous paragraphs, above all to the theological importance of conceiving this in interactive terms, than I am to this particular metaphor. Alternative metaphors might be suggested that meet the desiderata better. The key theological point is that the activity of the Spirit in relation to people should be conceptualized in a way that makes explicit the role of human receptivity in relation to the Spirit's activity.

8 Moral and Religious Intuitions

Finally, I want to suggest that conscience is a fruitful topic around which to focus discussion of the activity of the Spirit of God in relation to people. The cognitive processes underlying conscience can perhaps be understood in a hierarchical way, rather like that which has been developed for emotion, and which I have applied above to religious experience. At the lowest level, some reactions to transgression (or possible transgressions) seem to involve an almost reflex-like internalization of prohibitions. There is also a more propositional knowledge of what is supposed to be right and wrong, and a capacity for a discursive exploration of these notions. Finally, at the deepest level, there are moral intuitions arising initially at the inarticulate level of the implicational system, and which are probably particularly powerful in guiding conduct because people feel an instinctive obligation to follow that guidance.

Christian philosophy has often taken a rather naturalistic view of how conscience arises, emphasizing the importance of having a well-educated conscience. Through steeping ourselves in the guidelines of a moral community, such as the church, we develop reliable intuitions about how we should behave in particular circumstances. This emphasis on educating one's conscience can be found in many classic writings on the subject, such as Aquinas and Samuel Butler. It is also consistent with an important strand of contemporary moral philosophy, such as that associated with Alasdair MacIntyre, which emphasizes the role of the moral community.

There has been a parallel tradition which has seen conscience as the inner voice of God. There was a particular flowering of this tradition in some of the moralists of the eighteenth century. The interesting question is how we should understand the relationship between these two traditions about conscience. Some, such as Kenneth Kirk, would deny that conscience is a special moral faculty, or in any special sense the voice of God, and would see it simply as ourselves making moral decisions as best we can.[35] Of course, on this view, conscience may still be in accordance with the will of God, but insofar as that is the case, it would be because people, through the education of their consciences, have come to hold intuitions that reflect the will of God. On this view, it would not be necessary to assume that God had any *direct* influence on people's consciences. Others, while admitting the helpful role of the moral community and the education of conscience, might wish to allow for the active and direct work of the Spirit in shaping our intuitions about how to behave in particular circumstances.

This raises, in different terms, the issue addressed by Peacocke in the final section of his essay in this volume, where he considers the relationship between mediated and unmediated religious experiences. He is no doubt right to emphasize that there are no experiences that are wholly *un*mediated. All religious experiences, whatever their felt revelatory power, arise in the context of a particular cultural and social background, a particular personal developmental history and the memories arising from it, and a particular set of conceptualizations about the nature and will of God. William James's notion of "direct" religious experience, if understood as experience that bypasses all these contextual features, is simply not viable.[36] Also, as emphasized earlier in this essay, the physical brains of people are involved in their religious experience; there is no way in which "unmediated" religious experience could somehow bypass the physical brain and so stand outside the natural world.

However, this is not the end of the matter. Granted that all religious experiences and all intuitions of conscience are contextualized in these ways, the question still remains as to whether there might not also be religious experiences and moral intuitions that arise in part through individual communion with God. As Peacocke emphasizes, thoughts and intuitions can arise entirely through the ordinary processes of social influence, reflection on memories, etc., which are in accordance with the mind and purpose of God, rather in the way that people can discern, through a well-educated conscience, how to act in accordance with the will of God. However, this does not necessarily capture all that the Christian tradition would wish to say about God's influence on people. It is an important part of the tradition that the Spirit can "dwell" within people as a source of revelation and guidance. It is not clear that this influential presence of the Spirit is being adequately formulated if it is reduced entirely to externally-mediated influences that happen to bring people into attunement with God. Also, it is not clear, if God's influence in the world is allowed at all, that there is any particular problem with suggesting that there are ways in

---

[35] Kenneth Kirk, *Conscience and Its Problems* (London: Longmans Green, 1927).

[36] Lash, *Easter in Ordinary*.

which people's thoughts and cognitive processes can be drawn into attunement with those of God.

It is another interesting aspect of conscience that it relates to both thoughts and actions—the two are intertwined. Discerning what is right to do is a kind of enactive knowing. The process should not be seen as two sharply distinct steps occurring in sequence, first discerning the will of God and then doing it. Often the discernment is enhanced or revised in the process of beginning to act.

Here again the metaphors of "resonance" and "tuning in" to the Spirit of God may be helpful. They seem to place us in the right kind of position midway between extremes of acceptance and denial of the direct influence of the Spirit on people's thoughts and intuitions. "Resonance" does not suggest a "controlling" influence on human thought processes, because of its interactive emphasis and because it can accommodate the fact that all our thoughts inevitably arise out of our social and personal background. However, it also allows for the possibility of a facilitation or enhancement of thoughts and intuitions which are in accordance with the activity of the Spirit. As always, general providence is primary; any notion of special providence would be secondary. To be more specific, the first requirement is to have a scientifically-credible view of God's general guiding providence in relation to people's thought processes. What is being suggested is not the insertion by God of particular thoughts in people's minds, but a more general way in which people can allow themselves to be drawn into a way of thinking that is in accordance with the mind of God.

My guess is that, within a generation, we will have a fairly precise understanding in cognitive science of what Teasdale and Barnard call the implicational system, and its neural substrate. We will perhaps also understand the particular ways in which this implicational system can operate in accordance with the mind of God. That would be a major step forward in the scientific understanding of religious consciousness. It is much more difficult to see how, within foreseeable scientific development, we might begin to understand how there could be some kind of tuning of that system to the influence of the God. Nevertheless, we should not necessarily assume that the limited scientific understanding that we currently have, after only about 350 years of modern scientific research, will necessary include all the elements needed to conceptualize the activity of God. It remains a task for the future, in the light of developing scientific understanding, to formulate this activity more adequately. For now, it may be possible in the dialogue between theology and science to do no more than indicate where it might be fruitful to look.

A NEUROPSYCHOLOGICAL-SEMIOTIC MODEL
OF RELIGIOUS EXPERIENCES

Wesley J. Wildman & Leslie A. Brothers

1 Introduction

1.1 Goal

The goal of this essay is to present a richly textured interpretation of a large tract of the territory of religious experiences that we shall call *experiences of ultimacy*, a name that will be explained below. We develop this interpretation in two phases. First, we describe these religious experiences as objectively as possible, combining the descriptive precision of phenomenology informed by the neurosciences with a number of more obviously perspectival insights from psychology, sociology, theology, and ethics. Our hope is that the resulting taxonomy is compelling enough to suggest criteria for the plausibility of constructive efforts in theology and philosophy that depend upon an interpretation of religious experiences, including those in this book that attempt to speak of divine action in relation to human consciousness.

Second, we make two constructive ventures on the basis of this description. In the first, inspired by existing social processes used to identify authentic religious experiences, we describe a procedure whereby genuine experiences of ultimacy can be distinguished from mere claims to such experiences. This brings such experiences into the domain of public, scientific discussion as much as they can be, which is a great advantage from the point of view of encouraging more mainstream discussion of them by scientists and other intellectuals. The other constructive venture is a theory about the causation of ultimacy experiences. This is our attempt to evaluate claims made concerning the ultimate cause and value of experiences of ultimacy. The modeling procedure we adopt makes use of semiotic theory to plot not causal interactions themselves but rather their traces in the form of sign transformations— all terms that will be explained in detail later. In the language of semiotic theory, these causal traces take the form of richly intense sign transformations. This proposal keeps ontological presuppositions to a minimum by focusing on causal traces rather than on the nature of the cause itself. Nevertheless, it does offer a religiously or spiritually positive way of interpreting authentic ultimacy experiences, and at the end we offer a suggestion about the nature of the ultimate reality that might leave such causal traces.

1.2 Motivation

The motivation for the task we undertake here is primarily the intuition that religious experiences are important elements of human life, worthy of respectful and energetic interdisciplinary study. A word of explanation is required, however, because this intuition may seem obscure or trivial, depending on one's point of view. On the one hand, when religion is understood in the tradition of Emile Durkheim as the expression and codification of the most important cosmological and ethical commitments of a group,[1] individual experiences may seem irrelevant to the account of religion proffered. Yet appearances in this case are misleading: as Durkheim himself understood, without personal religious experience in some form, whether aberrant or not, the cohesiveness of religious groups and the motivation for underlying cosmological and ethical commitments remain unintelligible features of human life. On the other hand, a religious interpretation of human life in terms of categories such as sin and salvation, suffering and liberation, will be so apt to emphasize individuals that religious experiences will seem inevitably preeminent. The danger here, however, is that the complexity and diversity of religious experience and practice will be reduced to fit what a particular religion's belief structure can comprehend. Juxtaposing these two points of view leads to the conclusion that religious experiences are important in any analysis of human life and that many different points of view need to be integrated in order to achieve a properly balanced theory.

This assessment of the general importance of the study of religious experiences needs to be related to several other motivating factors. First, the increasing obscurity of scriptural, ritual, and theological language about divine action in recent centuries has drawn attention to the individual person as a possible locus for the action of God or gods. In fact, to the extent that divine action in the natural order has been eclipsed by scientific accounts of nature, divine action directly in relation to human consciousness can have the significance of a last resort for making sense of such language. This adds a sense of urgency to the investigation of religious experiences, especially among those who have had them.

Second, the neurosciences have largely succeeded through their analyses of brain structure and function in portraying that which is distinctively human as continuous with regularities and forms of complexity observed throughout nature. This generally accepted conclusion about human beings reconfigures the whole question of religious experiences by proposing explanations for them that are independent of the assumption that they are experiences of anything properly called a religious object.[2] The rise of the neurosciences does not make this reductionistic challenge philosophically different in kind than it was previously, but it does demand that theories of religious experiences should attend to the neurosciences.

Third, although neuroscientific accounts have focused on isolated brains, there is growing interest in the social capacities of the human brain. This research area suggests an approach to theorizing about religious experiences that exploits fruitful links between isolated-brain neuroscience and the various forms of communal wisdom that traditionally have been vital to the understanding of religious experiences.

1.3 Limitations

So much for motivation. Our goal must also be qualified by several practical considerations. First, research into the nature of religious experiences is still in its

---

[1] See Emile Durkheim, *The Elementary Forms of the Religious Life*, tr. from the French by Joseph Ward Swain (New York: The Free Press; London: Collier Macmillan Publishers, 1915).

[2] For an early and notable example of such a theory, see Julian Jaynes, *The Origin of Consciousness in the Breakdown of the Bicameral Mind* (Boston: Houghton Mifflin Company, 1976).

infancy in most respects. In particular, neuroscience—including cognitive neuroscience, the subdiscipline most pertinent to our current project—lacks a central theory capable of organizing the fragments of knowledge that we have. The field at present is a collection of part-concepts whose composition shifts with each new wave of experiments and interpretations.[3] Detailed neurobiological accounts are therefore premature: we can only make tentative suggestions, and nothing will get done without a sense of adventure. The same must be said of the phenomenology of religious experience. Disciplined and properly informed cross-cultural comparison has barely begun, and the means to determine agreement and disagreement between culturally bound descriptions of religious experiences remain obscure.[4]

Second, a cornerstone of our position is its neutrality. In the descriptive phase of the essay, we assume neither the reality nor the nonreality of that which is taken to be the object of an experience of ultimacy, and we take for granted neither the efficacy of belief in that object nor even the coherence of the idea. Subsequently, in the constructive phase of the essay, while we shall assume that ultimate reality leaves causal traces of a particular kind, we assume nothing about the nature of this ultimate reality; it could be anything from ontological emptiness to a supernatural God, from the self-grounding mystery of Godless nature to the wondrous divinity beyond being and not-being of the great mystics. We shall explain how this neutrality is possible below but state the two associated limitations here. First, there are some theological and existential-philosophical perspectives from which this posture of maximizing neutrality necessarily dooms our project because, it is held, ultimacy can only be discussed fairly if its reality and efficacy are fully accepted. We take this dictum seriously because it is the view of so many theologically serious viewpoints in the world's religions. We think, however, that it can only be evaluated empirically on the basis of the success or failure of projects that set it aside, as ours does. Second, our attempt to be as ontologically neutral as possible in the constructive phase of the essay avoids a self-defeating reductionism by making use of a philosophical framework drawn from semiotic theory (the theory of signs).[5] Some philosophical complexity is the inevitable result, but we try as much as possible to deal with the philosophical details in footnotes and only introduce them as they are needed toward the end of the essay.

---

[3] A similar analysis holds good, we think, for cognitive science. This is as true now—see Fraser Watts's essay in this volume—as it was a quarter of a century ago; see Allen Newell, "You Can't Play 20 Questions with Nature and Win: Projective Comments on the Papers of This Symposium," in *Visual Information Processing: Proceedings of the Eighth Annual Carnegie Symposium on Cognition*, William Chase, ed. (New York: Academic Press, 1973), 283–308.

[4] One component of this challenge is to develop cross-cultural comparative religious categories and the means to criticize and improve them. This is the goal of a series of volumes forthcoming from the Cross-Cultural Comparative Religious Ideas Project, directed by Robert Cummings Neville, Peter L. Berger, and John H. Berthrong, to be published by SUNY Press. The first of these, *The Human Condition*, is scheduled for publication in 1999. Subsequent volumes, to appear in 2000, are *Ultimate Realities* and *Religious Truth*.

[5] The elements of semiotic theory that we use are drawn especially from the pragmatic philosophy of the North American philosopher Charles Saunders Peirce, whose paleopragmatism (the apt designation of Robert C. Neville) is to be distinguished sharply from the neopragmatism of Richard Rorty. See section 7, below, for a more detailed account of the salient points.

Third, we acknowledge other difficulties: our analysis of religious experiences is not, in fact, independent of considerations in the philosophy of mind bearing on the ontological complexities of the mind-brain problem.[6] Nor is it independent of the various problems of consciousness, including the "hard problem" of first-person experience.[7] And we are forced to take a provisional stand on the notoriously controverted problem of defining religious experience. We shall assume that we can pursue our own line of investigation in spite of these and other complications.

1.4 Focus: "Experiences of Ultimacy"

Religious experiences include experiences in religious groups, as when worshiping, and experiences alone, as when meditating or in prayer. They may be mundane or sublime, wordlessly simple or replete with ideas. They include drawn-out periods of character transformation and spectacular episodes of conversion. This suggests too vast a diversity to describe all at once, so we need to define and name a target group of experiences.

The target group is determined by our interest in eventually developing a model that will be useful for discussing the ultimate causes and value of religious experiences (see section 7 below). We need to include experiences that religious people say are caused by God—whether correctly or mistakenly is unimportant at this stage. This narrow group of experiences conceivably might be called "God experiences."[8] This phrase is inappropriate for designating experiences within non-theistic religions, however, so we use the vaguer, more inclusive phrase, "experiences of ultimacy,"[9] which also expands the target group significantly.

Defining experiences of ultimacy more precisely is a complicated task, for two reasons. On the one hand, the way people describe their experiences crucially depends on the particular social and linguistic contexts in which the descriptions are used. On the other hand, we cannot know with certainty the contents of other minds,

---

[6] On the mind-brain problem, see Daniel C. Dennett, *Consciousness Explained* (Little-Brown and Co., 1991); Roger Penrose, *Shadows of the Mind* (Oxford and New York: Oxford University Press, 1994); and John R. Searle, *The Rediscovery of the Mind* (Cambridge and London: The MIT Press, 1993).

[7] On the "hard problem" of first-person consciousness, see David J. Chalmers, "Facing Up to the Problem of Consciousness," *Journal of Consciousness Studies* 2.3 (1995): 200–19, and idem, *The Conscious Mind: In Search of a Fundamental Theory* (New York: Oxford University Press, 1996); J. Levine, "Materialism and Qualia: The Explanatory Gap," *Pacific Philosophical Quarterly* 64 (1983): 354–61; and Thomas Nagel, "What Is It Like To Be a Bat?" *Philosophical Review* 83 (1974): 435–50.

[8] This is the terminology used in Michael A. Persinger, *Neuropsychological Bases of God Beliefs* (New York and London: Praeger, 1987).

[9] "Ultimacy" is a better category than "God" for registering the primary goal and object of a wide variety of religious traditions. Of course, the term "ultimacy" has to be construed sufficiently vaguely to comprehend the ultimate realities of religious traditions that think in such terms (such as most strands of the Abrahamic traditions and much of Hinduism), the ultimate paths or ways of religious traditions that subordinate questions of ultimate realities (such as strands of Buddhism and Hinduism), and the many ultimates of religious traditions that tend to avoid speaking of encompassing ultimates of either variety (such as strands of Chinese religion). In fact, most or perhaps all religious traditions thematize ultimacy in a variety of ways, ranging on one axis from ultimate realities to ultimate paths and on another axis from explicit to implicit formulations. When ultimacy is construed so as to take account of such variations, it is the optimal comparative category for our purposes.

so it is hard to know whether we are describing the same experiences even when we use identical descriptions within a single social-linguistic context. These considerations draw our attention to the hermeneutical circle connecting social-linguistic context and individual descriptions of ultimacy experiences.[10] While some might welcome relativism of descriptions as a way of protecting religious experience from scientific scrutiny, we treat it as a problem to be overcome. Our efforts can only be useful for questions about the causes and value of ultimacy experiences (the focus of section 7) if there is a way to determine, at least approximately, when and what sort of ultimacy experiences occur (the focus of section 6).[11]

Many attempts to define religious experiences have been made. We think most of them are flawed but we have found their insights quite helpful, as the following examples show. First, some definitions rely on phenomenological characteristics to circumvent the problem that people's descriptions are unreliable (for example, William James). This is wise, and we think that a sense of oneness with the divine and a sense of awe are good phenomenological markers for some experiences in our target group. Yet we cannot rely solely on a phenomenological approach to defining our target group because phenomenological reports are themselves subject to hermeneutical difficulties.[12] Second, some definitions focus on the irrational and usually spectacular elements of religious experience (for example, Rudolf Otto). This is useful because the phenomenological markers are easy to identify in those cases. However, we also want to include the more rational experiences surrounding the forming and changing of convictions and behaviors. Particularly interesting for understanding the causes and value of religious experiences are the sometimes mundane-seeming, sometimes spectacular experiences of conversion and character transformation, which typically involve both irrational and rational elements. Third, definitions focusing on individual experiences make obvious sense, and yet the role of the social-linguistic context is easy to overlook. We wish to pay close attention to the way social-linguistic contexts condition an individual's description of ultimacy experiences, for which a rich resource is the refined judgment of religious groups concerning the authenticity of claims to conversion and character transformation. Fourth, most definitions focus on what people are willing to call religious experiences, but we also want to include in our target group episodes in the lives of non-religious people who do not have the category "religious experience" at their

---

[10] This hermeneutical circle can be described by defining the social-linguistic context as the domain (1) in which experiences of ultimacy are described and redescribed, (2) in relation to which people form their expectations about experiences of ultimacy, (3) under the influence of which people learn how to use the words that will later help them describe their own ultimacy experiences, and (4) by means of which people's descriptions of their experiences are assessed, corrected, and regulated.

[11] It is important to note that detecting authentic ultimacy experiences is not merely an academic instinct imposed on religious practice. It is a pressing concern for religious groups as well, many of which have developed sophisticated methods of discernment to help make the judgments they want to make about the authenticity of religious experiences.

[12] The difficulties of too narrowly phenomenological an approach to delimiting the target group are as follows. First, this approach is precarious through its exclusive dependence on people's descriptions of religious experiences; phenomenological description requires skills in reporting that most people do not have. Second, the exclusively phenomenological approach to definition tends toward too narrow a definition, de-emphasizing many important features of religious experiences, especially those surrounding conversion and character transformation.

disposal. Nonreligious people sometimes describe their experiences in ways that lead religious people to call them "religious experiences."[13] Moreover, such experiences sometimes appear to be potent forces for character transformation. We conjecture that people's self-identification as religious or nonreligious is not an overriding consideration in determining the causes and value of religious experiences.

Having pondered existing definitions of religious experience, we are forced to concede that a precise definition of ultimacy experiences is probably out of the question. Nevertheless, there are several sorts of markers for ultimacy experiences: people's descriptions within social-linguistic contexts, phenomenological characteristics, the judgment of experts in religious discernment or of psychologists, the wisdom of generations encoded in theological and ethical traditions, and even neural signatures. These markers may not always be in complete harmony, as when a phenomenologically spectacular religious conversion is judged inauthentic by a religious group or when a person not affiliated with any religious group refuses to describe as religious an experience that utterly transforms his or her character. Nevertheless, such markers can still be used to evaluate putative experiences of ultimacy. In section 6 we shall give some examples of how this evaluation process might work. The point to be made here is that establishing a process of evaluating putative ultimacy experiences is equivalent to offering a dynamic definition for our model's target group of religious experiences. The resulting definition is dynamic in two senses. On the one hand, applying the definition in any given instance requires running through the process of evaluating the various markers for ultimacy experiences and remembering at the same time that there is a complex taxonomy of such experiences whereby different types are associated with different sets of phenomenal characteristics. On the other hand, the definition is not dyadic, excluding some experiences and including others. It is more like a set of targets, with the ultimacy experiences closest to the bull's eye for each type being those with the strongest agreement among markers.

Diagram 1 (see Appendix B) illustrates both the relation between ultimacy experiences and other experiences and the complex process of definition that we need to develop. The various considerations relevant to the description of ultimacy experiences are introduced in the next section and discussed in detail in sections 2–5. How all of this descriptive work contributes to a dynamic process of definition is described in section 6. The causal model of ultimacy experiences developed in section 7 is built on this descriptive foundation. As complicated as they are, we think experiences of ultimacy are delineated well enough for us to proceed with trying to describe them.

1.5 Components: Four Perspectives on Ultimacy Experiences

We gather the considerations we use for the description of ultimacy experiences into four groups. Two—the phenomenological and the social-psychological—refine our understanding of the fundamental dialectic between individuals and social-linguistic contexts; it is within that dialectical tension that the meanings of descriptions of ultimacy experiences are established. The other two components are less closely bound to the social-linguistic systems. One is neurology, which may in the future

---

[13] There are accounts of this sort in William James, *Varieties of Religious Experience: A Study in Human Nature* (New York: Longmans, Green, 1902).

contribute criteria to the task of assessing ultimacy experiences. The other is theological or ethical convictions that stipulate criteria for authenticity of claims to experiences of ultimacy in the form of correlations between such experiences and the behavior of those that have them. Theology sometimes also ventures to stipulate the specific causes of certain kinds of ultimacy experiences. We shall introduce each of these four components briefly and then devote the next four sections to a more detailed discussion of each.

First, phenomenological description of ultimacy experiences furnishes a thick description of their quality and relations to other events and experiences.[14] Phenomenological description depends on intensifying a linguistic system with new vocabulary and meanings, which allows experiences to be described with great nuance and precision. We may think of Rudolf Otto's phenomenology of numinous experiences (see section 2.2). Or we may think of Søren Kierkegaard's three-staged phenomenology of religious conversion and character transformation from the aesthetic to the ethical to the religious (see section 2.5). These and other phenomenologies, we take it, often induce strong feelings of recognition in those who read them; they often succeed in evoking assent when the reader is sensitive enough to grasp the enhancements of the linguistic system that the phenomenologist is trying to establish.

The second set of considerations derives from neurology. It is questionable whether brain states and processes can be correlated with personal descriptions of purported experiences of ultimacy at the present time, or ever. To the extent that correlations become possible, however, they would promise objective access to internal experiences through functional imaging and other measurements of brain activity, even as phenomenology promises objective access to internal experiences through disciplined cultivation of descriptive expertise. Though both neurological scans and phenomenological analyses are somewhat removed from the day-to-day use of linguistic systems to describe ultimacy experiences, both are relevant factors in the hermeneutical mix and presumably neurological considerations will become more important with time, even at the level of the individual religious person's self-consciousness.[15] Furthermore, neurological correlates could conceivably lead to criteria for "false positives" with the potential to weigh against the authenticity of

---

[14] We must provisionally set aside the philosophical commitment of phenomenologists such as Edmund Husserl to the possibility of achieving public, objective descriptions of internal conscious states through his phenomenological method. If this Husserlian claim is correct then the problem of other minds is essentially overcome and there is powerful evidence both for the autonomy of experiences of ultimacy and for the capacity of experiences of ultimacy to amend descriptions of them; it is not a matter of "hermeneutics all the way down" after all. Of course, even this result would say nothing about the cause of ultimacy experiences (though some phenomenologists would insist that this too could be determined), for their shared features may derive from the biological givenness of human beings or similar factors. But we cannot evaluate even this moderate claim adequately here and so must proceed by thinking of phenomenology as a disciplined development of part of a social-linguistic network so that that network becomes dense and sensitive enough to permit a properly trained person to make subtle discriminations among his or her experiences.

[15] This suggests a humorous image. Instead of demanding that group members handle poisonous snakes, speak in tongues, or give an enlightened answer to a koan, some religious groups might require specific sorts of brain activity as measured by functional imaging equipment. fMRI equipment in place of confessionals? While humorous, scenarios like this are surely not absurd or unlikely in the long term.

ultimacy experiences in spite of personal testimony. By contrast, if experiences of ultimacy turn out to be inconsistently realized or widely distributed across brain structures and variously expressed in brain processes, then they may have no obvious neurological correlates, and neurology would be of correspondingly little use as a criterion of authenticity. Of course, the situation is likely to be somewhere between these two extremes, but not enough is yet known to be sure how useful the neurological criterion will prove to be. And however well it serves as a criterion, it is an important element in any theory of ultimacy experiences.

Third, descriptions of experiences of ultimacy are greatly enriched by experts in psychology and sociology. These experts concern themselves less with the *thick* descriptions of experiences of ultimacy and more with the description of *typical* experiences of ultimacy, attending to how they cohere with other aspects of the human person by means of categories drawn from psychology, ethics, or spirituality. Understanding the processes of emotional and physical development in the typical human person, along with common aberrations, casts reports of experiences of ultimacy into a helpful light. Similarly, understanding the influences of religious groups on individuals allows experts to give nuanced descriptions of the complex social interactions within which many ultimacy experiences occur. Many religious people have at their disposal a vast database of first-hand and second-hand stories of ultimacy experiences in which the before and the after of the episode itself expose typical patterns of behavior. Of course, exceptions are unsurprising and even expected; expert psychoanalysts or religious advisors do not have privileged access to the experiences in question. But even exceptions have a kind of plausibility, perhaps due to thoroughly systematic ways in which typical patterns are broken. A person's chosen description of experiences of ultimacy and the meaning of that description in his or her social-linguistic context are profoundly influenced by such expert readings of the *typical* psychological and behavioral accompaniments of *typical* ultimacy experiences.

The fourth set of considerations is theological in character. Theological theories of ultimacy can be sufficiently detailed to permit stipulation of the psychological and behavioral correlates of experiences of ultimacy. For example, it is almost universally held in theological systems that experiences of ultimacy should transform people's character. The experience of samadhi in Buddhist meditation is supposed to make a person more caring toward other creatures and the experience of assurance in Christian piety is supposed to make a person unaccountably peaceful. These theologically-based beliefs are crucial in the operation of both individual spiritual direction and corporate discernment processes in religious groups, and we think they are also active in diffuse ways, perhaps also more generally ethical than specifically theological in character, in the secular analogues of discernment such as psychoanalysis. Of course, we might well say that theories of ultimacy of this kind should not count in forming our ideas of how experiences of ultimacy should affect people, but this would be an overreaction. Theological considerations function as a source of suggestions and hints as to what psychological and behavioral characteristics should be expected in the presence of an authentic claim of an experience of ultimacy. From this point of view, we would be justified in assuming that theological reflection and ethical theories over the centuries are well placed to give good hints, informed as they are by a wealth of individual and corporate experience. It is not the specifically normative character of theological or ethical reflection that makes these hints useful, therefore, but the long-term functionality of the theological or ethical theories themselves in religious and other groups.

The taxonomy of ultimacy experiences we develop draws chiefly from the phenomenological considerations, with crucial support coming from the neurologically important distinction between short-term and long-term episodes. Social-psychological and theological considerations play especially important roles in the process of distinguishing authentic ultimacy experiences (see section 6). Our model of the causation of ultimacy experiences (see section 7) has little direct use for phenomenological considerations, focusing instead on those that are neural, social-psychological, and theological.

2 Phenomenological Considerations

Phenomenology of religion is a diverse collection of partly descriptive, partly interpretive approaches to religious phenomena.[16] Phenomenological approaches to religious experience typically have been oriented to mystical states and conversions, which are familiar instances of what we are calling experiences of ultimacy. Phenomenologists have directed less attention to other kinds of religious experiences, such as corporate ritual experience and long-term character transformation, which can also be instances of ultimacy experiences.[17] In view of this emphasis, it is unsurprising that studies of religious experiences have often taken over distinctions generated in the phenomenology of religion. As useful as these distinctions may prove in some studies, we use phenomenological observations to divide the territory of ultimacy experiences in a way more congenial to exploring connections with the neurosciences.

We first distinguish ultimacy experiences on the basis of temporal extension because there seems to be a vast phenomenological difference between shorter and longer experiences. The phenomenology of discrete states that can be described as ultimacy experiences involves components having to do with sensory awareness,

---

[16] For a review, see Eric J. Sharpe, *Comparative Religion: A History*, 2nd ed. (La Salle: Open Court, 1986), especially chap. 10, pp. 220–50. Sharpe is careful to point out that the phenomenology of religion is basically an attempt at objective description of religious manifestations—places, people, actions, words—that respects the perspective of the religious person and that can help in the task of interpreting the nature of religion. It owes little more than a few key concepts to the philosophical phenomenology of Husserl, and its vaguely defined limits embrace numerous different methodological approaches. Sharpe's characterization is accurate so far as it goes, but the various methodological approaches he has in mind themselves have a history that lives on in their use within the phenomenology of religion. It follows that there is significantly more to the phenomenology of religion than simply objective description. These methodological approaches usually can be traced back to the needs of a discipline to which phenomenological techniques have been applied as a means to fuller understanding. Francisco J. Varela, Evan Thompson, and Eleanor Rosch identify several such methodology-defining allied disciplinary traditions that make extensive use of phenomenology: logical-philosophical analysis of human being, meta-analysis of patterns in existing theories (usually historical, sociological, or anthropological theories), and analysis of techniques used in clinical therapy. See *The Embodied Mind: Cognitive Science and Human Experience* (Cambridge: MIT Press, 1991), xvi–xvii. Theorists in the phenomenology of religion usually add to the basic goal of objective description the aims of one or more of these existing methods of applying phenomenological techniques.

[17] There are important phenomenological studies of the sacred, including sacred ritual and social transactions, which have some overlap with ultimacy experiences. See, for example, Gerardus van der Leeuw, *Religion in Essence and Manifestation*, 2nd ed. with a foreword by Ninian Smart (Princeton: Princeton University Press, 1964; tr. from the 2nd German ed. by Hans H. Penner; originally published 1933).

sense of self, presences, cognitions, and emotions. The phenomenology of extended experiences that can be described as ultimacy experiences divides into two classes. Dynamic processes of orientation and control help people maintain their relations to themselves, groups, and the wider world; these sometimes but not always fall under the ambit of ultimacy experiences. Gradual processes of transformation often take the form of experiences of ultimacy; these processes involve apparently lasting change in behavior, personality, and beliefs.

2.1 Discrete Ultimacy Experiences: Persinger

Michael Persinger includes interesting phenomenological characterizations of discrete ultimacy experiences in his book, *Neuropsychological Bases of God Beliefs*.[18] We find the book rhetorically unstable, with few links to the data he offers in support of his conclusions, few appropriate data in the articles to which he refers, and problematic patterns of interpreting his data.[19] In spite of the book's flaws, Persinger's extensive exploration of connections between temporal lobe function and religious experience leads him to a thoughtful characterization of what he calls God experiences, a phenomenological contribution worth quoting at length.

God Experiences are transient phenomena that are loaded with emotional references...

The God Experience exists for a few seconds or minutes at any given time. Multiple experiences can occur in quick succession. During this period, the person feels that the "self," or some reference indicating the "thinking entity" becomes united with or "at one" with the symbolic form of all space-time. It might be called Allah, God, Cosmic Consciousness, or even some idiosyncratic label. Slightly deviant forms include references to intellectual abstracts such as "mathematical balance," "consciousness of time," or "extraterrestrial intrusions." These phenomena are similar to mystical states and the more secular "peak experiences."

Usually the God Experience involves euphoric and positive emotions. The person reports a type of God high that is characterized by a sense of profound meaningfulness, peacefulness, and cosmic serenity. Invariably the state is perfused with references to reduction of death anxiety. It is defined as the anticipated extinction of the self-concept or "the thinking entity." During the God Experience, the person suddenly feels that he or she will not die. Instead, he or she will live forever as a part of subset of the symbol of all space-time. If the symbol is a father image, then the person expects to become a child of the father. If the symbol is "imageless," the person expects to become a part of the Universal Whole.

Sometimes God Experiences can have negative emotional valences. During these periods, the same sense of oneness is pervaded by anxiety and fear. It is the epitome of terror. These experiences rarely happen more than once, except in psychiatric patients; the consequences punish any further display. Labels applied to these experiences reflect the bad, aversive or generally evil components in the culture in which the person survives. Classic references involve "hell," "demon world," or the more abstract "nether world."

---

[18] Michael A. Persinger, *Neuropsychological Bases of God Beliefs* (New York and London: Praeger, 1987).

[19] See Persinger, "Religious and Mystical Experiences as Artifacts of Temporal Lobe Function: a General Hypothesis," *Perceptual and Motor Skills* 57 (1983): 1255–62; "Striking EEG Profiles from Single Episodes of Glossolalia and Transcendental Meditation," *Perceptual and Motor Skills* 58 (1994): 127–33; "People Who Report Religious Experiences May Also Display Enhanced Temporal-Lobe Signs," *Perceptual and Motor Skills* 58 (1994): 963–75; "Propensity to Report Paranormal Experiences Is Correlated with Temporal Lobe Signs," *Perceptual and Motor Skills* 59 (1994): 583–86; and "Death Anxiety as a Semantic Conditioned Suppression Paradigm," *Perceptual and Motor Skills* 60 (1995): 827–30.

They are not traditionally called God Experiences, although they are certainly derived from the same source of variance. The self, with respect to space-time and imminent dissolution (death), still dominates the experience.[20]

Persinger goes on to discuss God concepts and how God experiences and God concepts combine in God beliefs. His description of God experiences apparently derives from many interviews and clinical encounters with people who claim to have had them.[21] He strikes the main themes that recur in phenomenological descriptions of discrete ultimacy experiences: they involve modifications of sensory awareness, sense of self, sense of presences, cognitions, and emotions.

2.2 Discrete Ultimacy Experiences: Otto

In *The Idea of the Holy*, Rudolf Otto attempted to describe the irrational or supra-rational elements of religious experience.[22] He focused on what we are calling discrete rather than extended ultimacy experiences, calling them numinous experiences. He argued for the autonomy and uniqueness of numinous experience and he tried to show that it is involved in everything from faint religious stirrings to the most profound mystical experience. Otto characterized their two main features in the phrase "*mysterium tremendum.*" He described *tremendum* in terms of three elements: awefulness, overpoweringness, and energy or urgency. He described *mysterium* in terms of the wholly other and fascination.

Otto also discussed the means of expression of the numinous, including how it is awakened in one mind upon seeing its experience described or enacted by another.[23] He pointed out that it cannot be taught or described in such a way to "pass" it on but rather that there must be some independent experience that answers to the descriptions of it that are passed around the group. Thus it can be expressed directly only through an individual's encounter with holy places, holy events, and holy people. It can be expressed indirectly by making use of the ways we express feelings similar to those with numinous elements. Thus he spoke of art and language that convey terror and dread, responses that are capable of evoking numinous feelings of the *tremendum* kind. Under this heading he also mentioned its higher expressions: grandeur and sublimity. Under the heading of the *mysterium* Otto mentioned expression in the form of miracle; that which cannot be comprehended serves as analogy for the mysterium and is capable of evoking it. Under this heading he also treated the only half-intelligible language of devotion, including liturgy, ritual, and some music, as well as many other types of analogies.

One of the great strengths of Otto's work is his focused exploration of the emotional content of numinous experiences. This focus is also a weakness with respect to the desire for completeness of phenomenological descriptions. But even

---

[20] Persinger, *Neuropsychological Bases of God Beliefs*, 1–2.

[21] Ibid., xi.

[22] Rudolf Otto, *The Idea of the Holy: An Inquiry into the Non-rational Factor in the Idea of the Divine and its Relation to the Rational*, 3rd ed. (London: Oxford University Press, 1925; tr. from the ninth German ed. by John W. Harvey; first published 1917).

[23] See section 4, below, for a discussion of the function of mirror neurons. This can be thought of as one neurological consideration bearing on Otto's ideas about the awakening of numinous experiences acted on or described by one person in other people. This remains highly speculative, however, because mirror neurons have been studied primarily in relation to motor functions.

this brief summary of Otto's discussion permits an inference that the rest of the book justifies: discrete ultimacy experiences involve modifications of sensory awareness, sense of self, sense of presences, cognitions, and emotions. The characteristic elements of discrete ultimacy experiences appear in both Otto's and Persinger's very different analyses.

Another strength of Otto's work is his argument for the uniqueness and autonomy of the numinous. That is, he believes that the numinous element of discrete ultimacy experiences is objective enough to force particular descriptions and specific forms of concept stretching and cognitive breakdown to appear repeatedly and predictably across the various manifestations of religion. For example, key metaphors recur such as: encounter with a person, reason-transcending mystery, the power for salvation or liberation, peaceful presence, abysmal anxiety, and so on. We find this claim intriguing and return to it below in section 3. However, we think that far too little has been done by way of comparative phenomenology even now to draw Otto's conclusions with his confidence, and he drew them when almost no work had been done.[24]

2.3 Discrete Ultimacy Experiences: Elements

We might well draw other phenomenological descriptions into the mix.[25] We think the same five elements recur in those accounts. In summarizing the phenomenology of discrete ultimacy experiences we will make brief remarks about each of the five elements, indicating some of the variations within each element.

Sensory Alterations. Under this heading we would include perceptions that are incongruous with the current environmental situation. There may be the perception that the surroundings are suffused with light or otherwise perceptually different. There may be auditory or olfactory sensations as well as visions or hallucinations. Related to this category but distinguishable are percepts bearing on the sense of self or on the sense of a presence near oneself, or perhaps a nonlocalized presence (see below).

---

[24] There are a number of scholarly traditions in the study of religion that have explored the claim that the sacred or numinous causes the same symbols, myths, and ideas to recur in the world's religions, including most famously the comparative studies of mythology influenced by Carl Jung's theory of archetypes. See, for example, Joseph Campbell, *The Masks of God* (New York: Viking Press, 1959–68), 4 vols.: *Primitive Mythology, Oriental Mythology, Occidental Mythology,* and *Creative Mythology;* and many of the works of Mircea Eliade, including *Cosmos and History: The Myth of the Eternal Return* (New York: Harper, 1959; first English ed., 1954), *Myths, Dreams, and Mysteries* (New York: Harper, 1960), and *Images and Symbols* (New York: Sheed & Ward, 1961). Also influenced by Jung, and an important influence on this trend in religious studies, is structuralism; see especially Claude Lévi-Strauss, *Totemism* (Boston: Beacon Press, 1963; tr. from the French by Rodney Needham). On Jung himself, see Joseph Campbell, ed., *The Portable Jung* (New York: Penguin Books, 1976). The perennial philosophy makes the same claim on a different basis, though not independent of the Jungian emphasis on archetypes. See, for example, Huston Smith, *Forgotten Truth: The Common Vision of the World's Religions,* reprint ed. (San Francisco: HarperSanFrancisco, 1992; first ed., 1965); and Aldous Huxley, *The Perennial Philosophy* (New York: Harper & Brothers, 1945).

[25] See the phenomenological descriptions in, for example, James, *Varieties;* and Eugene G. d'Aquili and Andrew B. Newberg, "Religious and Mystical States: A Neuropsychological Model," *Zygon: Journal of Religion and Science* 28 (June, 1993): 177–99.

Self Alterations. A person may feel as if outside his or her own body. There may be a loss of the sense of the individual self as real or as the source of thought or will, and a sense of merger with the universe. There may be a feeling of union of the self with an entity such as God or the Infinite. There may be a sense of the enlarging of the self accompanied by powerful feelings of compassion and confidence. There may be a sense of the self as profoundly threatened by judgment or annihilation in the presence of a being of enormous power. There may be a sense of altered bodily functions or of the self being taken over by another being (see below).

Presences. In certain discrete states, a person may experience the sense of a presence felt as mysterious or awesome; this may have both positive and negative modulations. A person may feel the presence of nonphysical beings, either benign or evil, such as angels or demons. There may be a sense of being invaded, inhabited, or controlled by such beings.

Cognition. There may be a sudden sense of illumination or profound understanding. There may be a sense of increased awareness, or a sense of the unreality of the world. There may be a conviction of sin or weakness, or a sense of assurance of salvation or emotional and spiritual healing. There is a very important cognitive feature that invariably accompanies all the other phenomena of discrete states: "They are as convincing to those who have them as any direct sensible experiences can be, and they are, as a rule, much more convincing than results established by mere logic ever are."[26]

Emotions. Under this heading we would include intense feelings that are either incongruous with the current context or expressive of a social process that is itself incongruous with usual patterns, such as feelings of ecstasy, awe, dread, guilt, safety, or tranquility. There may be the experience of utter darkness or despair in the quest for mystical union, the mystics' dark night of the soul.

2.4 Extended Ultimacy Experiences: Berger

There are two classes of extended ultimacy experiences. The first concerns dynamic, socially embedded processes of orientation and control in relation to the cosmos, the social world, and one's self. For convenience, we shall call these processes *social ultimacy experiences*. The Durkheimian tradition of the social analysis of religion focuses on such processes but tends to downplay the individual religious experiences associated with them. Peter Berger's *The Sacred Canopy* is more balanced.[27] In particular, Berger blends the Durkheimian tradition with the sociology of knowledge and extends both of them in a direction that is at once more sensitive to individual experiences and more useful for theologically directed inquiries that seek to press questions of truth and causation in relation to ultimacy.

Berger assumes both that "every human society is an enterprise of world-building"[28] and that "all socially constructed worlds are precarious."[29] From these premises he analyzes the role of religion in society. World-construction is a dialectical process between individuals and their social context. Human beings first externalize their being in the world, whereupon the outpouring of themselves is

---

[26] James, *Varieties*, 72.

[27] Peter L. Berger, *The Sacred Canopy: Elements of a Sociological Theory of Religion* (Garden City: Doubleday, 1967).

[28] Ibid., 3; see chap. 1, "Religion and World-Construction."

[29] Ibid., 29; see ch. 2, "Religion and World-Maintenance."

objectified both in material social and economic structures and in immaterial ideas and culture. Finally, these objectivized realities are internalized by individuals, conditioning their activity and self-understanding. It follows that "the socially constructed world is, above all, an ordering of experience," a *nomos* that human beings must construct because, unlike other animals, they are not biologically equipped with any fixed such ordering; culture and socialization are necessary for being human.[30] Socialization is most effective when taken for granted. When it is, the meanings of the constructed nomos embrace the entire cosmos, yielding meanings for the fundamental questions of human life, a process in which religion plays the part of creating a sacred cosmos.[31]

> It can thus be said that religion has played a strategic part in the human enterprise of world-building. Religion implies the farthest reach of man's self-externalization, of his infusion of reality with his own meanings. Religion implies that human order is projected onto the totality of being. Put differently, religion is the audacious attempt to conceive of the entire universe as being humanly significant.[32]

The precariousness of social order is managed by socialization, as mentioned, but also by the resistance-limiting mechanisms of social control and, more subtly, by processes of legitimation. In legitimation, the social order is explained and justified with reference to ideas that are rendered plausible and even obvious by their having been already objectified in the dialectical process of social construction.[33] Among many mechanisms of legitimation, "religion legitimates social institutions by bestowing upon them an ultimately valid ontological status, that is, by *locating* them within a sacred and cosmic frame of reference."[34] At least as importantly, religious legitimation is capable of handling many marginal situations in which the *nomos* is threatened by ideas or activities not already managed by ordinary socialization.[35] These marginal situations are common, ranging from dreams and nightmares to hallucinations and intuitions, all of which were accorded ontologically real status in most cultures until recent times.[36] They also include discrete ultimacy experiences, whose religious legitimation serves the interests of maintaining the stability of social constructions of reality.

In this analysis, human experiences of the sacred orient individuals—in an enormous range of ways and not necessarily in religious contexts—within a cosmic environment. They can seem to confirm what religious beliefs assert about the cosmic meaningfulness of many other experiences and even of the social order itself.

---

[30] Ibid., 19.

[31] Ibid., 25.

[32] Ibid., 27–28.

[33] Ibid., 29.

[34] Ibid., 33; Berger's italics.

[35] Ibid., 42–43.

[36] Julian Jaynes makes a great deal of these marginal experiences, both positing a neural basis for them and developing a theory of religion on that basis. What he calls the bicameral mind involves the human left-brain with its speech centers in balance with the right brain in which the areas corresponding to left brain speech produce divine speech; he takes this divine voice to be a direct expression of the will to act. The bicameral mind has now broken down, he further supposes, leaving us with bicameral traces in many religious practices and the conscious entertaining of alternatives in place of direct action on the basis of divine voices. See his *The Origin of Consciousness*, especially pp. 84–125.

They also threaten the social order whenever they lie beyond the reach of the control achieved by ordinary socialization. Occurring outside of a context in which they are expected and explained, such experiences may upset the stability of the relationship between individual and society. Their occurrence in religious contexts, however, provides an effective means of controlling an individual's engagement with the social order. More than that, ultimacy experiences occurring in such contexts may even enhance social regulation by reinforcing processes of legitimation already active in religious groups: to experience personally is to confirm a group's legitimating claims. In these ways, therefore, ultimacy experiences orient individuals and, when occurring in an appropriately authoritative context, serve social interests of control.

While these effects are relevant to both discrete and extended ultimacy experiences, it is only in the context of an extended process of socially-guided interpretation of ultimacy experiences, which in many cases can themselves be regarded as extended ultimacy experiences, that the effects of orientation and control appear. For this reason, we classify the orientation and control dimensions of religious experiences as extended ultimacy experiences of the social type.

2.5 Extended Ultimacy Experiences: Kierkegaard

The second class of extended ultimacy experiences concerns gradual and chronic experiences of personal change or self-transcendence, such as Confucian self-cultivation, Christian sanctification, and possibly also character changes having little explicit connection with religious symbols and practices. Some conversions are of this extended type. For convenience, we shall call these *transformative ultimacy experiences*. While a vast literature on conversion clamors for attention here, we turn to Søren Kierkegaard's extraordinary phenomenology of the process of transformation associated with extended ultimacy experiences.[37] Intending his analysis as an answer to the great question, "What ought I do?" his answer famously subordinated moral sensibilities to religious ones in the third and final stage of an ongoing process of transformation driven by awareness of an intimate relationship with God. The rationally and ethically transcendent character of religious transformation has been noticed repeatedly and Kierkegaard's classic expression of it is worthy of summary.

The first stage of the religious-moral quest is the aesthetic. This is the search for sensual and intellectual pleasure. Kierkegaard argued that such a search eventually leads to boredom and then suicide and thus that there is an impulse to move to a form of life in which there is a conception of oughtness. The second stage is thus the moral or ethical stage in which we freely align ourselves with the moral law and make a determination to be good. Kierkegaard's arch-enemy was G.W.F. Hegel, who tried to synthesize the moral life and the aesthetic life; Kierkegaard admired Hegel's effort but judged it to be merely the highest form of aestheticism. Kierkegaard argued that

---

[37] Kierkegaard is usually neglected as an asset for descriptive tasks such as ours because he is so explicitly passionate an author. His analysis covers much more than conversion, however, and it captures dimensions of the process of religious transformation that most treatments of conversion miss. Moreover, it is the archetypal instance of objectivity of description achieved through passionate inwardness. See Søren Kierkegaard, *Fear and Trembling; Repetition*, ed. and tr. with introduction and notes by Howard V. Hong and Edna H. Hong from the 1st 1843 Danish ed. (Princeton: Princeton University Press, 1983); *Either/Or*, ed. and tr. with introduction and notes by Howard V. Hong and Edna H. Hong from the 1st 1843 Danish ed. (Princeton: Princeton University Press, 1987).

a jump is involved in moving from the aesthetic to the ethical and that we must simply choose. The third stage is the religious, in which we find ourselves driven to suspend ethical concerns in the name of fidelity to an awesome encounter with God (Kierkegaard called this a "teleological suspension of the ethical"). In the religious life, divine command is paramount and true love for God is expressed in the willingness to set aside moral habits and to respond to the divine command with purity of heart. If purity of heart is to will one thing, then for Kierkegaard its highest form is to will not the moral law but God.

Whereas Hegel and also Immanuel Kant took everything, even God, to be consistent with the moral law, Kierkegaard argued that the divine command is rationally unapproachable. The contrast between the moral and religious stages is movingly expressed in the discussion of Abraham and Isaac.[38] Abraham becomes for Kierkegaard the one whose life of faith (the religious stage) transcends moral categories through obedience to God—even rationally and ethically impeachable divine whims; morality derives from God, it does not rule God. According to Kierkegaard God has set us in a situation in which these choices (particularly in the movement from the second to the third stage) cannot be made rationally but are criterionless; this is essential to the life of faith. This is the brutal situation of human life and draws our attention to the fundamental character of decision: one's very soul depends upon it.

Kierkegaard's analysis of what we are calling transformative extended ultimacy experiences highlights the importance of choice and focuses on the existentially potent transformation of personality and character under the influence of profound, ongoing experiences of loyalty to and love for God. It also highlights the way that the transformation of people under the impact of extended ultimacy experiences induces new beliefs about themselves, about ultimacy (be it represented as God or something else), and about their own behavior and choices. Kierkegaard says comparatively little about the social embedding so characteristic of extended ultimacy experiences, which most other accounts, especially in the literature of conversion, stress.

2.6 Extended Ultimacy Experiences: Elements

Extended ultimacy experiences are typically less perceptually dramatic than discrete ultimacy experiences. They may occur in conjunction with episodes of the discrete states described above, however, and they may be strong enough that individuals feel as if they are more or less continually in communication with a deity and receiving assurance or direction in daily matters. Whether explicitly religious or not, we notice several recurring characteristics of extended ultimacy experiences: existential potency, social embedding, transformations of behavior and personality, and transformations of beliefs. All four elements seem important to various degrees in both the social and the transformation types of extended ultimacy experiences.

Existential Potency. Whereas discrete ultimacy experiences can occur in ways that may sometimes leave people wondering what happened and how it might be relevant to their lives, one of the hallmarks of extended ultimacy experiences is the direct existential relevance they are felt to have. The orienting and transforming dimensions of extended ultimacy experiences make this particularly clear.

---

[38] See Kierkegaard, *Fear and Trembling*, "Eulogy and Abraham," 15–23, and the subsequent discussion.

Social Embedding. Extended ultimacy experiences make little sense in isolation from a community within which they can be interpreted and by whose interpretation they are made existentially potent. The social embedding is effective in two directions, as we have seen. On the one hand, a person participates in the benefits of the community's interpretive and narrative power. Their experiences of ultimacy are channeled into and through that narrative framework and then focused into transformative potency or the need for orientation to cosmos, world, society, and self. Without this community, the person must self-generate the authority needed to make the assumptions expressed in such orienting and transforming processes plausible and effective, and very few people seem capable of doing that alone, if it even makes sense to do so.

On the other hand, a community mediates the wider society's need for stability (including control of marginal situations created by the occurrence of ultimacy experiences) by means of its participation in social legitimation processes. This control is exercised in a variety of ways. Sometimes social values are reaffirmed in the cosmically loaded narrative offered by the religious group; this is ubiquitous. When ultimacy experiences make that narrative existentially more vivid for individuals, the legitimation of linked social values is correspondingly strong. Other times, the effects of potentially socially disruptive sentiments and even critiques inspired by ultimacy experiences are controlled by being given limited expression within the religious group, releasing tension that otherwise might be socially explosive. This is the case, for example, with shamanic rituals, which are often performed in public: they help people let off steam, as it were, without threatening the social structure or calling its values too much into question.

Transformation of Behavior and Personality. The classic religious expression of behavior and personality transformation is permanent conversion. Conversion occurs when an individual orders his or her life in accordance with the felt reality of ultimacy experiences. Rarely is a conversion experience accomplished under the influence solely of a discrete ultimacy experience, but rather extended ultimacy experiences that result in conversion often occur in conjunction with the more discrete experiences.

It is important to note that, in the absence of discrete ultimacy experiences and a religious social-linguistic context, a nominally nonreligious conversion may take place in the form of character transformation, a combination of behavior and personality transformation. Character transformation is a staple of literature, a well-known example being the novel *Emma* in which the thoughtless young protagonist, at first chided by Knightley, gradually comes to assume moral responsibility herself.[39] Such accounts are usually punctuated by a crisis of remorse in which the protagonist perceives the whole of his or her existence up to that point as morally deficient and shameful.

What are the distinctive features of the new interpretive framework whose internalization marks character transformation? We can think of other instances in which someone is brought into a new interpretive framework—for example, when he or she is introduced to a school of philosophy; or when he or she joins the military and adopts its vocabulary, actions, and modes of thought; or when he or she is psychoanalyzed and adopts certain new concepts for understanding his or her own experiences and actions. In the kinds of transformations we are considering at the moment, by contrast, the interpretive framework has primarily to do with a moral

---

[39] Jane Austen, *Emma* (New York: Knopf, 1991; first published 1833).

order—Kierkegaard's movement from the first to the second stage. Now, the self and what is called the moral-social order are fundamentally related, as reflected by the designation of the self as "a location within the moral-social order."[40] Since the transformation in question has to do with participating in a new kind of moral order, the self must inevitably be changed.

A feature of the new kind of moral order in which a person participates, in the process we are calling character transformation, is that it seems to supersede the prior moral order in a recognizable way. For example, in *Emma* the moral order in which the heroine operated at first was one of trivial gossip, shallow amusement, attention to appearance rather than substance, and a disregard for the feelings of others. The question of ultimacy is raised because the movement to a new moral system in all such stories is not just a move to something different: it is a move to something we recognize as better or higher. This implies that there is something unique about the conceptual-linguistic system offered by the moral advisor, something that may be universally recognizable—just as accounts of discrete ultimacy experiences in various cultures are often cross-culturally recognizable. Such accounts are widespread in literature, which we take to mean that the topic of moral development is compelling to human beings regardless of their religious background. Furthermore, we note that such accounts can induce transformative effects in their readers: that is, a narrative depicting a character's introduction to a new moral system can itself promote moral transformation.

The kind of transformation of behavior and personality that we have been discussing thus far seems to encompass both nominally religious and nominally non-religious contexts. But the transformation expressed by Kierkegaard in the leap from the ethical to the religious is specifically religious. It rarely shows up in secular literature; the relativity of moral conventions is sometimes thematized, especially in existentialist literature, but the inevitability of anxiety and despair tend to be the lessons drawn rather than the possibility of a supra-rational, morality-transcending, transformative religious experience. The closest literary analogue for this kind of transformation may be outlaw heroes—not Robin Hood, who is essentially a moral prophet, but early Wild West outlaws such as Billy the Kid or Jesse James. They engage in a teleological suspension of the ethical, with the moral law being defied in service of their own gain. If allegiance to the moral law were suspended for the love of God instead of for personal profit, there would be strong affinity between such figures and Kierkegaard's truly religious person. In religious literature, by contrast, especially in the lives of the saints or in great mystical writings of many religious traditions, the possibility of the truly religious person in Kierkegaard's sense shows up lucidly again and again. Friedrich Nietzsche may have had something like this in mind when he spoke of the *Übermensch*: the person whose morality is autonomously generated out of a rich mystical sensibility rather than being merely a personal appropriation of extant social conventions.[41] Kierkegaard

---

[40] Rom Harré, *Personal Being: A Theory for Individual Psychology* (Cambridge: Harvard University Press, 1984).

[41] See Friedrich Nietzsche, *Thus Spake Zarathustra*, in *The Portable Nietzsche*, ed. and trans. Walter Kaufmann (New York: Penguin Books, 1954); and idem, *The Will to Power*, ed. and trans. Walter Kaufmann and R.J. Hollingdale (New York: Vintage Books, 1968). The interpretation of Nietzsche's *Übermensch* is complex; for a penetrating account of its mystical and warrior sensibilities, see Stephen Main, "Abyss Without a Ground: Nietzschean Spirituality and Self-Healing" (Chicago: University of Chicago, dissertation, 1999).

and especially Nietzsche tend to be overly optimistic about what individuals can accomplish independently of sustaining communities. More often, the transformation they describe depends upon a background community even when the transformation itself passes beyond the bounds of the usual for that community, as is typical.

Transformation of Beliefs. The nature of the beliefs that accompany behavior and personality changes distinguishes religious and nonreligious transformation. The content of a person's beliefs is heavily conditioned by the social-linguistic framework available to him or her from others; thus, religious beliefs are highly variable due to the many cultural contexts in which they arise. In any individual case, they also may depend on the phenomenological aspects of discrete ultimacy experiences, especially when they involve presences and unusual cognitions. In general, the beliefs that accompany religious transformation (conversion) may concern the individual's relation to a higher being or abstract principle—the placing of his or her finite existence into a meaningful context, the worth or value of other living beings, and the general meaning or purpose of the whole of the universe. These beliefs are usually intertwined.

3 Neurological Considerations

As mentioned above, those interested in the neural basis of religious experience are at the mercy of the stage of neuroscience's development during the period in which they are working.[42] As very little was known about the brain in James's time, he was not able to offer detailed neural hypotheses, which has undoubtedly contributed to the timelessness of his writing. Contemporary speculations on the neural basis of religious or mystical experiences tend to be freighted with neuroscientific part-concepts doomed to be left by the wayside (as opposed to tested and rejected) as neuroscience evolves. Nevertheless, while we await the arrival of a general scientific theory of brain function—and only such a theory can render speculations regarding psychological phenomena sensible—we may adumbrate certain links between the phenomena of ultimacy experiences and clinical neurological data.

These links support a neurological model of ultimacy experiences that we shall rely on in the causal model of section 7, but it is at best a tentative part-model. We try to stay in close contact with physiological knowledge experimentally derived from large numbers of clinical cases in order to reduce reliance on speculation, with the consequence that we can adumbrate our model in some areas while we are forced to give scant attention to other important areas. Other theorists have prized completeness in neurological model making more highly than we have; they have been willing to pay, and indeed have paid, the higher price demanded in the unstable currency of neurological speculation. Both approaches can be helpful, notwithstanding the predictable fate of any detailed neurological speculation at the current time. Relatively complete models such as that of Eugene d'Aquili and Andrew Newberg and that of James Austin[43] have the great virtue of indicating what might be possible even if their detailed descriptions of putative brain states that underpin mystical

---

[42] This view is also expressed forcefully in H. Rodney Holmes, "Thinking about Religion and Experiencing the Brain: Eugene d'Aquili's Biogenetic Structural Theory of Absolute Unitary Being," *Zygon* 28.2 (1993): 201–30.

[43] See d'Aquili and Newberg, "Religious and Mystical States: A Neuropsychological Model"; James H. Austin, *Zen and the Brain: Toward an Understanding of Meditation and Consciousness* (Cambridge: MIT Press, 1998).

experience place them far from anything that has been established empirically in mainstream neuroscience.[44] The tests of intelligibility furnished by relatively complete speculative models play an important role in subsequent theorizing.

A preliminary point about the relation between disordered brain function and discrete ultimacy experiences needs to be made. Although we have emphasized the similarities between certain types of disordered brain function and some of the phenomena of discrete ultimacy experiences, we do not imply that ultimacy experiences are a form of illness. Unusual mentation, presumably based on unusual brain function, does not imply that the resulting experiences are "wrong": consider mathematical geniuses or individuals with perfect pitch. Conversely, the fact that an individual has temporal lobe epilepsy does not rule out the possibility that she or he is also having ultimacy experiences. That determination would depend on additional criteria (see section 6 below).

3.1 Elements of a Tentative Neurological Part-Model

We begin here by introducing the various elements of our neurological model, which we cluster into three phases: activation, quality, and social-linguistic conditioning.

Neural Expression: Activation. In broad terms, we expect discrete ultimacy experiences to be correlated in family resemblance fashion with neuronal events occurring in medial temporal lobe regions, as has long been thought,[45] perhaps spreading to the hypothalamus, as speculated by d'Aquili and Newberg.[46] It is unlikely that the exact pattern of neural activity in discrete experiences of ultimacy is invariant from one individual to the next. Instead, there may be brain regions that are more or less typically involved in such experiences. If a description of total, real-time brain activity becomes available in the future, we would expect that the neural patterns corresponding to subjective experiences of ultimacy would be variable, bearing family resemblances to one another, with some structures—perhaps anterior temporal cortices or the amygdala—more frequently represented in such patterns than others. In any event, most discrete ultimacy experiences probably require transient activation of the amygdala and hippocampus. Such activation probably occurs spontaneously in normal individuals due to random fluctuations in neuronal activity. This is not to say that individuals could not train themselves to induce such activity: in animal models chronic stimulation produces permanent alterations of

---

[44] Rodney Holmes insists in his review of d'Aquili's work that there is not yet any scientific way to confirm much of what the model hypothesizes; see his "Thinking about Religion and Experiencing the Brain." Yet d'Aquili is quite correct that mainstream neuroscience has wanted little to do with brain imaging of religious experience both because it is hard to arrange mystical states on cue and perhaps because of a vague prejudice against religion. This has left the territory to research groups with an ideological agenda, whose results are typically ignored by mainstream neuroscience; d'Aquili and Newberg exclude them, as do we. Were this not the case, effective scientific evaluation of the speculative proposals of d'Aquili and Newberg and other theorists might be more feasible. See d'Aquili's reply to his critics in "*Apologia pro Scriptura Sua*, or Maybe We Got It Right After All," *Zygon* 28.2 (June 1993): 251–66.

[45] Persinger aptly calls them "temporal lobe transients"; see *Neuropsychological Bases of God Beliefs*. Also see Wilder Penfield and Phanor Perot, "The Brain's Record of Auditory and Visual Experience: a Final Summary and Discussion," *Brain* 86 (1963): 595–702; Jaynes's discussion of Wernicke's area, as well as Penfield's and Perot's results in *The Origin of Consciousness*, 107–12; and almost every other neurological study of religious experience.

[46] D'Aquili and Newberg, "Religious and Mystical States."

connectivity in medial temporal structures more readily than in any other brain areas. Whether induced intentionally or not, there may be precursor experiences—certain kinds of concentration or preoccupation—that dispose toward these events.

The neural expressions of social and transformative extended ultimacy experiences are poorly understood and doubtless extremely diverse. In relation to the dimension of control, d'Aquili and Newberg hypothesize a set of neural schemas that underlie the detection of and the striving imaginatively to complete causal sequences of events. They further speculate that these schemas underlie much of the human need of religion for the crucial task of controlling the environment.[47]

Neural Expression: Quality. With regard to discrete ultimacy experiences, there are neurological data relevant to each of the phenomenological elements discussed earlier. With regard to sensory alterations and emotions, experiential phenomena occurring in discrete epileptic episodes, correlated with abnormal electrical discharges, have some similarities to phenomena described during discrete mystical or religious experiences in non-epileptic individuals; see section 3.2 on temporal lobe epilepsy, below.[48] With regard to cognition, data are scarce. However, the sense of conviction that attends ultimacy experiences may be explicable in neural terms; the relevant data are discussed in section 3.3. With regard to self-alterations and sense of presences, neurological data on alterations of person experience are most thought provoking; see section 3.4 below.

In relation to the phenomenal qualities of extended ultimacy experiences, we have little to say specifically about the neural underpinnings of the quality of existential potency. Of course, insofar as this involves cognitive certainty, the process of global semantic matching (mentioned in section 3.3) is relevant. Data on chronic personality changes due to temporal lobe pathology exist, however, and these are important for understanding transformation of behavior, personality, and beliefs; see section 3.5 below. The entire model is also relevant here: it shows how cognitive-somatic-emotional experiences might lead to a revision in stored,

---

[47] D'Aquili and Newberg call these schemas cognitive operators:

The cognitive operators we are referring to handle abstraction of generals from particulars, the perception of abstract causality in external reality, the perception of spatial or temporal sequences in external reality, and the ordering of elements of reality into causal chains giving rise to explanatory models of the external world, whether scientific or mythical. Briefly, the inferior parietal lobule on the dominant hemisphere of the brain, the anterior convexity of the frontal lobes primarily on the dominant side, and their reciprocal neural interconnections have been fairly definitively shown to account for causal sequencing of elements of reality abstracted from sense perceptions. The operation of cross-modal transfer, which is specific to the function of the inferior parietal lobule, is particularly implicated in causal sequencing. For convenience we refer to the anterior convexity of the frontal lobe, the inferior parietal lobule, and their reciprocal interconnections as the *causal operator*. Thus the causal operator…organizes [a] strip of reality into what is subjectively perceived as causal sequences back to the initial terminus of that strip. In view of the apparently universal human trait, under ordinary circumstances, of positing causes for any given strip of reality, we postulate that if the initial terminus is not given by sense data, the causal operator automatically generates an initial terminus."

Therein lies the connection to religion, as well as a bold attempt to specify the neural basis for what Immanuel Kant called the transcendental illusion. See d'Aquili and Newberg, "The Neuropsychological Basis of Religions, or Why God Won't Go Away," *Zygon* 33.2 (1998): 190–91.

[48] Of course, there are important differences between the phenomena of epilepsy and experiences of ultimacy. One is that ultimacy experiences are much more likely to be positive in tone, whereas the emotions experienced in complex partial seizures are more usually negative, though dread is fairly common in both cases. Another difference is the stereotypically repeated character of complex partial seizure experiences. This, however, does not prevent us from learning from similarities, where they exist.

generalized representations of self and world. The social embedding of extended ultimacy experiences has neurological connections as well; see section 3.6, which concerns brain functions that allow individuals to participate in the elaboration of social-linguistic systems.

Social-Linguistic Conditioning. Through their interconnections, the brain's neurons form a dense network that functions as a social-linguistic milieu for the interpretation and integration of novel experiences. The social-linguistic conditioning of ultimacy experiences begins with the process of global matching (already mentioned) and continues in enormously complex ways to allow individuals to participate in the performances and narratives of the larger social world, and to incorporate these into neural semantic structures. Neurological data relevant to social-linguistic conditioning are discussed most directly in section 3.3 below.

The remainder of section 3 discusses the five important classes of relevant neurological data alluded to in this introduction. Most of the data we treat derive from careful study of many clinical cases. One part of our account—global matching—is more speculative. It derives from a well-articulated theory of how the hippocampus and neocortex interact during normal learning; the theory is based in part on experimental evidence and in part on neural net simulations of semantic learning. The data we discuss involve no necessary religious content but rather bear on general processes for making sense of any current episode or information.

3.2 Temporal Lobe Epilepsy and Discrete Alterations of Experience

Deep within the anterior end of the temporal lobe in each hemisphere, below the cortical surface, lie two phylogenetically ancient structures: the amygdala and the hippocampus. Under normal conditions, the amygdala links incoming, highly processed sensory information to somatic outputs through its connections to the hypothalamus, to brainstem centers also involved in visceral control, and to primitive motor centers in the basal ganglia. When the amygdala is artificially stimulated by external electrical sources in human patients, somatic events that accompany emotions—sensations of tightness in the chest or piloerection ("goose bumps"), for example—are often produced. Visible signs such as pallor or fearful expressions may be seen. Various emotions, usually unpleasant, are often reported. The hippocampus is known to play a significant role in memory processes: disruption of its normal activity may cause failure to encode events. Because the hippocampus and amygdala lie adjacent to one another and are interconnected, they are often considered as a functional unit in investigations of the clinical effects of stimulation. It is also possible that the cortices immediately surrounding them, on the ventral and medial surfaces of the temporal lobe, are activated by electrical discharges occurring in these deeper structures. Cognitive experiences such as déjà-vu (the feeling that what is currently before one has been seen before) frequently result from stimulation of the medial temporal lobes, as do unpleasant emotions and brief mnemonic episodes.[49]

In animal models, the amygdala and hippocampus have very low thresholds for the induction of spontaneous seizures in response to chronic, low-level electrical stimulation. In clinical populations, these structures are often the source of complex

---

[49] See Eric Halgren, R. Walter, D. Cherlow, and P. Crandall, "Mental Phenomena Evoked by Electrical Stimulation of the Human Hippocampal Formation and Amygdala," *Brain* 101 (1978): 83–117.

partial seizures (CPS)—seizures characterized by cognitive, affective, or psycho-sensory symptoms, with or without motor automatisms. Unusual experiential phenomena are correlated with repetitive electrical discharges in the temporal lobes. The following descriptions are summarized from a review by G.W. Fenton.[50] Sensory phenomena in CPS can occur in any modality or in several together. Formed visual hallucinations may be simple and static, or intricate and progressing in time. An example of the latter is the image of a man carrying a cane accompanied by a dog. Another patient reported seeing irregular colored triangles replaced by the hallucination of a robber coming after him with a gun. Usually such experiences, and the ones described below, are repeated almost identically each time the patient has a seizure. Vertiginous hallucinations vary from simple sensations such as rotation to more complex sensations such as floating. Illusions can also involve any sensory modality. Objects may appear larger or smaller than they are. Shapes or sounds may become distorted. A limb may feel as if it does not belong to the patient or it may seem detached.

Any emotion may occur as a seizure phenomenon. The quality ranges from a crude undifferentiated welling up of feeling intruding on the patient's consciousness and unrelated to anything in the immediate environment, to highly refined feelings related to ongoing events in the environment. Fear is the most frequent. Pleasurable experiences are rare (sudden feelings of ecstasy, elation, happiness, serenity, or relaxation) but do occur. Unpleasant emotions that cannot be identified by the patient are not uncommon. The physiological basis of pleasurable versus unpleasurable experiences in CPS is unknown.

Under the category of cognitive symptomatology occurring in CPS, Fenton describes three subgroups. The first is ideational. The most common of this type of symptom, according to Fenton, is forced thinking—that is, the subject is incapable of resisting or putting out of his head some repetitive thought. The thought itself can be subjective, as in an idea such as death or immortality; it can be objective, as in a fixation upon a phrase read or heard before the attack; or it can be unidentifiable and impossible to recall after the attack. The second category is dysmnesic. Illusions of memory are common in CPS. These include déjà-vu, déjà-entendu, and déjà-vécu (these are encompassed by our usual use of the term déjà-vu), which are illusions of familiarity, and jamais-vu, jamais-entendu, and jamais-vécu, which are illusions of unfamiliarity. Other cognitive experiences in CPS include disturbances of time perception and the feeling that the world is not real (derealization). An additional cognitive phenomenon worthy of note in CPS is depersonalization; that is, the feeling that one's self is not real or that one is seeing one's body from an outside location. We discuss this below, in section 3.4.

There are theories regarding the anatomical basis of somatic, affective, mnemonic, and cognitive experiences arising from temporal lobe dysfunction. Somatic and affective phenomena are ascribed to the activation of amygdala efferents, such as those to the brainstem and hypothalamus mentioned above. Mnemonic phenomena (memories) are thought to be evoked when either the amygdala or hippocampus activates more widespread cortical areas in which networks of neurons encode records of sensory experience. The cognitive experiences are less well understood, however. According to some theories, the

---

[50] See G.W. Fenton, "Psychiatric Disorders of Epilepsy: Classification and Phenomenology," in *Epilepsy and Psychiatry*, E. Reynolds and M. Trimble, eds. (New York: Churchill Livingstone, 1981), 12–26.

hippocampus matches current experience with previously encoded episodes: possibly, if a match is falsely created, the experience of déjà-vu could occur. M-Marsel Mesulam speculates that unusual neuronal discharges in medial temporal structures might disrupt the normal "balance between affect on one hand and perception and thought on the other."[51] He considers phenomena such as déjà-vu and feelings of unreality to be a combination of sensory and affective experience.

The occurrence of derealization in CPS suggests that the conviction of the undeniable reality of ongoing experience—our usual stance towards the flow of events—depends on temporal lobe mechanisms. Experiences of déjà-vu teach us that the sense of familiarity can go awry; similarly, derealization phenomena suggest that the sense of conviction of reality can go awry as well. Just as the sense of familiarity can be either inappropriately missing (jamais-vu) or inappropriately present (déjà-vu) due to altered neural activity, the sense of conviction of reality might be affected by altered neuronal activity in the temporal lobe. We speculate that the conviction of reality, to which we now turn, depends on a process of global semantic matching that takes place in the temporal lobes.

3.3 Semantic Processing of Discrete Experiences

Semantic memory refers to stored information that is impersonal, and includes knowledge of words and their meanings, knowledge about objects and their interrelationships, and also general information about the world (e.g., knowing the meaning of the word *generous*, knowing the capital of France). Episodic and semantic memory are considered to be closely related and to interact with each other continuously.... Thus semantic knowledge is at least partly built up from information first acquired via episodic memory. Conversely, episodic memories have to be interpreted within the framework of existing semantic knowledge.[52]

Neuropsychological investigations show that the hippocampus is responsible for recording each new experienced event; however, permanent storage takes place gradually as a result of changes in the temporal cortex. According to one model, "repeated reinstatement of the hippocampal memory results in an accumulation of subtle neocortical changes, allowing the new memory (either episodic or semantic) to be integrated gradually into existing neocortical networks."[53]

The following clinical case, a patient with a progressive deficit in semantic memory, highlights the centrality of semantic processing and indicates its probable anatomy.[54] When the patient presented for evaluation, he had a five-year history of word-finding problems and recent problems with word comprehension. As his comprehension continued to decline, his everyday functioning became impaired. "For example, on one occasion, A.M. put orange juice in his lasagna and on another,

---

[51] M-Marsel Mesulam, "Dissociative States with Abnormal Temporal Lobe EEG: Multiple Personality and the Illusion of Possession," *Archives of Neurology* 38 (1981): 176–81; the quotation is from 181.

[52] A. McKay, P. McKenna, P. Bentham, A. Mortimer, A. Holbery, and J. Hodges, "Semantic Memory Is Impaired in Schizophrenia," *Biological Psychiatry* 39 (1996): 929–37.

[53] J. McClelland, B. McNaughton, R. O'Reilly, "Why There Are Complementary Learning Systems in the Hippocampus and Neocortex: Insights from the Successes and Failures of Connectionist Models of Learning and Memory," *Psychological Review* 102 (1995): 419–57, cited in Kim Graham and John Hodges, "Differentiating the Roles of the Hippocampal Complex and the Neocortex in Long-Term Memory Storage: Evidence from the Study of Semantic Dementia and Alzheimer's Disease," *Neuropsychology* 11 (1997): 77–89.

[54] This is summarized from Graham and Hodges, "Differentiating the Roles."

brought the lawnmower up to the bathroom when he was asked for a ladder." Brain imaging revealed marked atrophy of the inferolateral temporal lobes bilaterally, the left more than the right. Subsequent studies have confirmed an association between atrophy of the temporal lobe cortex, especially on the left, and semantic dementia.

Schizophrenic patients may also have relatively severe compromise of semantic knowledge. D. Tamlyn and others showed that of sixty schizophrenic patients, "nearly a quarter made significant numbers of errors on sentences like *rats have teeth* and *desks wear clothes*."[55] In this context, A. McKay and coauthors suggest that a "hyperfunctional" semantic memory could explain delusions—a person's "knowing" (that is, believing) things that are untrue.[56] We develop this suggestion somewhat differently as follows.

For background to our proposal, we return to the hippocampus. "Within the hippocampus itself, we assume that the event or experience is represented by a sparse pattern of activity in which the individual neurons represent specific combinations or conjunctions of elements of the event that gave rise to the pattern of activation."[57] These patterns would have arisen from activation of neocortical areas representing features of the experience, ultimately feeding into the hippocampus via its primary input source, the entorhinal cortex. The patterns are considered to be codes for the conjunctions of features that make up an experienced event.

In analogy to data compression schemes used for computer files, the information contained in neocortical patterns is thought to be redundant, and thus compressible. Fewer synapses are needed for storage of the information in the hippocampus than in the neocortex. The compressed version is called a "summary sketch." Compression is assumed to occur from the neocortex to the hippocampus, with decompression going the other way. If there are several way stations going in and out—and there are, including the perirhinal cortex and the parahippocampal gyrus—then compression-decompression can be sophisticated.

When a new event takes place—consisting of internal, somatic sensations as well as external ones—it would first be sparsely represented in a pattern of hippocampal connections, then decompressed in temporal cortical regions surrounding the hippocampus for transfer to neocortex. The transfer is essentially an interaction between temporal cortical patterns and extant widespread neocortical patterns in more primary sensory areas. Connectionist models suggest that the throughput to the neocortex is straightforward if global characteristics of the event pattern are already shared to a large extent with representations in the neocortex. However, when new input is at odds with what is already stored, widespread alterations in the overall performance of the neocortical network result.

We can recast the hippocampal-neocortical interaction as follows. In normal circumstances, the continuity of current events with representations of prior events—a match not at the level of "has this single event occurred before?" but at the more global level "is this event consistent with all I know?"—takes place seamlessly, the ongoing, unnoticed internal response being, "yes...yes...yes." This "yes" is another way of describing a straightforward decompression—a good enough fit—between

---

[55] D. Tamlyn, P. McKenna, A. Mortimer, C. Lund, S. Hammond, and A. Baddeley, "Memory Impairment in Schizophrenia: its Extent, Affiliations and Neuropsychological Character," *Psychological Medicine* 22 (1992): 101–15. Quoted in McKay et al., "Semantic Memory."

[56] See McKay et al., "Semantic Memory."

[57] McClelland et al., "Why There Are Complementary Learning Systems," 423–24.

currently formed patterns of synaptic activity in the hippocampus and pre-existing, more widespread patterns in surrounding cortices. When an event occurs that does not match what we know about the world, it is usual to experience the event as unreal, at least transiently.

We speculate that ongoing global matching is the basis for a background feeling of the reality of current experience that is generated by temporal lobe processes. We furthermore suggest that global matching is a personal semantic process, meaning that it depends on accumulated knowledge of the way events, objects, and the self normally relate. Every complex, unique experience can be globally matched if it can be made consistent with some region of the total semantic network that the individual has constructed. It is unlikely that a patch of temporal cortex (corresponding to areas that are atrophied in patients with semantic dementia) contains all the semantic information and relations that an individual possesses. It may be, however, that such a patch of cortex is critical in the decompression process, indexing the web of relationships that the person has been building throughout his or her life by accessing the bits and pieces represented elsewhere in widespread regions of cortex.

We can imagine what would happen if decompression cortices began to behave anomalously, as they might in the context of spreading electrical activity from the nearby hippocampus or amygdala. They might falsely send signals that declare, "Current experience has successfully been matched and incorporated into the global semantic network." Or they might declare the opposite. In the latter case, the person might have a feeling that the event just then occurring is unreal. In the former, he would decide that his current experience, however unusual, is real. He would then be faced with the subsequent problem of incorporating a bit of "real" experience into a network with which it is not compatible. The unusual but real-seeming experience could be denied access to larger networks—walled off in some way—or it could force changes in the rest of the semantic economy.

We do not know what occurs when a real-seeming bit of experience must be incorporated into a global semantic network with which it is incompatible. It is possible that the brain might go into a state of widespread, heightened activity when it is necessary to update and revise widespread semantic networks in light of an anomalous experience. Such widespread revision would probably produce heightened metabolic activity throughout the brain as large numbers of synaptic connections are modified, a state that might be experienced as positive. Anecdotally at least, when persons who are working on a difficult problem suddenly see a new way to look at it, a way that forces revision of many of their previous assumptions, the experience may be described colloquially as a "rush."

However, not all unusual experiences that are accepted as real succeed in forcing widespread semantic re-organization so as to become integrated into the individual's total experience of the world. Mesulam described a series of twelve patients with clinical or EEG signs of temporal lobe epilepsy, some of whom developed dissociative disorders (commonly known as multiple personality disorders), some of whom had delusions of possession, and one of whom had elements of both. Based on his analysis, he speculated, "It is conceivable that autonomous mental events that originate in the nondominant hemisphere are more likely to lead to dissociative states, whereas those that originate in the hemisphere dominant for language may be more likely to be adopted as part of the self."[58] There has been no further research to shed light on his theory.

---

[58] Mesulam, "Dissociative States," 181.

Referring to seizure events, Fenton states, "It is important to note that this altered content of consciousness constitutes an intrusion upon the patient's ongoing stream of awareness. No matter how vivid, complex or 'real' the ictal experience, the patient recognizes that it is an experience imposed upon him."[59] In other words, Fenton holds that most CPS experiences are recognized as nonreal by the patients. When an individual has a series of experiences that occur in stereotyped fashion many times, it is to be expected that he or she would learn to label these as seizures. However, it is apparent that at least some epileptic individuals do embrace their experiences as real. We speculate that what makes the difference between cases such as those described by Fenton and those described by Mesulam is how the seizure activity affects the anterior temporal cortex, and hence the global matching process that underlies the conviction of reality.

Discrete experiences of ultimacy probably involve the automatic attempt to match the unusual experience with the person's total global semantic network. At this point, the experience could be rejected as "not real," in which case it would not achieve the status of ultimacy. To be an experience of ultimacy, the event must be stamped with what James called "conviction." In analogy to altered experiences of familiarity, we believe the conviction of reality can occur anomalously when neural discharge spreads to the cortical areas that perform global semantic matching. In general, however, the threshold for a match would depend on the contents of previously stored memories and concepts, as well as on the activation of temporal cortical decompression mechanisms. In a religiously acculturated individual with previous ultimacy experiences, the neural threshold for a conviction of the reality of any given ultimacy experience might be quite low. In a person with no previous religious experience or concepts, the unusual experience might have to be "pushed through" by activation of temporal lobe global matching mechanisms; conditions favoring large-scale plasticity of semantic representations in more posterior cortices would also favor global matching in such a case. When a match results due to any of these factors, the individual not only might have (for example) a feeling of extraordinary tranquility and a loss of the sense of the individual self, but he or she would also have the sense that "*This*—however unusual it seems—*is real*."

Although we can imagine how alterations in the conviction of reality might be effected by mechanisms such as the one proposed here, we do not thereby pre-judge discrete experiences of ultimacy and attendant convictions of reality simply as a breakdown of the neural machinery for accurate judgment. In both everyday experiences and ultimacy experiences, the brain's activity in attempting global matches is likely to be at least as "artful" (to use the terminology of ethno-methodologists) as social practices are in establishing consensual reality. To see brains as either true or flawed mirrors of nature denies the continuous processes of engagement that characterize the relations between brains and the worlds in which they operate. Global matching must depend on neural processes underlying decompression, in interaction with the representations already available in other cortical areas for the assimilation of the current experience, representations that are themselves subject to alteration by the experience.

In the terminology of religious reports of discrete altered states, it is not unusual for an individual to have first a complex sensory experience (such as the conviction of a presence together with a feeling of awe or ecstasy) succeeded closely by the revision of all previous understandings of a certain kind. That is, the previous

---

[59] Fenton, "Psychiatric Disorders of Epilepsy," 17.

understanding of garden hoses and stop signs, for example, is unaffected, but previous understanding at higher levels—of the nature of the self and its relation to the universe, of the general principles expressing the nature of the world—is revised, the revision being accompanied by a heightened sense of intellectual perception. In a discussion of conversion similar to the present account, Warren Brown and Carla Caetano postulate that a revision in the brain's semantic networks based on novel religious experience produces a sense of excitement and joy at having discovered a new schema with a better fit to one's life experiences.[60]

3.4 Alterations of Person Experience

The normal human brain attributes subjective life to bodies according to certain rules. The basic rules are (1) one self per body, which "owns" the body and is located in it in a peculiar way, and (2) one identity per mind/body unit. P.F. Strawson termed this mind-body unit the "person."[61] As we shall see, when brain function is disturbed, these rules are broken. Put differently, the affected individual cannot make sense of his or her altered experience of "person" except by changing the rules.

Neurologic disorders may cause alterations in the sense of self; these are known collectively as misidentification syndromes. The brain lesions responsible for these symptoms vary and are often diffuse rather than localized. A survey of the literature on brain injury and misidentification suggests that neural pathways linking representations of the body in the parietal lobes with more anterior temporal structures such as the amygdala, especially in the right hemisphere, must be damaged in order for misidentification to occur.[62]

Occasionally, misidentification syndromes are related to temporal lobe epilepsy, as in depersonalization phenomena of CPS, in which one feels as if one's self is not real, or as if one's body is regarded from some outside location. There are other interesting and more chronic symptoms affecting the sense of subjectivity. In such syndromes, the person can feel as if his or her mind may have been located to another body; or that other minds are taking over his or her own body. Affected people may feel that others around them have had other minds substituted for their "real" ones—that is, that impostors have taken over the bodies of familiars. Alterations of the perception of self in the first person are pertinent to the self-dissolution phenomena that sometimes occur in discrete ultimacy experiences.

The phenomenon of experienced "presences" in discrete ultimacy experiences, by contrast, pertains to other minds. A presence is the representation of another person without the representation of the body. In a sense it is the converse of the illusion of other bodies represented as being without minds, such as zombies or persons with "alien" minds, common percepts in misidentification syndromes. In either case, whether it is the mind that is missing and the body present, or the other

---

[60] Warren Brown and Carla Caetano, "Conversion, Cognition, and Neuropsychology," in *Handbook of Religious Conversion*, H.N. Maloney and S. Southard, eds. (Birmingham, Ala.: Religious Education Press, 1992), 147–58.

[61] See P.F. Strawson, "Persons" in *Individuals: An Essay in Descriptive Metaphysics* (London: Methuen and Co., 1959), 87–116.

[62] Regarding the neural basis of misidentification syndromes, see Hadyn Ellis, "The Role of the Right Hemisphere in the Capgras Delusion," *Psychopathology*, 27 (1994): 177–85; and Leslie Brothers, *Friday's Footprint: How Society Shapes the Human Mind* (New York: Oxford University Press, 1997).

way around, the third-person identity is lost. A mind without a body must be experienced in a different way than ordinary persons, as a different kind of being. Furthermore, in mystical or religious experience, it is usual for a presence to be only vaguely or not at all localized in space.

3.5 Chronic Personality Changes Related to Temporal Lobe Pathology

A syndrome of chronic characterological features has been described in persons with temporal lobe epilepsy. Since it is present in the absence of acute electrical activity (the "ictus"), it is termed interictal personality.

> The syndrome includes the following features: increased concern with philosophical, moral or religious issues, often in striking contrast to the patient's educational back-ground, an increased rate of religious conversions (or strongly justified, rather than casual, lack of religious feeling), hypergraphia (a tendency to highly detailed writing often of a religious or philosophical nature), hyposexuality (diminished sex drive sometimes associated with changes in sexual taste), and irritability of varying degree.[63]

David Bear reviewed previous studies of sixty-nine patients who had developed psychotic symptoms on the average of fourteen years after the onset of seizures as follows:

> Affective disturbance was "shown by all patients," most frequently a deepening of emotion... and preserved affective intensity. Delusional ideas appeared in 67 out of 69 patients, mystical religious conceptions being extremely common. Paranoid feelings and explanatory systems justified a diagnosis of paranoid schizophrenia in 46 patients... Hallucinations occurred in 63 patients, typically consisting of formed visual images or conversational phrases experienced with intense emotional significance (e.g., a vision of Christ on the Cross in the sky, the voice of God saying, "You will be healed, your tears have been seen."[64]

In a subsequent paper, Bear and his coauthors wrote,

> The summary traits which most powerfully differentiated temporal lobe epileptics from a mixed psychiatric group were excessive interpersonal clinging (viscosity), repetitive preoccupation with peripheral details (circumstantiality), religious and philosophical preoccupations, humorless sobriety, tendency to paranoid over-interpreta-tion, and moralistic concerns.[65]

The authors also remarked on these patients' "propensity to write extensively—diaries, notebooks, novels, or biographies," noting that this writing often has a cosmological or moral tone. (While Bear's descriptions are widely accepted, not all authors agree with his characterizations of the interictal personality.[66]) To explain the

---

[63] Norman Geschwind, "Behavioural Changes in Temporal Lobe Epilepsy," *Psychological Medicine* 9 (1979): 217–19; quotation is on 217.

[64] David Bear, "Temporal Lobe Epilepsy—A Syndrome of Sensory-Limbic Hyper-connection," *Cortex* 15 (1979): 357–84; quotation, 363. Reviewed were E. Slater and P.A.P. Moran, "The Schizophrenic-Like Psychoses of Epilepsy: Relation Between Ages of Onset," *British Journal of Psychiatry* 115 (1969): 599–600; E. Slater and A.W. Beard, "Schizo-phrenia-Like Psychoses of Epilepsy," *British Journal of Psychiatry* 109 (1963): 95–150.

[65] David Bear, K. Levin, D. Blumer, D. Chetham, and J. Ryder, "Interictal Behaviour in Hospitalised Temporal Lobe Epileptics: Relationship to Idiopathic Psychiatric Syndromes," *Journal of Neurology, Neurosurgery, and Psychiatry* 45 (1982): 481–88.

[66] For example, see Dan Mungas, "Interictal Behavior Abnormality in Temporal Lobe Epilepsy," *Archives of General Psychiatry* 39 (1982): 108–11; David M. Tucker, R. Novelly, and P. Walker, "Hyperreligiosity in Temporal Lobe Epilepsy: Redefining the Relationship," *Journal of Nervous and Mental Disease* 175 (1987): 181–84.

interictal personality, Bear speculated that recurrent seizure activity in medial temporal lobe structures causes aberrant synaptic connections to form, resulting in what he termed "sensory limbic hyperconnection." He theorized that these hyperconnections give rise to an overinvestment of perception and thought with affective significance.

Recently, in a preliminary study of two epileptic patients with religious preoccupations, V.S. Ramachandran and his colleagues detected unusually intense autonomic responses to religious images, while other images that usually provoke autonomic activity—such as sexual and violent material—produced less response than in normal subjects. These findings suggest that if hyperconnectivity is responsible for features of the interictal personality, it acts selectively: such new circuits might produce "new peaks and valleys in the patients' emotional landscape."[67] Although the new landscape might yield pronounced religious feelings, this does not prove that there are circuits potentially dedicated to religious belief in normal brains. It is possible, as Ramachandran points out, that certain more general-purpose emotional circuits are simply conducive to religious experience when selectively potentiated.

We consider below, in section 7, the question as to whether the interictal personality belongs within the category of extended experiences of ultimacy.

3.6 Neurological Considerations Relevant to Sociality

Social participation enters into our neural description of ultimacy experiences at several levels. First, the phenomena of ultimacy are rendered intelligible and meaningful to the individual experiencing them by means of the social-linguistic systems of his or her group. We explored above how specialized groups, using their particular systems, interpret aspects of what might be theologically termed divine action in the form of experiences of ultimacy. Second, under the heading of extended ultimacy experiences, we have considered moral transformations of the individual occurring, for example, in dialogue with a mentor. We have said that the fundamental feature of character transformation is the ability to enter into a proffered social-linguistic framework, consequently changing the self by participation in a new moral order. In both cases, social participation is essential.

The neurobiology of human social participation is beginning to be understood.[68] In the course of primate evolution, certain structures that had linked olfaction with social behavior became much more diverse in their connections and differentiated in their architecture. As ancestral primates became diurnal instead of nocturnal, used vision to receive social signals and expressive faces to send them, and lived in more complex social groups, these brain structures correspondingly received input from more extended cortical regions and were able to generate more intricate behavioral responses. A specialization for social cognition appears to involve brain areas that process faces—especially the cortex of the temporal lobes—and deeper structures such as the amygdala and orbital frontal cortex, together with those parts of the frontal cortex that evaluate complex and rapidly shifting contexts. The sophisticated deployment of these circuits is best seen in the demanding face-to-face situation known as conversation.

---

[67] V.S. Ramachandran and Sandra Blakeslee, *Phantoms in the Brain: Probing the Mysteries of the Human Mind* (New York: William Morrow and Company, 1998), 188.

[68] See Brothers, *Friday's Footprint*.

The brain structures that subserve social participation are conduits for the transmission of symbolic systems between individuals. Defects in their function would prevent acculturation of the individual by preventing social participation.

Terrence Deacon comments that social attention is crucial for the acquisition of symbols, and discusses the role of the prefrontal cortex in recruiting attention to social stimuli.[69] He also gives great emphasis to the role of the prefrontal cortex in constructing the distributed mnemonic architecture that supports symbolic reference—not just in relation to language, but in relation to human cognition in general, for cognition requires the ongoing construction of novel symbolic relationships.[70] We saw earlier that the matching of current experience with stored semantic memory is likely to involve the temporal lobe cortex. How the functions of this cortex may relate to the prefrontal symbolic system hypothesized by Deacon is still not known. Nevertheless, we have the outlines of a neural system that enables social communication in the first instance and storage of acquired semantic categories subsequently. Such a system would make pedagogy possible, and probably must be intact in order for characterological transformation under the influence of social learning to occur.

Even less well understood, but mentioned here for the sake of completeness, are innate propensities for imitation, seen within minutes of birth in human infants. The rapid and seemingly automatic spread of certain behaviors in groups has been observed in many species of animals, as well as in human beings. Social contagion is observed in human infants in the phenomenon of so-called "contagious crying" that occurs in nurseries. Clues to the neural mechanisms for such behavior may reside in "mirror neurons" studied in monkeys. These "subjectless" neurons fire in response to certain actions both when they are viewed by the animal subject and when performed by it.[71] Certain discrete ultimacy states such as trances and mystical ecstasy seem to be facilitated by group participation. These, then, constitute a further instance of the role of social participation in generating experiences of ultimacy.

4 Social-Psychological Considerations

Experiences of ultimacy can be described from many points of view, each involving characteristic terminology and usually presupposing the social-linguistic framework of some social context.[72] These contexts and vocabularies may be vernacular or professional, religious or secular, theological or nontheological. We need to examine as many thought systems as possible for observations relevant to the multi-faceted model of ultimacy experiences that we propose in section 7. In practice, however, there are two limitations. On the one hand, we have space only to discuss a few lenses through which human beings observe themselves and which can be used to

---

[69] Terrence Deacon, *The Symbolic Species: The Co-evolution of Language and the Brain* (New York: Norton, 1997), 272.

[70] Ibid., 266.

[71] G. Rizzolatti, L. Fadiga, V. Gallese, and L. Fogassi, "Premotor Cortex and the Recognition of Motor Actions," *Cognitive Brain Research* 3 (1996): 131–41.

[72] We have found the sociology of religion important for understanding the social-psychological considerations relevant to ultimacy experiences, particularly works influenced by the sociology of knowledge such as Peter L. Berger and Thomas Luckmann, *The Social Construction of Reality: A Treatise in the Sociology of Knowledge* (Garden City: Doubleday, 1966); and Berger's *The Sacred Canopy*.

describe ultimacy experiences (usually under other descriptions, of course). On the other hand, we are limited by not being participants in all such systems. This means that, for any given thought system, we may only turn up a small sample of ideas and observations relevant to experiences of ultimacy. In principle this limitation can be overcome through deeper participation, but we have dealt with it by trying to limit ourselves to systems with which we have some familiarity.[73]

4.1 Psychoanalysis

A number of insights from psychoanalysis have bearing on an adequate understanding of the role of social-linguistic systems in mediating ultimacy experiences.

Early Development. First, beginning with Sigmund Freud, psychoanalytic theorists have postulated psychological developmental stages through which children naturally progress. The idea of developmental stages provides a richly descriptive framework for characterizing the various ways in which self, other, and the world may be experienced. Adults are viewed as the products of developmental processes, with highly individual and more or less successful outcomes. Freud's scheme of stages of libidinal development[74] has been succeeded by others such as those of Margaret Mahler's separation-individuation scheme[75] and Heinz Kohut's focus on narcissism and the self.[76] William Meissner[77] and Ana-Maria Rizzuto[78] have thoughtfully elaborated some connections between such stages and the possibilities for experiences of faith. These later psychoanalytic theorists have tended to remain neutral to the question of the cause of experiences of ultimacy, while providing rich psychological frameworks within which the potently transformative effects of ultimacy experiences can be understood.

Subject and Object. Second, D.W. Winnicott has written on the transitional object, that part of experience that is in between being self and not-self. In a telling sentence, he says, "In the rules of the game we all know that we will never challenge the baby to elicit an answer to the question: did you create that or did you find it?"[79] The position suggested by this quotation has interesting implications for interpreting personal experiences of ultimacy because it provides a way of understanding how

---

[73] We think that sociology of knowledge and other sociological lines of analysis would offer especially important perspectives on ultimacy experiences. For example, there appear to be significant correlations between socio-economic status and the types of ultimacy experiences that typically occur, suggesting that economic and class analysis could be quite fruitful.

[74] Sigmund Freud, *Introductory Lectures on Psycho-Analysis, Part III, General Theory of the Neuroses*, in *The Standard Edition of the Complete Psychological Works of Sigmund Freud*, J. Strachey, ed. (London: The Hogarth Press, 1963; first published in 1917).

[75] Margaret S. Mahler, *The Psychological Birth of the Human Infant: Symbiosis and Individuation* (New York: Basic Books, 1975).

[76] Heinz Kohut, *The Analysis of the Self: A Systematic Approach to the Psychoanalytic Treatment of Narcissistic Personality Disorders* (New York: International Universities Press, 1971).

[77] William W. Meissner, *Psychoanalysis and Religious Experience* (New Haven: Yale University Press, 1984).

[78] Ana-Maria Rizzuto, *The Birth of the Living God: A Psychoanalytic Study* (Chicago: University of Chicago Press, 1979).

[79] D.W. Winnicott, "The Use of an Object and Relating Through Identifications," in *Playing and Reality* (New York: Basic Books, 1971), 89.

ultimacy experiences could be both within the self and from outside. Faith can be thought of as the suspension of the question, "Did you create that (revelatory experience) or did you find it?"

Winnicott's ideas have been further elaborated by Christopher Bollas in the concept of a transformational object, "[a person, place, event, ideology] that promises to transform the self."[80] This concept resonates with temporally extended experiences of ultimacy in the form of conversion or character transformation mediated through a relationship with symbols, rituals, or other persons. As is the case with other psychoanalytic thinkers, these theorists illumine the transformative power of ultimacy experiences. But the power of the ideas of transitional objects and transformational objects to explain personal transformation also indicates something about the social-linguistic entanglements of such experiences: they are crucially linked both with the internal world of the person and with their social-linguistic milieu. The objects of ultimacy experiences function as transitional and transformational objects.

Spiritual Guide. Third, although the term "spiritual guide" is alien to the vocabulary of psychoanalysis, the relationship between the analyst and analysand has received much attention, with results that might enrich a more traditionally theological framework concerned with the role of a spiritual mentor or a non-theological scheme of moral development involving a mentor. The key point is that "The analyst at the outset and throughout the work functions not simply on what he observes the patient *is,* but on what he both infers and implies the patient *might become,* that is, someone with a capacity for realizing further ego growth."[81] On the basis of this insight, a number of concepts have been advanced to capture how the analysand's relation to the analyst produces meaningful change: working alliance, therapeutic alliance, transference, transference neurosis, the holding environment, the transformational object, and others. Radmila Moacanin has also drawn a parallel between the classical guru, or "spiritual friend," and the psychoanalyst.[82] These insights of psychoanalytic theorists are most relevant to temporally extended ultimacy experiences. The "spiritual guide" serves as a trusted source of wisdom by which the social-linguistic environment of the one undergoing transformation is enriched and extended in efficacious ways. This underlines the importance for transformation ultimacy experiences of both flexibility within social-linguistic frameworks and established wisdom about appropriate patterns of social-linguistic change. Personal transformation without social-linguistic flexibility is impossible, and change without the regulation of established wisdom is precarious.

Discernment. Fourth, some of the psychoanalyst's activities are similar to religious discernment (see below for a discussion of the latter). Presentations of clinical cases often center on critical evaluative moments in which the analyst discerns that a bit of unconscious material has been brought to light. Here is a typical example, regarding a patient whose elder brother had died when he was three and who had been unable to reach the analyst by phone during a previous missed session. The analyst writes, "I pondered and puzzled, wondering why on earth he was

---

[80] Christopher Bollas, "The Transformational Object," *International Journal of Psychoanalysis* 60 (1979): 97–107; quotation is on 14.

[81] W. Poland, "On the Analyst's Neutrality," *Journal of the American Psychoanalytic Association* 32 (1984): 283–99; quotation is on 296, italics in original.

[82] Radmila Moacanin, *Jung's Psychology and Tibetan Buddhism: Western and Eastern Paths to the Heart* (London: Wisdom Publications, 1986), 56–58.

relieved once he had got through to my secretary. Then the truth struck me, 'Ah, of course, he was relieved that I was not dead'. I communicated this to him and he immediately assented. We both realized how great was his anxiety that I would suffer the fate of his elder brother."[83] Now, this example does not involve ultimacy experiences, but analytic discernment is a basic preoccupation of the psychoanalytic community that applies to all patient experiences, including ultimacy experiences. Analytic discernment presupposes both that there are unconscious truths to be discovered and that it is not so easy to figure out what those truths are. Criteria for having discovered unconscious truths include not only the assent of the patient, but also his or her subsequent behavior, dreams, and other communications. The psychoanalyst's role as an expert in discernment underlines the hermeneutical complexity of the social-linguistic systems within which ultimacy experiences must be interpreted. This complexity makes discernment vital in the identification of ultimacy experiences either within religious groups, in the ongoing conversations with spiritual advisors, or in self-evaluation. The meanings of ultimacy experiences are thus typically far from obvious and require as much discernment as does the identification of unconscious truths in the psychoanalytic context.

Based on all that has been said, it seems to us that ultimacy experiences cannot be understood in isolation from a rich appreciation of human dependence on the conceptual-linguistic conditions of the social environments in which people live and change. The previous points express various aspects of this social-linguistic conditioning from a psychoanalytic perspective. But both discrete and temporally extended ultimacy experiences call for a discussion of the provenance of ultimacy experiences in psychoanalytic (as well as other) terms. Psychoanalysis is relevant to this issue in at least two ways.

Role of an External Force or Power. First, one of the key tenets of psychoanalysis is that the unconscious is dynamic, producing effects on conscious life and behavior. In this sense it acts as an external power over which the individual has no control. James wrote, "since on our hypothesis it is primarily the higher faculties of our own hidden mind which are controlling, the sense of union with the power beyond us is a sense of something, not merely apparently, but literally true."[84] The connection James envisaged between the dynamism of the unconscious or "hidden mind" and the religious idea of "higher" control is fascinating and important. On the surface it may seem to be unduly reductive of ultimacy experiences, but we think this would be a hasty conclusion. To identify the unconscious as a locus of ultimacy experiences merely indicates in psychoanalytic terms a part-condition for ultimacy experiences to occur without thereby also demonstrating that there can be no authentic encounter with ultimacy mediated by unconscious dynamics.

Experiences of Loss of Self. Second, experiences of depersonalization or merger states take place in psychoanalytic treatments. These experiences are typically rendered in narrative form using such concepts as drives, self-states, or regressions to earlier developmental stages. An example is J.M. Masson's account of the "oceanic feeling"[85] and other psychoanalytic accounts of the sense of reunion with

---

[83] Neville Symington, "Psychoanalysis: A Servant of Truth," in *The Analytic Experience* (New York: St. Martin's Press, 1986), 15–24. Quotation is on 18.

[84] James, *Varieties*, 503.

[85] J.M. Masson, *The Oceanic Feeling: The Origins of Religious Sentiment in Ancient India* (Dordrecht: D. Reidel, 1980).

an omnipotent force.[86] Here we see a correlation between psychoanalytic and religious categories that once again raises the specter of reductionism. The same argument as in the previous point also applies here, however: to render religious descriptions of ultimacy experiences in psychoanalytic categories is not to reduce the former but simply to redescribe them. Such redescriptions and correlations are the way that religious language about ultimacy experiences is connected with other social-linguistic spheres. Such connections are to be desired for the intelligibility they bring to religious categories. To achieve the dual result that religious categories are utterly superfluous and actually misleading through invoking the influence of imaginary entities would require more than just convincing translations between social-linguistic frameworks. In fact, we do not see how any amount of psychoanalytic theory would be capable of establishing such a result.

4.2 Life-Stage Psychology

There are a number of insights into ultimacy experiences to be gained from the psychology of life stages.

Adult Development. Eric Erikson extended the psychoanalytic idea of developmental stages into adulthood. Two such stages are generativity and ego integrity. "Generativity is primarily the interest in establishing and guiding the next generation or whatever in a given case may become the absorbing object of a parental kind of responsibility." A failure to achieve generativity results in stagnation. Ego integrity involves "a post-narcissistic love of the human ego—not of the self—as an experience which conveys some world order and spiritual sense, no matter how dearly paid for...The lack or loss of this accrued ego integration is signified by fear of death."[87] These concepts are useful for describing a gradually achieved personal transformation in conjunction with experiences of ultimacy.

Spiritual Development. Stephen Happel and James Walter[88] cite the work of Lawrence Kohlberg on cognitive-moral development[89]; they show that James Fowler's description of stages of faith development is a further elaboration of Kohlberg's account.[90] Following Fowler, they conclude that religious conversion depends on innate ontogenetic structures that unfold in an invariant way during development.[91] Both Erikson and Fowler offer insights into human development over the course of life that are important for understanding temporally extended ultimacy experiences. Specifically, we learn that the meaning of temporally extended ultimacy experiences shifts with age and spiritual experience. These are further factors modifying the social-linguistic conditions for the understanding and expression of temporally extended experiences of ultimacy.

---

[86] Otto Fenichel, *The Psychoanalytic Theory of Neurosis* (New York: W.W. Norton, 1945), 40.

[87] E. Erikson, "The Eight Stages of Man," in *Childhood and Society* (New York: Norton, 1950), 231.

[88] Stephen Happel and James Walter, *Conversion and Discipleship: A Christian Foundation for Ethics and Doctrine* (Philadelphia: Fortress Press, 1980).

[89] Lawrence Kohlberg, *Essays on Moral Development, Vol. 1, The Philosophy of Moral Development* (San Francisco: Harper and Row, 1981).

[90] James Fowler, *Stages of Faith: The Psychology of Human Development and the Quest for Meaning* (San Francisco: Harper and Row, 1981).

[91] Happel and Walter, *Conversion and Discipleship*, 54–60.

Death. Some kinds of discrete ultimacy experiences can be usefully framed within the vicissitudes of adult development, especially the approaching end of life. Persinger thematizes the issue as follows:

> God Experiences... are precipitated by personal crises, such as the loss of a loved one (real or imagined) or the confrontation of an insoluble problem. Certainly the greatest insoluble problem is the anticipation of self-extinction. Death anxiety increases in incremental steps as the person ages and approaches the latter portion of life. God Experiences proliferate during these periods and may even occur as death-bed episodes. The God Experience is followed by a remarkable anxiety reduction and a positive anticipation of the future.[92]

4.3 Evolutionary Psychology

The basic tenet of evolutionary psychology is that the human mind evolved in response to the demands of a hunting and gathering way of life so as to increase reproductive fitness through avoiding predation, assisting kin, finding a mate, and so on.[93] Moral behavior is explained in this theory by the need to promote relationships based on reciprocity, with the effect that the individual represses selfish behavior in favor of altruistic behavior. Evolutionary psychologists speculate that there are universal deep structures for moral beliefs due to common patterns of kin and reciprocity relationships in the environment of evolutionary adaptation. They also invoke self-deception, akin to the psychoanalytic concept of repression, as a means for disguising to oneself one's own selfish interests, in the service of making selfless acts appear more convincing to others.[94]

These ideas have relevance for both discrete and temporally extended ultimacy experiences. On the one hand, the evolutionary-psychological inducements for having certain beliefs and behaviors may be sufficiently strong in some cases to predispose people to discrete ultimacy experiences that are capable of forging or solidifying such adaptive beliefs and behaviors. This is speculative but highly probable in view of the development in at least human beings of a rich emotional life, one of whose effects is precisely the predisposing of people to certain beliefs and behaviors. Discrete ultimacy experiences would seem to be nothing more than a special case of this more general process. On the other hand, the insights of evolutionary psychology are important for understanding the processes of character transformation and increased integration into the moral-social order that are characteristic of temporally extended experiences of ultimacy. Complex social organizations will be most adaptive when stable conformation of individual members to social rules and commitments is achieved. In such social environments, conversion and character transformation naturally take on special significance.

---

[92] Persinger, *Neuropsychological Bases of God Beliefs*, 2.

[93] Leda Cosmides, J. Tooby, and J. Barkow, "Introduction" in *The Adapted Mind: Evolutionary Psychology and the Generation of Culture*, J.H. Barkow, L. Cosmides, J. Tooby, eds. (New York: Oxford University Press, 1992), 3–15.

[94] R. Nesse, A. Lloyd, "The Evolution of Psychodynamic Mechanisms," in *The Adapted Mind*, Barkow, Cosmides, and Tooby, eds., 601–24.

5 Theological-Ethical Considerations

A number of theological or ethical perspectives are relevant to ultimacy experiences.[95] Ultimacy experiences are often referred to in stories with theological or ethical overtones (the story of the Buddha's calling and enlightenment, for example). Ultimacy experiences are also presupposed by some theological or ethical concepts (such as Zen's satori or Judaism's repentance). These narrative and conceptual ways of invoking ultimacy experiences are theologically and ethically loaded especially for the groups within which such narratives and concepts play key roles. This makes the experiences presupposed in the narratives and concepts expected and intelligible in those groups. Moreover, people typically describe their ultimacy experiences in terms of the narratives and concepts of their group and the suffusion of such narratives through the shared practices of a group may help to induce the occurrence of particular ultimacy experiences. All of this occurs in part because these narratives and concepts probably encode with tolerable accuracy, in the specialized language of the group, the group's collected wisdom about the way ultimacy experiences occur and the sorts of transformations their occurrence can induce. In what follows, we discuss these dynamics in relation to theological concepts and narratives (5.1), ethical concepts and narratives (5.2), and the way experiences are expressed in language (5.3). Finally, we comment (in 5.4) on the almost ubiquitous yet highly varied processes of discernment that regulate the application of theological and ethical ideas to the ultimacy experiences of individuals within a religious group.

5.1 Theological Concepts and Narratives

Specifically theological concepts active within the social-linguistic environment of a religious group are assigned meanings usually by means of narratives that express the typical experience of group members. These concepts and narratives frequently presuppose that ultimacy experiences of the discrete or extended variety should occur on particular occasions as the narrative is lived out or in particular ways as dictated by the narrative's key concepts. One example drawn from Christianity will suffice to make the point concrete.

The narratives and concepts associated with the Christian understanding of salvation indicate that a process of salvation typically is accompanied by a number of discrete and extended ultimacy experiences. There should be a sharp consciousness of guilt for past weakness and sin, along with a sense of being invited by God (specifically by Jesus Christ or the Holy Spirit) to confess such weakness and sin, an overwhelming sense of peace associated with belief in the biblical promise that God forgives through Jesus Christ when true confession is made, a felt need to be with other Christian people and to be sustained by the sacrament of the Eucharist or Holy Communion, an unaccountable increase in love and tolerance for other people, and a powerful urge to share what has occurred with those both familiar and unfamiliar with it.

Any given experience may vary from this narrative statement of the typical process in two ways. On the one hand, some of the implied ultimacy experiences may occur for some people but not others. In the process of group discernment (see

---

[95] George Ellis in his essay for this volume takes up a similar theme in relation to aesthetic experiences and experiences of love (we classify these as socially extended ultimacy experiences bearing especially on orientation).

5.4 below) of authentic salvation, for instance, conformation to the *overall sweep* of the narrative typically is deemed more important than the report of a particular, spectacular ultimacy experience. On the other hand, some groups value certain specific types of experiences that others value less, and the narratives vary across groups accordingly. So Pentecostal Christian groups typically (and thus with variations) expect salvation to be accompanied by glossolalia (so-called "speaking in tongues"); Pietist Christian groups typically emphasize the confession and behavioral change phases; Catholic Christians typically stress the role of the sacraments; and Evangelical Christian groups care more than most about the desire to share the good news of salvation with others—and there are significant variations within each of these types of groups.

In this way, the concept of salvation is stabilized and spreads through a religious group, though with the variations noted. To the extent that there is a shared history of commentarial or devotional literature or a body of narratives enshrined in a common sacred text such as the Bible or universally used rituals, a theological idea can spread to many different groups and acquire different nuances in each context.[96] The history of an idea, its spreading and its variations, is the essential background for trying to make sense of theological descriptions of ultimacy experiences. With this background in place, many theological concepts and narratives can be richly informative about ultimacy experiences and the ways they are described. Without this background, theological language cannot be penetrated very far and analyses of ultimacy experiences depending on theological concepts will be limited by the perspective of those concepts with no way either to discern the nature of the limitation or to overcome it. This is a daunting caveat on the usefulness of theology for interpreting ultimacy experiences because the spread and modification of theological ideas is a complex subject. Much work has been done in this area, however, so it is safe to say that, with care, theology can offer genuine insights into the character of ultimacy experiences.

5.2 Ethical Concepts and Narratives

Many ethical concepts intimate ultimacy experiences in much the same way that some theological concepts do. The Confucian account of the virtues of human-heartedness (*jen*) and ritual propriety (*li*) involves such an ethical narrative.[97] There are staple literary themes with ethical import, such as the dramatic crisis of conscience or the process of character transformation described in *Emma* (see p. 363, above). Other examples might be drawn from political or activist groups such as Greenpeace or Amnesty International, in which the social texture is rich and focused enough to support complex narratives involving stages of commitment and risk in the furthering of the group's aims. The movement through these increasingly demanding stages of commitment is accompanied by changes of values and self-understanding; many of these can be described as transformation ultimacy experiences. Moreover, discrete ultimacy experiences may occur as this process of

---

[96] This way of thinking about the spread of ideas is akin to epidemiological analysis of the spread of diseases; this analogy is exploited in Dan Sperber, *Explaining Culture: A Naturalistic Approach* (Oxford: Blackwell Publishers, 1996).

[97] Note, however, that many contemporary interpreters of Confucianism, especially those indebted to the vision of Neoconfucianism, readily treat these ethical concepts and the narratives framing them as essentially religious.

increasingly risky and challenging commitment proceeds; testimonials from within such groups suggest that they do.[98]

There is an important double difference between theological and ethical narratives. On the one hand, *narratives* expressing ethical concepts are less clearly defined and less widespread within and across cultures than theological narratives. This means that specific narratives will typically be less useful for furnishing ethical descriptions of ultimacy experiences than is the case with theological descriptions. Also, ethical descriptions will tend to use ethical terms in a more *ad hoc* way. On the other hand, ethical *concepts* are more deeply embedded in the language and practices of groups across cultures than most theological concepts. Every culture has variations on the conceptual themes of good and bad, loyalty, kindness, generosity, and honesty because these come with any form of social togetherness. Theological concepts are culturally more refined and diverse, by comparison; the cross-cultural diversity in conceptions of divinity or salvation is extreme. That means that ethical narratives are actually needed less to focus ethical concepts than theological narratives are needed to give meaning to theological concepts. It also means that ethical descriptions of ultimacy experiences will tend to draw less on highly structured narratives and more on universally recognizable ethical ideas, even if those ideas vary in content from context to context.

This double difference offers an advantage and a disadvantage to the analyst of ultimacy experiences. The advantage is that understanding the complexities of the spread and modification of ideas is a less crucial precondition for making use of ethical descriptions of ultimacy experiences than is the case for theological descriptions. The corresponding disadvantage is that it is harder to tie down precisely what is meant in an ethical description of an ultimacy experience without a highly structured narrative to guide interpretation. It follows that the insights offered by ethical descriptions will tend to be helpful in a relatively vague way for understanding ultimacy experiences.

Finally, ethical perspectives on ultimacy experiences are most directly relevant to understanding extended ultimacy experiences, especially of the transformation kind, for obvious reasons. They are relatively less useful for aiding interpretation of discrete ultimacy experiences where moral status is frequently unclear. This constitutes both a limitation on the contribution of ethics to interpreting ultimacy experiences and an indication of its great strength.

5.3 Expressing Experiences of Ultimacy in Language

Behind the scenes in what has been said so far about the contribution of theology and ethics to understanding ultimacy experiences is a famous problem that has haunted the phenomenology of religion for years: the development of criteria capable of detecting when dissimilar reports describe phenomenologically similar religious experiences. So dramatic is this problem in relation to theology that, at present, there is every reason to expect neurally and phenomenologically identical experiences to be describable in such different theological terms that the lurking identity would remain undetectable. Theorists have advanced methods proposing to circumvent this

---

[98] The occurrence of ultimacy experiences in individual members of groups with striking ethical commitments and potent ethical narratives is worthy of close examination; we have not investigated literature on this subject to discover if anything along these lines has already been attempted.

limitation on ordinary human communication, from Husserl's philosophical phenomenology to Daniel Dennett's heterophenomenological method.[99] Carrying out the methods is difficult at best, however, and the problem only seems to grow more pointed as cross-cultural knowledge deepens.

This problem of relativism of description of religious experiences is compounded in theological and ethical contexts by the fact that the specialized concepts presupposing or referring to religious experiences (for example, conversion, samadhi, salvation, and human-heartedness) are themselves hard to compare. This second kind of difficulty in comparison is no easier to manage than the phenomenological difficulty just mentioned and, predictably, methods for attempting to manage it have proliferated.[100]

We have several remarks to make about the relativity of description of ultimacy experiences. First, the issue is complex and this complexity must be faced if a theory of ultimacy experiences is to have standing in a broad public of intellectuals. In relation to narrower social contexts, such as a particular theological-intellectual tradition within Hinduism, some of the approaches to comparison that focus solely on the theological categories used for describing ultimacy experiences do seem to succeed in exposing such descriptions to judgments of similarity and difference.[101] The same can be said for some of the approaches to comparing descriptive phenomenological accounts.[102] The problem is sharpest and most threatening when the intellectual context of discussion is broadly interdisciplinary and cross-cultural.

Second, theories about the neural expression of ultimacy experiences in conjunction with improving technologies for mapping brain activity promise a heretofore underappreciated means of testing claims that dissimilar descriptions refer to similar experiences. Such testing would be far from simple, as it must presuppose correlations between neurology and the phenomenology of experience that are in fact likely to be related no more strictly than in family resemblance fashion. Nevertheless, research indicates that such correlations exist and that a few areas of the brain tend to be involved in discrete ultimacy states, so neural scans should in time have a statistically relevant contribution to make.

Third, theories about religious language have a role to play here, the more so when they join considerations from neurology and the sociology of knowledge, as well as other disciplines such as comparative ethology and linguistics. Religious groups adopt patterns of description of ultimacy experiences under several

---

[99] See Daniel C. Dennett, "A Method for Phenomenology," in *Consciousness Explained* (Boston: Little, Brown and Company, 1991), 66–98.

[100] For a survey of such methods, see Wesley J. Wildman with Robert Cummings Neville, "How Our Approach to Comparison Relates to Others," in *Ultimate Realities*, Neville, ed. (Albany: State University of New York Press, forthcoming, 2000).

[101] The book cited in the previous note is the second of three volumes in a series devoted to testing a particularly sophisticated comparative method; the series is the published output of the Cross-Cultural Comparative Religious Ideas Project, already mentioned in footnote 4. This project presents strong evidence on behalf of the effectiveness of a comparative method that establishes a dialectic between religious texts and experts, on the one hand, and a network of vague categories, on the other. This dialectic both allows the vague comparative categories to be specified by more concrete categories and systematically subjects the entire scheme to correction by concrete details in an ongoing process of deepening comparison.

[102] The most realistic and promising of these is Dennett's heterophenomenological method, which is really the formal statement of a process of comparison that is already employed in many contexts; see Dennett, "A Method for Phenomenology."

influences, including the internal logic of the concepts and experiences with which they have to grapple, and the historical circumstances that condition key decisions taken. And most such processes of adoption develop against the background of neurally and socially conditioned ranges of possible options. Because of this, many thinkers have been able sensibly to advance analyses of distinctive patterns of religious speech or other activity that lay the groundwork for the effective detection of the similar under the guise of the differently described.[103] Such interdisciplinary work is relatively new and plagued by problems of arbitrariness of analysis, but there is every reason to expect it to improve rapidly as more scholars reach across disciplines for promising resources.

Together, these observations indicate that the relativity of description of ultimacy experiences—phenomenologically, culturally, communally, theologically, and ethically—while daunting, is not the impasse that it can seem upon first encounter.

5.4 Spiritual Discernment

Many religious groups embody considerable wisdom with regard to the authenticity of both discrete and extended ultimacy experiences. Discernment is not magic; it does not afford otherwise impossible access to the minds and experiences of an individual. And be that as it may, the effectiveness of discernment can be accounted for without the supposition of supernatural guidance. Traditionally, its operation involves a wise and experienced mentor or teacher, or the collective wisdom of a group, in dialogue with an individual who reports a profound or unusual religious experience. At minimum, it is a regulative process that refines the application of special words used in that group's social-linguistic context to describe such experiences. Sometimes discernment also involves the evaluation of information or instructions whose origins the individual might attribute to his or her experience. Typically it is a caring, pastoral process from which the individual and discerners alike emerge wiser, more experienced, and better adjusted.

The social implementation of these discernment processes varies widely, as does their regulative effectiveness. Examples include the teacher-student relationship in some Hindu schools such as Vedanta and in most forms of Buddhism, the relationship between Christian believers and their confessors or spiritual advisors, the benevolent authority of groups of Jewish Rabbis over the groups they serve, and the relationship between the community and an individual in groups such as the Quakers. Discernment processes are evident almost everywhere religion is found. While they are typically absent in the (usually explosive) phase of a religion's infancy, they tend to be established quickly. As discussed above, there are arguably

---

[103] One example especially relevant to the analysis of ritual experiences is Eugene G. d'Aquili, Charles D. Laughlin, and J. McManus, *The Spectrum of Ritual: A Biogenetic Structural Analysis* (New York: Columbia University Press, 1979). An example pertaining to the connections between the architecture of religious buildings and religious experience, and relying mostly on arguments from hemispheric dominance, is James B. Ashbrook, *The Human Mind and the Mind of God: Theological Promise in Brain Research* (Lanham: University Press of America, 1984). More generally, see Carol Rausch Albright and James B. Ashbrook, *The Humanizing Brain: Where Religion and Neuroscience Meet* (Cleveland: Pilgrim Press, 1997); Laurence O. McKinney, *Neurotheology: Virtual Religion in the 21st Century* (Cambridge: American Institute for Mindfulness, 1994); and David Porush, "Finding God in the Three-Pound Universe: The Neuroscience of Transcendence," *Omni* 16.1(1993). Little has been done specifically in relation to theological modes of speech, but see Wesley J. Wildman, *Speaking of Ultimacy*, forthcoming.

even secular forms of discernment, such as the patient-analyst relationship in psychoanalysis. The fact that these secular forms of discernment exist shows that discernment processes take their rise in the first instance not from religious interests but from ordinary processes of social-linguistic regulation by which groups try to make sure important descriptions are applied to experiences correctly.

In what sense is discernment a regulative process? People wanting to describe their ultimacy experiences will tend to use the terminology that their social-linguistic context makes available to them, as they understand it. They can only really discover the aptness of their descriptions by imaginatively or actually trying them out and seeing what happens. They may or may not revise their description; that will depend on their group's reactions and how much they care about those reactions. For example, a lazy, narcissistic person is unlikely to be believed when he or she claims to have received a revelation from a divine being in the absence of radical personality changes and a plausible angelic message. A skeptical reaction might force a change in the description offered or it might marginalize a person intransigent about his or her claim. In either case, the complex discernment relation serves to regulate current and future descriptions of purported ultimacy experiences.

Discernment is a complex dialectical process and it is not always successful in creating agreement among all parties. But it is remarkably successful just the same and the natural question is: Why? Other examples of regulative social-linguistic processes exist, including those operative when a child is learning how to apply names to observable objects. These processes work because mistakes can be corrected. But how does the regulative dialectic operating in the description of ultimacy experiences work? Since experiences are not observable in the same way that physical objects are, how can mistakes be corrected? Might not the situation for ultimacy experiences be one of "hermeneutics all the way down"? That is, could completely different experiences receive identical descriptions and still pass muster in the discernment processes? At one level, we cannot answer these questions because the problem of other minds is not going to dissolve; not even the prospect of neurological correlates for mental events removes the problem in its abstract philosophical form. Yet, as we try to show in section 6, many factors are relevant to assessing putative ultimacy experiences, and together they make less plausible the logically possible option of utter relativism of descriptions of ultimacy experiences. Within religious groups, the wisdom about typical and atypical ultimacy experiences built over centuries and activated in discernment processes is the key to producing such agreement as exists in judgments of authenticity.

There are two observations to be drawn for our interpretation of ultimacy experiences from the perspective of spiritual discernment. First, spiritual discernment is a crucial factor in the linguistic environment of religious groups because such groups typically exercise considerable influence over their members' beliefs and the modes of expression of those beliefs. Thus, religious people tend to describe their experiences of ultimacy in terms likely to satisfy the discerning eye of their group. Because of this, second, they are also likely to value some types of discrete experiences more than others, some descriptions of experiences more than others, and some patterns of transformation more than others. Communities that value mystical states will have members who value them and describe ultimacy experiences in such terms. Groups that stress the virtue of long-term, stable character transformation will promote such temporally extended ultimacy experiences in their members as well as appropriate descriptions.

Of all methods for assessing ultimacy experiences, religious discernment claims the most extensive experience and the richest traditions of interpretation, much more extensive than nonreligious branches of psychology or phenomenology. Thus, discernment is at present a crucial factor in evaluating the emerging candidates for neurological and phenomenological markers of ultimacy experiences. Openness to normative judgments in religious contexts also makes possible a move from judgments about *typical* ultimacy experiences to judgments concerning their *authenticity*—a characteristic important for the constructive venture of section 6.

6 Procedures for Identifying Typical and Authentic Ultimacy Experiences

6.1 A Description of Typical Ultimacy Experiences

Our description of ultimacy experiences constitutes a detailed taxonomy (section 2 on phenomenology) in conjunction with a neurological part-model (section 3) and a number of more explanatory considerations (social-psychological in section 4 and theological-ethical in section 5). This descriptive apparatus is useful for determining when a *typical* ultimacy experience has occurred. In traditional religious practice, working taxonomies conditioned primarily by informal phenomenological and theological perspectives have dominated the interpretation and evaluation of ultimacy experiences. Our taxonomy gives greater weight to the neurological and social-psychological perspectives than they have traditionally received. The taxonomy is recapitulated in Appendix A, which also serves as a reader's guide to the entire argument.

We wish to stress the idea that there is a spectrum of both discrete and extended ultimacy experiences. Our taxonomy describes *typical* ultimacy experiences, and it is not the case that they just occur or not in binary fashion. At the center of an imaginary target there is an ideal type, recognizable by traditional theological criteria; richly describable in consistent ways by, say, a psychoanalyst or an evolutionary psychologist; with brain activity known to be strongly correlated with what are usually accepted as ultimacy experiences; and with experiential phenomena embracing several of the categories discussed earlier (for discrete states, sensory alteration, self-alteration, presences, cognitions, emotions). Other discrete experiences in the ultimacy spectrum would possess some of the features listed above, but in some categories might lack relevant features or even possess contradicting features. There would be controversy, then, as to whether to accept such experiences as being ultimacy experiences. In addition, there would be experiences lying at the edges, acknowledged simply as belonging to the ultimacy spectrum, but as being only partial in character.

Furthermore, characteristic features of ultimacy experiences permit the distinguishing of types. The two main types we have identified are the discrete and extended ultimacy experiences. We further distinguished between social and transformation ultimacy experiences within the extended type and we hypothesize that clusters of features recur with such frequency that it makes sense to distinguish other subtypes. For example, within the discrete type, feelings of oceanic peace and wideness of compassion are frequent consequences of certain meditative techniques. Meditation produces other experiences, but the peaceful, compassionate state seems to be so frequently encountered that it can usefully be designated a subtype. We have been careful not to elaborate subtypes both because the systematic survey data needed to identify them do not exist and because the descriptive apparatus needed to create meaningful survey instruments is unstable, mostly due to the need for further development of stable cross-cultural descriptive categories and of the neurosciences. We can predict, however, that these subtypes would be related to one another in family resemblance fashion and that each would have its own distinctiveness, with typical instances closer to the center of its own circular target than marginal instances. This is precisely the situation that obtains in the diagnosis of neurological disorders, and the prominent role of brain processes gives us every reason to expect that ultimacy experiences, especially of the discrete kind, will be similar.

Our description of typical ultimacy experiences has produced more than a multifaceted taxonomy. Because we have paid close attention to the relation between social-linguistic environments and the way that individuals describe purported ultimacy experiences, we have a working model of the complicated hermeneutical transactions that occur between individuals and groups in the having and describing of ultimacy experiences. This model is useful both for understanding why ultimacy experiences are described the way they are and for evaluating putative ultimacy experiences.

With regard to the way typical ultimacy experiences are detected, the application of our descriptive taxonomy can be likened to medical diagnostic procedures that also distinguish conditions based on typical sets of features. The cross-cultural aspect of this analogy is particularly apt. For example, an individual presents with the complaint, "It hurts when I take a breath." A Western-trained doctor might note that the person is also coughing and breathing somewhat rapidly. Were he to have a conversation with a colleague, they would likely agree that pneumonia must be ruled out. In a different diagnostic culture, the clinician and her colleagues might consider that there is an imbalance of qi. In the first setting, a set of objective data would be sought: the temperature, the number of white cells in the blood, a radiographic picture of the chest. In the second setting, other data such as detailed characteristics of the pulse and the appearance of the tongue would be gathered.

Within the Western diagnostic framework, there is an ideal type—pneumonia—that has certain co-occurring subjective and objective features. In other cases with the same subjective complaint, the white cell count and x-ray will be silent, leaving the clinician with the cough, dyspnea, and pleuritic pain. He or she will consider pneumonia to have been ruled out and may diagnose a viral illness with pleuritis. In still another case, a person may have no complaint but be found to be increasingly weak and lethargic. A physical exam and blood cell count might raise the suspicion of pneumonia, subsequently confirmed by x-ray. Various combinations of subjective and objective phenomena are interpreted both by the individual and by relevant experts.

Modern Western medicine is but one hermeneutical approach to diagnosing bodily ailments. It is not uncommon to find detailed and recognizable descriptions of certain illnesses in the literature of other cultures and eras. The observation that tubercular symptoms may remit at high altitudes, for example, was established long before the pathogenic bacillus was identified and its oxygen requirements understood. The association between a rich diet and gouty attacks was known before the pathways of uric acid metabolism had been elucidated. Many observations and varying vocabularies in the course of time can be made consistent within an enlarged framework. Nevertheless, gray areas and ambiguities will always remain, stimulating further investigative and hermeneutic efforts.

This extended analogy from medicine expresses the diagnostic procedure made possible by our description of ultimacy experiences. It also illustrates the flexibility of the descriptive taxonomy with respect to managing complex or partial instances of ultimacy experiences. It even suggests how sufficiently detailed descriptions of ultimacy experiences might be used to relate culturally distinctive patterns of description. But there is an important disanalogy here as well. Western medicine has a sophisticated causal theory of pneumonia that is more detailed, less arbitrary, and better able to be tested and improved than competitive causal theories such as those based on *qi*. This is so even though Western patterns of diagnosis and treatment are inferior to those of non-Western medicine in some instances. To this point we have offered no causal theory for any type or subtype of ultimacy experiences, but only an extended, multi-faceted description that makes possible a diagnostic procedure. We return to the question of a causal model for ultimacy experiences in section 7.

6.2 Detecting Typical Ultimacy Experiences

We have described a procedure for identifying typical ultimacy experiences using our descriptive taxonomy and using a working model of the complicated hermeneutics of the social-linguistic processes that condition the having and describing of ultimacy experiences. All this constitutes a rich framework for articulating why one does or does not accept an experience as belonging to the category of ultimacy and whether or not a putative ultimacy experience would be deemed typical. The *authenticity* of such experiences is another matter, to which we turn in section 6.3. As a demonstration of how the procedure for detecting *typical* ultimacy experiences might be applied, let us evaluate some possible examples of ultimacy experiences.

The Interictal Personality. We refer here to the chronic personality traits observed in some temporal lobe epilepsy patients (described above). From the phenomenological side, the salient feature is not discrete experiences (sensory alterations, self-alterations, presences, etc.) but an ongoing interest in religious and philosophical matters together with a moralistic attitude, intensified emotions, and a "viscous" interpersonal style. Thus, the temporally extended phenomenological category is pertinent here. What is notable, as regards the elements of that category (existential potency, social embedding, transformation of behavior and personality, and transformation of beliefs), is a dissociation of beliefs (which are pronounced in these individuals) from behavior and character changes of a moral and ethical nature; it is not reported that these individuals are more altruistic towards others or more selfless than prior to their illness. To summarize the phenomenological dimension of this type of individual, then, we find only one element of the several that we deem typical of extended experiences of ultimacy.

Moving to the neurological domain, we can surmise that this personality is based on accumulated episodes of aberrant temporal lobe activity. Perhaps our putative markers, involving medial temporal structures and the lateral temporal cortex, would be indicated in brain images. Moving next to the social-psychological dimension, different experts (neurologists, psychoanalysts, and evolutionary psychologists) might frame the phenomena differently. As we have seen, Bear and Ramachandran understand the phenomena as a form of limbic-sensory hyperconnection. Neurologists generally do not find these patients particularly pleasant to deal with, as they tend to be overly concerned with detail and to prolong encounters, disregarding cues directed towards terminating a conversation (viscosity). A psychoanalyst would regard such a patient as an obsessive character—again, with connotations of rigidity and insensitivity to others. Evidence of unconscious aggression might be discerned.

In sum, what might appear to be a religious conversion taken solely on the face of the individual's expressed interests does not seem particularly convincing when viewed from these multiple perspectives. Hermeneutically speaking, we might relate it to an "ideal" conversion as we would relate pleuritis to pneumonia: they are not the same even though they share a few features.

Discrete Temporal Lobe Phenomena. Dostoevsky's *The Idiot* contains the following passage describing a complex partial seizure.

> His mind and his heart were flooded with extraordinary light; all his uneasiness, all his doubts, all his anxieties were relieved at once; they were all merged in a lofty calm, full of serene, harmonious joy and hope. Since at that second, that is at the very last conscious moment before the fit, he had time to say to himself clearly and consciously, "Yes, for this moment one might give one's whole life!" then without doubt that moment was really worth the whole of life.[104]

From a phenomenological point of view, some complex partial seizures indeed are remarkably similar to the discrete experiences generally accepted as religious. If they are reflected on and lead to changes such as those described under the temporally extended category (behavior, character, and belief change), then they would conform to an elaborated phenomenology traditionally associated with ultimacy experiences. From a social-psychological perspective, again, different experts might view these episodes in various ways. In a psychoanalytic setting, a person with the discrete experiences just quoted might be understood as regressing to an infantile state of bliss and security which in his real infancy had been traumatically torn away. In the context of a life-stage psychology, if a person were struggling with a terminal illness an experience such as the one above might be considered a defense against an impossible struggle and its attendant despair, or a letting-go of both. We can imagine still other hypothetical interpretations arising in other social-psychological settings: it is not obvious, though, that experts in all fields would have cogent explanations to offer.

From a neurological perspective, such an individual might very well have the neurological markers of ultimacy experiences discussed above. In the future it might be possible to show that brain activity indistinguishable from the transient temporal lobe discharges seen in epilepsy is an invariable concomitant of discrete religious experience, equivalent to a positive x-ray in pneumonia. In the meantime, we would say that this case stands much closer to the ideal type of discrete ultimacy experiences than the case of the interictal personality does to the ideal type of extended ultimacy experiences.

Sex and drugs. In some cultures, alteration of emotion and the experience of self are sought through sexual arousal or the use of drugs. Drugs can also be used to generate hallucinations that are held to be spiritually significant. In the instances we have in mind, the use of arousal or intoxication is highly ritualized and is contextualized by a group's articulated religious or metaphysical beliefs. To decide whether these are ultimacy experiences, we would apply our several criteria. We would want to know how closely the phenomena conform to the five elements of discrete ultimacy phenomena, and whether they bring about chronic changes conforming to the four elements of extended ultimacy experiences. We would ask whether these

---

[104] Fyodor Dostoevsky, *The Idiot* (New York: Bantam Books, 1981; first published 1869), 218–19.

manipulations produce the neural hallmarks of ultimacy experiences. Probably the group in which these practices occur would make the richest social-psychological interpretations and relevant observations, but other observations and interpretations would also be possible.

Remorse and Transformation. Although psychoanalytic treatment focuses on unconscious conflict rather than on undesirable behavior, in a successful treatment patients alter their previous behavior, sometimes radically. For example, a man who could not keep a job because of his arrogance and who had alienated two wives by his narcissistic demands had been in treatment for several years when he began to contrast some new, unselfish behavior with his past patterns. As he became fully able to see himself as he had been, he spent many sessions in a state of deep remorse, bemoaning the damage he had done in his relationships, even weeping. He made attempts at rapprochement with his estranged adult children and was deeply moved by their forgiveness. He became able to tolerate the demands of a job, even though his position was much less prestigious than in the past. As he described his new capacity for acts of kindness towards others and patience with himself, he realized he felt very much at peace and actually happy.

Let us subject this account to our fourfold criteria. Phenomenologically, it does not contain any discrete states relevant to ultimacy. The account is strongly marked by a change in behavior and character, unaccompanied, however, by a belief system involving concepts of ultimacy. From a social-psychological viewpoint, psychoanalysts of various schools might give rich accounts—for example, that he had developed the capacity to serve as a self-object for others once he no longer required them to be narcissistic self-objects for him. An evolutionary psychologist might note the negative consequences of his previous selfish behavior, arguing that they had caused a shift in strategy to altruism. From a neurological perspective, we would expect neural signatures to be absent. From a theological/ethical point of view, it might be noted that the person had passed into the second of the three Kierkegaardian stages of transformation. We might conclude that this vignette bears some of the signs of ultimacy, but lies at too far a distance from the ideal type to be counted as an ultimacy experience in the usual sense of the phrase.

We might think, however, that this person had potential for moving closer to a recognizable experience of ultimacy, a Kierkegaardian third stage, in contrast to the interictal personality, for example. This is, in effect, a prediction. Medical diagnosis is useful to the extent that it predicts the future course of events. In the absence of interventions, a diagnosis of pneumonia predicts a different course than a diagnosis of viral-associated pleuritis. As our understanding of ultimacy experiences increases, it should allow predictions as to course and outcome to be made and tested.

6.3 Detecting Authentic Ultimacy Experiences

Detecting typical ultimacy experiences does not require normative categories such as authenticity. There is no concern with the reality of the purported ultimate object of such experiences in this detection, and no judgment is made about what is better or worse in any religious sense. Within various narrower social-linguistic frameworks, limited normative judgments are made, as when a sociobiologist points out the serendipitous outworkings of an adaptive behavior or a psychoanalyst notes that unhappiness is caused by denial of trauma because denial engenders obsessive repetition of psychic processes in situations akin to that in which the trauma first occurred. But there is no comprehensive judgment made about the authenticity of ultimacy experiences in our diagnostic assessments of putative instances of them.

The conventional wisdom seems to be that there is no meaningful way to discuss richer normative judgments of authenticity in scientific contexts; such matters should be left to the discernment procedures of religious groups. We demur. Judgments of authenticity are not much different than judgments of what is typical. The added element is merely the willingness of a social-linguistic context to stipulate what *ought* to occur. Psychologists may try to refrain from normative "ought" language, confining themselves to the "if-then" logic defined by the patient's own convictions—"if such-and-such is an important goal to you, then so-and-so is the way to achieve it"—as though therapy were essentially a means-ends technology. In practice, however, psychologists often care about their patients and identify with their goals of happiness or peace of mind, making a community of at least two people with shared goals and a clear sense of what *ought* to happen. Sociobiologists can probably afford to be stricter about avoiding normative language, but impartiality is only a limited virtue. Sympathetic, explicitly partial identification with others is the very stuff of friendship, community life, and caring behavior, and thus has its own virtues.

Where psychologists tend to walk the line between scientific objectivity and passionate concern—losing balance while walking this line is an occupational hazard, in fact—other social contexts promote full-blooded commitment to the value of ultimacy experiences. In such contexts, along with an amassing of experience about what is typical, there is a concern to recognize and cultivate *authentic* ultimacy experiences. This passionate embrace of comprehensive normative categories can be described in the same way that any complex relation between a group and its members is described. Such descriptions can be impartial even if the judgments of authenticity being described are thoroughly, adventurously biased by a passionate sense of what *ought* to happen.

In practice, many religious groups manifest a double concern with the authenticity of ultimacy experiences. On the one hand, they promote authentic ultimacy experiences by lavishing honor and encouragement upon those who have them, especially in virtue of the deeper forms of wisdom and maturity achieved by means of such ultimacy experiences. Such cultivation is one of the expected commitments of a religious group. On the other hand, the discernment procedures of many religious groups are designed to parse inauthentic from authentic ultimacy experiences, as described above in section 5.4. Besides applying accumulated experience to judging whether a purported ultimacy experience is typical, such discernment also relies heavily on a background of assumptions about what is in fact occurring in a typical ultimacy experience. For example, whereas folk religions encourage trance states and revere those who achieve them, the greatest respect is reserved for those shamans who subsequently exhibit behavior confirming the group's background belief that a trance state involves a spiritual journey in which the traveler receives wisdom capable of solving real problems in the community.

The interpretation of ultimacy experiences underlying such criteria for authenticity is effectively a causal model. It follows that judgments of authenticity can be described impartially. However, they cannot be confirmed from beyond the social-linguistic community within which those judgments are made without evaluation of the underlying causal interpretation of ultimacy experiences. Any evaluation of a group's causal model is complicated by the fact that it will be expressed in the distinctive language of that social-linguistic context. That makes comparing and testing such causal models so tricky that the scientific literature in religious experiences tends to stay in the domain of the typical and to avoid altogether the fuzzier territory of the authentic. Nevertheless, in section 7 we shall propose our own causal model that we hope is helpful for analyzing and comparing other causal models and for evaluating judgments of authenticity with some degree of impartiality. That causal model also offers us the opportunity to venture some ideas about the ultimate causes and value of ultimacy experiences, albeit within the constraints of our metaphysically relatively neutral approach.

6.4 An Orientation to the Future

It is worthwhile noting that considerations from neurophysiology can have little *direct* relevance for causal theories of ultimacy experiences (and so for theories of divine action) at the present time. There is of course the obvious obligation to express such theories in ways that are responsive to the general relevance of the neurosciences for understanding human beings. But the situation seems to be one of significant independence of the two realms of discourse as far as direct relations are concerned, with indirect relations possible because of mutual connections to general theories of human nature. Our approach to diagnosis of typical ultimacy experiences is likely to become more effective with the passage of time, however, for three reasons.

First, phenomenological accounts of religious experience will become more refined as advances are made in comparing descriptions of experiences from diverse cultures. This sort of comparison is in its infancy, yet it is the most promising line of solution to the problems associated with deciding when phenomenologically similar episodes are being described in different ways and when similarity of descriptions is misleading.

Second, research in neuroscience does not yet furnish us with any correlations between neural activity and phenomenology definite enough to identify a specific neural signature of discrete ultimacy events of any sort. We have said that such neural signatures would probably include activity pervading the medial temporal regions and extending into the anterior temporal cortices, perhaps especially in the dominant hemisphere. These tentative suggestions are based on data from temporal lobe epilepsy and our speculations regarding semantic processing and social participation. But they are pointers towards a hopeful future, not firm criteria in the present.

Third, our use of social-psychological categories drawn from nontheological frames of reference is meant to open up the possibility that experiences recognizably related to ultimacy can occur without being so labeled. Although descriptive and explanatory frameworks differ between religious and nonreligious contexts, it is probable that some of the same phenomena are being described. Moreover—and linking the first two points—neural markers of ultimacy experiences could support this conjecture and thereby furnish an intriguing basis for the comparison of experiences across cultures and social-linguistic systems. Such markers could also supply a partial test of the ability of phenomenological descriptions of ultimacy experiences to register these similarities.

It follows that an orientation to the future is crucial both for the development of better correlations between neurology and other types of descriptions of ultimacy experiences, and for correcting the theoretical underpinnings of the model as they take shape at any given time.

7 The Causes and Value of Ultimacy Experiences

7.1 The Problem with Modeling Causes

We have passed from describing ultimacy experiences (sections 2–5) to the problem of detecting them (section 6) and now we turn our attention to the intriguing issues surrounding their causes and value. For now we concentrate on causes; we turn to value in section 7.5, below.

Models of the causes of ultimacy experiences underlie the judgments made in religious groups about the authenticity of putative ultimacy experiences, as we saw in section 6.3, above. Unfortunately, we can make little use of such folk models because, though they take account of a rich mass of testimonial data, they are too narrow in scope, limited by the usually unexamined convictions of the group, uninformed by outside experts, and oblivious to neurological considerations. By contrast, our investigation of the causes of ultimacy experiences has to take account of all the interpretative perspectives offered in the descriptive phase of this essay, except of course that phenomenology is of little use, being specifically neutral to questions of causation.

The idea of investigating the causes of a class of phenomena is problematic. For the sake of argument, let us suppose that we can leap over the high hurdle of internal variation within the class of phenomena; then we could treat the class of phenomena as defining a single question about causes rather than a horde of distinct questions. Our task then would seem to involve finding the best causal model for the phenomena. To do that we would have to construct the best model we can, making sure that it is capable of being compared with competitors and that it surpasses them in descriptive richness and explanatory effectiveness. That is a complicated but thinkable task and tasks like it are carried out all the time in the natural and social sciences.

What does a causal model look like? Put simply, causal models describe how causes and effects are arranged in chains or networks of events by giving a theoretical and sometimes mathematized account both of the linkages between events and of what events ought to be observed. In order to give a causal model of cosmic ray detection, for example, a series of events, some observable and some hypothetical, is described. Successive particle decays are postulated to occur with probabilities consistent with going theoretical models from high-energy physics and then constraints on observable effects are deduced that make the model vulnerable to correction through experiments. The same is true for deterministic causal models of colliding billiard balls, planetary motion, or the cleaning mechanism of a machine that washes clothes.

Such a model is certainly possible for ultimacy experiences, though there would be no mathematizing of linkages or observable events involved. Consider two such models. Let us suppose, for the sake of argument, that Ultimacy is thought of as an intentional being with causal powers. We could include such a causal factor in one model, making of it a complex hypothesis in favor of intentional divine action as a causal factor in ultimacy experiences. Alternatively, we could exclude that factor and develop a model advancing the hypothesis that the causes of ultimacy experiences can be accounted for without reference to an intentional divine causal agent. Then we set the two hypothetical models to the task of accounting for data and watch to see what happens. The better of the two wins the day and must take on yet other hypothetical models of the causes of ultimacy experiences.

This is a simplified version of what happens in the natural and social sciences but it is enough to show that finding the best causal model of ultimacy experiences is an intelligible goal providing that the competing models produce sufficiently different predictions to enable the data to discriminate between them. Unfortunately, centuries of wrangling over the issue of divine action have proved fruitless, essentially because the infamous slipperiness of language about God and divine action seems to make falsifiable predictions unachievable. This situation is not a matter of intellectual dishonesty on the part of those throwing their weight behind the model that affirms an intentional divine agent, though some are quick to leap to that conclusion and subsequently dismiss theology as a serious form of intellectual work. It is rather a consequence of the intrinsic difficulty of the problem itself. The actions of an intentional divine being—as with any causal agent—are only predictable to the extent that quite a lot is known about the nature of this divine being. But the sacred texts of most traditions, when they portray an intentional divine being, typically insist that the divine intention is inscrutable, often in flagrant contradiction to human expectations, and impossible to predict. Moreover, the events in which divine action is said to occur are dense with meanings and subject to all kinds of more or less persuasive interpretations. All this makes it virtually impossible for the data of ultimacy experiences to drive a wedge between our two models, pushing one into the superior position.

It is tempting to yield to the positivist instinct and jettison the divine-action model on the grounds that it is too arbitrary, too suspiciously convenient, too hard to test and so too difficult to render meaningful, and weighed down by an extra hypothesis (intentional divine action) that seems superfluous in explaining all of the scientifically admissible data about ultimacy experiences. Theologians certainly have displayed plenty of self-conscious embarrassment about this problem during the last two-hundred years. Like not being invited to a party where everyone who is anyone is having fun, theologians defending divine action are out of fashion in modern and postmodern intellectual circles. Yet fairness demands that the difficulties in showing the sufficiency of the alternative, more minimalist view for explaining ultimacy experiences be squarely admitted. The minimalist model is a more progressive research program and so more persuasive, in one way, yet the nature of the subject matter demands that we hesitate to apply the usual criteria for deciding theoretical debates. So the debate wanders along, reeling and rolling from criteria for theory choice to the peculiarities of models involving God and back again, never getting very far, and driving everyone crazy. A few people resolve the unbearable tension by taking the positivist escape route but that seems to have all the leap-of-faith character of the decision to affirm the action of an intentional divine being. The debate seems futile.

In light of this analysis, there is an obvious problem with trying to develop a causal model for ultimacy experiences. Deciding between competitive models has to be possible in order to make building causal models of ultimacy experiences a worthwhile activity, yet these decisions seem infuriatingly difficult to make. The best that can be done, it would seem, is to invent causal models as artists express themselves on canvas, and then leave all of the causal models hanging up for view in the philosophical analogy of an art gallery. Even if ruling out the weakest models is possible, deciding on the best model is an arbitrary and highly personal exercise that can never win rational consensus. Well, that may indeed be the best outcome but we think there might be a more promising way. It is not that the impasse facing causal models of ultimacy experiences can somehow be evaded after all this time. No, the history of the question weighs too heavily for that. The point is rather that there is an alternative to modeling causes in the usual way.

We propose to model ultimacy experiences by tracing sign-transformations rather than causes and effects. While this approach is less familiar and requires some explanation, we claim for it three significant virtues. First, it is more impartial than traditional causal models, a virtue deriving from the fact that it does not require the specification of ontological causes and effects. Indeed, the ontological status of causes can be treated as a question to be answered in this approach, making it ideal for modeling ultimacy experiences. Second, the model employs a strategy of vagueness with regard to irresolvable ontological questions in a constructive way that facilitates a nontraditional resolution to debates over the causes of ultimacy experiences. Third, the model is as useful for discussing the value as it is the causes of ultimacy experiences, enabling us to deal with both issues at once. We do not claim that this approach to modeling ultimacy experiences gets us further than the debate over causal models has gone but only that a change of theoretical scenery can't hurt and that the vaguer, more impartial way we propose for framing questions about causes sidesteps many of these futile debates and stays closer to intelligible data.

The approach of using sign transformations to model various complex process has enjoyed limited success within semiotic theory for many years and it is particularly apt for modeling phenomena related to neural processes. Moreover, sign transformations are the ideal way to express and manage the double fact that ultimacy experiences arise from brain activity that has been reflexively subjected to an interpretive process in a social-linguistic environment, and that this semiotic conditioning in turn depends on brain function.

Any model makes simplifying and organizing assumptions. Independent arguments can be offered for such assumptions but they are most convincing when a model using them proves to be effective. Most of the assumptions of our model are gathered into what we are calling its philosophical-semiotic framework, a specifically philosophical elaboration of semiotic theory. The task of section 7.2 is to lay out this framework and to explain how it allows our model to be both neutral to, and yet well suited for investigating, the causes and value of ultimacy experiences. In section 7.3 we elaborate the model specifically in relation to the relevant considerations from the descriptive phase of our essay. Throughout these two sections, we make extensive use of footnotes to elaborate on technical points so as to keep the main text as lucid as possible for the general reader. The model is applied to a discussion of the causes of ultimacy experiences in section 7.4, including whether they demand explanation in terms of divine action, and to an analysis of theological claims for their truth, value, and importance in section 7.5. We conclude in section 7.6 with some speculative yet plausible and metaphysically modest suggestions about the nature of the ultimate causes of ultimacy experiences.

7.2 The Model's Philosophical-Semiotic Framework

In the most general terms, semiotics is the theory of signs. As a technical discipline, it has applications in intellectual ventures as diverse as communication theory, developmental psychology, and anthropology. We adopt a philosophical view of semiotics that is indebted to the early North American pragmatists, especially

Charles Peirce.[105] Signs are treated abstractly as anything that can be taken to stand for something else.[106] The way signs *stand for* other signs in certain respects is analyzed in semiotic theory.[107] When one sign is taken to stand for another, we speak loosely of *sign transformation* of the object sign into the interpretant sign, intending to suggest that signs flow from one to the next, each standing for the previous one in some respect. The entire flux of signs in all its complexity registers whatever underlying causal processes there might be, though other relations are registered in the semiotic flux as well.[108] Yet it does so without assuming anything about the ontology of those causes (if indeed there are causes; on some South Asian metaphysical theories there is no underlying causation). When we speak of causation, we usually have in mind causal chains that are particularly significant for our interests; in fact, causation is much more like a surging river than a single thread of string. In the same way, the semiotic flux is enormously complex, and we pay attention to pieces of it selectively as circumstances and interests demand.[109] By focusing on

---

[105] For convenient access to Peirce's view of pragmatism (he called it pragmaticism to distinguish it from the pragmatism of William James; Robert Neville has called it paleopragmatism to distinguish it from the neopragmatism of Richard Rorty), see Part II of Philip P. Wiener, ed., *Charles S. Peirce: Selected Writings* (New York: Dover, 1958). For his view of semiotics and logic, see the compilation in "Logic and Semiotic: The Theory of Signs," 98–119; and other essays in *Philosophical Writings of Peirce*, Justus Buchler, ed. (New York: Dover, 1955; originally published in 1940). The official edition of Peirce's works is Charles Hartshorne and Paul Weiss, eds., *Collected Papers of Charles Saunders Peirce*, 6 vols. (Cambridge: Harvard University Press, 1931–35).

[106] Semioticians have elaborated subtle distinctions among many types of signs. The most basic is the three-fold distinction between icon, index, and symbol, but sophisticated taxonomies have been produced in semiotic theory by paying close attention to the astonishingly diverse ways in which things stand for other things.

[107] This is a simple formulation of the basic semiotic structure. More precisely, the basic semiotic structure is four-fold in its logical form: (1) an object-sign (2) interpreted by (3) an interpretant-sign, with the interpretation occurring (4) in a particular respect. This structure is highly abstract, applicable quite generally to complex and simple processes, in every kind of system, including the neural, physiological, and cognitive. Structures of this kind are important elements of theories, illuminating common structural features of disparate objects and processes. Without it, a theory of religious experiences is likely to be arbitrary and unprincipled. With it, complex communication against the background of sophisticated cognitive habits can be analyzed as readily as physiological processes or neural signaling, all using the same semiotic framework. That is what our model requires.

[108] Peirce's semiotic theory is compatible with a causal theory of reference in which reference is understood to be achieved causally rather than merely "thought" in some way independently of the physical, causal world. Accordingly, sign relations of reference that are registered in the semiotic flux actually reflect causal processes. Other relations between signs, such as merely associative relations, can also be thought of as actually causal, given the complex social and neural connections that underlie the imaginative connections we make. It follows that the semiotic flux is more closely reflective of underlying causal processes than might seem to be the case at first.

[109] Alfred North Whitehead in *Process and Reality: An Essay in Cosmology* (New York: The Macmillan Co., 1929) articulates a theory of causation that is explicit about the enormous, multi-directional complexity of causal relations. A few thinkers influenced by Whitehead have found in semiotic theory a convenient way to express a similarly rich theory of the semiotic traces of causation, though without what they take to be the metaphysically over-determined and so unduly speculative character of Whitehead's theory. See, for example, Robert Cummings Neville, *Creativity and God: A Challenge to Process Theology* (New York:

signs and how they transform in a complex flux of signs, reflecting underlying causal processes, this philosophical version of semiotics facilitates a model of ultimacy experiences that can neutrally frame questions about the ontological status of supposed underlying causes, including "ultimacy." At the same time, our philosophical version of semiotic theory is critically realist and thus well suited for asking questions about the causes and value of ultimacy experiences and the reference and truth of language about ultimacy.[110]

With regard to human experiences of all kinds, our philosophical version of semiotic theory especially focuses on the ways we enter into the flux of signs—in fact, semiosis flows through us and all our interactions just as the river of causation (on common views of causation) embraces all that we are and do and think as human beings. The name we use for this participation in the encompassing semiotic flux is "engagement." This concept is the key to understanding how our version of semiotic theory can be critically realist, thereby enabling our model to frame questions of causation and value in helpful ways. The task of the remainder of this section is to explain engagement, showing how it brings to our model the virtues we claim for it. We will do this by asking and answering three questions.

First, in more detail, what do we mean by engagement? Semiotics takes with great seriousness the symbolic character of human experience, thought, and communication. We constantly take one thing to stand for another, understand some things by means of other things, mean some things in speaking of other things, and so on. In doing this we are interpreters who wield signs as tools for engaging a world of objects. Engagement through signs is especially obvious in language use,[111] but if we look more closely at nonlinguistic activity we see that it is pervasive. For example, a skilled dancer engages the world in magnificently shaped ways, and that engagement can be analyzed in terms of sign processing. Information pours into the dancer's senses about the dancer's body and the dancing surface, and that flood of information functions as signs for interpretation by highly trained habits, producing with incredible efficiency a further flow of signs that move the body in subtly adjusted ways. Again, listening to or making music can be understood as engagement through a richly textured flux of signs, in some respects cognitive, in others not. It might seem odd at first to interpret every kind of engagement with the world in terms of signs and sign processing—the more so because it seems we could speak of engagement more vaguely without ever mentioning signs—but the payoff is a powerful tool for analyzing structures and process of all kinds. Philosophers

---

Crossroad Publishers, 1980); and idem, *The Truth of Broken Symbols* (Albany: State University of New York Press, 1996).

[110] Semiotic theory often focuses narrowly on signs and sign relationships, abstracting from meanings (the domain of semantics) so as to focus on pure structure as much as possible. This is not a limitation intrinsic to the theory of signs but merely one of the ways this theory has been used. Since we are concerned with meanings, reference, truth, causes, and value, we move beyond this narrow sort of semiotics. We indicate this in our choice of names. For example, we speak of a philosophical version or application of semiotic theory and of a "philosophical-semiotic" foundation for our causal model of ultimacy experiences.

[111] This is one of the points made by Deacon in *The Symbolic Species*; he argues that the human capacity to engage the world through signs coevolved with the brain and was the necessary evolutionary precondition for language.

appreciate that kind of generality and are willing to use specialized terminology to get it.[112]

Second, how does the concept of engagement help our model to be critically realist about the world we engage? The case for critical realism is quite simple: the strongest argument for the reality of our world is our ability to engage it in every sense—to move within it, sense it, talk about it, change it. Of course, the world is not known directly in its reality but only indirectly *as engaged*; that is the point of speaking of *critical* realism. Nevertheless, signs must facilitate engagement with the real world in some way, whether loosely—as in the case of a hallucination—or tightly, as in the apprehension of the sound of a twig snapping in a lonely forest. Most signs are complex so that, even in the loose case, they engage the world

---

[112] Another scalable concept useful for general analysis of structures and processes is the schema, a concept developed by Michael Arbib (see his essay in this volume for a fuller discussion of schemas). Because several chapters in this volume refer to schemas, we pause to compare the schema structure with the basic semiotic structure we use here. On our understanding of it, a schema is a conceptual tool modeled after objects in computer programming architecture. An object represents an operation that functions in relation to many different yet structurally similar input conditions and produces different yet structurally similar outputs; in the context of computer science, an object has the same set of input and output variables. Objects are useful when they represent repeatedly occurring processes, thereby allowing a much larger process to be conceived analytically in terms of many simpler component subprocesses, each represented as an object. Objects are more useful when they can be reused in the analytical description of many superficially different large-scale processes; when the same objects recur, superficially different processes are shown to have common sub-structural elements. In the same way, in the philosophical context, the schema is a concept whose instances are reusable. A particular neural subprocess might be represented as a schema, for example, and then that schema reused in the description of many different large-scale neural processes. Sound pattern recall might be used in modeling both hearing and speaking, for instance. Or an often-repeated physiological movement such as a smile could be represented as a schema and then that schema reused in descriptions of numerous larger social transactions, such as expressing enjoyment or ice-cool hatred. The reusability of a schema at the neural level is a strategic boon when it comes to computer modeling of brain processes. Sub-models for reusable neural schemas then become building blocks for more adventurous modeling efforts, with the details of each schema sub-model now helpfully hidden from view at the higher level. Likewise, schema reusability is an important component in the modeling of any complex process, from economic systems in the social sciences to cell systems in biology.

Reusable objects are still more useful when they are also scalable, which means that their structural features are applicable at many different levels of complexity. Scalable objects are relatively rare in programming architecture, the usual goal of good program design being reusable objects, but scalable objects do exist and they are as powerful as they are abstract. Scalable concepts are also important philosophically. For example, the concept of an object itself is highly scalable, as is the concept of a schema, as is the basic four-fold semiotic structure. It is their scalability that makes the schema and the basic semiotic structure so useful for the general analysis and modeling of complex processes. The difference between the semiotic structure and the schema structure lies essentially in degrees of complexity. The semiotic structure is simple and therefore extensively applicable; that makes it better suited for tracing the semiotic transformations necessary for the adequate framing of epistemological and ontological questions about the reference of language and the causes of experience. The schema structure is typically far more complex (except, of course, that the semiotic structure is one instance of a schema) because of the need for a structure of consistent input and output conditions; it is far less generally applicable, accordingly. But the schema concept is ideal for registering the structurally invariant features of complex processes, which makes it useful in model construction.

through multiple channels. Think, for example, of the rich way that Macbeth's hallucination of the dagger, its handle turned towards him, signified his actions in the real world. Considering both the sound of a snapping twig and the hallucination in this light illustrates how unappealing is a forced choice between naïve realism and the nonreferential relativism of hermeneutical circularity. The better way is to understand signs as making possible diverse forms of engagement with a real world. There is nothing more real or more basic than engagement with the world; that is the way the world shows up for us.[113] The upshot of all this for experiences of ultimacy is crucial. We are not forced to choose between a blunt realism about ultimacy and a hermeneutical disengagement from reality. In our approach, the distinction between what there really is and what people think there is turns out to be less

---

[113] The complete argument for these points is complex and too long for this context. We may deepen the analysis one level, however, by considering how reference emerges in the sort of philosophical semiotics we use. To begin with, the argument here is deliberately reminiscent of Kant; engagement is the pragmatic-semiotic solution to—actually, a dissolution of—the famous dual problems of linguistic reference and ontological realism pressed into the modern philosophical consciousness by Kant; see *Critique of Pure Reason*, trans. Norman Kemp Smith, 2nd ed. (New York and London: Macmillan, 1933). Reference is, in effect, an emergent property of a semiotic flux. Analysis of the objects of our interpretation shows that the signs by which we interpret have as their objects other signs. In one sense, then, from the semiotic point of view, interpretation is a matter of "signs all the way down." The virtue of this point of view is the forcefulness with which it brings to our attention the pervasive reality of signs and socially embodied systems of signs. Thus, semiotics has the potential to furnish an extremely general description of reality in which "signs as interpretants of other signs" becomes the universal mode of analysis, both within and beyond the realm of human interpretation.

If semiotics suggests "signs all the way down," however, where are the philosophical resources needed to speak of a world to which language refers, a world in which we act, a public world of objects and events, of meanings and values? Some philosophical perspectives happily capitalize on the apparent impossibility of reference to an external world in semiotic theory's infinite regress of signs. We go another way, however, following Peirce's emphatic combination of semiotic theory and a critical realist view of the physical world that is congenial to the natural sciences. The key to this step is always, under one description or another, the fundamental category of engagement. Accepting that the reference of signs to other signs helps us to engage the world in which we live disarms what otherwise might appear to be a vicious infinite regress. Peirce expressed this colorfully: "everything that is present to our minds... appear[s] as a sign of ourselves as well as a sign of something without us... just as a rainbow is at once a manifestation both of the sun and of the rain." (This is the characterization of Milton Singer, including Peirce's rainbow quotation, from "Signs of the Self: An Exploration in Semiotic Anthropology," *American Anthropologist* 82(1980): 485–507.) Because we engage the world so richly, the reality of a world can be affirmed without having to pretend that the extraordinarily complex process of signs referring to other signs somehow just cuts out at some link along the chain. This view thus embodies an impressive refusal to oversimplify the (appropriate and optimal) messiness of reference in human symbol systems while dissolving the problem of reference to objective reality. To put the point cryptically, reference to what there is to be referred to is explained for free by means of the category of engagement without having first to settle what exists to be referred to in the first place. We note that this approach is quite compatible with a correspondence account of the meaning of truth and a combination approach to *criteria* for truth, in which correspondence, coherence, aesthetic, and pragmatic criteria all may play roles in helping to detect the true, with coherence criteria usually playing the leading role. (Note that some might expect a pragmatic theory of truth to emphasize pragmatic criteria but this is to think of William James too much and Peirce too little.)

significant than the distinction between what can be effectively—truly as well as efficaciously—engaged and what cannot be effectively engaged.

Third, how does the concept of engagement enable our model of ultimacy experiences to be neutral as to the question of the reality of ultimacy? Experiences of ultimacy are forms of engagement in the world. We do not need to settle the question of whether there is an "ultimate reality" causing experiences of ultimacy before we can legitimately speak of the experiences themselves. Nor must we decide the question of the reality and efficacy of ultimacy before we can begin examining people's reports of ultimacy experiences. On the contrary, we settle our metaphysical inventory questions about the world and ultimacy in and through our engagements with reality, which in our case means in and through the process of developing a model of ultimacy experiences.[114]

Within our philosophical-semiotic framework, therefore, we are justified in speaking of experiences of ultimacy while remaining formally neutral to the reality and efficacy of ultimacy. None of this requires surrendering a critical realist account of nature or our determination to ask and answer questions about the causes and value of ultimacy experiences.

7.3 Ultimacy Experiences Viewed in Terms of Sign Transformations

Within this philosophical-semiotic framework, it is possible to relate the relevant components of our description of ultimacy experiences—the neural, the social-psychological, and the theological-ethical—under the rubric of sign transformation. We view these components as aspects of reality registered by an encompassing semiotic flux, within which signs refer to other signs, constantly transforming our interpretations from one sphere to others.

More concretely, consider the relation between the neural and the social-psychological. Let us begin with patterns of neural firing—'objects' (of interpretation) in semiotic vocabulary, which of course should not be confused with physical objects, even though our language constantly blurs the distinction. When a particular pattern impinges on another set of neurons it becomes a 'sign' to be interpreted. The action of the second set of neurons, in response to the original pattern, is the 'interpretant'. This action in turn becomes a sign and so forth, until the interpretant arises at the level of somatic effectors—for example, as movements of the muscles. Once again, muscular movements become signs. To the individual moving the muscles, proprioceptive sensory input is relayed back to sensory areas of the brain as a sign. To other individuals who may be observing the movements, the visual

---

[114] It is possible to extend semiotic terminology to all processes and objects and thereby to make semiotic categories metaphysically fundamental, in which case the real world engaged with signs is just the real world of signs. Alternatively, it is possible to insist that the real world's engagement is well described by semiotics but only at levels of nature complex enough to speak of interpretation in something like the usual way. *Engagement* receives different interpretations in these two cases and in others, as do the key words: "interpretation," "world," and "real." We do not take a position on such debates here as it is unnecessary for articulating our model of religious experiences. It is enough for us to make use of semiotic theory's frank acknowledgement of the endless complexity of systems of signs and to insist on some form of critical realism by means of the category of engagement. We leave the fundamental metaphysical inventory questions (is reality at bottom probability distributions? particles? energy? ideas? signs? Peirce's firstness, secondness, and thirdness?—see Peirce, "The Principles of Phenomenology" in *Philosophical Writings of Peirce*, 74–97) to those who think they can be answered and want to try.

sensory patterns produced become signs to be interpreted within their brains. If you stick your tongue out "at" someone, you are not just performing a motor act. In the terminology of George Herbert Mead, you are invoking—for yourself and others—a public system of signs in which there is consensus as to the effects that gesture should produce.[115] The same invocation of a public system of signs occurs when I utter a sentence. Networks of interrelated signs may be demarcated into separate territories that define communities, which vary in their interpretations of human acts. The observation of a public gesture in turn transforms the neural activity of the observer, creating a two-way intercourse of signs. Furthermore, all individuals who are not so estranged that they are incapable of discourse, or unconscious, are automatically and continually engaged in social semiosis. As we type, we are engaged in the social practice of "writing"; we might be wearing clothes that signal our status as at home or in the office; and so on. Certain neural events arise as interpretants of the keys beneath our fingers and the words before our eyes. These in turn act as signs to produce interpretants in the form of further patterns of neural activity, conditioned by patterns previously stored. In sum, the link between neural events and social-psychological events is a flow of semiosis, with successive transformations that lead from neural events, conditioned by the semiotic environment of the brain, to social events, conditioned by the semiotic environment of a particular society.[116] Neural events and social life can be thought of as dimensions of reality that show up together within a single, continuous flux of signs.

At this point, metaphysical choices multiply. There do seem to be levels of reality ranging from the universal and simple to the special and complex. We all believe this because our attempts to study reality as we engage it give rise to scholarly disciplines that in some respects relate to each other in hierarchical ways. But how should this hierarchy of levels be understood metaphysically? We may turn to an emergentist physicalism, to Whiteheadian naturalism, or to dimensionality metaphors for articulating what we mean. Semiotics can play a complementary role to these metaphysical constructions, one that is relatively independent of them. Specifically, more complex levels of reality are registered in (or just are; we are not deciding that question here) a denser flux of signs. Density in this sense can be quantified in principle by counting the ways that signs function as interpretants for and within that which is complex; the density of the semiotic system is the way we know that we are engaging more complex levels of reality. In this way, we continue to evade settling metaphysical inventory questions while preserving a natural way to speak of the relationships between types or domains or dimensions or levels of reality. This allows us to extend the discussion above of the semiotic linkage between the neural and the social to two higher-level components of our model: the social-psychological and the religious-ethical.

---

[115] George Herbert Mead, *Mind, Self and Society from the Standpoint of a Social Behaviorist* (Chicago: The University of Chicago Press, 1934).

[116] We find Teske's view one-sided when he states, "Our spirituality resides, not in the finitude of our individual biology, but in a historically and culturally emergent symbolic world," because our neurologically based capacities make the emergent symbolic world possible. See J. Teske, "The Spiritual Limits of Neuropsychological Life," *Zygon* 31.2 (1996): 209–34. Likewise, we find unconvincing those accounts of language and mental life that focus solely on neurology and neglect the necessity of social environment for the establishment and function of semiotic systems. Just as individuals operate within a semiotic milieu, and just as signs are more generally conditioned by the network of signs in which they emerge, so neural events are shaped and constrained by the brain's semantic networks.

First, the social-psychological considerations reflect the ways that the semiotic flux of which we have been speaking relates individual subjectivity to social-linguistic environments. Everyday consciousness is permeated with the sense of a standpoint, the subject. The subject may be present explicitly, as in self-consciousness, or implicitly, as in intentional thought and even some kinds of habitual actions; but we mean to exclude the sorts of habitual and mystical awareness in which no sense of subjectivity is present. A complete account of the subject cannot be derived without both signs originating in neural events and signs interpreted within a special sign-system arising from the social milieu. We take it that the latter sign-system, which includes what we may designate as the "person system," exists because of its compatibility with brain function and its usefulness for organizing societies of individuals. Thus, the experiencing self is neither solely brain-derived, nor solely socially derived.

To account for the experiencing subject, then, we must consider semiotic processes arising both in the brain and in the social milieu, thereby making connections between the individual subjectivity and both the neural and the social-psychological. Body-related neural activity includes the firing patterns that encode movement and the patterns that register sensory events arising within the body and at its peripheral receptors. The boundedness of bodies by their skins, and the fact of their separate spatial positions, are signs derived from interactions with the physical world and other bodies. Neural processes transform bodily activity in the context of a group into a sign—a unique bodily identity—without implying subjective identity. Bodies do not impinge on one another only through their spatial requirements, however. In certain highly social species, they deploy motor acts, even secretions, that function as signs. Such signs produce interpretants in the form of responsive behavior in other individuals. In primate species, social signs are registered and interpreted through a stream of neural semiosis that is well developed, particularly in apes and human beings. In humans, a rich neural semiotic capacity related to the gestures of others may have set the stage for a step into a still more intense kind of semiosis, the person-system. The key in every case is the semiotic richness of the system, which depends in turn on the physiological achievement of critical levels of semiotic density and complexity. For example, the capacity of primate brains for social signaling appears to have accelerated; in addition, human brains are immense compared to those of our closest primate relatives. The sheer amount of wiring permits unimaginably large numbers of semiotic transformations, social and non-social alike. The evolution of very complex societies, in turn, provides a remarkably dense system of signs within which individual brains operate. Within such dense and sensitive systems, signs relating to one's own and other bodies could be interpreted in new ways.[117]

---

[117] We step back to note that the subject is a special representation within awareness that appears to be the location of awareness itself. Thus, it is preeminently a spatial concept. In an essay on Kant, Strawson demonstrated that the intuition that we each possess a singularity of subjective experience—the certainty that my experience is mine and not someone else's—derives from an antecedent notion, that of the singularity of embodied persons. See P.F. Strawson, *The Bounds of Sense: An Essay on Kant's* Critique of Pure Reason (London: Methuen and Company, 1966): 162–74. There are empirical criteria for that notion, namely, the existence of individual bodies that have unique identities. Attributing subjecthood to ourselves, then, is an interpretant of individual bodily identity. We saw above that individual bodily identity was itself an interpretant of neural patterns that encode bodily experience. Strawson termed it the illusion of a purely inner and yet subject-referring use for "I."

Second, the religious-ethical can be understood semiotically as an extension of what has just been said about the social environment in relation to the individual subject. The flux of signs becomes denser and more complex under the impact of increasing complexification of both brains and social milieus, but it is not just a quantitative increase in engagement of the world that results. There are also qualitative increases, as demonstrated in the difference between the subjectivity of higher primates and its apparent absence in plants and simpler animals. It is reasonable to suppose that complex semiotic systems and complex brains permitted forms of engagement with dimensions of reality, including questions of value and purpose, that formerly were not possible. The expression of religious and ethical concerns is, like subjectivity, a form of the intensification of engagement with reality. In fact, we suggest that the semiosis of subjectivity, with its accompanying density and complexity, itself constitutes a necessary basis for the intensification of semiosis that we experience in morality and religion.

We argued that the social system of concepts and rules related to persons is external to the brain on the one hand but compatible with it and dependent on it on the other. We said that a nascent phenomenology of separate bodily individuals interacts with neural and social semiosis to produce the subject. Similarly, one might think of the ultimacy dimension as external to human subjectivity, but compatible with it and dependent on it. We know that neural ensembles can be set into motion by language. Can human experience be engaged, in an analogous manner, by experiences of ultimacy? If so, we would expect the dimension of ultimacy to deepen every aspect of the semiotic flux we have described. In fact, the description of ultimacy as the depth dimension of reality seems singularly apt here.[118]

Deep engagement with reality in the forms of morality and religion is both incontestably important in the history of human life and, necessarily on our critically realist point of view, indicative of reality. An explanation that does justice to the richness and potency of the moral and religious dimensions of life—including ultimacy experiences—is therefore required. This is so regardless of how the causes of religious and ethical engagement are finally assigned; the traces of those causes in the form of semiotic transformations themselves demand an adequate explanation. Even a world in which all religious beliefs and all attributions of divine action are fundamentally mistaken remains astonishing, intriguing, and terrifying enough to demand at least the semiotic richness, the poetry and puzzled awe, though perhaps not the dogmatism or exclusivism, of the world's spiritually-oriented traditions. It may even be that literally mistaken yet symbolically referential rituals and dogmas and the forms of life they sponsor can advance profound engagement in the wild semiotic flux of the religious and moral dimensions of life. We do not thereby insist that religious beliefs are mistaken but we are led to ponder their *usefulness even if*

Strawson also pointed out that the concept of person derives from our linguistic community. See "Persons" in *Individuals: An Essay in Descriptive Metaphysics* (London: Methuen and Co., 1959): 87–116. He said that a person is the union of mental life and body and pointed out that the concept necessarily has both first-person and third-person ascriptive uses. While the deployment of the basic person concept appears universal, local cultures have unique ways of characterizing persons that are narrated and enacted dramaturgically by individuals. Thus, at both the level of the subjectivity concept, and the level of the socially constituted person, there is intimate mutual interplay between phenomenological and social psychological signs. Each is mediated through complex neural semiotic activity.

[118] This is the terminology of Paul Tillich; see *Systematic Theology*, vol. 1 (Chicago: University of Chicago, 1951).

mistaken by reflection on the sign-flux within which human beings live and move and have their being. This is the practical implication of what it means to say that religion and morality are intense forms of engagement with the world even while remaining neutral as to the reality and efficacy of putative religious objects to which the cause of such realms of life is sometimes attributed. Semiotic complexity is required for such engagement regardless of the causal story offered for established religious and moral experiences and beliefs.

Individual subjectivity, social groups, and brain processes are connected to one another through semiosis. They may be thought of as the vertices of a triangle whose sides denote bi-directional semiotic transformations. All the transformations taken together constitute the experiencing subject-in-a-social-context. In one respect, therefore, our view stands in contrast to the concept of a hierarchical model in which phenomena at each level are made to give rise to phenomena at the next, "higher" level. Hierarchical models may be epistemically and metaphysically helpful in some ways but the semiotic perspective represents matters differently and, we think, helpfully: the neural, social, and individual territories are mutually bound and mutually determining through semiotic interfaces. And deepening it all are the extraordinarily rich semiotic transactions that we call ethical and religious, of which experiences of ultimacy are the most direct, personal manifestations. The structured character of the semiotic flux to which our model draws attention is pictured in Diagram 2 (see Appendix B).

In saying that ultimacy experiences are peculiarly rich and deep forms of engagement with the world, we are making a couple of other suggestions. First, because ultimacy experiences are defined in the semiotic model as rich and deep forms of engagement and defined in mostly phenomenological fashion in the descriptive section of the essay, we are hypothesizing that these two definitions coincide. That is, wherever exceptionally deep and rich engagement is found, there will also be found ultimacy experiences in the ways we described them, and *vice versa*. Now, such experiences may not be especially religious in character; that is the purpose of the misalignment between the box denoting ultimacy experiences and the circle denoting conventional religious experiences in Diagram 1 (see Appendix B). But they are ultimacy experiences all the same. We think that the experiential source of religion is in part the deepest and richest elements of our engagement of the world; but only some of those are suitable for inclusion in domain of officially recognized religious experiences because of the other social interests of religions. That said, it is interesting to note that virtually every kind of rich and intense form of experience that occurs to human beings is recognized as religious in some cultural context. Second, we are positing a distinction between richness of engagement and focus of cultural and individual attention. While that to which we attend may involve our rich engagement in the world, it is often the case that there is a lot of talk about some unimportant and relatively superficial phenomena and a lot of silence about and even inattention to exceptionally important phenomena. Thus, the amazingly large number of genocides in the last two hundred years is rarely spoken of and actually unknown by most people whereas in some communities the state of grass lawns is an intense topic of conversation. We think that there is a phenomenological difference between focused attention and intense engagement—one that is actually fairly easy for a person to detect given enough time.[119]

---

[119] We are grateful for communications with Patrick Mcnamara about this essay and especially for his request for clarification of this point.

This philosophical-semiotic framework allows our model to be neutral but what, we must ask, is the specific content of this model? After all, semiotic transformations, even when good stories are told about them (as we tried to do above), remain fairly vague; for example, you cannot actually individuate or count them. Apart from the possibility of using the model to frame debates about causes in neutral fashion, therefore, it might seem that nothing has been gained by way of specific content. But something has been gained, though it is a modest gain. The positive content of the model has two aspects. First, the relations between individual subjectivity, social groups, and brain processes have been spelled out in some detail in the descriptive taxonomy. Sign transformation is the cement that holds these considerations together and represents them as mutually codetermining factors in conditioning ultimacy experiences. In this way the model brings that descriptive material to bear when it is used to evaluate theories attempting to penetrate beyond the semiotic flux of ultimacy experiences to the underlying causes. Second, the model stresses depth and richness of engagement as the hallmark of ultimacy experiences. This is a consideration for which any grander theory of the causes of ultimacy experiences would have to account.

7.4 The Causes of Ultimacy Experiences

On the basis of this model of ultimacy experiences, can we say whether in fact Ultimacy causes ultimacy experiences? We have been able to speak of ultimacy experiences in terms of the density of the semiotic flux we inhabit without having to decide whether there is anything ultimately corresponding to "ultimacy." That was the plan: the usefulness of the model was envisaged to lie not in making claims about the causes of ultimacy experiences but in framing questions about such claims. How, then, does our model help to frame and evaluate claims about the causes of ultimacy experiences? In two ways, we think, and we describe them in what follows.

First, with the background of centuries of futile debate about divine action in mind, we are working on the assumption that there is little point in setting up a bunch of competitive causal models for ultimacy experiences and then playing them off one another in attempt to make best sense of the data; the odds of success are remote at best. The odds of success might be increased, however, if we were to use the semiotic model as a coordinator of data to aid in selecting among competitive causal models of ultimacy experiences. We wouldn't really need the full power of the philosophical semiotic framework to carry this off, perhaps, but the model would be more useful than its function merely as an inanely overbearing reminder that the best explanation has to make sense of all the various perspectives we elaborated in sections 2–5. So, how would this work? For the sake of simplicity, let us deal with this question in terms of the action of an intentional divine being, which is one kind of divine action. The positing of action by an intentional divine being in explaining the causes of ultimacy experiences occurs frequently in religious and theological contexts, whether it be an experience of divine comfort, the sense of the presence of Jesus, an unaccountable peace given as divine gift in response to the faithful worship of Allah, a moment of inspiration that creates or confirms a belief, the hearing of divine voices and the seeing of heavenly visions, an event of psychological healing following the invocation of the power of God over the natural world, or a process of character transformation springing from a sense of divine forgiveness for past wrongs. Models affirming and rejecting the intentional divine action thesis would be assessed against the other levels of description, such as the neural and the sociobiological. The usual frustratingly slippery questions would emerge. Is the theological account consistent with the other insights? Must the theological account be deemed merely the higher-order, heavily coded description within a specialized semiotic environment for an experience that is more adequately described in other terms (say, psychological and neural—or Madhyamaka Buddhist, for that matter)?

It might seem that the semiotic model takes us no further than did the ordinary demand that successful models should account for all of the perspectives on ultimacy experiences organized in the descriptive taxonomy—the phenomenological, the neural, the social-psychological, and the theological-ethical. But that is not quite the case. While the semiotic model does not stipulate strict rules for answering questions about which description of the causes of an ultimacy experience is true or more accurate, it does impact the process of debate in three helpful ways. First, all candidate accounts of the actual causes of ultimacy experiences would have to be able to explain the fundamental datum of the structured pattern of sign transformations described in our model (again, refer to Diagram 2 in Appendix B for a summary of that structure)—this is, in effect, a new and sturdy constraint on models. Second, the model's neutrality to questions of causal provenance and truth facilitates its usefulness for conducting debates among experts with very different views about the reality of divine action. People can discuss their models in a common language without simply repeatedly stumbling over the conflicting presuppositions of their explanations. Finally, the model's foundation of detailed, integrated descriptions of ultimacy experiences establishes an important criterion for debate among competitive accounts: the candidate explanations have to *demonstrate* their compatibility with all of the various levels of description. It is not acceptable for candidates merely to *assert* their compatibility with the various levels of description, nor is it possible without penalty to withdraw from the debate for fear of being uncompetitive. The critically realist emphasis in the model's philosophical-semiotic framework implies a common referent of the various sign-descriptions, even when those descriptions are from diverse social-linguistic contexts. The bland assertion of compatibility by virtue of some perspectival move—"they see it their way, we see it our way"—is, accordingly, unintelligible or indistinguishable from mere wishful thinking. If such perspectival compatibility actually obtains, then the semiotic framework implies that it can be explained in detail, at least to some degree.

These three contributions mark a modest but real advance in conducting debates about the causes of ultimacy experiences. These debates, we must remember, may in fact be irresolvable—to return to the image used above, we may simply be refurbishing the art gallery's décor with the switch to philosophical-semiotic conceptuality. We cannot determine that in advance, however, so it is as well to have a relatively neutral and data-rich framework within which to conduct such debates and a slightly enlarged set of criteria to which appeal may be made in distinguishing inferior from superior causal models of ultimacy experiences.[120]

---

[120] Note that these constraints apply to all theories of divine action in relation to human experience, of which this volume contains a number. For example, concepts such as whole-part causation or primary-secondary causation are invoked in various places to express how God's action might be compatible with other descriptions of ultimacy experiences. We judge such speculative concepts to be an advance over the mere redescription of the problem in terms of such concepts as supervenience, as usually defined. (But see Nancey Murphy's essay in this volume in which she promises more on behalf of her definition of supervenience than merely the potential for redescribing an established problem.) However, our model not only invites such speculative attempts to demonstrate the compatibility of a *few* levels of

In connection with this, we note that the challenge of reductionism that threatens to make superfluous God-descriptions of the provenance of ultimacy experiences is made neither harder nor easier by our model, but merely is posed happily. With all of the levels of description in place, the full complexity of the problem of reductionism can be appreciated: it is nothing other than the problem of intelligibly coordinating the various levels of descriptions of ultimacy experiences that our model incorporates. And that problem must be pursued using a variety of strategies. We shall not venture far into this territory but we do make the following suggestion. Centralizing the concept of engagement in the semiotic underpinnings of our model shifts the burden of proof in the reductionism debate by bluntly demanding that *all* parties pay careful attention to *all* data relevant to the task of describing that which is engaged and the means by which it is engaged. There is no question in our minds that neglect of relevant data is one of the crimes perpetrated by the hasty assumption of unjustifiably reductive accounts of religious experiences (and there is more than one way to be hastily reductive!).

The second way of bringing our semiotic model to bear on the evaluation of causal models is, we think, far more interesting. Were it not for this second way, we would not have bothered with all of the philosophical-semiotic overhead. The starting point is to notice that the semiotic model is not a causal model. Rather, it is vaguer than traditional causal models in an important respect. Yet it stays close to the data, making it empirically more responsible than traditional causal models of religious experience. This strategy of developing an empirical model that remains vague about ontology is a deliberate attempt to fit the model to the lines of debate, making it vague where the debate seems least tractable and highly specific where the debate seems to promise the clearest answers. The result is the plotting of a course around the edge of the untraversable ontological swamps, all the while staying on the solid ground of relatively uncontroversial data. In short, pursuing this strategy delegitimates debates about whether or not real contact with some sort of Ultimate occurs in religious experiences. In exchange, there is an emphasis on the issue of whether that which is *actually engaged* in ultimacy experiences supports assertions made about ultimacy in theological accounts. For example, is that which is *actually engaged* in ultimacy experiences rightly described as a personal God, as a mysterious natural force, as the hangover of childhood developmental frustrations, or as a meaningless artifact of temporal lobe transients? This is by no means an easy problem but it is a far different and more interesting problem than the irresolvable and often data-censoring debates about reductionism in relation to religious experiences. It clumps all the intractable ontological questions into one corner while organizing ways of answering the more resolvable questions. This approach even sets up an environment for dealing with ontological debates in a systematic, comparative way—and there is no question that systematic comparative metaphysics will be the venue for the best of such debates in our era of multicultural awareness. The key to the effectiveness of the philosophical-semiotic model is the fact that it is vague in all the right places. We indicate below (section 7.6) the sorts of speculations that we think our model permits when it is used to interpret ultimacy experiences on its own terms rather than being limited to playing referee for

---

description; it also insists that *all* of the various levels of descriptions are registered in any theoretical portrayal of their compatibility and resists the mere assertion of compatibility of those levels of descriptions.

competing causal models in the traditional but futile game of deciding if a divine being ever intentionally does anything in particular.

7.5 The Value of Ultimacy Experiences

Once the whole array of pertinent data is allowed to have its effect on the interpretation of ultimacy experiences, rather than a subset of it determined by the unreflective or ideological embrace of reductionistic assumptions, a balanced discussion of the value of ultimacy experiences is possible. Admittedly, the theological problems posed by a model such as ours are no easier to manage just because the data set is richer, as we pointed out above. Theological claims about the value of ultimacy experiences are easy to contest, accordingly. In several other respects, however, the value of ultimacy experiences is easier to make out.

First, the transformative efficacy and sheer emotional color of ultimacy experiences makes them valuable. People often change their lives dramatically because of ultimacy experiences, almost always in the direction of greater contentment and emotional maturity. What reams of argumentation and hosts of exemplars are often powerless to achieve, ultimacy experiences can induce almost instantaneously. The data confirm this resoundingly.

Second and more abstractly, the semiotic density expressed in the having and in the social meaning of ultimacy experiences indicates that the environment of human life is astonishingly rich. Indeed, it is rich enough to make ultimacy experiences important and maybe necessary reminders of how quick we are to oversimplify our lives, to trim the interest, to dull the color. And to what end? Ultimacy experiences drive into awareness our tendency to flatten out our life for the sake of the comfort of predictability and the appearance of safety; this awareness helps us to become more realistic and adaptable.

Third, and connected with this, ultimacy experiences are valuable interruptions of the socially programed character of much of our lives. The legitimation structures of society aim for pacific steadiness in all social transactions, in the name of controlling our intriguing environment. But ultimacy experiences are capable of casting social transactions into new and perhaps uncomfortably bright light. They relativize social norms and assumptions in such a way as to open up space for critique. As a result, ultimacy experiences time and again have been wellsprings of transformation and reform. Unfortunately, they have also been the triggering events for violence and enthusiastic neglect of important balancing perspectives. But that there is a positive side at all to the social effects of ultimacy experiences indicates a notable virtue.

All of these benefits are incontestable—at least from all but the most obscure and anti-social perspectives—no matter how mistaken the interpretations attached to the causes and theological significance of ultimacy experiences may finally prove to be. But none of these virtues is even discernible unless explanations of the extraordinary richness of life that is expressed in ultimacy experiences prize fidelity to the data to be explained, a scientific virtue too often neglected in the study of religious experiences.

7.6 One View of Ultimacy and Ultimacy Experiences

We began with neutral description of ultimacy experiences and ended with a model capable of neutrally framing debate about the actual causes of ultimacy experiences. We just now let our hair down a little and argued that ultimacy experiences are valuable in important ways even if theological theories about their origins and nature prove to be mistaken. In concluding we shall throw caution to the wind and offer a modest suggestion about what the Ultimacy really is that we actually engage in ultimacy experiences. That a psychologist-neurologist and a philosopher-theologian are capable of agreeing on something like this is remarkable, to us at least, but it just so happens that we do agree and so we take a moment here to express our considered opinion.

We have said that the world shows up as *something in particular* in our engagement with it and, therefore, that Ultimacy is really showing up in ultimacy experiences. The manner of Ultimacy's showing-up is a spectacularly rich and dense pattern of sign transformation that links enormous amounts of our experience together and brings mutually enriching significance to them all. We have said that this may be explicable in many ways, but we have resolutely rejected thinly reductive accounts that deny such rich density of meanings; they are insufficiently empirical and, by leaving out the good stuff, commit the sin of being boring. That leaves us with a wondrous world, as monstrously terrifying at some times as it is blissfully peaceful at others. And our aim is to say what sort of Ultimate lies behind these astonishing ways that ultimacy shows up for us. We know what Buddhism says about this, by and large: there is nothing lying behind this showing-up of ultimacy; the point is to journey on an ultimate path toward enlightened acceptance of the world in its wondrous actuality. We know what serious theisms say about it, too: God in various conceptions is what shows up in this most amazing way and God is even more terrifying and wonderful than ultimacy experiences suggest. We even know what serious forms of evolutionary naturalism say about it: nature, needing no creator being and subject to no ultimate purpose, is richer and more wondrous than we can imagine; all we experience so far is what our evolved neural and social capacities enable us to register—it only gets more intense.

We think that each of these three is roughly correct about what it affirms and insufficiently imaginative about what it denies. The evolutionary naturalists and the Buddhists will forever disagree over nature but the former are right about the evolutionary preconditions for our current state and the latter are right about the importance of a path in which enlightened acceptance of the wonders and challenges of life is prized. Neither group tends to have much sympathy for any sort of theism, but that may be because theists typically speak about God in ways that needlessly conflict with the insights of evolutionary naturalists and Buddhists. The mystics of theistic traditions do not make that mistake, however. They describe an Ultimate that defies all descriptions, whether as physical nature, as a divine being, or as a path of enlightenment. Our social and neural capabilities presumably allow us only primitive forms of engagement with this most profound mystery even now. After all, it must be as subtle and profound as the creative potential of nature. When in the process of evolutionary development this mystery began to be consciously perceived in a rich flux of sign transformations, it is probable both that social-linguistic systems already existed that were capable of expressing that encounter to some extent and that these experiences forced adaptations in those social-linguistic systems. Even at their most sophisticated, however, social-linguistic expression of ultimacy experiences has always and only been from a particular point of view. If simple flowers and complex personal relationships defy systematic description, then how much more will our descriptions of Ultimacy refract into uncountably many perspectives? This Ultimate is the power of being in one conceptuality, a morally unfamiliar drive for enrichment and complexity in another, and a passionate lover all-consuming in its demands in yet another. None of these descriptions will do, yet the stirrings in the mysterious depths of reality continue to leave traces in the semiotic flux we inhabit and especially in ultimacy experiences. We can no more remain silent about this enormous strangeness than we can undo the evolutionary development that gave us the capacity for subjectivity, sociality, culture, language, and rich intensity of meaning.[121]

Religion is right at a deeper level than most of its theological claims and some of it practices suggest: there is something out there and in here. We think many scientists sense the same thing in studying the wonders of nature, even when they elect not to use religious categories in describing those wonders.[122] This ultimate *something* that we sense in the depths of nature is not much like a personal God. It is not a causal force independent of the rest of nature. It does not reflect human moral categories very closely. It is not especially amenable to cognitive investigation because inquiry quickly trips up on the phenomenon of conceptual refraction described above. It is not even much like a being. And yet it is not nothing, either, even if Buddhists are right to say that it is somehow indeterminate or empty. It *is* real, and it is doubtless more wondrous and strange than our best and worst guesses. This is the hypothesis that best makes sense of the basic data, including the data of religious experience. It is a proposal vague in the right way and one with which more elaborated theories of Ultimacy (such as doctrines of God and divine action) should strive to be consistent. It is a modest hypothesis, metaphysically minimalist, realistic about the conflicting descriptions of Ultimacy found in the theological claims of religious groups, and bearing little resemblance to folk religious ideas. But it is a powerful idea, well attested by mystical traditions worldwide, as congenial to the natural and social sciences as it is to religion, and well matched to the amazing facet of human life that we call ultimacy experiences.

---

[121] There are obvious ways to extend this view in the direction of religiously and philosophically viable versions of naturalism. A number of recent books advocate such forms of religious naturalism. For example, see Charley Hardwick, *Events of Grace: Naturalism, Existentialism, and Theology* (Cambridge and London: Harvard University Press, 1996); Gordon D. Kaufman, *In Face of Mystery: A Constructive Theology* (Cambridge: Harvard University Press, 1993); and Robert Cummings Neville, *The Truth of Broken Symbols*.

[122] For a recent example of this, see Ursula Goodenough, *The Sacred Depths of Nature* (New York and London: Oxford University Press, 1998).

Appendix A: Reader's Guide

Appendix B: Diagrams

Diagram 1. Identifying the Target Group—Ultimacy Experiences

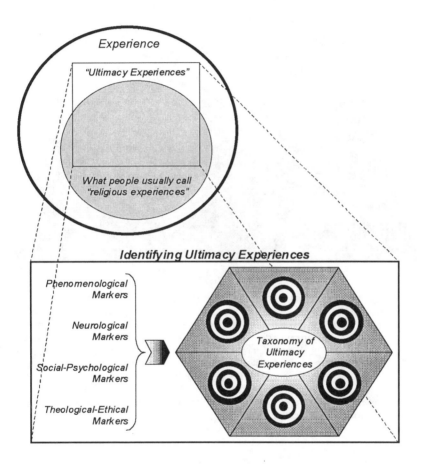

(1) Ultimacy experiences include much of what people usually describe as "religious experiences" but also other experiences not usually described as "religious." (2) Several sorts of markers help to identify the various types of ultimacy experiences. (3) Purported instances of ultimacy experiences approximate the ideal types to greater and lesser degrees (represented by the targets at the center of each type in the taxonomy).

Diagram 2. The Structure of Sign Transformations in Ultimacy Experiences

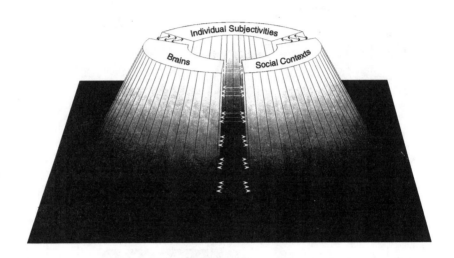

(1) Sign transformations (represented by arrows) flood back and forth among individual subjectivities, social contexts, and brains. (2) Sign transformations are greater in number in ultimacy experiences than in ordinary experiences (represented by arrows deeper into the emergent structure of human experience). (3) This binds all three domains together more richly and intensely, enabling deeper reaches of reality to be registered, in ways that are (a) difficult to express cognitively, and (b) frequently transformative in their effects. (4) These are structural features of causal traces (sign transformations), not causes themselves. Hypotheses about the actual causes of ultimacy experiences must account for these structural features.

IV. CONTRASTING REFLECTIONS ON THE THEOLOGICAL CONTEXT

Michael A. Arbib
George F.R. Ellis

CRUSOE'S BRAIN: OF SOLITUDE AND SOCIETY

Michael A. Arbib

1 Introduction

In 1986, Mary Hesse and I published *The Construction of Reality*.[1] A major theme of the book was the reconciliation and integration of an epistemology based on mental schemas and brain mechanisms "in the head" (Arbib) and an epistemology addressing the creation of social schemas by a community (Hesse). This collaboration led me to view brain mechanisms within the context of the animal's or human's social interactions. However, the bulk of my computational modeling of the brain continued to focus on sensorimotor coordination, on how the brain may extract parameters from sensory stimulation to shape the organism's behavior. Indeed, the bulk of research on neuroscience—and cognitive science more generally—continues to focus either on components of the brain or on the behavior of an isolated animal or human rather than one engaged in social interactions. It was in reaction to this bias of neuroscience in general that Leslie Brothers published her book, *Friday's Footprint: How Society Shapes the Human Mind*, to stress the vital role of social interaction in the evolution and function of animal and human brains. The book provides an extensive and fascinating literature review and builds on Brothers' experience both as a neurophysiologist recording from the brains of monkeys observing "social stimuli" and as a psychiatrist interacting with her patients. Brothers notes that the image of Robinson Crusoe as an isolated individual alone on a desert island embodies the "isolated mind metaphor" typical of much contemporary neuroscience. She stresses, however, that it is a mistake to view Crusoe as truly isolated before he saw Friday's footprint in the sand. Indeed,

> constructing Robinson Crusoe as an isolated mind...dependent only on the manipulation of the nonhuman natural world...[ignores] the history of socialization he brought with him to his exile...[and] the extensive socialization that produces organized thought and behavior in every human being, socialization for which our brains are innately prepared.[2]

The brain is able to restructure itself for many different tasks, whether social or not, but Brothers focuses on the former both because she views it as critically important for understanding the human brain and human experience, and because it has so often been ignored.

Brothers stresses the human brain's inborn mechanism for generating and perceiving "person," a construct that assigns subjectivity to individuals. She discusses the particularities of persons, selves, and the social order, and how these are continually mediated and defined in conversation—a behavior for which our brains have evolved by elaborating on mechanisms for more ancient forms of communication. The crucial point is that

---

[1] Michael A. Arbib and Mary B. Hesse, *The Construction of Reality* (Cambridge: Cambridge University Press, 1986), based on our Gifford Lectures in Natural Theology given at the University of Edinburgh in 1983.

[2] Leslie Brothers, *Friday's Footprint: How Society Shapes the Human Mind* (New York: Oxford University Press, 1997), 68.

The network of meanings we call culture arises from the *joint* activities of human brains. This network forms the living content of the mind, so that the mind is communal in its very nature: It cannot be derived from any single brain in isolation.[3]

Brothers defines a "person" as a being with a mental life, an "owner" of conscious subjective experience, and holds that the person concept—the union of a mental life with a body—is logically primitive in that the concept of a mental life derives from the concept of person, not the other way around.[4] For Brothers, we perceive a "person" *automatically* when we perceive a human body, obligatorily endowing it with mental life in the same way we obligatorily endow words with meaning. However, it is a general perceptual necessity, not a specifically social one, to go beyond the "facts" to endow something—whether an object, a movement, or a person—with meaning beyond the immediate sensory input. Given this, what she calls person-representations could very well be called person-schemas, making contact with the schema theory discussed below.

As Brothers insists, to be a person is to belong to a network of persons, the "social order" which is made up of shoulds and oughts and other moral categories. Strawson emphasizes that person-predicates such as "in pain" or "depressed" necessarily have *both* first-person and third-person uses, but I would stress that we can only learn to use such "subjectivity words" if we can link them to our subjective states—either because someone else infers our mental state and gives us the word for it, or because we infer that a (possibly remembered) mental state corresponds to a behavioral manifestation which can be compared to the state of others for which the word is used. (Note that this can include remembered causes and reasons, and not just the current state or behavior.) In other words, we learn the words for a core set of subjective states by matching, or having matched, our overt expression of these states with those of others. In these cases, one learns to use person-predicates (descriptions of subjective mental states) *after* having subjective mental states in the first place to pin them to. Brothers notes that it can also be the other way around, that the availability of person-predicates in our milieu is what enables us to "have" subjective states. She further notes (personal communication):

> There are three parts to refined counting and person-rules, the rule system that includes subjective experience. (1) Both are created by a shared linguistic-conceptual system. The system is particularly stable and universal because (2) once it is in place, it works so well on a social level. [Like traffic lights in this country, or roundabouts in other countries, both counting and assigning subjectivities permit us to get certain kinds of business done in an orderly way, the more so to the degree that everyone signs on.] It is also universal because (3) the computations involved correspond in some way to what the brain's hardware can do, making it easy to learn and automatic once learned. [See Deacon on the idea of language as a coevolved parasite whose host is the human brain.[5]] Now, once such shared social systems are in place, our tendency is to naturalize them: our experience is read as simply a mirror of "what's out there," i.e., numbers and subjective experiences.

---

[3] Ibid., xii. According to George Ellis (personal communication), viewing the mind as communal in its very nature reflects a strong stream in African traditional thought, where it has been given the name Ubuntu, encapsulated in the phrase "people are people because of people." He notes that Augustine Schutte has written on this recently in his *Philosophy of Africa* (Cape Town: University of Cape Town, 1998).

[4] Brothers, *Friday's Footprint*, chap. 1, following P.F. Strawson, *Individuals: An Essay in Descriptive Metaphysics* (London: Methuen and Co., 1959).

[5] Terrence W. Deacon, *The Symbolic Species: The Co-evolution of Language and Brain* (New York: W.W. Norton & Company, 1997), 111.

Brothers argues that the person concept is no arbitrary result of cultural learning. Rather, human beings are biologically prepared to subscribe to the concept of person just as we are biologically prepared to learn a language. In *Friday's Footprint*, Brothers offers data from primate studies for this biological substrate, but says all too little about the way in which cultural evolution provides subtle and amazingly diverse variations on a biological theme. To rectify this, I shall use language as a model system—for we know that the human brain differs from that of other primates in its neural readiness for language, yet we also know that languages vary immensely from culture to culture, providing an "external reality" which each child must internalize if it is to become a member of a given language community. In a later section, I shall review data on the mirror neurons of the monkey brain, suggesting that they provide a basis for the biological evolution of language.

The biological substrate of the concept of person must be an important element in any discussion of personhood. The schema theory below advances the idea that perception in general "goes beyond the data given," and that, in particular, we have "person schemas" which usually integrate and thus subsume diverse sensory features. I further argue that our theory-laden perception has to do with our being embodied—our perceptual schemas coevolve and codevelop with motor schemas as they guide our interactions with the world. Section 3 attempts to show how social interaction and ultimately any further evolution must build on this.

Section 4 will then take a "great leap upward" by seeking to relate (somewhat loosely) the treatment in section 3 to the interaction between the brain and society as the individual comes to know his or her world and to ponder the question of divine action. Is there God? Or is God only a projection of our minds? I re-examine this discussion in the light of the new insights gained from the framework offered by the three levels of schema theory—social, "basic," and neural—to which I now turn.[6]

2 Basic Notions of Schema Theory

This section provides a brief introduction to schema theory, summarizing points made at greater length in the section also titled "Basic Notions of Schema Theory" in my companion essay in this volume, "Towards a Neuroscience of the Person." That essay supplies references and details missing from the present summary. Schema theory explains behavior in terms of the interaction of many concurrently active processes for recognition of different objects and the planning and control of different activities. These basic units are called *schemas*. Schema theory has been developed at three levels:

1. Basic Schema Theory: Schema theory *simpliciter* has its basic definition at a functional level which associates schemas with specific perceptual, motor, and cognitive abilities and then stresses how our mental life results from the dynamic interaction—the competition and cooperation—of many schema instances. It refines and extends an overly phenomenological account of the "mental level."

---

[6] In addition to the debts to Mary Hesse and Leslie Brothers that are made clear in the above paragraphs, I would also like to express my thanks to Giacomo Rizzolatti for the great contribution to this article made by our collaboration as exemplified in the material on "From Mirror Neurons to Language" and to the late Jane Hill for the enrichment of the linkage of individual and social schemas provided by her model described in the section on "Acquiring a Language." Finally, I acknowledge with gratitude the many thoughtful comments—some reproduced here, others whose beneficial effect is less explicit—on an earlier draft of this essay by participants in the conference on which this volume is based.

2. Neural Schema Theory: William Stoeger notes that at a very low level schemas do not presuppose either symbolic reference or abstract concepts, but they do always seem to presume some selection, integration, or "binding," and evaluation of diverse perceptions. At certain higher levels, they obviously presuppose both symbolic reference and abstract thought—for example, in social and religious schemas. Some schemas cannot, therefore, be determined just from the "outside"; they require "inside-outside" correspondence and reference. He asserts:

> To avoid only facile explanations of mental and personal phenomena in terms of schemata, it is crucial, it seems to me, to distinguish carefully among the different types and levels, and the presuppositions each involves.... A key question in neuroscience at the moment is how [perceptual] integration occurs—how neural processes are marshaled to yield a single integrated perception, intention or response. This seems to imply a basic level of interiority, or subjectivity—even in primitive organisms—which must be explained. Or perhaps this integration constitutes it. Even if we carefully and exhaustively correlate behavioral components and perceptions with patterns of brain states in various areas and parts of the brain, this question is not answered. We can know that various functions of a car are correlated with certain engine parts without having any idea how those components work to yield such functions. This seems to be where we are at with respect to the brain and perceptual and mental functions (personal communication).

My companion essay notes that a variety of functions exists for which an account at the neural level is now available. In relation to this issue, that essay charts the "downward" extension of schema theory to form *neural schema theory*, in which we seek to understand how schemas and their interactions may indeed be played out over neural circuitry—a basic move from psychology and cognitive science as classically conceived (viewing the mind "from the outside") to cognitive neuroscience.[7] Moreover, it is already clear that in many cases schema integration requires no subjectivity in the sense suggested by Stoeger. Rather, many schemas develop together (whether on an evolutionary or individual's time-scale) so that basic neural interactions (excitation, inhibition, modulation) serve to integrate the schemas into a harmonious "schema assemblage" without invoking any processes that involve subjectivity. The essay also stresses the many open problems confronting cognitive neuroscience, the neural basis of interiority and subjectivity among them.

3. Social Schema Theory: In the second part of this essay, I move "upward" to develop *social schema theory*, an attempt to understand how "social schemas" (collective patterns of behavior in a society) may provide an external reality for a person's acquisition of schemas "in the head" in the sense of basic schema theory. Conversely, it is the collective effect of behaviors which express schemas within the heads of many individuals that constitutes, and changes, this social reality.

Earlier work on schema theory showed how to describe perceptual structures and distributed motor control in terms of schemas (composable units of brain function/neural activity) which include perceptual and motor schemas. For example, one *perceptual* schema would let you recognize a house; in doing so, it might provide

---

[7] Three books of mine bear on this move: Michael A. Arbib, *The Metaphorical Brain 2: Neural Networks and Beyond* (New York: Wiley-Interscience, 1989) provides an overview of how to model the brain; Michael A. Arbib, ed., *The Handbook of Brain Theory and Neural Networks* (Cambridge: MIT Press, 1995) is a large compendium embracing studies in detailed neuronal function, system models of brain regions, connectionist models of psychology and linguistics, mathematical and biological studies of learning, and technological applications of artificial neural networks; while Michael A. Arbib, Péter Érdi, and János Szentágothai, *Neural Organization: Structure, Function, and Dynamics* (Cambridge: MIT Press, 1997) integrates modeling, anatomy, and physiology in a comprehensive view of neural organization.

strategies for locating the front door. The recognition of the door (activation of the perceptual schema for door) is not an abstract end in itself—it helps activate, and supply appropriate inputs to, *motor* schemas for approaching and opening the door.

Schemas are *functional* units. In neural schema theory, we seek to go the further step of studying the neural implementation of the schemas. Our initial functional definition of a schema may change as we work out its implementation, revealing details that escaped our attention on superficial examination. A schema is both a store of knowledge and the description of a process for applying that knowledge, synthesizing cognitive (not necessarily propositional) and procedural types of knowledge into one. As such, a schema may be instantiated to form multiple active copies called *schema instances*. For example, given a schema that represents generic knowledge about a chair, we may need several active instances of the chair schema, each suitably tuned, to subserve our perception of a scene containing several chairs. Schema instances may well be linked to others to form *schema assemblages* which provide yet more comprehensive schemas.

Schemas are modular entities whose instances can become *activated* in response to certain patterns of input from sensory stimuli or other schema instances that are already active. The *activity level* of an instance of a perceptual schema represents a "confidence level" that the object represented by the schema is indeed present; while that of a motor schema may signal its "degree of readiness" to control some course of action. The activity level of a schema may be but one of many parameters that characterize it. Thus a schema for "ball" might include parameters for its size, color, and velocity—properties we might notice when we see a ball or play with it, or put differently, properties that would be observed at a level of detail appropriate to our skill and interest rather than at the level of highly precise measurements.

Schema theory provides a language for studying *action-oriented perception* in which the organism's perception is in the service of current and intended action rather than (though not exclusive of) processing stimuli to which the organism provides unintended responses. A crucial notion is that of *dynamic planning*: the organism is continually making plans—in the form of schema assemblages called *coordinated control programs* which combine perceptual, motor, and coordinating schemas—but these are subject to constant updating as perception signals obstacles or novel opportunities. In particular, action-oriented perception involves passing parameters from perceptual to motor schemas: for example, perceiving a ball instructs the hand how to grasp it. However, schema assemblages and dynamic planning ensure that behavior seldom involves direct relationships of a behaviorist, stimulus-response simplicity; rather, context and plans help determine which perceptual clues will be sought and acted upon.

Schema theory can also express models of language and other cognitive functions. There is a tendency (though not a necessity) to root such models in action and perception. Schema theory sees behavior as based *not* on inferences from axioms nor on the operation of an inference engine on a passive store of knowledge. Schema theory explains behavior as resulting from *competition* and *cooperation* among schema instances (in other words, interactions which, respectively, decrease and increase the activity levels of these instances) which, due to the limitations of experience, cannot constitute a completely consistent axiom-based logical system.

A schema network does not, in general, need a top-level executor since schema instances can combine their effects by distributed processes. This may lead to apparently emergent behavior due to the absence of global control. As such, the concept of schema goes well beyond the phenomenal fact that we have concepts, for example, of a ball. It makes explicit aspects of "concepts" that might be lost in other accounts. For example, it integrates perceptual schemas (how to recognize a ball) with motor schemas (what to do with a ball) through the parameter-passing mechanism, but also expresses likely and unlikely patterns of co-occurrence through the patterns of competition and cooperation that develop within the schema network.

In any particular study of brain and mind, schema theory separates a part of the schema network from its connections inside and outside the body and thus must do damage to the plenitude of lived experience. However, our job as cognitive scientists is only to *chart* the territory of mental life in order to establish the major phenomena and their relationships, not to provide the full-scale map or to replace a life richly lived by the running of some computer program. Schema theory seeks the benefits of holism while avoiding the disadvantages attendant upon any approach that precludes analysis of processes into constituents with distinct contributions to the whole.

Schema theory is also a learning theory. In a general setting, there is no fixed repertoire of basic schemas. Rather, new schemas may be formed as assemblages of old schemas; but once formed a schema may be tuned by some adaptive mechanism. This tunability of schema-assemblages allows them to start as composite but emerge as primitive, much as a skill is honed into a unified whole from constituent pieces. My approach to schema theory thus adopts the idea of the Swiss genetic epistemologist Jean Piaget that the child has certain basic schemas and basic ways of *assimilating* knowledge to schemas, and that the child will find at times a discrepancy between what it experiences and what it needs or anticipates. On this basis, its schemas will change, *accommodation* will take place. It is an active research question as to what constitutes the initial stock of schemas. Much of Piaget's writing emphasizes the initial primacy of sensorimotor schemas, where other scientists study the interactions between mother and child to stress social and interpersonal schemas as part of the basic repertoire on which the child builds. For example, Colwyn Trevarthen argues that the human brain has systems which integrate interpersonal and practical aims from the first few months after birth.[8] His argument is based on close observation of young infants interacting with other persons and with objects, but it also has support from studies of human brain growth. In any case, we can see that right from the beginning, the child has schemas—whether in terms of personal relationships or in terms of hunger and comfort—in which the (perhaps unconscious) knowledge of how to do something is inextricably intertwined with the knowledge of what to do. To the extent that we have skills to do something, we make an implicit value judgment that this something is worth doing. This has useful resonances with Brothers' observation that to be a person is to belong to a social order which is made up of shoulds and oughts and other moral categories, but it also stresses that the child's apprehension of the social order must be built up schema by schema.

3 Social Schemas, Language, and the Brain

I will now introduce *social* schema theory and use its terminology in presenting a specific model of a specific socialization process, namely the acquisition of language. The last part of this section describes some novel insights into the evolution of the brain mechanisms of language. As such, it is a detour in terms of

---

[8] Colwyn Trevarthen, "The Primary Motives for Cooperative Understanding," in *Social Cognition: Studies of the Development of Understanding*, George Butterworth and Paul Light, eds. (Chicago: University of Chicago Press, 1982), 77–109.

preparing the reader for section 4, but it is an important contribution to the theme of relating neural activity to social interactions.[9]

3.1 Social Schemas, Briefly

To move from the basic schema level of analysis to an understanding of the human individual, we will need to study the coherence and conflicts within a schema network that constitutes a personality, with all its contradictions, and the holistic nets of social reality, of custom, language, and religion. It is interesting that Marvin Minsky chose the word "frame" for his AI account of schemas for such social situations as a birthday party while Erving Goffman, quite independently, chose the title *Frame Analysis* for his sociological analysis of the way in which people's behavior depends on the frame, or social context.[10] One example: A patient enters the doctor's office, and the doctor asks "How are you?" and the patient replies "Fine, thanks." After they are both seated, the doctor again asks "How are you?" and now the patient replies "Doctor, I have this terrible pain." What changed? The action moved from the "greetings frame" to the "doctor-patient frame." As the name "frame" suggests, frames tend to "build in" from the overall framework, while schema theory is more generative. Rather than represent a birthday party by a single frame or script with specific slots to be filled in, schema theory forms such a representation as an assemblage of schemas for salient objects and actions.

Schemas have, in fact, been applied to the interactions of individuals in social context. A bottom-up view begins from neurochemistry and the cellular level, to layers and columns of neurons, and from there to cortical areas and behavior, whereas a top-down view descends from the social through the behavior of the individual to lower levels. Schemas provide the bridging level for a "science of the person." In *The Construction of Reality* Hesse and I contrast the individual's schemas with *social schemas*—regularities expressed in social behavior which constitute an external reality for individuals without necessarily being codified in any one place or expressed in any one mind. The coherence of a society with its many actors as perceived by the individual can be codified in each individual's schemas. But these are based on different samples of societal regularities and so can be in discord with what is demanded by other members of society. In some cases the order perceived in the complexities around one gives one the ability to interact happily with the world; in other cases, it leads to tension.

Each of us has very different life experiences on the basis of which our schemas change over time. Each of us thus has our knowledge embodied within a different schema network. Thus, each of us has constructed a different worldview which each of us takes for reality. This observation is very important as we try to reconcile the schemas of the individual and those of society. With this, we have come to understand the diversity of possible schemas. We have seen that they may rest on individual style and yet be shaped by the social milieu. A network of schemas—be it an individual personality, a scientific paradigm, an ideology, or a religious symbol

---

[9] The preliminary draft of this chapter had an extensive analysis of data and arguments on, for example, "Social Interactions and Emotion" from Leslie Brothers' *Friday's Footprint,* but it was agreed that the material unbalanced the central argument of this essay. I hope to have a chance to publish the material elsewhere.

[10] Marvin L. Minsky, "A Framework for Representing Knowledge," in *The Psychology of Computer Vision*, Patrick H. Winston, ed. (New York: McGraw-Hill, 1975), 211–77; Erving Goffman, *Frame Analysis* (New York: Harper and Row, 1974). •

system—can itself constitute a schema at a higher level. Such a great schema can certainly be analyzed in term of its constituent schemas, but—and this is the crucial point—once we have the overall network, these constituents can find their full meaning only in terms of this network of which they are a part.

According to schema theory, to make sense of any given situation we call upon tens or hundreds of schema instances in our current "schema assemblage," whereas our lifetime of experience—our skills, our general and social knowledge, our recollection of specific episodes—is encoded in a personal "encyclopedia" of hundreds of thousands of schemas, enriched from the verbal domain to incorporate the representations of action and perception, of motive and emotion, of social and nonsocial interactions of the embodied self. It is in terms of this encyclopedia that I would offer a naturalistic account of the self, embodied in space and time.

With this background, I now turn to the reconciliation of the schemas of the individual and of society. How is it that we as individuals come to be members of society, and to what extent can we, having become members of a society, provide a critique of it? Mary Hesse and I have shown how networks of schemas could integrate the manifold experiences of the embodied self—where the self is not an isolated actor, but participates in rich social interactions. Our task, then, is to extend schema theory from a description of the mind of the individual to a schema theory that addresses the apparent reality of social forces and institutions, including such social phenomena as language, ideology, and religion.[11] Before doing so, I stress again that schema theory involves three separate enterprises: (1) *Basic Schema Theory*: the development of schema theory as a dynamic model of (not necessarily conscious) functions in the mind of an individual; (2) *Neural Schema Theory*: the analysis of how our understanding of schemas may be enriched when we seek to understand their neural implementation; (3) *Social Schema Theory*: the extension of schema theory "the other way" to social forces—a separate (and much less developed) research program. "Social schemas" are posited as "thermodynamic entities" in the sense that they represent the collective effect of behavior governed by related schemas (in the sense of basic schema theory) in the individuals of a community. Basic schema theory thus provides the bridging level between neural schema theory and social schema theory.

In attempting to develop a social schema theory, it is appropriate to chart a tension within schema theory between the current formal models of limited phenomena (whether expressed in computer terms, mathematical equations, or—in the extension to neural schema theory—neural networks) and the richer description of human experience that we have in our everyday language. Our job here is twofold: not only to provide explicit accounts of cognitive processes where we can, but also to understand the limitations of those accounts. And so we always exist in the tension between the charted and the unknown. In sensing that tension, some may argue that the current limitations will in due time be removed by further scientific research, while others may argue that the limitations are not temporary but are irremovable in principle. But here I want to examine the extent to which the somewhat individualistic account of schemas can be extended to address social realities.

Consider the question: "How does one become a member of a particular community?" whether the community be a language community, a social community,

---

[11] This section is based in part on chap. 2 of Michael A. Arbib, *In Search of the Person: Philosophical Explorations in Cognitive Science* (Amherst: University of Massachusetts Press, 1985), and on chap. 7 of Arbib and Hesse, *The Construction of Reality*.

or a religious community. Or put differently, "How does the individual acquire the schemas that constitute, or construct, her social reality?" To proceed, I must distinguish between a schema as an *internal* structure or process (whether it is a computer program, a neural network, or a set of information-processing relationships within the head of the animal, robot, or human) and a schema as an *external* pattern of overt behavior that we can see when we look at someone "from the outside." Reading Jean Piaget, one finds ambiguity as he switches without warning from talking about the schema as a structure inside the head, which explain a child's behavior, to talking about the schema as though it were something that the psychologist can observe.

The distinction between the schema as internal and external can be more formally understood by looking at the theory of finite automata.[12] Very simply, a finite automaton is just a machine that receives, one at a time, symbols from some fixed set of inputs and produces, one at a time, symbols from some fixed set of outputs. What makes the theory of automata interesting is that these are not just stimulus-response automata. The present input symbol alone does not determine what the present output will be; a particular stimulus does not elicit a fixed corresponding response. Rather, the automaton exhibits *sequences* of behavior, with sequences of inputs corresponding to sequences of outputs.

1. The *external* description of such an automaton characterizes how it behaves by specifying for each sequence of inputs the corresponding sequence of outputs, given that it starts in some standard state. This would correspond to the psychologist looking at the child and deciding that the next sequence of activity constitutes an example of a particular schema (as externally observed), a grasping schema, or a suckling schema, or the schema for object permanence.

2. The *internal* description of the finite automaton explains its behavior in terms of a finite set of states, together with a specification of what happens when each input arrives, namely how the automaton changes state and, in changing state, which output it emits. The reason we get different immediate responses to a given input is that different histories of input can drive the automaton to drastically different states.

The task of the cognitive scientist with those schemas defined as part of the individual (as distinct from social schemas) is to infer, from externally observed regularities, internal structures that can provide an internal state explanation of them. But, of course, this raises problems. Within automata theory itself, there was the problem of knowing that the observed sequences of behavior were each initiated when the machine was in some standard state.[13] But if you are observing people from the outside, how do you decide that they are in comparable states when you look at their behavior on different occasions? There is also the problem of deciding which pieces of behavior correspond to one schema and which correspond to another. Moreover, when do we ascribe a behavior to a specific schema, and when do we ascribe it to a larger coordinated control program of which it is part? Note that here I take for granted the critique of behaviorism which insists that behavior cannot be explained without the invocation of internal states, whether linked by programs, as are the internal states of a computer, or by the dynamics of a neural network. What I stress is that in trying to analyze the behavior of animals or humans, we will often start with the observation of sequences of behavior, and this raises two questions: (1)

---

[12] For an elementary exposition, see chap. 2 of Michael A. Arbib, *Brains, Machines and Mathematics*, 2nd ed. (New York: Springer-Verlag, 1987).

[13] Again, see Arbib, *Brains, Machines and Mathematics* for details.

are the sequences best seen as the expressions of one schema or of several; and (2) to the extent that we isolate overt behaviors characteristic of a single schema, how can we then make explicit the internal mechanisms (states and state transitions) which underlie that "external" schema? Recall also Piaget's stress on the change of schemas through accommodation. Much effort in neuroscience now looks at neural mechanisms of "synaptic plasticity" which provide the basis for such processes of schema change. My own concern with the interaction of multiple schemas raises additional questions for which cognitive neuroscience has as yet few answers.

The problems of disentangling the multiple schemas which constitute behavior, however, are not just problems that the cognitive scientist faces in observing and rationally analyzing behavior. They are also problems that the child faces, perhaps unconsciously, when it comes to interact with its world. The child is in some sense trying to find out how to behave in such a way that it will achieve what it wants, that it will not be punished, and that it will gain some pleasure from its interactions with the world. We suggest that the child is trying to go from observed patterns of behaviors to internal patterns which provide appropriate representations. For example, we may see Sigmund Freud's notion of *identification* as the child's process of extracting from the parents' behavior schemas for behaving like the parent. To some extent this helps the child's growth, yet it also causes problems when the child incorporates into its schemas behavior which the parents abrogate for themselves, punishing the child for exhibiting them. This leads inevitably to tensions in schema acquisition and mental development, showing that Piaget may be too sanguine in his view of the child progressing through stages of increasingly coherent abstraction to make sense of more and more experiences. New schemas need not always just generalize or enrich old schemas. Even though parts of schema acquisition may reduce inconsistencies and seek out greater generality, inconsistencies may remain.

Where Piaget provides an almost utopian vision of reflective abstraction[14] yielding coherent schemas of ever greater generality, Freud reminds us that consistency is not an attribute of humanity. For Freud, the superego provides the mental representation of society. It embodies the first identification with the parents. The blocking of aspects of identification with the parents becomes the basis of what society says one can and cannot do—yielding the tension, experienced in the sense of guilt between the demands of conscience and the actual attainments of the ego. Schema theory places as much stress on competition between alternative schemas as it does on cooperation among those that can cohere within a particular situation. Freud tells us that this is not simply a computational abstraction but a part of what we must confront as we try to understand the darkness of the human soul.

Whether in the use of language or in the expression of ideology, the shared schemas within the individuals of a community may provide patterns of behavior that can provide regularities sufficient to allow the child to build schemas which will internalize for that child the patterns of the community. We continue to contrast the schemas outside, the observable patterns of behaviors, with those inside, the schemas as processes within the individual's head. In the following section on "Language Acquisition" I outline a specific model of how the regularities of language in the child's environment may provide the data whereby the child acquires the "schemas in the head" that enable it to produce and understand utterances in the language of its community.

---

[14] The idea of *reflective abstraction* is developed by Evert W. Beth and Jean Piaget, *Mathematical Epistemology and Psychology*, trans. W. Mays, (Dordrecht: Reidel, 1966).

To repeat, the crucial distinction here is between the individual's schemas *about* the society, which the individual "holds" in his own head—schemas which embody his knowledge of his relations with and within society—and what Hesse and I call a *social schema*, a schema which is held by the society *en masse*, and which is in some sense an external reality for the individual. The primary sense of "schema" was as a unit of mental function, which neural schema theory then seeks to relate to distributed patterns of neural interaction. The notion of "social schema" is an *extension* of that schema theory. It addresses the fact that entities like "the law" or "Presbyterianism" or "the English language" are not exhausted by any one individual's stock of schemas but are constituted by a "collective representation,"[15] which is experienced by each individual as an external reality constituted by patterns of behavior exhibited by many individuals as well as related writings and artifacts. Hesse and I explored ways in which individuals respond to a social schema to acquire individual schemas which enable them to play a role in society—whether as conformists, or as rebels who change the social schemas which define society. Such change may involve a process of critique whereby individual experience and social schemas are engaged in a process of accommodation in which either or both classes of schema may change. Note, then, that no individual schema may exhaust the social schema—in the case of "the English language" we each know words and grammatical turns that another does not know. As the child comes to internalize the language, she too creates internal schemas that constitute an idiolect as she learns both general lexical and syntactic patterns from those around her, but also picks up some idiosyncracies but not others.

The next section on language acquisition will illustrate what constitutes a social schema, in distinction from a schema within any one person's head, and how such a schema can affect what an individual does. As individuals come to assimilate communal patterns, they will provide part of the coherent context for others. But the commonality of social behaviors still leaves space for discordances between the individual and the community, as we shall see when we look not at language but at ideology as a social structure.[16]

3.2 Schemas and Language

3.2.1 Acquiring a Language

We will see below that mirror neurons may provide a crucial ingredient in the set of neural mechanisms whose evolution makes the human mastery of language possible (though, as the observations by Doreen Kimura will make clear, the evolutionary pressure may be more for generic mechanisms for "dynamic planning," say, than for language *per se*). However, it requires several years for a child endowed with these neural mechanisms to become fluent in a specific language. The present section reviews one specific model of language acquisition to exemplify the way in which

---

[15] Here, I am adapting a term from Emile Durkheim's *The Elementary Forms of the Religious Life*, trans. J.W. Swain, (London: George Allen and Unwin, 1915).

[16] George Ellis notes two essential references on the way society influences individual thought and vice versa: Peter L. Berger, *Invitation to Sociology* (New York: Doubleday-Anchor, 1963); and Peter L. Berger and Thomas Luckmann, *The Social Construction of Reality* (New York: Doubleday, 1966), which addresses how the individual acquires schemas that construct their social reality and how society constitutes what is in some sense an external reality for the individual.

individuals may come to interiorize the social schemas that define the community to which they belong. To this I add that the very diversity of human language makes it clear that, whatever the extent of biologically based universals that may unite human languages, most of what defines any specific language is rooted in a cultural, rather than biological, process of historical evolution.[17]

The concept of "English" is an example of Ludwig Wittgenstein's notion of "family resemblance."[18] There is no single, standard definition of English, either as to grammar or vocabulary. Rather, a speaker of English is someone who speaks a language intelligible to enough people who agree that they speak English. How does the child extract patterns from the utterances of a community of mutually intelligible speakers to form schemas which interiorize his own version of the language?

To be competent in English is, I would take it, not to interiorize some formal grammar, but to have the ability to communicate with others who speak English. By this statement, I mean to counter Noam Chomsky's view that the grammar is the essence of language. By contrast, I stress communication as primary[19] and note that it leads to development of a system whose regularities can be *described* by a grammar whether or not that grammar plays a causal role in the mechanisms of language production and perception. A Chomskian approach to language acquisition would suggest that the child's innate schemas incorporate a "universal" grammar, including such general categories as noun and verb, as well as certain constraints on the relationships between these categories.[20] On the contrary, I am suggesting that in the beginning was *not* the word, nor even the syntactic category. An alternative model of language acquisition, due to Jane Hill, starts not from innate syntactic categories and constraints but from the observation that the child wants to communicate and likes to repeat sounds at first and, later, words and sentences.[21] However, when a two-year-old child "repeats" a sentence, she does not repeat the sentence word-for-word, nor does she omit words at random. Rather, the child's behavior is consistent with the suggestion that she already has some schemas in her head and that an active schema-based process is involved in assimilating the input sentences and generating the simplified repetitions.[22]

To focus her research, Hill studied a two-year-old girl responding to adult sentences either with a simple paraphrase or with a simple response. The child was studied once a week for nine weeks to provide a specific database to balance the general findings in the literature. Intriguingly, her utterances changed every week. There was no such thing as "two-year-old language" to be given one general model.

---

[17] For an historical view of language change, I particularly recommend Robert M.W. Dixon, *The Rise and Fall of Languages* (Cambridge: Cambridge University Press, 1997).

[18] Ludwig Wittgenstein, *Philosophical Investigations*, trans. G.E.M. Anscombe, (Oxford: Blackwell, 1953).

[19] In no way is this meant to belittle the power of language as a medium for thought.

[20] See, for example, David Lightfoot, *The Language Lottery: Toward a Biology of Grammars* (Cambridge: MIT Press, 1982).

[21] Jane C. Hill, "A Computational Model of Language Acquisition in the Two-Year-Old," *Cognition and Brain Theory* 6 (1983): 287–317; Jane C. Hill and Michael A. Arbib, "Schemas, Computation and Language Acquisition," *Human Development* 27 (1984): 282–96; Michael A. Arbib, E. Jeffrey Conklin, and Jane C. Hill, *From Schema Theory to Language* (Oxford: Oxford University Press, 1987).

[22] Frederic C. Bartlett, *Remembering* (Cambridge: Cambridge University Press, 1932).

Since the child was different every week, the model had to be one of microchanges, in the sense that every sentence could possibly change the child's internal structures. This is "neo-Piagetian" in that it builds on Piaget's ideas of schema change, but analyzes it at a fine level rather than in terms of fixed developmental stages.

At birth the child already has many complex neural networks in place which provide "innate schemas" which enable it to suckle and to grasp, as well as the ability to feel pain and discomfort so as to learn that to continue a certain action in some circumstances and to discontinue another action in others is pleasurable. Note, however, that to say a schema is innate is not to imply that the adult necessarily has that schema. Once we begin to acquire new schemas they change the information environment of old schemas so that they can change in turn.

The issue is not whether or not language rests on some innate substrate, but rather it concerns the nature of what that (highly adaptive) substrate provides. We know that certain subsystems of the brain have to be intact for a person to have language and that language is degraded in specific ways by removing certain portions of the brain. What is at issue is to determine what it is that the initial structure of the brain gives to the child. Does it give it the concept of noun and verb? Does it give it certain universals concerning transformational rules? Or does it give, rather, the ability to abstract sound patterns, to associate sound patterns with other types of visual stimulation or patterns of action? Hill's model suggests that, at least for certain limited portions of a child's linguistic development, innate patterns of schema change can yield an increasing richness of language without building upon language universals of the kind postulated in Chomsky's universal grammar.

Hill posited a grammar based on child language at the two-word stage and kept it free of any characterization of the more adult grammar which will emerge, but is not yet present. The grammar consists of templates which represent relations. In the beginning every template consists of one invariant word, the relation word, and one slot with an example slot-filler, for example, "want milk," in which "want" is the relation word and "milk" is the slot-filler. Given such a template, Hill's model shows how the child may come to produce an entire set of two-word sentences in which any object of her desire for which the child knows the lexical label may be substituted for the word "milk." Thus she may express "want doll," "want blocks," "want juice," the number of utterances being limited only by her vocabulary for objects which she may want. The slot will be found to permit a small set of slot fillers, based on the meaning of the relation word. The model begins by forming a different template for every individual relation word. The concepts encoded in the templates will express relations of interest to the child which are related to her needs or which describe instances of movement or change which attract her attention. As the child relates fragments of adult utterances to situations of interest to her in her environment, templates for expressing the stated relations are added to the grammar, and the items in the lexicon are tagged according to the way in which they might combine with relation words. In this way word classes based on potential for use are constructed, and it is by means of the classification process that templates are generalized.

The "two-year-old schemas" of the model were basic schemas for words, basic schemas for concepts, and the basic templates which provided a grammar marked by a richness of simple patterns the child had already taken from experience, rather than the grand general rules that we would find in the grammarian's description of adult language. And what was built in were not grammatical rules but rather processes whereby the child could form classes, try to match incoming words to existing templates, and use those templates to generate the response. In particular, the model explains how categories such as "noun" and "verb" might arise through the developmental aggregation of words into diverse classes, rather than being imposed as innate categories within a biologically specified universal grammar.

Hill's model thus provides a set of innate mechanisms that could drive acquisition of a certain body of language. However, these mechanisms do not explain how language eventually becomes nested or recursive in the sense that certain structures repeatedly incorporating simpler forms of that structure in increasingly elaborate constructions. Hill outlined what those mechanisms might be, but she did not study them in depth. It does not appear that the elaboration of the model would force one to build in the structures that Chomsky claims to be innate. However, it would also be premature to claim that Hill's model is already sufficiently strong to invalidate Chomsky's claims about innateness. Current work will relate Hill's model to recent work in psycholinguistics and connectionist modeling to bolster the type of schema-theoretic view of language acquisition—and thus of socialization—offered here.

3.2.2 Language as Metaphor

In this section I turn from language acquisition to a view of language which sees it as inherently metaphorical—so that literal and metaphorical readings of a sentence are part of a continuum, rather than forming a strict dichotomy. The view of language presented by Hesse and myself in *The Construction of Reality* is that language provides partial expression of a richer network of schemas, which is in turn the partial crystallization of the true richness of lived experience, and that it provides a means of seeing science and human experience (both quotidian and transcendent) within a unifying framework.[23] In some approaches to language, each word is viewed as having its own literal meaning, or perhaps a finite set of different literal meanings. To find out what a sentence means, one then has to compose those individual literal meanings. In this approach, metaphor is seen as somewhat aberrant, an exception to be found in poetic language. A metaphoric meaning is then a distortion of literal meaning and is only to be sought when the literal rendering of the sentence fails. But our schema-theoretic viewpoint suggests that meaning is extracted from a dynamic process which is virtually endless. A sequence of words is always to be seen as an impoverished representation of some schema assemblage. When we engage in conversation, we engage in a process of interaction which may tease out more and more of the meaning rooted in this assemblage, and may even change the meaning as the conversation proceeds (regardless of whether that conversation is an actual dialogue or an internal "conversation" in one's own exploration further and further into the network of one's own internal meanings).

---

[23] While Hesse and I have focused on the role of schemas, others have taken a more "external" approach (recall our earlier distinction between internal and external schemas), most notably put forward in the work of George Lakoff and Mark Johnson. Theo Meyering notes (personal communication): "Their theory extends the theme of the metaphorical nature of natural cognitive modeling to metonymic models and so-called radial categories, and includes an extensive critique of classical categorization theory and the Mind-as-Machine paradigm. I believe there may be interesting points of contact between schema theory with its action-oriented perspective and the 'experientialist' paradigm that Lakoff and Johnson oppose to the cognitivist paradigm of classical AI. They regard the embodied character of the mind as essential for a proper understanding of the nature of mentality." Certainly, the embodied nature of mind is central to schema theory, and the relations between the Arbib-Hesse and Lakoff-Johnson accounts of metaphor are ripe for further analysis.

John Marshall (personal communication) comments that "If you are overheard plotting to 'kill' someone (who then turns up dead with your fingerprints on the knife), I'm not sure that the Judge will sympathize with your views that 'all language is metaphorical'." Certainly, a sentence may have a "literal" meaning in terms of skimming off the most common set of associations from the related schemas. But there is no dividing line that sets off the literal from the metaphorical. In all cases, the schema or discourse provides an entry into a schema network. Perhaps "literalness" is then simply a measure of the speed with which we can break off the exploration. When we try to interpret a poem, we may see ourselves as going indefinitely deeper into this network of meanings, both in articulating the knowledge implicit in the schemas that provide the core of the interpretation and also in exploring the network of associations that takes us further and further out into the richness of experience.

Our understanding of this continuity between the literal and the metaphorical can allow us to extend the field of our science as we enrich our vocabulary for talking about that to which science cannot yet do justice. As we extend our perspective, we will come to look at science as something not limited to the purely physical sciences, even in the extended form of the information-processing sciences which include schema theory. Rather, it reaches out into the moral sciences and the *Geisteswissenschaften*, forming a continuum between those studies anchored in the quantifiable and those anchored in the richness of lived experience.

We must continue to explore the creative tension posed by the apparent incompatibility of these extremes. However, I do not want this openness to dialogue to be seen as suggesting an acceptance of the view that physical science is "just one more myth." Mere myths cannot build jet planes that fly, or probe the properties of matter to enable us to build electronic circuitry. Science is driven by the pragmatic criterion of successful prediction and control.[24] More generally, we can see science as bringing into concordance socially agreed-upon methods of observation with socially shareable types of formal explanation, all making sense to the individual who tries to measure experience against these standards. These general criteria still apply when we move to the less quantifiable concerns of cognitive science. We may expect our person-reality to be changed in the light of careful science and our science to be changed and extended as we assimilate more of our human experience within the continuing dialogue of two-way reductionism.[25] Let me close this section by noting a concern of Theo Meyering (personal communication):

> By the time we get to "language as metaphor"... [the] notion of schema is expected to do much too much work, it seems. If it is stretched too far, it threatens to become vacuous: facilitated reflex arc, routine behavior, internalized skills, metaphor, implicit knowledge, inarticulable background knowledge, socio-cultural models, ideology, religion—from the amoeba to Bach, from moving a bacterial flagellum up to inventing the Art of the Fugue—perhaps all of this should be seen indiscriminately as the mere exercise of (lower- or higher-order) schemata? Is there any fish this net can't catch?

To this I reply: No, I do not include amoebae—but schema theory *must* be protean if it is to develop into a truly adequate model of the mind (and related "support processes"). Consider, by way of comparison, how protean the notion of "program" is for computation.

---

[24] Mary B. Hesse, *Revolutions and Reconstructions in the Philosophy of Science* (Bloomington: Indiana University Press, 1980).

[25] See fn. 18 in my companion essay in this volume.

3.3 From Mirror Neurons to Language

As noted earlier, this section is a detour in terms of preparing the reader for section 4, but it is important as an example of relating neural activity to social interactions. I summarize data on what are called "mirror neurons" in order to develop an hypothesis on the possible role of this observation/execution matching system in the evolution of human brain mechanisms for language.[26]

Grasping movements play an important role in both social and nonsocial behaviors. Specialized mechanisms in the primate brain link perceptual schemas for objects to the motor schemas for grasping them.[27] In the monkey, parietal area AIP (anterior intraparietal cortex) and ventral premotor area F5 form key elements in a cortical circuit which transforms visual information on intrinsic properties of objects into hand movements that allow the animal to interact appropriately with the objects. Giacomo Rizzolatti and his coworkers discovered "mirror neurons," a *subset* of the grasp-related premotor neurons of F5.[28] Mirror neurons are active not only when the monkey grasps an object (as do other F5 neurons) but also when the monkey observes meaningful hand movements made by the experimenter or another monkey, such as placing objects on or taking objects from a table, grasping food, or manipulating objects. Mirror neurons, in order to be visually triggered, require an interaction between the agent of the action and the object of it. The majority of mirror neurons are selective for one type of action, and for almost all mirror neurons there is a link between the effective observed movement and the effective executed movement. If the human or other monkey is moving its hand toward an object, and the shape and orientation of that hand are appropriate to that object, then those mirror neurons will fire which also fire if the observing monkey were itself to grasp such an object in such a way. The simple presentation of objects, even when held by hand, does not evoke the neuron discharge, nor does a movement of the hand which is not directed towards an object.

The mirror neurons, then, constitute the set of all F5 neurons with the property that each will be activated for both observation and execution of the same specific range of grasping behaviors. We thus say that area F5 includes an *observation/execution matching system*: when the monkey observes a motor act that resembles one in its movement repertoire, a neural code for this action is automatically retrieved. This code consists in the activation of a subset of the F5 neurons which discharge when the observed act is executed by the monkey itself.

---

[26] The present section is based on two papers written with Giacomo Rizzolatti: Michael A. Arbib and Giacomo Rizzolatti, "Neural Expectations: A Possible Evolutionary Path from Manual Skills to Language," *Communication and Cognition* 29 (1997): 393–424; and Giacomo Rizzolatti and Michael A. Arbib, "Language Within Our Grasp," *Trends in Neurosciences* 21 (1998):188–94.

[27] Marc Jeannerod, M.A. Arbib, G. Rizzolatti, and H. Sakata, "Grasping Objects: The Cortical Mechanisms of Visuomotor Transformation," *Trends in Neurosciences* 18 (1995): 314–20.

[28] G. Di Pellegrino, L. Fadiga, L. Fogassi, V. Gallese, and G. Rizzolatti, "Understanding Motor Events: A Neurophysiological Study," *Experimental Brain Research* 91 (1992): 176–80; Vittorio Gallese, L. Fadiga, L. Fogassi, and G. Rizzolatti, "Action Recognition in the Premotor Cortex," *Brain* 119 (1996): 593–609; Giacomo Rizzolatti, L. Fadiga, V. Gallese, and L. Fogassi, "Premotor Cortex and the Recognition of Motor Actions," *Cognitive Brain Research* 3 (1996): 131–41.

An observation/execution matching system *in humans* was demonstrated by PET experiments (studies which image areas of the human brain to show which areas are more active when the subject is performing one task rather than another) which showed that grasp observation significantly activates the cortex of the left superior temporal sulcus, of the left inferior parietal lobule, and of Brodmann's area 45.[29] What is most intriguing is that Brodmann's area 45 is part of Broca's area, which is most commonly characterized as an area for speech.

What can we make of the similarities between Broca's area and F5? There is a general consensus that animal calls and human speech are different phenomena. Among the many aspects that differentiate them, such as the strict relation of animal calls (but not of speech) with emotional and instinctive behavior, there is also a marked difference in the anatomical structures responsible for the two behaviors. Animal calls are mediated primarily by the cingulate cortex together with diencephalic and brain stem structures.[30] Speech is mediated by a circuit whose main nodes are the classical Wernicke's and Broca's areas of the dorsolateral cortical convexity.[31] The data reviewed above on the observation/execution matching system suggest that what determined the speech development in the lateral cortical circuit was that a part of this circuit, and Broca's area in particular, was endowed before speech evolved with the capacity of recognizing actions made by others. It is obvious that there is a great conceptual difference between recognition of gestures made by the actor with the communicative intent of conveying some message to the observer, and gestures such as picking up a piece of food which are made without perceptible communicative intent. No claim is made that the mirror neurons in monkeys which have been described above are themselves part of a communication system; rather, the homology with Broca's area leads to the hypothesis that the noncommunication F5 system of some hominid ancestor of the human provided the evolutionary precursor of the human Broca's area, with its crucial role in human language. The mechanisms that allow the recognition of hand actions without communicative intent must, in principle, be no different from those for recognition of facial-mouth movements having a communicative intent (for example, monkey "lipsmacks") and those for recognition of "phonetic gestures." It is only the input-output coupling which is different, not the recognition process.

Two sets of data reinforce this view. The first is based on neurophysiological findings concerning neurons coding face-mouth movements in F5. Preliminary experiments carried out in Rizzolatti's laboratory (unpublished observations) showed that in F5 there are face-mouth neurons that become active when the monkey

---

[29] Giacomo Rizzolatti, L. Fadiga, M. Matelli, V. Bettinardi, D. Perani, and F. Fazio, "Localization of Grasp Representations in Humans by Positron Emission Tomography: 1. Observation Versus Execution," *Experimental Brain Research* 111 (1996): 246–52; Scott T. Grafton, M.A. Arbib, L. Fadiga, and G. Rizzolatti, "Localization of Grasp Representations in Humans by PET: 2. Observation Compared with Imagination," *Experimental Brain Research* 112 (1996): 103–11.

[30] Uwe Jürgens, "Primate Communication: Signaling, Vocalization," in *Encyclopedia of Neuroscience*, George Adelman, ed. (Boston: Birkhauser, 1987); Paul D. MacLean, "Introduction: Perspectives on Cingulate Cortex in the Limbic System," in *Neurobiology of Cingulate Cortex and Limbic Thalamus: A Comprehensive Handbook*, Brent A. Vogt and Michael Gabriel, eds. (Boston: Birkhauser, 1993).

[31] D. Frank Benson, "Classical Syndromes of Aphasia," in *Handbook of Neuropsychology*, vol. 1, François Boller and Jordan Grafman, eds. (Amsterdam: Elsevier, 1988), 267–80.

makes "lipsmacking" movements. Some of these neurons also fire when the monkey observes similar movements made by another individual. It thus appears that the mirror mechanism is not limited to actions devoid of communication purposes such as grasping movements, but also mediates communications between individuals.

The second set of evidence comes from phonological studies of Alvin Liberman and others.[32] On the basis of empirical findings and general considerations of plausibility, they formulated what is usually referred to as the "motor theory" of speech perception. According to this theory, the "objects" of speech perception are not the sounds, but the phonetic gesture of the speaker. These gestures are represented in the brain as invariant motor commands. The phonetic gestures are "the primitives that the mechanisms of speech production translate into actual articulator movements, and they are also the primitives that the specialized mechanisms of speech perception recover from the signal."[33] The common link between sender and receiver is therefore not sound, but motor pattern. In both speech perception and gesture recognition the basic mechanism thus appears to be that of matching the neural activity resulting from observation of a gesture with that underlying its execution. Since the meaning of the latter is "known," the gesture is, to a first approximation, recognized.

The force of the phrase "to a first approximation" is to emphasize that, since phonemes may be hard to distinguish in isolation, hearing a sound may well rest on cooperative computation between representations at multiple levels, including the lexical and the phrasal, before contextual constraints yield a coherent interpretation of an utterance.[34] In this spirit, note, for example, that some Japanese speakers who acquire English late enough may be unable to hear the difference between "lid" and "rid" yet can pronounce them distinctly. For such people, the network for "gesture recognition" has been stabilized for the Japanese phonemic repertoire before English is learned. Thus, although careful attention to mouth shape and tongue placement can be used to form an "l" and "r" distinguishable by English speakers, the gesture recognition system of the Japanese speaker is by then insufficiently plastic to acquire this distinction. On the other hand, this Japanese speaker is well able to understand English when context can settle whether "l" or "r" occurred in a particular utterance.

The sequence proposed for the evolution of language is the following: facial gestures first, then manual gestures, and finally speech gestures.[35] Facio-visual communication is used in a wide variety of social circumstances by nonhuman primates. It is produced by both males and females, regardless of their rank, and is accompanied by eye contact. Note that this specific relation between the sender and the receiver of the message may be absent during emission of vocal calls—the latter

---

[32] Alvin M. Liberman and M. Studdert-Kennedy, "Phonetic Perception," in *Handbook of Sensory Physiology, Volume VIII: Perception*, Richard Held, Herschel W. Leibowitz, Hans-Lukas Teuber, eds. (New York: Springer-Verlag, 1978); Alvin M. Liberman and I.G. Mattingly, "The Motor Theory of Speech Perception Revised," *Cognition* 21 (1985): 1–36; Alvin M. Liberman, F.S. Cooper, D.P. Shankweiler, and M. Studdert-Kennedy, "Perception of the Speech Code," *Psychology Review* 74 (1967): 431–61.

[33] Alvin M. Liberman and I.G. Mattingly, "A Specialization for Speech Perception," *Science* 243 (1989): 489–94.

[34] For example, see the early analysis of neural computations underlying language in Michael A. Arbib and D. Caplan, "Neurolinguistics Must be Computational," *Behavioral and Brain Sciences* 2 (1979): 449–83.

[35] Rizzolatti and Arbib, "Language Within Our Grasp."

may be directed to the whole tribe rather than to a single individual. The use of manual gesture marks an important evolutionary step because it allows the possibility of introducing a third, new element in the basically dyadic communication based on facial gestures. By using manual gestures the sender may indicate to the receiver the position of a third person or object, and may represent the speed and direction of a movement. These possibilities are very limited with visuo-facial communication, even if facial movements are accompanied by head movements.

Both facial and arm gestures can be used only when lighting conditions are suitable. Sound may overcome this drawback, but the advantage the basic vocalizations bring to communication are rather limited. They simply stress a facial message. The advantage of associating sound to gesture becomes far more important when manual gestures convey information on objects and events. The appearance of such flexibility was one of the most important evolutionary steps toward speech. We suggest that the ability to use sound to reflect the richness of manual communication then followed, at which stage speech took off. Note that Broca's area has the neural structure for all the above discussed gestures: face movements, oro-laryngeal movements, and hand-arm movements. Moreover, and crucially, it is endowed with an observation/execution matching system.

The hypothesis that the language specialization of the human Broca's area derives from an ancient mechanism related to the observation/execution of motor acts is made more plausible by the fact that language can be exhibited in the brachio-manual-visual domain of sign language. Doreen Kimura argues that linguistic and motor control functions evolved in an interactive manner, and that it would be extremely inefficient to duplicate the motor control mechanisms in a separate language system.[36] I want to suggest that the issue is, rather, one of "partial separation." In spite of the strong spatial component necessary for the production of sign language, the basic communicative function is, as for speech, localized in the left hemisphere and, basically, in the same areas where speech is localized in individuals without auditory deficits.[37]

Finally, I have adopted a "case structure" grammar for analyzing hand actions, being careful to distinguish (1) "case structure" as studied in the monkey: a search for the neural representations of perceptually guided actions, from (2) "case structure" as studied in the human: a search for the neural representations of linguistic structures.[38] However, by relating "uttering a sentence" to "achieving a goal" I am reducing the "evolutionary distance" between action and utterance, thus increasing the plausibility of the overall hypothesis. The details of our "pre-linguistic grammar" of action in the monkey brain are beyond the scope of this essay. However, note that the problem of transforming an underlying case structure into one of the many possible sentences which expresses it may be seen as analogous to the nonlinguistic problem of planning the right order of actions to achieve a complex goal. Suppose the monkey wants to grab food, but then—secondarily—realizes it must open a door to get the food. In planning the subsequent action, the order food-

[36] Doreen Kimura, *Neuromotor Mechanisms in Human Communication* (Oxford: Oxford University Press, 1993).

[37] Helen J. Neville, D. Bavelier, D. Corina, S. Padmanabahn, V.P. Clark, A. Braun, J. Rauscheker, P. Jezzard, and R. Turner, "Effects of Experience on Cerebral Organization for Language: fMRI Studies of Hearing and Deaf Subjects," *International Journal of Psychology* 31 (1996): abstract no. 434.

[38] Arbib and Rizzolatti, "Neural Expectations."

then-door must be reversed. This is like the syntactic problem of producing a well-formed sentence, but now viewed as going from "ideas in the head" to a sequence of words/actions in the right order to achieve some communicative/instrumental goal. Moreover, my analysis of the "grammar" of actions made explicit that the neural principle underlying organization of action is that of a combinatorial system. Each of the "classical" F5 neurons codes a discrete motor act (for example, precision grip, whole hand prehension, wrist rotation).

In summary, a well-formed sentence is like a well-formed action plan. Leslie Brothers (personal communication) observes that the idea that we understand one another's sentences analogously to the way we understand one another's purposeful motor acts is borne out by our ability to finish someone else's sentence for him. We know what the goal is when we hear the beginning of the act. In any case, by relating the neural substrate of language—an important medium for human social behavior—to a possible evolutionary precursor for which a body of neural data and computational models is accumulating, I have offered a contribution to the program of *Friday's Footprint* of developing a neuroscience of sociality.

4 Ideology and Religion

The model of the acquisition of language outlined earlier delineated a path from social schemas (defined by the collective behavior of a given social group, in this case their use of a shared language for communication) to individual schemas viewed as constitutive of different patterns of behavior (the schemas acquired by the child in becoming a member of the language community). The following discussion of the evolution of the neural basis of language then provided a useful detour, exemplifying the linkage of "external" schemas to the neural mechanisms which provide their "internal realization." I will now use my understanding of the relation of social schemas to "schemas in the head" to provide a brief discussion of ideology and religion, linking the knowledge of individuals to the society around them. In schema theory, the individual's knowledge is inseparable from his or her evolving schema network. Thus, there is no such thing as sure knowledge. Certainly, there are schemas that embody what we now take to be true. But schemas also include representations that the naturalist would not want to call knowledge. There are false but useful models of the natural world, ideal types of human behavior and social order that may never be realized, including utopias, heavens or hells; there are also schemas for the beautiful, the true, and the good, which determine the way in which we behave, whether or not we believe they represent some external reality.

I earlier qualified the Piagetian view that schemas adapt to yield increased coherence with Freud's view that schema change could lead to conflict as well as to coherence. New schemas that give us further control of some new aspect of our experience may lead to tensions and conflict with what we already know within other contexts. On the social scale, special interests are inevitable and cannot be circumvented, reminding us of Ambrose Bierce's definition, in *The Devil's Dictionary*, of an egotist as "a person of low taste, more interested in himself than in me." In the following section, I offer a schema which sees social systems as necessarily provisional, as necessarily open to critique through the process whereby the patterns of a society, no matter how well established, are continually internalized by new members of the community, who bring individual experience to bear. Because of that experience, these members may in time generate critiques of the societal patterns, the *social schemas*.

In the final section I will turn to the social schemas of theistic religion, offering arguments—consistent with but *not* implied by schema theory—for the view that God may be a "projection" rather than an "external reality." In introducing the earlier section on language, I noted that as individuals come to assimilate communal patterns, they will provide part of the coherent context for others. But I also stressed that the commonality of social behaviors still leaves space for discordances between the individual and the community. This becomes particularly clear when we look not at language but at ideology as a social structure, and I shall exemplify this by looking at a secular schema that stands in contrast to the theistic schemas espoused by many Christians and other believers. I do *not* claim that schema theory can tell us whether or not God exists as an external reality. Rather, social schema theory extends an account of cooperative computation of schemas in the head, with strong links to the data of neuroscience, to embrace lessons from the philosophy of science and the sociology of knowledge in order to understand how individual schemas relate to social schemas. With this I have provided a bridge between accounts of social change and an understanding of mental or neural mechanisms for individual and social behavior. However, if this theory is to be successful, it cannot dictate one single ideology or, in particular, one single system of religious belief.[39] Rather, it must let us understand why some hold that God is a transcendent reality, whereas others view God as a social construct. What do I mean by "must let us understand"? First, schema theory's "natural territory" of creating a computational theory of the mind presents a lifetime of challenging science even without any contact with theology. However if one accepts the challenge of seeking to understand the relation between neuroscience and divine action, then (I argue) schema theory provides as good a bridging language as any, and the diversity of belief in the nature of God and in divine action then becomes a challenge for schema theory.

Schema theory cannot answer complex questions in a simple way. Each hard problem requires major effort for its solution. Thus in no way do I claim that I have been able to apply schema theory to answer how it is that some of us hold that God is a transcendent reality, whereas others view God as a social construct; all I have done below is to open a discussion of some of the issues that such a challenge to schema theory must confront. Moreover, schema theory itself is just one more perspective on "socio-cognitive neuroscience." I have tried to use this perspective to create an epistemology from which one can analyze and compare other perspectives. However, there is no claim that this perspective is absolute. Given the whole thrust of this article, to admit to such perspectivism is not self-defeating. Schema theory recognizes the need for making prejudgments, and it allows for criteria in terms of which this perspective can be compared with others. In this spirit, I offer schema theory not to provide recipes for absolute value but rather to help us understand the diversity of value systems and to help us analyze how they interact and how they may change.

---

[39] Robert Russell asks whether this view implies "that religious beliefs are *only* ideologies, or whether they merely have an ideological component or dimension?" "I would agree," he continues, "with the claim that there is an ideological dimension to religion, as there is to science, etc., but not with the more reductive claim that you seem to be implying here." I use ideology here in the OED's fourth sense: "A systematic scheme of ideas, usually relating to politics or society, or to the conduct of a class or group, and regarded as justifying actions, especially one that is held implicitly or adopted as a whole and maintained regardless of the course of events." In view of this I would think it fair to say that science is less ideological than religion (but that's another debate!).

4.1 Ideology, Social Schemas, and Critique

The time has come to look at ideology from the perspective of schema theory. I begin by first ridding ideology of its pejorative connotation, accepting that an ideology is not to be seen as imposed upon us by history or as inherently bad, but rather as a necessary expression of social interaction. There have to be social schemas to bind the members of a society together, to allow them to live together, even in the simple case of agreeing which side of the road to drive on. This realization that an ideology is not necessarily repressive but is necessary for social cohesion does not deny that a critique can be made of each ideology.

In his analysis of religion Emile Durkheim rejected Karl Marx's one-way explanation of religion in terms of economic superstructure.[40] Instead he focused attention on the multiple interactions between different social institutions—the religious, the economic, and others—even though one can look at them as relatively autonomous. He stressed that the effectiveness of a religion depends in part on the effectiveness of its rituals; and this, in turn, depends on the effectiveness of belief. The social system must be psychologically internalized; a clarion call for schema theory! We need a theory of the internal workings of the symbols of an ideology or religion. We need to study individual beliefs not in isolation, but to see how they reinforce each other and to see their relation to the external world. We have to understand that the justification for a social schema comes not only from social cohesion but also in terms of the natural world, personal emotions and relations, and social history.

Building on our earlier discussion, I now want explicitly to view an ideology as a "large" social schema—outside the individual, but not immutable. As emphasized in the previous discussion of language acquisition, it is something that the child comes to as an external reality and internalizes in order to become a member of society; but the way in which the ideology is internalized does not preclude a critique. This is consonant with Hesse's philosophy of science.[41] In becoming a member of a scientific community, one must learn how to make observations and one must learn the techniques of theory building. But in learning these, one comes to gain a sufficiently rich expressive language so that one not only understands what observations and theory came to coherence in the work of earlier scientists; one also can provide a critique of extant theory in terms of one's own observations and thus contribute to eventual changes in theory.

When I was an undergraduate, my friends and I were speculating about the idea that computers in the home would support direct democracy—people would be able to log in each night and vote their decisions about the pressing issues of the day. One friend destroyed the whole idea by just saying "Decisions, decisions, decisions." None of us has the time to learn enough to make all the decisions necessary for our complex society to work. We want to be involved in particular issues of special concern to us, and we want the checks and balances of democracy so that we can apply some high level constraints to the overall nature of decision making. But to be involved in making a decision on every twist and turn in the course of society could only overload us. Similarly, I would suggest that it is a tool of mental economy to view social roles and social schemas, including ideologies, as necessary *and* as necessarily provisional. In fact, if schemas are provisional, as the Piagetian approach

---

[40] Durkheim, *The Elementary Forms of the Religious Life*.

[41] Hesse, *Revolutions and Reconstructions in the Philosophy of Science*.

to schemas has suggested, then the more reflective amongst us will almost necessarily form a critique.

Let us see the bad and the good of what happens when one forms a critique. When a society embodies a repressive ideology, someone who dares to form a critique may be subject to political or religious persecution, trouble holding a job, ostracism, exposure to ridicule, troubled social relationships, and so on. If someone learns that his view of society is out of joint with society itself, that can mean that society is disarrayed, or that the person is disarrayed, or both! However, in some cases, we have a body of individual experience, or we belong to a subgroup in society that has sufficient cohesion, and we persist, we become rebels, the critique continues. In other cases, the costs prove too great, and we come in due time to repress these elements of critique. Then either these elements fade away and we become a happy member of society or they continue as a festering sore. But if one does not view ideology as necessarily a bad thing, one can see that the critique can proceed in other ways. One may indeed be persuaded that the ideology is beneficial; or one may see that, given the current state of society, the benefits of change are outweighed by the likely costs of trying to bring that change about. On the other hand, critiques can contribute to the evolution of social institutions. Society does change, and while some of these changes can be seen as accompanied by great costs, others can be seen as progressing in a relatively comfortable way. Schemas are our set of operational means both for acting and for changing things.[42]

The language in which we exercise our critique is inevitably value laden. As a child, we start with the interests of easing discomfort, or of getting enough to eat and drink. As our schemas grow, they always interact with schemas embodying personal comfort and social interaction. We grow up within a society. By the time we have reached an age and level of education in which we could provide a critique, much of the vocabulary in which the critique is expressed is already laden with the values of that society. This is inevitable, even if the language is not as biased as George Orwell's NEWSPEAK. One might say cynically that an ideology succeeds to the extent that it excludes conflicting schemas from consideration by making their reality appear untenable. But no matter how holistic, integrated, or encompassing an ideology or other symbol system may be, the network is open to other experience and this may either lead to evolution of the system or to rejection of the system.

Nancey Murphy commented (personal communication):

> "I think it is important to see theological systems as entirely comparable to other sorts of knowledge, so I am happy with your arguing that religious beliefs are major schemas or ideologies. I think you need to make it clearer than you do why in this case there is special reason to distinguish between social construction and reality depiction, whereas your schema theory itself seems to hold these two sides of the epistemological account together so nicely."

I do not argue that "in this case there is special reason." Rather, I would say it is a central tenet of schema theory to analyze the mechanisms whereby social construction and reality depiction are dynamically interlinked. It is because many "realities"

---

[42] George Ellis (personal communication) observes that the necessity for some kind of ideology in order for society to cohere is developed in the useful article by D.F. Aberle, A.K. Cohen, A.K. Davis, M.J. Levy, and F.X. Sutton, "On the Functional Pre-requisites of a Society," *Ethics* 60 (1950): 100–11. Peter Berger, in *Invitation to Sociology*, develops the concept of ecstasy precisely to refer to the individual breaking out of the conceptual constraints given to him or her by society and hence becoming a potential agent for change.

are socially defined rather than "physical" that Hesse and I developed the notion of social schema—our schemas are not shaped solely by perceiving doors and how to open them, or by apples and when to eat them.

I have already noted Brothers' observation that to be a person is to belong to a social order which is made up of shoulds and oughts and other moral categories, while stressing that the child's apprehension of the social order must be built up schema by schema. I have already argued that there cannot be one true ideology, but it may be possible to understand why people have the sense that they should live in one way or another. Our moral and ethical systems may be purely social constructs, bound to change from generation to generation, but there may be biological constraints on social organization. In my view, neuroscience can address the moral domain in original ways because, to some degree, it is possible to get "ought" from what "is." Human evolution has led to an exceptionally adaptive animal, with an extremely complex nervous system, that can not only adapt to its environment, but also tends to adapt the environment to itself. More strictly, science (for example, anthropology) may analyze diverse moral systems, but one cannot infer a specific moral system on the basis of the scientific method. My suggestion is that neuroscience may help us understand the constraints which shape moral norms. At the same time, I would suggest, paradoxically, that moral norms are *contrary* to the functioning of the brain! We need norms only to the extent that behaviors are not the automatic result of biologically evolved patterns of brain function. However, schema theory can address the issue because it shows how learning about objects is inextricably bound up with learning about practices concerning them, and these in turn involve what one ought to do with them.

4.2 The Social Construction of God [43]

Finally, I relate all this discussion of the interaction between brain and society as the individual comes to know her world to the question of the nature of God. Is God an *external reality* that we dimly apprehend? Or is God only a *projection* of our social schemas? In *The Construction of Reality*, Hesse and I asked whether God is more like gravitation or embarrassment: although our theories of gravitation have changed from the Greeks to Newton to Einstein, we believe that even if humans never existed, gravitation would still endure. On the other hand, embarrassment is a real presence in our lives and plays a socially important role, even if different generations and societies "shift the goal posts." It depends upon our physiology as well as on our cultural background—but if there were no humans, there would be no embarrassment. Embarrassment is a physiologically real phenomenon, but we also understand that different societies have totally different constructs as to what it is to be embarrassed. There is no great embarrassing thing with a capital "E" out there that is the trigger for embarrassment. Therefore, if you accepted that God is like embarrassment, you would see God as being a human social construct; but if you accept the view that God is like gravitation, then God with a capital "G" does indeed exist whether humans are here or not. [44]

---

[43] This section is something of a palimpsest—my original draft is overwritten with the comments of many readers, and these in turn are overwritten by my replies. I thank all these commentators, and hope that readers will find the resultant composite of interest.

[44] Nancey Murphy notes that the distinction between "physical" and "social construction" is unsatisfying, since most schemas are socially constructed and reality-depicting. I agree that

Tom Tracy notes that here I address two different, though connected, questions:

(1) You offer an interpretation of theism (and other comprehensive interpretive views) in terms of schema theory, treating such systems of belief and practice as "large social schemas."

(2) You ask whether God is an "external [mind-independent] reality" or just a projection.... It may turn out to be illuminating to analyze theism as one socially constructed schema among others available to us in constructing/interpreting our experience. It does not follow from this, of course, that theism is merely a projection that fails to represent (however incompletely) a mind-independent [transcendent] reality.

Indeed, schema theory does not support a claim for or against theism. As Hesse and I showed in *The Construction of Reality*, schema theory is neutral—all it can do is give a critique of certain forms of theism and atheism, but it cannot prove or disprove the reality of God. Moreover, many commentators have questioned the above dichotomy. George Ellis asks: "There may well be an element of projection occurring: but is there nothing but projection?" Stephen Happel states:

Perhaps it is not an either/or question. Could it be true that God *is* and that humans interpret/construct the identity/meaning of divinity in an emotional and cognitive dialogue? Moreover, it is possible that God is "like" both kinds of objects—and like neither?

Anne Clifford, too, asks:

Is the question necessarily an either/or one? Is it possible to say that God *is* (something akin to Thomas Aquinas's "pure act") and also laden with historically shaped projection that is both analogical and metaphorical in kind?

Wesley Wildman goes even further, asserting that

the undeniable reality of projection in religion looks very different in the context of a nuanced treatment of the way language refers. Specifically, I think the neat distinction between successful reference and socially-supported and even functional projection that you rely on (cf. gravity versus embarrassment) pretty much collapses.

First, I acknowledge that I do not have the resources for a new, scholarly approach to projection. Second, the dichotomy is more subtle than the commentators may have noted: the "gravitation" alternative acknowledges that our understanding of gravitation has changed from Newton to Einstein and continues to be colored by the "projections" of current scientific theory. Clifford believes that

persons relating to other persons involves projection. If God is viewed as more than merely a principle of cosmic creation or of created order (a deist god—*deus ex machina* of the eighteenth century) and is invested with personal attributes then projections that are emotion laden seem to be part of the package. Human beings have a propensity for projecting notions onto their personal acquaintances.

Exactly. All our perception is mediated by schemas, and thus always involves a constant compromise between our current beliefs and the current sensory evidence

---

there is no strict dichotomy. Rather I am planning to define ends of a continuum. For example, chairs are physical objects, but their exact delineation (e.g., chairs versus stools) may vary from society to society. On the other hand, a character in a familiar legend (e.g., Ulysses) may have some tenuous linkage to historical events and familiar human behaviors, but nonetheless is "more like embarrassment than gravitation." Murphy poses the question: "Why think of religious schemas as falling at the extreme of total construction. And even if they do, how do they relate to other pure constructions, like justice?" My view is that a specific religion is very much a social construction developed by a community or communities without there being a divine reality of which it is the imperfect expression, whereas theists would see the reality of God as central to their definition.

(itself a most imperfect sample of the other person).[45] As one sees from different religions, there are different theories, different perceptions of what God is—again raising the question of whether this is because there is a God who is, in essence, unknowable; or whether different religions project God upon the search for answers to diverse social needs. In the rest of this section I will pursue a number of arguments in support of the latter view.

To start with, I offer the somewhat implausible suggestion that God is like the style of a city. For example, the town of Santa Fe has, at any time, a distinctive style, both as a present livable city and surroundings, and as a set of cultural artifacts— museum pieces, famous buildings, books, urban myths, etc. Each person who moves to Santa Fe imbibes some of this. In the way people design a new house or live their lives, they "update" that style, and both preserve and change the legend. But what is the "real" Santa Fe? We may, in the fashion of the *via negativa* of theology, strip off the apparent inessentials in search of the true, underlying essence. However, the excavation of different levels of ancient Paris below the Louvre takes us further away from the "real" Paris, not closer to it. Imagine Italy before tomato sauce! If we imagine London without double-decker buses, or the Tower of London, or if we peel off the layers we do *not* get closer to the "real" London. The Londons of Henry VIII and of Charles Dickens both color our perception of the late twentieth-century city. But it is the interaction of history, current actuality, and individual experience that creates "our" London. We may "believe" in London, we may "love" London, but it is a totality as much in the minds of fellow enthusiasts as in the current day city. The idea that God is like the style of a city, then, suggests that we view God not as an external reality but rather as the hypostatization of a culture-laden yet personally enriched experience to a set of questions about how life is to be lived. Then major religions are like large, history-rich, highly populated cities, and God is like the "spirit of the city," which is our projection on the base of social experience, not an "external reality" in the sense of gravitation.[46]

On this basis, some aspects of a religion are "as valid as any other"; others are open to a broader ideological critique of the kind outlined in the previous section. It was only in 1983 that the northern and southern branches of the Presbyterian Church of the United States were reunited, healing a split caused one hundred years ago because people were able to read the teachings of Jesus Christ as either supporting slavery or rejecting it, depending on which part of the country they lived in.

The process of dialogue or schema change leads at best to a process of learning in terms of current criteria, rather than to absolute progress to the right text, the right science, or a true understanding of God.

People mean many different things when they speak of God. There is God the Creator, who created the universe. There is God the Father, who somehow looks after and protects us, and perhaps related to that, though not necessarily the same,

---

[45] Note the discussion of Brothers' view of "person" given earlier in this essay, as well as Jeannerod's essay on mental states in this volume.

[46] This discussion reminded Leslie Brothers of the following: "…the grammar of a living language must always be open to the linguistic novelties and the sheer increment of vocabulary that is the fate of every living language:…[I]f theology is grammar, then there is the task, always pertinent, of learning to extend the rules, the order, the morphology, of godliness over the ever-changing circumstances" (Paul Holmer, "The Grammar of Faith" in *The Grammar of the Heart: New Essays in Moral Philosophy and Theology*, Richard Bell, ed. [San Francisco: Harper & Row, 1988], 8).

is God the Answerer of prayers. For those who think of the divine arrow of time, there is God who will resurrect us at the end of the physical universe, in a New Creation. There is a God who presides over Heaven and who fights the devil.

Many of those scientists who would consider themselves religious hold a view of God that is less concrete than any of these properties, a God who is beyond our experience but who imbues our experience with its meaning. Many other scientists hold a different, Spinozian view of God as the underlying Order in the universe. As a scientist, my faith is that I can make progress continually in finding an explanatory system for a whole body of apparently unrelated facts. Therefore, there is a sense that there is an Order in the universe that can allow understanding. The following observation by Happel seems to me consistent with this view:

> That the world is intelligible (mathematically comprehensible) is a datum of your experience to which as a scientist you give yourself. This is not only because you as a scientist are part of the world you explore, but also because you continue to trust that this intelligibility will go forward. The "religious" question is, Why is this so? Is there a basis for trusting this order? A scientist's willingness to give him or herself to the continuing ordering of the universe is a heightened example of the trust in ordinary language most people have. That we use metaphors [whether in ordinary speech, science, or theology] to express the meaning/trust in this Order does not mean that they have only emotive value. They have heuristic, cognitive values as well by which we continue to explore... the truths [under certain conditions] of the claims [that]... they lay out.

And one may choose, with Spinoza, to use the word "God" to name that Order, but God in this sense need not come anywhere close to the God of religious faith in the sense of Judaism or Christianity or Islam, where there really is a God or a Christ or an Allah, who in some sense cares for us and has something to do with our lives. The Spinozian God or the God who created the Big Bang and then sat back—the God of the Deists—is irrelevant to our lives. Clifford responds that

> I cannot grasp why the God of Order... is presented in opposition to the God of the Bible.... the Hebrew Bible's Wisdom tradition concerns itself with people discerning and living in accord with the order God intended for creation.

The distinction I had in mind was that (1) the God of Order is one who created the laws of nature and the initial conditions of the universe and then left nature to take its course. This God may or may not have anticipated certain aspects of this "unfolding" but the vagaries of chance in evolution are such that even God could not have ensured the emergence of Christian morality. There is no scope for divine action, including divine acts of redemption, since the Creation. By contrast, (2) the God of the Bible is a God of divine action: for Jews, this God formed a covenant with Israel and for millennia expressed his pleasure and displeasure with his chosen people. For Christians, the Resurrection is the supreme example of divine action... or so it seems to me.

My concern is that if one is to say one believes in God, one may have intimations of the reality of that God based on one or more religious traditions (even the phrase "the God of the Bible" has no fixed denotation) or on a nonreligious tradition—God the underlying Order of the Universe. I accept that the scientist proceeds on faith that apparently diverse phenomena will reveal an underlying pattern to the diligent investigator; I am not convinced that attaching the word "God" to such a faith adds intellectual clarity. One may experience "a harmony with the universe" but these experiences do not imply that one "really" is in harmony with the universe (it is a big place). Rather, such phrases are metaphors to express feelings for which we have few words.[47] Great works of art, and a variety of substances, can sweep us in this direction. The problem arises when believers take their feelings as evidence for the "reality depiction" of their God-schemas, or, *a fortiori*, when people seek in their experiences evidence of a God who can rule us and force us to act in ways which justify cruelty to others in the name of God. Problems also arise when the distortions caused by drug-taking are instead seen as an opening to a greater reality, as in Aldous Huxley's *Doors of Perception*. Having said this, I confess that our culture owes much to visionaries such as William Blake whose wild visions force us to change our own perspectives, whether or not we can accept their "reality."

Ellis provides the following comment on my view of religion:

> Your discussion of the nature of God tackles real and important issues, but why do you (i) ignore the central theme of Christianity [by suggesting humans are insignificant to God], (ii) apparently propose that all believers trust in a cruel God, and (iii) suppose that drug-taking is important to true religion? What purpose is served by setting up these straw men and then demolishing them? While the historical role of the Church is clearly a major problem in many respects, (a) there has also been a very positive side to the role of the Church in history that is ignored here, and (b) there are many religious people who reject the kind of authoritarian religion referred to here in favor of a kenotic one, and whose worldview is not so easily demolished.

This is an unfortunate misreading of a discussion of the challenges posed by the very diversity of human beliefs about the nature of God. The comments to which Ellis addresses (i) were in the context of the view of God as Order (as above), as distinct from God as Father, etc.; those addressed by (ii) formed a reflection on such historical phenomena as slavery (they apply as well to the Expulsion and the Spanish Inquisition), raising the issue of which elements of a religious tradition are indeed "valid," while (iii) was occasioned by my comments on drug taking in relation to experiences of "apparent transcendence," not in relation to religion. As Happel notes:

> Just as there are distorted notions of the order of the universe, which require intellectual correction, so there are distortions in religious interpretations (for example slavery was justified, as you note, to speak of only one element of barbarism in religion). These too require corrections.

I have never doubted that there are admirable versions of Christianity and Judaism and Hinduism and Buddhism. The issue is: What are the truth claims of these diverse systems? They cannot all be true in all respects. In particular, is a belief in God as an external reality a necessary basis for their moral virtues; and, if it is, does such virtue imply that the belief is "true"?

I cannot avoid talk of "reality." We know that our description of the physical universe is imperfect. Nonetheless (as in our talk of gravitation) most scientists still speak of a "world out there" against which our science strives to build approximations. Over a broad range of experiences we can agree on the objective reality of perception and action: "Yes, that is an apple; yes, he did cut it in two with a knife." At another stage, there are experiences that we agree upon as having an objective content, even though we are not sure what happened, "It happened so fast," "I was looking the other way," and so on. At the other extreme, there are personal reactions

---

[47] I am unable to develop links here between the present informal remarks and the deeply thoughtful treatment of "experiences of ultimacy" by Wesley Wildman and Leslie Brothers in this volume. Their choice of the term "experiences of ultimacy" rather than "religious experiences" has a two-fold meaning: such experiences may or may not be linked to a specific religious tradition, and a religious community may interpret (and cultivate) these as being experiences of an Other whether or not it is an Other that is indeed being experienced.

of love or disgust, for example, which we may trace either to our social milieu or to our own very personal history. In this way, each of us has not only a reality of perception and action rooted firmly in the physics of the world around us, but also a personal "reality" which appears completely real to us even if it cannot be experienced by others, a reality of pain and pleasure, of grief and joy, of insight and incomprehension. Within this personal reality, we can define the spiritual quest—but the definition of that quest and the terms in which it is couched vary from person to person, from social group to social group. As Brothers comments:

> The set of questions about how life is to be lived speaks to the... [personal] reality but not... [that treated by physical science]. Existentialists have characterized science as miserably limited when held against the second reality (which they would put first). Others appreciate science but hold that it does not properly reach to the problem of how to live. Its grammar, we might say, is different.

One of the great challenges to Christianity is that of theodicy, which is, essentially, the question of why, if God is so good, the world is such a place of travail. In particular, why is there evil? The story of the Fall from the Garden of Eden is a very powerful story: human beings were created in a certain blessed state, but because they have free will, they were able to fall from grace; but now, contingent on the acceptance of the gift of divine grace, they have the chance to redeem themselves, so that no matter what the vicissitudes of the current situation, whether the Holocaust comes, whether Rome falls, there is still meaning beyond our personal death, even beyond the death of our entire village, or the extinction of our civilization. For Ellis, this brief comment on theodicy fails to take seriously the central message of Christianity: the voluntary acceptance of suffering by Christ on behalf of humanity, so that in the end God's answer is to say, I am not asking you to accept anything that I have not also accepted for myself. My counter-point is that this is *an* answer to the problem of evil which some—but not all—theists have found compelling. Of course, if God does not exist, then the problem does not arise.

Many people start from a belief of the immanence of God in human affairs, and some—but not all—such theists also accept scientific findings. Conversely, while many nonbelievers simply dismiss religious belief as mere superstition, others seek to understand what psychological imperatives might lead some people to embrace religion as they tried to answer important questions about human existence. I am in no position to prove that a secular viewpoint is superior to a religious view. I can simply say that through my own body of experience I have come to find myself with a schema network that is secular and that, as I have exposed it to a dialogue with various religious traditions, I have been able to account (to my own satisfaction) for certain aspects of the human condition that some people would find to require a transcendent view of the person. Following Freud's notion of projection, but not his myth of the primal horde, I see ways in which a group of people within space and time might come to the historical belief in God, even though I do not believe that a God-schema itself is a manifestation of a transcendent reality. This section has sketched some of the considerations leading to, or related to, my secular view. The aim was not to convert the devoted theist to this view, but rather to suggest that the theistic view should not be taken for granted.

The previous section links the present personal reflections to the essay as a whole, offering a framework for charting the dialogue between the individual and society, whether or not that society is rooted in a religious tradition. Ellis notes that the argument relating schema theory to society...applies equally validly to atheists and agnostics as to believers: it equally explains why there can be atheists if there really is a transcendent God, as why there can be believers if there is not.

This point is also made in *The Construction of Reality*.[48] My concern here is to stake out the atheistic position so that the reader of this volume will not blandly assume that Christianity is a given, and that the only issue is to relate neuroscience to divine action. In any case, belief in God is part of a social schema, and it is the nature of the grounding of that schema in "the world" that is of concern here.

Many of the Christians represented in this volume express their openness to the wisdom of other religious traditions, not only Judaism and Islam, but also Hinduism and Buddhism. But the teachings of the latter systems are in so many ways divergent from "the religions of the Book"—the polytheistic robustness of Hinduism, the etiolated sense of the Godhead in Buddhism, reincarnation rather than resurrection— that I am left with little sense of how the religious pluralist can truly view religious systems as God-given, rather than as fruits of cultural evolution whose lack of mutual accommodation must be overcome if the human race is to progress beyond its current period of international and internecine strife.

I thus ask that Christian theologians address the diversity of theologies that Christians have offered in light of an attempt to create a science of the person in which religious diversity, extending far beyond Christianity or even monotheism, is a central fact. In such a science of the person, neuroscience may be central, or it may instead be ancillary to the mental or the social sciences. Schemas may prove to be more relevant than neurons in providing the necessary abstractions for understanding people and the religious and other beliefs which define them, even as we come to understand better the ways in which schemas may reflect both the neural structure which underlies them as well as patterns of learning which dramatically transform the "pre-experiential" patterns of neural activity. I have traced a path from social schemas (defined by the collective behavior of a given social group) to individual schemas viewed as constitutive of different patterns of behavior, and sketched a path (with more detail provided in my companion essay and the references cited therein) from these "external" schemas to the neural mechanisms which provide the "internal realization" of these schemas. I have illustrated these notions through a consideration of the acquisition and evolution of language. In so doing, I have begun to show how social interaction may exert a top-down influence on the microstructure of the brain. The resulting challenge for neuroscience is to understand how such research topics as neural plasticity and brain circuits for emotion may enter into a scientific analysis of how large scale social schemas are related to schemas at other levels and to the neural structures that embody them.

---

[48] Chap. 11 presents the belief system based on the Christian Bible as constituting a "Great Schema" which grounds the type of view of reality to which Hesse subscribes, whereas chap. 12 presents the type of "secular schema" which I advocate.

INTIMATIONS OF TRANSCENDENCE: RELATIONS OF THE MIND AND GOD

George F.R. Ellis

1 Introduction

This essay assumes the need for a coherent approach to Cosmology, understood in the large sense: that is, an integration of our understanding of daily life, the world, and the universe.[1] This implies some kind of integration of scientific and everyday views of life, including views of neurology and psychology as well as cosmology.

In order to develop foundations of such a position satisfactorily, it is proposed here that one should adopt a philosophy that takes a strong view of religion as well as of science, and features a kenotic understanding of religious reality and practice, focused on and valuing generosity and sacrificial love. The purpose of this essay— apart from making the implicit claim that, contrary to much present opinion, such an integration is both possible and desirable[2]—is to propose two major propositions. First, such an understanding can, in line with age-old traditions, profitably adopt an approach to transcendence that, as well as comprehending specific numinous events,[3] envisages a broad-ranging understanding of everyday-life as providing experiences of ultimacy.[4] Second, then, a very traditional Christian view[5] can be vindicated which makes sense to the ordinary believer, and which can make sense also in relation to present-day science, including neuroscience. However, such a position does not reject other faiths that adopt a Cosmological understanding of a fundamentally similar nature. This essay does not claim to solve the problems of apologetics or of religious pluralism; rather, it aims to outline briefly one approach which has the potential to be developed into a viable position. I should state at the outset that this essay presupposes that language about the mind is irreducible, regardless of the findings of particular approaches to neuroscience or cognitive science (see the argument in section 5.3 below). On this view, cosmology must include the reality of mental experience; views that do not accommodate these realities are rejected as inadequate.

---

[1] See George F.R. Ellis, *Before the Beginning* (London: Bowerdean/Marion Boyers, 1993), hereafter *BTB*, for a motivation as to why a satisfactory viewpoint needs more than is attainable from a purely scientific analysis, as for example in physical cosmology.

[2] See Ian Barbour, *Religion and Science: Historical and Contemporary Issues* (San Francisco: Harper, 1997), for the range of possible relations between science and religion, and Nancey Murphy and George F.R. Ellis, *On The Moral Nature of the Universe: Theology, Cosmology, and Ethics* (Minneapolis: Fortress Press, 1997), hereafter *MNU*, for one detailed proposal as to how integration might be achieved.

[3] Cf. the essay by Wesley Wildman and Leslie Brothers in this volume.

[4] It also allows a concept of revelation that is not founded on the presupposition of a divine intervention that suspends temporarily the laws that govern nature's processes.

[5] Those Christian approaches that are tyrannical or cruel in their practice, or fundamentally dogmatic in their doctrine, are rejected.

2 Foundations: Models of Reality and Multiple Causality

The model of understanding adopted here is consonant with what is presented in many of the other essays in this volume. In brief:

2.1 Models as Our Method of Understanding

Our basic method of understanding, whether formalized or not, is modeling reality in a variety of ways. Complex scientific models give good understanding of specific aspects of physical reality, while metaphors are an essential tool in approaching higher-level understanding.[6] These models encapsulate the essentials of generic principles that apply in particular situations, specific applications involving parametrized and named instantiations of generic models ("a large marmalade cat named Orlando"). This is sometimes carried out explicitly (that is, the modeling process is identified and labeled as such) but often implicitly (mental models are routinely employed as a basis for understanding without being explicitly labeled as such).

There is always a tension between universal principles and the particular situations in which they are applied. The specific circumstances applying in any particular case give it specific identity, and still allow universal principles to apply; the related underlying questions are whether the model is appropriate to the particular situation at hand (a meta-question that cannot be answered fully by use of the model itself), and whether all significant particular circumstances have been taken into account by the model used.

Two specific issues are of fundamental importance in relation to the use of models. First, of necessity any model can only give a *partial representation* of the objects, actions, and situations it represents; it cannot adequately represent the full complexity of reality, let alone fully reflect its underlying existential nature.[7] Despite this, models are often mistaken for the reality they represent, leading to confusion and misleading arguments. The implication of this partial representation is the need for multiple models in order to get an adequate representation of reality. Secondly, and indeed to some degree as a consequence of this, there will be *multiple accounts of causation* in all real situations. Many different features will act in concert to produce any specific result; and focusing on a single causal feature and related description (thus taking all the other relevant factors for granted) will inevitably be misleading.

2.2 Hierarchical Structuring of Models

A multiple account of causation becomes particularly important in the context of the hierarchically structured models we must use in order to represent adequately complex systems in general, and the human mind and body in particular. There will be different layers of emergent order and meaning that apply simultaneously, and one

---

[6] Ian Barbour, *Myths, Models, and Paradigms: The Nature of Scientific and Religious Language* (New York: Harper and Row, 1974); Janet Martin Soskice, *Metaphors and Religious Language* (Oxford: Clarendon Press, 1992). See also Stephen Happel's and Arthur Peacocke's writings on metaphors in the volumes of this series.

[7] For a discussion of this partial nature of all models of reality, see for example Soskice, *Metaphors and Religious Language*. The amount of detailed (microscopic) information lost in any coarse-grained (macroscopic) description of the system is a measure of the entropy of that representation.

can choose to examine specific levels of description, or to contemplate the relations between different levels, in particular to examine the (partial) reduction of higher-level features to descriptions in terms of lower-level features, and the way the higher-level features supervene on the lower-level functioning.[8] It is clear that the higher-level descriptions cannot be fully reduced to those at the lower levels; indeed, the language used for the higher-level descriptions is simply inapplicable to concepts available at the lower levels.[9]

While scientific models give very effective but partial models of lower (functional) levels of the hierarchy, events at the higher levels are causally effective at their own levels, and can to a great degree be described without reference to lower- level properties. Clearly this is true in a practical sense (neither a motor mechanic nor a zoologist nor a brain surgeon has to be an expert in elementary particle physics). As explained by Silvan Schweber and Phil Anderson, this remarkable feature results because the underlying physics displays the renormalization group properties apparent in solid state physics, and allows statistical analysis such as that which underpins the success of statistical mechanics.[10]

Thus causal processes can operate simultaneously at the different levels in such systems, allowing multiple-level descriptions which must be compatible with each other but which cannot fully be reduced to each other. (Even in the much-touted case of statistical mechanics and thermodynamics, the attempt at causal reduction is only partially successful because the problem of the arrow of time has never been satisfactorily solved.) This compatibility is possible because both bottom-up and top-down causation takes place.[11] When we try to integrate lower-level (physical) descriptions and higher-level descriptions (inclusive of concepts of purpose and meaning), it is a mistake to think of causes at different levels as competing alternatives, when there are in fact multiple causes operating simultaneously at the different levels. Nevertheless, there are *compatibility conditions* between the different levels that must be satisfied if such multiple-level descriptions are to be coherent; the workings envisaged at the different levels must not be incompatible with each other. It is important to consider the nature of such conditions, and under what circumstances they may be satisfied.

2.3 Highest-Level Models—Theories of the World

The foregoing understanding suggests that when looking at the highest levels of causation, we should regard the theological and naturalistic perspectives as complementary. There is a need for consonance between these different levels but not reduction. At these higher levels, myths[12] and stories (the Greek tragedies,

---

[8] See Nancey Murphy's essay in this volume, and the references therein.

[9] See *BTB*; and Robert J. Russell, Nancey Murphy, and Arthur Peacocke, eds. *Chaos and Complexity: Scientific Perspectives on Divine Action* (Vatican City State: Vatican Observatory; Berkeley: Center for Theology and the Natural Sciences, 1995), hereafter *CAC*.

[10] Silvan S. Schweber, "Physics, Community and the Crisis in Physical Theory," *Physics Today* 1993 (November): 34–40; Phil Anderson, *A Career in Theoretical Physics* (Singapore: World Scientific, 1994).

[11] See Arthur Peacocke, *Theology for a Scientific Age* (Minneapolis: Fortress Press, 1993); and *MNU*.

[12] Joseph Campbell, *The Power of Myth* (New York: Anchor Books/Doubleday, 1988); Hans Blumenberg, *Work on Myth*, trans R. M. Wallace, (Cambridge: MIT Press, 1985).

Shakespeare, Dostoevsky, etc.) are a means of understanding the world and the human predicament and of learning how to cope with it: they are effective models at the higher (meaning) levels of the hierarchy.

This does not undermine or deny science; but it is essential to recognize that a strictly scientific description—effective at the lower levels—is simply unable to comprehend meanings at this kind of level.[13] Use of the term "myth" or "metaphor" does not mean that anything goes: these models have to satisfy criteria of effectiveness of a suitable breadth, which means they must in some sense adequately correspond to reality (or "tell the truth"[14]). If they do not, they will mislead those using them in potentially dangerous ways. These criteria will refer to cultural and psychological issues, but also to physical and environmental realities. A successful model will reflect these externalities accurately, in particular enabling reliable predictions of future situations that lie within its range of competence.[15] In this sense it will be a true representation of the real universe (at its appropriate descriptive level), which can be tested and confirmed as reliable. Hence, while we cannot obtain "truth" about ultimate reality, we can obtain effective *local truth* in terms of laws applicable to particular levels of description (for example, Newton's laws of motion, applicable with great accuracy when relative speeds of motion and gravitational field strengths are small in appropriate units; the Second Law of Thermodynamics, representing the inviolable degradation of energy to unusable forms in the circumstances presently holding on the Earth).

2.4 The Tension of Inner and Outer Views

A specific important case where we have multiple models is the case of consciousness, where there is a major contrast between the inner (or subjective) and outer (or objective) views of human beings, including ourselves. In these two quite different types of view, the same person or mind is viewed as a subject or object respectively. The languages, modes of expression, and methods of analysis are quite different, yet they refer to the same being.

In a sense this is the major problem facing us in understanding consciousness: the mind can be viewed from the external (scientific) side and from the internal (personal) side. We do not yet have a good way of relating these two views, which are in tension with each other both philosophically[16] and in practice (in medicine, for example, surgeons notoriously tend to treat patients as objects, while a family doctor's major role is to treat them as subjective and individual persons). Some approaches in effect try to solve this by rejecting the subjective view, labeling it an illusion; the viewpoint I adopt here is that any such move will be a major error, excluding from our analysis one of the most important sources of data we have available. Thus I strongly support a *two-view* model, regarding both views as valid, relating to quite different (although strongly related) aspects of the same conscious being. This does not imply that the reality to which they refer is dichotomous, but

---

[13] See *BTB, MNU*.

[14] A realist cognitional theory permits metaphors to refer to the world and to tell truth.

[15] Each model will come with a range of applicability stating what phenomena it represents, the level of description at which it applies, and the circumstances under which it is reliable.

[16] R. Scruton, *An Intelligent Person's Guide to Philosophy* (London: Duckworth, 1996); Thomas Nagel, *Mortal Questions* (Oxford: Oxford University Press, 1982); David Chalmers, *The Conscious Mind* (Oxford: Oxford University Press, 1996).

rather that our views of that reality have a dual (but related) character.[17] The defense against a simple reductionism is that the concepts involved in consciousness are not directly describable in terms of neurological concepts, and we do not have correspondence rules giving reliable relations between these kinds of description—they are indeed incommensurable.

2.5 Constructing and Testing Theories

Much has been written about the way theories (that is, complexly structured models of some domain) are developed, tested, and modified on the basis of experience and experiment.[18] Only two points will be made here.

First, in the process of building theories, the basic problem we face is *the hidden nature of reality*. This means that within the realm of valid (tested and tried) data, we have to distinguish between *revelatory data* (pointing at the true nature of reality) and *misleading data* (tending to hide the true nature of reality—as they are easily conceptualized in misleading ways). The revelatory data are the basis used to construct a good theory, which must then also be able to account for any apparent counter-data[19] in order to give a believable comprehensive view. Counter-data will mislead if one tries to use them as the basis for theory construction; and the problem is to know which is which. This is not obvious; a classic example is the evidence from daily life experience that led to Aristotelian dynamics (objects only keep moving if you keep pushing them), accepted as correct for over a thousand years before Isaac Newton saw that these data are in fact misleading; the revelatory data are the way a thrown ball or an object on ice keeps moving despite no force being exerted to keep it moving. The true theory (Newtonian dynamics) is counter-intuitive precisely because (due to the effects of friction) it is contradicted by daily experience; this experience misleads us.

Secondly, in order to determine the range of data relevant to testing a proposed theory, one needs first to decide clearly on *the scope intended for the theory*, and then to take into account *the full range of theoretical concerns and associated data that are relevant* in the light of this scope. This is particularly important in the context of high-level, broad-ranging theories, and here the previous comment is important: it may not always be clear what the relevant data are. Nevertheless it is clear that the highly specialized training of academics and intellectuals does not serve us well in this regard, for it leads to a lack of awareness of the wider contexts of high-level theories, and so a resistance to a broader investigation of that context. Examples include biologists who are unwilling to consider the implications of different laws of physics for biology and so cannot see the question underlying investigation of the Anthropic issue in cosmology; and neuroscientists who are unwilling to contemplate the possible importance of quantum phenomena for their field, despite the fact that the physical functioning of the brain is patently founded in quantum mechanics—assuming it is founded in physical laws.

---

[17] Indeed it may well be that we cannot easily "see" both views at the same time, just as in the famous visual illusions where one cannot simultaneously see the two possible interpretations of an ambiguous image—one has to switch between them.

[18] Nancey Murphy, *Theology in the Age of Scientific Reasoning* (Ithaca, N.Y.: Cornell University Press, 1990); *MNU*.

[19] This would include the existence of contrary viewpoints to that proposed.

These issues are important in the attempt to construct a wide-ranging Cosmology; the full data needed for this purpose include scientific and everyday data, and those one might characterize as of a spiritual or transcendent nature. Thus, I aim to incorporate—in principle—all these data.

3 A Provisional Thesis

The basic viewpoint considered here, very briefly outlined in section 1, has been laid out in depth in the book *On the Moral Nature of the Universe.*[20] What follows sketches just enough of that view to be a basis for the sequel.

3.1 A Consistent Transcendent Worldview

My proposal is that a kenotic theological and ethical view provides a world-picture that is consistent with all levels in the hierarchical structure of reality, giving the foundation for an overall worldview that accepts the scientific vision of the evolving universe and the evolution of life, but also takes everyday human experience and concerns seriously. Here, "kenotic" means a freely giving, joyous, kind, and loving attitude that is generous and creative, and—if needed—willing to give up selfish desires and to sacrifice on behalf of others, but in a humble way, avoiding the pitfall of pride, and doing this all in the light of the love of God and the gift of grace.[21] Kenosis is not in opposition to the power of God: it *is* the ultimate nature of the power of God, leading to the triumph of the Resurrection, which is the outcome and reward of perfect kenosis in accord with the paradoxical statement, "he who would save his life shall lose it; he who loses his life for my sake and the Gospel's will save it."

On this view, kenosis is an overall key to the nature of creation,[22] with an underlying theological basis: this is the nature of the Universe because it is the nature of God the Creator. Thus the Anthropic question in cosmology is answered by a modern version of the traditional design argument: a transcendent God designed laws of physics so as to enable the emergence through processes of self-organization of ethically aware beings—including the Son of God—able to be aware of the nature of God[23] and able to respond to moral issues in a kenotic manner. Developing this thesis in depth strongly suggests that inclusion of ethics is needed to give a satisfactory understanding of the science-theology-cosmology relation, and leads to a characterization of the branching hierarchy of the sciences discussed in depth in *MNU*, with theology as the topmost level of both the natural and human science branches.

This viewpoint thus takes a strong position on there being a transcendent reality underlying the existence and functioning of the universe as we know it. A series of

---

[20] See *MNU*, and also the American Academy of Religion (1997 Meeting) discussion essays by Christoph Lameter, Richard O. Randolph, Robert J. Russell, and Kirk Wegter-McNelly; reprinted in *CTNS Bulletin* 18.4 (Fall 1999).

[21] K. M. Cronin, *Kenosis* (New York: Continuum, 1992); William H. Vanstone, *Love's Endeavour, Love's Expense* (London: Darton Longman and Todd, 1977).

[22] But it is not the only one: as emphasized by Anne Clifford and Stephen Happel, concepts such as *covenant* are also of major importance.

[23] See the discussion below of the causal joint that could allow this to happen. Robert Russell has emphasized that this same causal joint would allow God to act within the process of evolution at the genetic level. I acknowledge this possibility without necessarily embracing it.

questions arise that need examination in developing this proposal further. I briefly outline some of them, indicating which will be considered in the rest of the essay.

3.2 Evidence for This Higher-Level Vision

A comprehensive epistemology underlying this view is presented in *MNU*, including proposals for testing the effects of adoption of a kenotic ethic as a foundation for understanding social issues. The counter-evidence to this overall position is also considered there, along the lines intimated in section 2.5 above, by identifying often quoted apparent counter-evidence (the existence of evil, particularly in relation to the history of the church) as misleading data, and identifying relevant revelatory data as including the life of Christ[24] and the lives of devotion and acts of courage and generosity of many others, specifically Mahatma Gandhi and Martin Luther King, Jr., whose lives demonstrate the viability of a kenotic approach.

In *MNU* the argument is based on consideration of the metaphysics of cosmology together with evidence and considerations of a moral and ethical nature. The further questions are: What is the full realm of data that can be used in addition to this? What is left out in this focus on morality alone? What further approaches to transcendence might usefully be taken into account? This will be developed below in section 4, where I consider intimations of transcendence in daily life, inevitably raising the related issues of discernment and scepticism. This will then be related to the nature of the mind in section 5.

3.3 Religious Pluralism and the Variety of Viewpoints

A broad defense of this kenotic position is given in *MNU* based on the methodologies of Carl Hempel, Imre Lakatos, and Alasdair MacIntyre. Apart from the issue of evil (dealt with to some degree in *MNU*), counter-evidence comes from pluralism in moral and religious viewpoints—why are there other worldviews if this is the correct one?—as well as from the broader scepticism about *any* realism that is evident in postmodern worldviews (despite the fact that such a position is to a considerable degree self-defeating[25]).

A thorough response cannot be attempted here because of lack of time and space; however, I note that some relevant perspectives from a scientific viewpoint (specifically relating to the theory of relativity) can be helpful to some degree. There can be no claim that certainty can be attained, but it can be claimed that a position of the kind suggested here can be rationally defended and can be shown to be supported by a considerable body of data. This will still leave other viable options

---

[24] See my "The Theology of the Anthropic Principle," in *Quantum Cosmology and the Laws of Nature: Scientific Perspectives on Divine Action*, Robert J. Russell, Nancey Murphy, and C.J. Isham, eds. (Vatican City State: Vatican Observatory; Berkeley, Calif.: Center for Theology and the Natural Sciences, 1993), hereafter *QCLN*, 367–405, and John Haught, "Darwin's Gift to Theology," in *Evolutionary and Molecular Biology: Scientific Perspectives on Divine Action*, Robert J. Russell, William R. Stoeger, and Francisco J. Ayala, eds. (Vatican City State: Vatican Observatory; Berkeley, Calif.: Center for Theology and the Natural Sciences, 1998), hereafter *EMB*, 393–418.

[25] See for example Peter Berger, *A Rumour of Angels* (New York: Anchor Books/ Doubleday, 1990), particularly chap. 2, "Relativising the Relativisers."

open, and so it is not claimed that this position is necessarily true or is the only defensible position.[26]

3.4 Theoretical Issues

The theoretical issues that need consideration include the compatibility conditions between the different levels of explanation that must be satisfied if such a multiple-level description is to be consistent, as well as the consistency of an evolutionary account of the origin of the structures of reality.[27]

For the present discussion, the issue is what if anything is implied for this kenotic worldview by the neurosciences and our current understanding of the brain. Here is where one particular compatibility condition will be identified. If the channels of transcendence discussed in section 4 are genuine intimations of transcendence, what implications might there be for the mind and brain? This will be considered in section 5, together with the closely related issue of the existence of free will. Here is where the issue of the nature of divine action comes to the fore.

3.5 Practical Implications

Finally, a kenotic worldview such as is considered here has substantial practical implications for how life should be lived. A range of issues arise, including:

1. How can one live a life consonant with the kenotic nature of the universe (thereby simultaneously testing and verifying its true nature)?
2. How does one handle the issues of pride and humility that underlie a kenotic approach and are closely related to its paradoxical nature?
3. How does one create the right balance of social action, communal and private contemplation and prayer, and intellectual analysis and investigation, in a well-ordered kenotically-based life?

These issues will not be pursued further here; however, it would be incorrect not to comment on their significance in the overall scheme.

4 Domains of Intimations of Transcendence

The concern in this section is to consider the full range of data relevant to the viewpoint proposed. One can point first to philosophical issues such as the design argument in its modern form of the Anthropic principle.[28] However we can also ask, What are the channels by which intimations of transcendence are available to us through our personal experience, in addition to those supported by rational argument? Historically this has largely been seen as occurring through experiences of ultimacy that are specifically spiritual or religious in their nature (as discussed by Wesley Wildman and Leslie Brothers in this volume). But this is evidence available through experiences that are in some sense out of the ordinary: they are certainly not

---

[26] The arguments for uncertainty about ultimate matters are given, for example, by John Hick, *Evil and the God of Love* (New York: Harper and Row, 1996), and are briefly summarized in my "The Thinking Underlying the New 'Scientific' Worldviews," in *EMB*, 251–80.

[27] The evolutionary issues (i.e., evolution of mind and brain) will not be considered here, because that was the focus of the previous volume in this series.

[28] See the essay by Ellis in *QCLN*; and Nancey Murphy, "Evidence of Design in the Fine-Tuning of the Universe," *QCLN*, 407–35.

experienced by everyone. We seek additionally to characterize such data that may come from ordinary rather than extraordinary experience.

In *MNU*, moral life was presented as such a channel; this was chosen as a focus because it is particularly compelling to many at the present time. However, it is not the only such route. Indeed there is a tradition in the Christian religion which sees daily life as providing such experiential evidence of transcendence, this being the viewpoint of some nineteenth-century liberal Protestants and of some contemporary theologians, in particular Arthur Peacocke and Ian Barbour. It is also evident in other traditions, such as in the writing of the Buddhist Thich Nhat Hanh.[29] In recent times this feature has been well argued *inter alia* by C.S. Lewis,[30] Peter Berger,[31] and George Gorman,[32] who describe various ways that transcendence is experienced as reality.

This section considers intimations of transcendence of the kinds characterized by Lewis, Berger, and Gorman. I will argue that it is not just the existence of these kinds of experience that is persuasive, but rather that in each case *there is more there than is strictly needed* from a functional viewpoint. It is this over-abundance that is the convincing intimation of transcendence to those who experience it and are open to it. Further, there is a subjective-objective dimension of the hermeneutics[33]: there is a "more" (an *excessus*, in the writings of Karl Rahner and Martin Heidegger) in the data. But then there must be a similar *openness* or space for disclosure in the subject so that the intimations of transcendence can be effectively received. I argue below that this openness may have its origins at a lower level in the structures of matter.

4.1 The Miracle of Creation

The first intimation of transcendence is the existence of the Universe, the world, and life, thematized as the Anthropic question: How can it be that the Universe exists and is structured so as to allow, perhaps even prefer, the development of the living things we see, including intelligent beings?[34]

At one level this is a scientific and metaphysical issue, as has been discussed in the literature on the Anthropic question (see *QCLN* for an extensive discussion). As such, it is indicative but far from conclusive: it is persuasive to some, but empty to others. It is a rather complex intellectual argument; indeed, to fully appreciate its import one must have a good understanding of physics, chemistry, biochemistry, thermodynamics and entropy, and of the hierarchical structuring and engineering problems that have been solved in the design of living things. However, this somewhat austere intellectual approach does not really comprehend the sense of

---

[29] Thich Nhat Hanh, *The Miracle of Mindfulness: A Manual on Meditation* (New York: Rider/Random House, 1991), 12.

[30] C.S. Lewis, *Mere Christianity* (New York: Collier Books, 1962).

[31] Berger, *A Rumour of Angels*.

[32] George H. Gorman: *The Amazing Fact of Quaker Worship* (London: Quaker Home Service, 1979).

[33] I am indebted to Stephen Happel for this comment.

[34] John Barrow and Frank Tipler, *The Anthropic Cosmological Principle* (Oxford: Oxford University Press, 1986); *BTB*; *MNU*; and essays by Ellis and Murphy in *QCLN*.

awareness of the miraculous nature of creation that some people have, as is beautifully illustrated, for example, in the writings of Hanh.[35]

This sense of awe and wonder is awakened by contemplation of the natural world: mountain views, sunsets, flowers, animals, the heavens, and in particular by the miracle of human life (contemplating babies, for example). What is meaningful in this sense will be specific to each person. I find it in eagles soaring above the mountains, owls gliding through the forest, sunbirds feeding from flowers, blue disas growing amongst the heathers and proteas on Table Mountain, in the sunsets on Noordhoek beach. All of this is a result of that order in nature, expressible as mathematical laws and resulting in an intelligible and beautiful universe, which allows the existence of life. When truly understood, the nature of creation is intellectually awesome and can be taken as an intimation of transcendence; indeed, this is what some modern design arguments seek to formalize.

However, the point here is that this can be seen as an intimation of transcendence even without understanding those physical laws and intricate philosophical arguments; it is a perfectly valid reaction to comprehend this excess in the nature of creation at an intuitive level.[36] The significant point has been made above, in the introduction to this section: what has been accomplished in creation is *more than is necessary*; instantiating conditions for the universe to exist need not have led to existence of living beings; and instantiating conditions for living beings to exist need not have led to such beauty and variety. What we see is more than the necessary minimum.[37] This is what is comprehended in an intuitive way by the sense of awe and wonder at creation and at the intelligibility and beauty of the universe. This is related to the language of the sublime in aesthetics (see below).

This may be disparagingly referred to as merely an emotional reaction, and indeed it certainly does have an emotional component; but the "merely" is wrong, for it is a far more holistic response than that appellation indicates. It is an overall appreciation of what exists, with both intellectual and emotional components. Its major feature is that kind of questioning of the taken-for-granted nature of everyday life[38] which is at the heart of the realization of the true mystery of existence, and hence of its metaphysical dimension. This interpretation is in keeping with the overall worldview proposed in section 3—with kenosis embodied in the concept of God's creating the Universe and then fully respecting the nature of the created order.

---

[35] Hanh, *The Miracle of Mindfulness.*

[36] In terms of the hierarchy of explanation, one is looking at the truth embodied in the higher levels of structure without being concerned about how they are embodied in lower-level structures. While an understanding of these further features leads to a more comprehensive understanding, it is not needed in order to have a valid understanding and appreciation of the higher levels.

[37] How is this compatible with the idea of kenosis, of giving up, which could perhaps be taken to imply that God would provide only the necessary minimum? The point here is that God restrains his/her own direct intervention in favor of imposing physical laws that have their own total internal autonomy, and respects that autonomy. The extraordinary creativity of the Creator is evidenced in the self-organizing character of matter that allows and results in the superabundance indicated here as a gift to animals and human beings. The kenosis in this respect is the voluntary self-limitation of God; the recipients of the benefits of that act of kenosis are living beings in the physical universe, which in turn delights God.

[38] This breaking out of the taken-for-granted socially conditioned frame is what is characterized by Peter Berger as "ecstasy" in *Invitation to Sociology* (New York: Doubleday, 1963).

This interpretation points to a different, more fundamental order underlying our perceived physical reality. Thus it provides an intimation of transcendence.

4.2 Moral and Ethical Issues

The arguments in the foregoing cases are similar in essence to the one from morality, but refer to different aspects of life. The case of ethics is argued in detail in *MNU*, and is closely related to issues of meaning, for ethics (the choice of highest-level goals for our behavior) is concerned with the *telos*, or purpose of life.

In this sphere, too, we experience more than the necessary minimum. Thus, while one is indeed concerned here with justice and sympathy, generosity and giving, kindness and empathy, a kenotic ethic goes far beyond this to sacrifice and love of a different order. This is illustrated, for example, by Mother Theresa, Mahatma Gandhi, and Martin Luther King, Jr., whose lives emulate the life of Christ and the Disciples. In the midst of pain and evil they see the potential for change and good in everyone, and are prepared to unlock that potential by making voluntary sacrifices on behalf of the other, even those perceived as an enemy.[39] So doing mirrors the nature of God as revealed in the life of Jesus and particularly in his voluntary acceptance of suffering. The kenotic nature of God is echoed in the behavior towards others that must follow if one fully acknowledges that there is "that of God" in everyone.[40] The sense of "ought" here is the sense of the highest-level of persuasive action, rather than that of compulsion or coercion. This is what gives a kenotic ethic its peculiarly generous and compelling nature ("compelling" not in the sense of coercion, but in the sense of resulting in assent to its high-level nature and rightness and in a consequent desire to follow that way).

This kind of attitude and action is far more than what one could easily explain satisfactorily by sociobiology alone, or justify by a utilitarian approach to ethics alone, even though that can make stringent demands on one. Indeed, as has been persuasively argued by Murphy, the attempt to explain ethics in either way simply does not do justice to the true nature of morality: it explains it away rather than explaining it.[41] The kind of kenotic behavior demonstrated in depth by exceptional people, and less fully mirrored in the lives of many, can again be seen as an intimation of the nature of transcendence underlying reality, based in the nature of God. This kind of action has a self-validating nature that invokes almost universal recognition.

4.3 Creativity

A further intimation of transcendence is given by the creativity of humankind. This creativity is evidenced in daily life, in economic achievements, in artistic endeavors, in scientific discovery, in technical achievement, and in team and individual sport (and requires a kenotic aspect for success, according to Vanstone[42]).

---

[39] See *MNU*; Ellis in *QCLN*; Vanstone, *Love's Endeavour, Love's Expense*; and Cronin, *Kenosis*, for further development of this concept.

[40] See G. Hubbard, *Quaker by Convincement* (Harmondsworth: Penguin, 1976).

[41] Nancey Murphy, "Supervenience and the Nonreducibility of Ethics to Biology," in *EMB*, 463–89.

[42] Vanstone, *Love's Endeavour, Love's Expense*.

Again one can claim that we see far more here than is strictly necessary for survival. The way people are prepared to sacrifice on behalf of their creative activity does not make sense if one takes a simply utilitarian view of life. As proposed by Philip Hefner's concept of men and women as "cocreators" with God,[43] human creativity mirrors to some degree the awesome creativity of God, and the effort and delight of our engagement in creativity results from this link. Here we are seeing another intimation of transcendence.

A particular aspect of this is humor and its relation to the unexpected. The prevalence of humor is a striking feature of human life. It helps to relieve the tension and stress of life, including the ultimate tension of relating to the finiteness of life. There is of course a substantial element of creativity in seeing and displaying humorous confusions, for example of inside and outside views of human being, or in the questioning of the continuation of present roles and conditions, which brings into focus the pathos of the human condition. Comic strips such as Peanuts and Calvin and Hobbes are widely recognized as having such a quality; to some degree it is present in much humor. Peter Berger argues that the incongruities of humor are hints of a transcendence underlying the taken-for-granted appearance of everyday life, because such incongruities can throw into question this taken-for-granted experience and hint at something further behind them.

4.4 Aesthetics: Beauty as Revelation

Closely related to this is the effort and devotion that many put into both creation of and appreciation of art of all kinds. Appreciation of the beauty of the natural world has as a counterpoint both the appreciation of the beauty created by human beings and the endeavor to create such beauty; and this is a truly creative act.[44]

We see this in all forms of art: painting (Rembrandt), sculpture, music (Bach, Debussy, Ravel), ballet, plays (Shakespeare, *Les Miserables*), novels (Dostoevsky, Saint-Exupéry), architecture (Gaudi, the Mendel church), photography (Walker Evans), handicraft (pottery, weaving, baskets, clothes, etc.; Shaker furniture). Some artists are prepared to dedicate themselves to a great degree to their creative vision, and the public often responds by being prepared to support and appreciate those artistic endeavors which it sees as reflecting a haunting, transcendent beauty. Again the specific art that will have this quality for specific people is a very individual affair (I have put in some names above that some will agree with and others will not); there is a great deal of cultural dependence in aesthetic taste. Nevertheless the basic quality of the experience of beauty as revelation of a transcendent quality may well have a rather universal character. It certainly has this nature in some specific cases for some individuals.

4.5 Love and Joy

Closely related to the proposed worldview is the way that unreserved and undeserved love occurs and mirrors the envisioned nature of God.[45] This love reflects joy in the

---

[43] Philip Hefner, "The Evolution of the Created Co-creator," in *Cosmos as Creation*, Ted Peters, ed. (Nashville: Abingdon Press, 1989).

[44] We also see other things in some art; I do not claim all art is of this kind, but rather that there is a major body of artistic work that has this nature.

[45] Berger, *A Rumour of Angels*; C.S. Lewis, *The Four Loves* (New York: Harcourt Brace, 1971).

value of each person, so clearly seen in the relation of mother and child, but also often in the relation of lover and loved. Indeed this is made explicit in the musical *Les Miserables,* based on the novel by Victor Hugo, where the last song contains the expressive phrase: "To love another person is to see the face of God."

This again reflects the overall theme proposed: such human love is an image of and intimation of divine love, made possible because there is that of God in everyone. While some of this is fully comprehensible on a socio-biological basis, much is not; for example the great love given autistic children by their parents exceeds what is logically necessary (and is indeed inefficient and unwanted from an evolutionary viewpoint). It can validly be interpreted as another intimation of transcendence.

4.6 Specifically Spiritual

Finally, there are the traditional approaches to transcendence provided by specifically spiritual practices and experiences.[46] These approaches provide experiences of ultimacy (cf. the essay by Wildman and Brothers in this volume) that are the interpretative keys to the rest of the experiences outlined here. There is a vast amount of data here concerning experiences that range from the intense and strongly visionary to the quietly devotional. The practices include prayer and contemplation and the practice of the presence of God, potentially leading to an ongoing awareness of the immanent presence of transcendent reality. This evidence raises a series of issues in terms of the human relation to reality, for devotional contemplation without consequent action in the world is partial and incomplete. The question is how best to enact such a response. Specific issues are (1) the need for ethical behavior, based on serious contemplation of the proper nature of such a response, and (2) the fundamental issue of pride and humility. The latter is a core issue in the relation of the Creator and the created. The temptation is to seek to be God (as opposed to being like God). This is the sin of idolatry, which often has disastrous consequences in practical terms; indeed, it underlies much of the evil in the world. Its most common evidence is a claim either to be infallible or to have infallible knowledge.

The specific revelation that is part of this experience makes explicit—on the understanding proposed here—both the creative action of God and the kenotic nature of the true Christian vision, made real and embodied in the life of Christ.[47] Similar support comes from other faiths, for example from Gandhi's vision and practice of *ahimsa* and *satyagraha*. Thus these are to be seen as revelations of the nature of God, given by him to enlighten us.

4.7 The Overall Picture

The overall picture then is as follows: basically, there is more than is needed in each of these areas. There is an experience of a gift. The issue is not just the existence of each of these concerns or experiences, but the unreasonable joy they provide, and the

---

[46] See G.S. Wakefield, ed., *The Dictionary of Christian Spirituality,* (London: SCM Press, 1983) for a comprehensive description (which, despite the title, includes other faiths). I have considerable reservations about the way the word "spiritual" is often used, and Wildman and Brothers' "experiences of ultimacy" might be better; nevertheless I find the term useful, provided it is used in a way consistent with the overall position of this essay.

[47] For example, see William Temple, *Readings in St. John's Gospel,* (London: MacMillan, 1961); Ellis in *QCLN*; Haught in *EMB*.

unreasonable amount of effort that people are prepared to put into them, for example devoting their whole lives to art or music. They illustrate the importance of the theme of "ecstasy" described by Peter Berger, namely, those visions, experiences, and actions that illuminate and transcend our daily lives by questioning its taken-for-granted (socially conditioned) apparent nature, and by pointing to something deeper behind it.

These considerations do not provide a logical proof or unquestionable evidence for the view proposed; rather, they give a pattern that one either recognizes or does not recognize as valid. Imagination is way of approaching and understanding these issues, with fable and myth as teaching tools helping people to understand and relate to them. Logical analysis and teaching can help, but they only take one a part of the way to the deeper understanding that comes from a fully fledged comprehension of the underlying pattern that is hidden in and can be comprehended through these experiences.[48]

The issue of discernment is critical in assessing this kind of overall view and the kinds of data referred to in this section because there are traps on every side, both in terms of the approach used (the kind of theory adopted) and in terms of what data will be accepted as reliable. In each case the need is to consider how to distinguish illusion from reality (as is done by the scientific method within its area of compe-tence). Consequently a reliable process of discernment is crucial with regard to both the standpoint taken and the data accepted.

Important as this topic is, I will not pursue it further here because other essays in this volume (particularly that by Wildman and Brothers) deal with the topic in some depth. I simply comment that assessment within any faith community of both evidence and counter-evidence and of overall patterns of integration relies on general plausibility and consonance with the overall intellectual and experiential position of that community and quality of outcome ("by their fruits ye shall know them"), together with specific tests in terms of communal tradition and experience, authorized texts, and considered response by authority figures or others in the community acting in concert. Assessment of this kind of broad scheme can only take place within a faith community, precisely because it deals with issues outside the competence of science to decide[49]; so issues of judgment, discernment, and faith inevitably arise, whether one considers the proposal put forward here, or any of its rivals (for example, atheism, Marxism, or one of the "scientific religions").[50] However, the difficulties in determining the degree of verisimilitude of an explanatory scheme does not undermine the realist ontological position adopted here; they merely underscore the need to apply an adequately developed epistemology (cf. the approach in *MNU*).

---

[48] See William E. Placher's *Unapologetic Theology: A Christian Voice in a Pluralistic Conversation* (Louisville, Ky.: Westminster/John Knox, 1989), chap. 8, for more on such pattern recognition.

[49] For example, see George F.R. Ellis, "Modern Cosmology and the Limits of Science," *Transactions of the Royal Society of South Africa* 50.1 (1995); *BTB*; Ellis in *EMB*.

[50] Actually such issues and a consequent need for discernment arise also within the sciences to some degree, particularly at the cutting edge where the nature of future developments is not yet clear, and they arise in relation to determining the boundaries of science. However, I will not pursue these controversial topics here.

Given that understanding, the proposal put forward here is both intellectually viable and supported by substantial data that are compelling to some communities. The proposal is summarized in the following figure:

Intimations of Transcendence

Patterns of understanding: More than the necessary minimum

Aesthetics:	*Love and Joy:*
The beauty of	the Father/Mother
transcendence	companion/friend

Spiritual awareness
awe, wonder
creation, Transcendent
reality/Immanence

Creativity:	*Ethics:*
Cocreators,	Right and wrong,
inspiration,	Justice and
humor, ability	forgiveness

Figure 1. Intimations of Transcendence through a variety of domains of perception and experience, experienced as a "gift."

Finally, it is important to note that we cannot control this kind of awareness: it is something given to us as a gift, often as a surprise. It is a gift of grace which we can hope for, but cannot command. However, we can try to create circumstances where we will be more likely to attain such experiences, by various kinds of preparation (for example, meditation, prayer, or spiritual reading), and by arranging events where access to such awareness is likely to be easier than otherwise. There are rites of passage in most societies marking major developmental stages in life that often relate them to higher meaning and purpose; the ceremonies and rituals of the church, particularly the sacraments, are designed precisely in order to facilitate the awareness of transcendent reality behind everyday life and to relate this to purpose in our own lives. The various forms of such ceremonies and rituals will be more or less successful in doing so, depending on one's culture and individual life experience; in effect it is an experimental question for each one of us to determine which route is most successful in allowing and enhancing such awareness, but always with a cautionary approach that tests and reviews what is happening. This is where personal discernment is crucial, greatly facilitated by a constant awareness of the possibility of being misled or going down wrong alleys and of self-deception, and a constant testing by looking at the fruits of any view or standpoint: where does it lead?

This cannot be pursued further here, important as it is, except to make one comment: the true process of developing and increasing this awareness is a long-term process of developing specific thought patterns and attitudes, which ultimately amounts to character change,[51] and indeed true religious experience has this nature. Attempts to hurry the process by taking short cuts (using drugs, for example) will almost always fail.

5 Mind and Brain

The issue to be considered in this final section is how all of the above relates to the theme of this volume: that is, what, if any, are the implications of mind/brain theory for this viewpoint, and *vice versa*?

In an experiential sense, the mind inhabits the body, watching it, receiving sensations from it, and exerting partial control over it.[52] In some sense we experience the mind as existing in and through the body, working according to its own logic, and trying to control the body, but to some degree experiencing the body as an external reality (albeit very closely connected), with its own laws of behavior and almost its own volition. Thus the process of learning to control the body and find out how it functions is one of our major developmental tasks in life, made very conscious, for example, in the training of athletes. The same arises in relation to some of the functions of the brain, as is evident in the whole process of learning. For example, one of the tasks of a pilot flying an aircraft on instruments is to exert a conscious counter-intuitive control of the rational mind over instinctive reactions. The practical task is to try to be aware of and balance internal and external views of what is happening (cf. section 2.4), and thereby arrive at a decision as to how to act (including how to try to train and control the mind and brain). The intellectual task however is to show why any kind of language about top-down causality or mental emergence makes sense in the light of cognitive science and neuroscience—given that our experience indicates that this must be possible.

5.1 Functions of Consciousness

As an overview of the high-level mental processes discussed in the essays in this volume, and the role of schemas in them,[53] I have developed the following diagram (figure 2) showing these various functions as a kind of flow diagram.[54] This is a high-level (phenomenological) model of functions of consciousness. In brief:

1. Data Input comes from the senses and is evaluated by Perception on the basis of an Awareness of the present Situation, updating current schemas in the Present Frame, and (utilizing information from Long Term Memory) activating new parametrized and named Frames or Schemas where necessary (for example, "that's a big black cat named Felix"). In addition to associating models of behavior with objects (animate and inanimate), enabling reasonably reliable extrapolation to future situations from present ones, a crucial feature of schemas is the ability to link concepts and modifiers together and to give the whole a new reference label, enabling construction of hierarchically ordered structures exhibiting properties such

---

[51] See *MNU* for a discussion of how ethics is related to character change.

[52] John C. Eccles, *How the Self Controls Its Brain* (Berlin: Springer, 1994).

[53] See Michael Arbib's essays in this volume; his "schemas" seem to me much the same as the frames that cognitive scientists have been discussing for some time; e.g., see A.M. Aitkenhead and J.M. Slack, *Issues in Cognitive Modelling* (New Jersey: Erlbaum, 1985).

[54] As in the work of Jay Forrester *et al.*: discussed, for example, in *Managerial Applications of System Dynamics*, Edward B. Roberts, ed. (Cambridge: MIT Press, 1981).

Functions of Consciousness

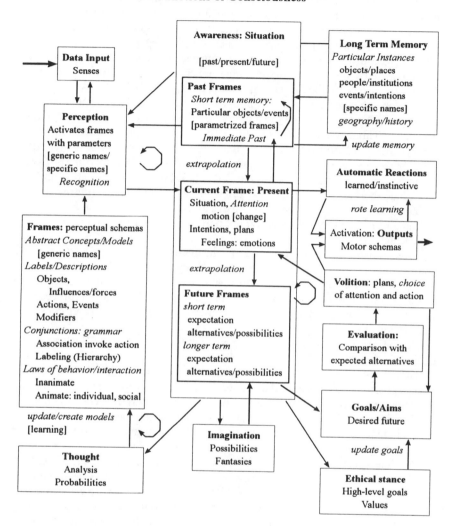

Figure 2. A phenomenological view of the functioning of consciousness, as enabled by the neuronal structure and connections in the brain. Note that this is what Chalmers characterizes as the psychological concept of mind[55]; no attempt is made here to relate this to the hard problem of the phenomenal concept of mind (the way any of this is consciously experienced as a mental state). The main cycles identified are the perception cycle (top left), the analysis cycle (bottom left), and the intentions/volition cycle (middle right).

---

[55] David Chalmers, *The Conscious Mind* (Oxford: Oxford University Press, 1996).

as abstraction and inheritance that characterize object-oriented languages.[56]

2. Awareness of the current Situation comprehends the Present Frame, immediate Past Frames, and anticipated Future Frames, all continually being updated. The Past Frame together with interpreted Input generates the new Present Frame, updating current parametrized schemas, activating new ones, and discarding old ones, as appropriate; the old Present Frame becomes part of the immediate Past; the anticipated and possible Future Frames are continually updated on the basis of the new assessment of the Present. As this happens, the available stock of (abstract) Frames or schemas may also be updated as necessary to accommodate new experience (in particular re-assessing probabilities of various schemas), through a process of Thought and analysis that proceeds partially consciously and partially unconsciously. There may also be an updating of the current stock of Automatic Reactions as needed (this can be partially consciously controlled). As time proceeds, the immediate past in the Past Frame is displaced and becomes part of the less immediate short term memory, this in turn being sorted and cleared (during sleep?), with the relevant memories either being discarded or incorporated in filtered form in Long Term Memory (as some form of parametrized, named, and dated schema representation).

3. Automatic Reactions (learned and instinctive) send messages to activate output schemas; some kind of competition and integration takes place with activation messages sent from the Intentional will, resulting in the activation of motor schemas that is the active output of the brain. Conscious choice can dominate this process, but this is not guaranteed (instinctive reactions can sometimes overpower conscious action choices).

4. The conscious activation messages result from a process of Evaluation and choice, leading to the chosen current Intentions. The Evaluation is based on a process of assessment of the expected Future Frames (if all proceeds unchanged) as against possible future frames (that may result if some conscious choice is made and implemented). This assessment in turn is based on the current stock of Aims and Goals, which are updated as time proceeds, and are constrained and shaped overall by the individual's Ethical stance (embodying at least in implicit form the *telos* or higher-level meaning understood by the individual).

5. The Expected Future Frames are based on extrapolation from the present (on the basis of the stored laws of behavior and the parameters associated with the instantiated present frames); while the Alternative Possibilities are generated by a process of Imagination that can be more or less constrained by the requirement of corresponding to those behavioral laws. One of the tasks of Evaluation is to assess the probabilities of these various possible outcomes; another is to consider them against the hierarchy of goals, and the values implicit or explicit in the Ethical Stance. These values are revised in a long-term process that is partially conscious.

5.2 Consistency with Neuroscience

The above is a phenomenological (high-level) description; the functional issues indicated arise in a slightly different form in relation to each sensory capacity and each broad area of functioning (language, locomotor skills, social skills, etc.), as well as in relation to the *hard problem* of how this can all be bound together to give the

---

[56] G. Booch, *Object Oriented Analysis and Design* (Menlo Park: Addison-Wesley, 1994).

experience of consciousness.[57] This is all made possible through the physics and biology of the functioning of the brain. The *neuroscience issue* is how all this may relate to the physical structure of the brain, remembering that the functioning and plasticity of the mind are to some degree mirrored in the physical structure and plasticity of the brain (the pattern of neuronal connections evolving to facilitate the pattern of thought that becomes habitual through multiply-repeated effectuation). Numerous questions arise, specifically:

1. Brain *functioning* in each individual must allow for all these capacities. The brain's microstructure must at least permit them, but it may do far more; it may reflect them or even facilitate them:

   Q1: To what extent is each functional capacity made possible by specific neural mechanisms and/or connections, and how is this done so as to allow mental emergence and top-down causation?

   Q2: In particular, how is the hierarchical structuring of sense, thought, and action implemented?

   Q3: Which (if any) of these functions are non-algorithmic, and how are they achieved? For that matter, how are the algorithmic functions implemented?

2. Brain *development* in each individual must permit/facilitate all these functions. They are partially derived from heredity and partially acquired developmentally in a social context:

   Q4: To what extent is each of these capacities available through initial in-born hard-wiring (determined by developmental processes controlled by the information stored in DNA)?

3. The *evolutionary history* of the body and brain in the species must explain the historical development of these capacities, passed from generation to generation through the genetic information primarily stored in DNA. This is a form of long-term top-down action by the environment on the microstructure of the molecular biology of the cell, thereby in turn influencing the microstructure of the brain through developmental processes:

   Q5: How did this occur, in regard to each functional capacity? In particular, to what degree did social structure affect that development, as well as environmental factors?

Other contributions to this volume consider several of these questions in more detail. I note here simply that this functioning is based in the physical hierarchical structuring of brain and the associated levels of emergent order[58] allowing hierarchical levels of control of the body by the brain.[59] The challenge in each case is to find an answer to these questions that is compatible with the multiple levels of structure and explanation required in a full theory of consciousness and the mind.

5.3 The Range of Data

As regards the main questions addressed in this volume, one of the key issues is the data to be considered and the range of concerns they incorporate. We need explicitly to consider this issue: namely, *what is the range of data that will be taken into*

---

[57] Chalmers, *The Conscious Mind*; see also John R. Searle, *The Mystery of Consciousness* (London: Granta Books, 1997) for an enlightening presentation of competing views.

[58] See Alwyn Scott, *Stairway to the Mind* (New York: Copernicus, 1995).

[59] See Stafford Beer, *Brain of the Firm* (New York: Wiley, 1981).

account and regarded as in need of explanation? As remarked in section 2.5, this will largely determine the nature of the theory that results; viewed the other way round, if we wish to investigate a theory of a broad scope (for example, that proposed in section 3), then we must look at data of a broad enough scope to be relevant in examining the validity of that theory.[60] If we consider and try to explain only restricted data, we will inevitably end up with a "thin" theory. In the present context, there are two specific issues: (1) consciousness and free will, and (2) the intimations of transcendence considered in section 4. I will consider these in turn.

Two quite different theoretical options can be chosen in relation to consciousness and free will, (which I take to be closely associated). One of them is based on theory, with a specific example being the following:

A colleague of mine believes that because of the operation in the mind of physically based bottom-up causation, free will is an illusion. However, in discussions about how to run an astrophysics institute, he comments that someone is responsible and someone else irresponsible. Such concepts are seriously truncated if one accepts his proclaimed stance on free will. I suggest that the fact that he uses these kinds of words in a meaningful way shows that *in practice* he believes in the reality and meaningfulness of free will, no matter what his proclaimed theoretical stance may be; the evidence of experience overrides his theoretical predilections when it comes to practical matters.[61] Similar issues arise in relation to consciousness: some theoretical approaches in effect discount it because they cannot satisfactorily accommodate it; but those theorists taking such a stance must necessarily treat consciousness as real when living their own lives.

The second option is to take this kind of experiential data seriously, accepting it at face value. Indeed one can regard the existence of free will and consciousness as among the best-tested bits of evidence that will ever be available to us. Hence one can make it a criterion of acceptability of any theory of the brain that it must be able to satisfactorily account for these phenomena; if it does not do so, it will be discarded. Thus this view, which is the one I adopt, takes the existence of free will and consciousness as essential experiential data, rather than as theoretical concepts that can be discarded or explained away if one's theory does not accommodate them. This then determines that theories are regarded as acceptable only if they can accommodate these data.

The philosophical reason for doing this is that we otherwise ignore fundamentally important data. Cognitive science and neuroscience then must be consistent with these data by allowing rather than denying them. This approach is consistent with and indeed implicit in the phenomenological picture of the functioning of consciousness sketched above; I will simply assume that it is possible to devise an understanding of the physical operation of the brain through neuronal signaling and connections that is consistent with this demand (if this is not true, then our understanding of physics and/or cognition and neuroscience is inadequate[62]).

---

[60] This does not mean that we assume the validity of these data to start with, but rather that we cannot sensibly investigate it by the use of too restricted a range of data.

[61] It can be argued that he does not necessarily affirm freedom and responsibility in the same strong sense as I do here. More formal, persuasive arguments against the compatibility of determinism with free will and responsibility are given in Peter van Inwagen, *An Essay on Free Will* (Oxford: Oxford University Press, 1983).

[62] If pushed I will take the unpopular view that the operation of consciousness and free will may well be tied to the not yet understood foundational issue of measurement processes in

The same essential issue then arises in relation to the various intimations of transcendence outlined in section 4. One option is to consider all of these experiences to be misleading and based purely on psychological projection; they reflect *only* inner needs and desires, and do not truly reflect any features of an external reality. This is a consistent position, and indeed it is not impossible that it is correct.[63] The alternative is to take at least some of these experiences at their face value: to assume that they are what they seem to be, and to make it a requirement of an adopted theory that it can accommodate them understood this way. This leads to a different class of theories than those attained under the first assumption; these are "strong" theories from the religious viewpoint, as they take religious experience seriously. Each class of theories can then be examined to see how coherent it is in terms of its inner logic, and how adequate it is in explaining the full range of relevant data, which must of course include these kinds of experiences. (The one theory will see them as delusion and the other as real; but the point is that each of them must give some account of these data.) The case for the first class has been made by many (for example, see Arbib's "Crusoe's Brain" in this volume); here, I will pursue the second option (developed in *MNU*).

A wide range of data is considered in section 4. How one regards them will depend on one's life experience. However, one can make a good case that, whatever people may say about moral relativism, in fact (as in the case of free will) they believe there is a real meaning to morality, for this is evidenced in the way they lead their lives.[64] Insofar as this is so, it can then reasonably be interpreted as reflecting a real feature of the universe—it is truly an intimation of transcendence; and this then arguably gives some evidence of a Creator with a moral nature (for this is the most plausible origin of the existence of true morality[65]).

Thus one can propose that at least one of the intimations of transcendence mentioned in section 4—when one looks into it in depth—carries substantial weight at the present time.[66] This then makes it reasonable to suggest that the others may have weight too—or at least it is sensible to include in one's investigations a theory which makes that supposition, and then to see how satisfactory that proposal is in other ways. The theory outlined in sections 3 and 4 (based on the argument in *MNU*) is one such theory which is coherent in its own right, and takes seriously all these data. It is therefore also a candidate for serious consideration.

5.4 The Issue of Revelation

Included among the intimations of transcendence are specifically religious experiences and special revelations, as discussed, for example, by Wildman and

---

quantum theory; cf. Roger Penrose, *The Emperor's New Mind* (Oxford: Oxford University Press, 1989); E. Squires, *Conscious Mind in the Physical World* (Bristol: Adam Hilger, 1990); Eccles, *How the Self Controls Its Brain*. However, even if this is not needed or viable, as claimed by some of its critics, the main point made here remains.

[63] Although to put them all down to temporal-lobe disorders is not plausible; if a large fraction of the population suffer from a "disorder" then it becomes the norm!

[64] See Ellis, "God and the Universe: *Kenosis* as the Foundation of Being," *CTNS Bulletin* 14.2 (1994); *BTB*; *MNU*.

[65] *BTB*; *MNU*.

[66] Which of them can have efficacy in any era will depend on the intellectual situation in that era.

Brothers in this volume (and by Denis Edwards in *CAC*), and this is central to the proposed viewpoint. The interpretation of the experiences discussed in section 4 as true intimations of transcendence can only come, in part, from the experiences themselves; it comes partly from interpretational keys—specific *revelatory events*— that help guide us to their true interpretation by placing them in their true context, not in a coercive way, but by offering a vision one can either accept or not, according to one's free choice. Thus they are fully consistent with the kenotic view offered earlier. Without such revelatory events, we simply have natural theology, which by itself is unconvincing. In the end, this is why the life of Christ was necessary: this intervention was needed precisely in order to provide such an interpretative key and hence transformation of understanding.[67] However there are also revelatory events in the life of ordinary believers as well as in the lives of the saints; the nature of these events—in some sense specific intimations of transcendence—has been discussed in many places.[68]

The occurrence of these events (as discussed in depth in my essay in *CAC*), requires some means of communication or inspiration to be available[69] in a way that is consistent with the laws of physics, but in a way that does not mean we are pitting God against natural explanation; rather, we are seeing God acting within natural processes and respecting their nature. Neither evolutionary mechanisms nor history by themselves supply that channel (if they could, there would be no need for specific revelation). In terms of the model of mental functioning presented in section 5.1, this requirement would represent a causal joint of some kind that allows the feeding of extra information, or at least intimations of understanding, into the mind, this being an effective mode of action of God in the world. Without this, any conclusions that are derived are completely internally generated and so are indeed simply internal projections, without any necessary external referent. The concept of multiple causality described above allows for these experiences to be explained, at the appropriate causal level, as psychological projections; but there must additionally be an element of verificatory or revelatory data to link the mind to an external transcendent reality if the whole view put forward here is to be persuasive.

It is true that we receive data of a revelatory nature through traditional religious and spiritual experiences of a social nature: church or temple, priests or gurus, the Bible or the Koran, and these can serve to provide the interpretative keys needed to complete the scheme. But this just displaces the problem, for the issue then is how those who provided this interpretation (for example, those who wrote the Bible) themselves came to this position. We have access to them through various texts, rituals, etc.; but how did they attain their revelatory views? At least some of them

---

[67] I accept here that other religions provide avenues towards understanding the nature of transcendence in a powerful and meaningful way, and provide helpful interpretations; nevertheless I see a kenotic Christian vision as giving the most complete embodiment of the kenotic view put forward here.

[68] See for example, Wakefield, ed., *Dictionary of Christian Spirituality*; Denis Edwards, *Human Experience of God* (New York: Paulist Press, 1983); Hubbard, *Quaker by Convincement*.

[69] The data can of course be conceptualized in this way without the suggestion of a causal joint, as pointed out by Theo Meyering; but then the question is how to justify that conceptualization. At some point a truly revelatory event is needed, or else the conceptualization is based on human imagination and nothing else; there is no reason to believe it reflects in a reasonably adequate way the nature of reality.

need some revelatory input that relates more or less directly to ontological reality or else the problem remains. Somewhere in the course of evolutionary development of the brain, such input is required for the whole to be more than a purely human construct; and this requires the possible existence of a revelatory channel or causal joint such as is proposed here. In any case I would wish to require this because of the vast data of spiritual experience of many people that has been tested by processes of discernment in religious communities that have an overall kenotic approach to life. Some of this has a revelatory nature which on the proposed interpretation deserves to be taken seriously at its face value.[70]

This revelatory causal joint does not control what goes on in our brains in the sense of constraining our options, but rather offers us visions or intimations of how things are and could be, and of how to act, experienced generally as pre-conceptual intimations[71] or as an "inner light,"[72] which can then be freely accepted or rejected by the person involved. Thus this action by God is of a kenotic rather than coercive nature; indeed, it is consistent with Murphy's view of noncoercive divine action at the quantum level and all the way up (see *MNU*)—the action takes place in a way consistent with the microscopic physical nature of matter and also with the free choices available to humanity. As is evident from the discussion above, and is stated elsewhere in this volume, much religious experience is not mystical; but then this experience cannot be correctly interpreted without some revelatory action assisting humanity in seeing what that correct interpretation is. Hence the need for the causal joint allowing specific revelatory experiences (in classic Christian terms, allowing action of the Holy Spirit as a divine inspirer without violating the laws of physics). Indeed if this is not possible, we are in danger of invoking a God who is useless in the sense implied by Arbib. I suggest this is not the case.

Making this assumption, the existence of such a causal joint imposes a *compatibility condition* on the viewpoint presented here in order that it be consistent (cf. section 2.2), namely that *the functioning of the mind and brain allow such a revelatory channel*. As in the case of free will and consciousness, in this approach we take seriously the validity of the intimations of transcendence discussed in the previous section, and require of a satisfactory theory of the mind and brain that it be consistent with this body of data. The ways this might happen were discussed extensively in my essay in *CAC*, and I will only give a brief summary of my view here.

Quantum uncertainty provides one way this channel could function (see the arguments by Tracy, Murphy, and Ellis in *CAC*, where the issue is considered in depth). I recognize that this account is highly speculative and that most physicists reject the idea. I speculate that this rejection is probably because they take quantum uncertainty to be an ontological reality rather than an epistemological one; but that of course is highly disputable. I see the answer to lie in something like a hidden variable theory, with all that entails, such that the uncertainty is not real in an ultimate sense. It is "real" for physics simply because physical manipulation is unable to control that hidden variable, which is, however, accessible to God. Hence

---

[70] As explained above, one can take this position when forming a strong theory to be compared with other theories, such as those that discount all this experience as illusory and manufactured without any external transcendent referent.

[71] Denis Edwards, *Human Experience of God*.

[72] Hubbard, *Quaker by Convincement*.

when God acts the quantum event is in fact determined, even though quantum processes are indeterministic from a purely natural point of view. This action is effected in such a way as to give reliable macroscopic physical behavior, as tested by the experiments of physics—and as required by the faithful nature of God in his creation of a reliable physical universe.[73]

The further issues, then, would be where specifically in the brain that uncertainty obtained, and how it is amplified to macroscopic level. The answer to the first question is *everywhere*—all matter is subject to quantum theory! (and it is logically sound to conceive of God as acting everywhere all the time, as emphasized for example by Murphy in her writings). Hence it is perfectly consistent for large arrays of neurons to be influenced in a coherent way in the timing of their firings in a way such as to allow an holistic response from brain regions where this occurs. In this case the distinction between micro- and macro-level causation made in the previous paragraph falls away—consistent micro-level causation in all atoms in the neurons in some region of the brain is equivalent to macro-level causation there, and these are simply different descriptions of the same phenomena. No amplification mechanism is necessary. However, it is also possible—and perhaps more economical—to assume that such a causal input is more localized, for example taking place specifically in the microtubules in neurons (where quantum effects could play a critical role in terms of cognition—as proposed by Roger Penrose[74]), and that amplification of individual quantum events takes place until they have a macroscopic effect. That this could be possible is shown both by Penrose's investigations of the possible role of quantum coherence within microtubules,[75] and by the fact that the eye is able to detect individual photons. It is here that it is possible that neural systems with sensitive dependence on initial conditions (and hence perhaps evidencing "chaotic" behavior) could be significant—and this is something that can be searched for by neuroscientists.

This is a "tough" proposal in the same sense that the neo-Darwinian paradigm is tough—it explains a vast range of data through a paradigm that is compatible with all the data and locates itself at the heart of current physical explanatory theory in a way that re-interprets a central feature in a new way. It is precisely because it is tough in this sense that it is difficult for many to accept it. It responds to Arbib's concern ("Can the God of Kenosis be the God of Initial Conditions who kowtows to the laws of physics?") in that here God is active in all consciousnesses—and could even be active in all evolution—without ever violating the laws of physics, *precisely because these laws have been designed to allow such continuous specific action*; and this feature is what physics perceives as quantum uncertainty. Because they all ultimately function in a way that is based on quantum physics (if their function is indeed based in physics), all neurons allow the possibility of such action, while still functioning in the way determined by neuroscience. The neuroscientists are not out of business, but they cannot put the theologians out of business.

Clearly this proposal is controversial. However it is achieved, there is a need for a channel of revelation in order to allow for agreement with a large body of specifically spiritual experience that may reasonably regarded as valid. Much of that experience may be regarded as interpretative rather than embodying a specifically

---

[73] See my essay in *QCLN* and *MNU* for further discussion of the need for this reliability.

[74] Roger Penrose, *Shadows of the Mind* (Oxford: Oxford Univ. Press, 1994), chap. 7.

[75] Ibid., sec. 7.6.

revelatory message—but at least some of such events some of the time must have that character if the whole is to have a causal link to an external referent rather than being simply internally-based invention. The mechanisms briefly discussed indicate that this idea is not necessarily in conflict with present-day physics; indeed, the idea appears to be compatible with physics in terms of causal mechanisms operating in complete consistency with known physical laws. A strong religious theory along the lines outlined in this essay will therefore take as one of its fundamental presuppositions the assumption that such a channel exists and underlies the interpretative validity and hence the efficacy of the tested intimations of transcendence that have been discussed in section 4 (which we take at their face value). Here "tested" means that they have been through a process of discernment that has subjected them to critical review (and hence those regarded as invalid after such a process of testing have been rejected). The issue of discernment remains central, and one can ask if neuroscience can assist here. Can one relate any neuroscience correlates to more traditional tests of discernment? Specifically, what are the discernment processes for distinguishing between temporal lobe disorders and religious experiences, if there are any? (cf. Wildman and Brothers in this volume).

5.5 The Challenge: The Issue of Consistency

The challenge is to find an overall explanation for each of the phenomena to be accounted for, with multiple levels of causation and with consistency between the descriptions at each level. How is it possible,

1. for the mind to function through physics/biochemistry and the design of the brain?
2. to develop this capacity in each individual through a developmental process?
3. to develop this capacity in the species through a process of evolutionary development?
4. to understand this as part of an overall process of revelation of the true nature of the universe?
5. to see these different levels of explanation as all functioning simultaneously and, in a harmonious, way reinforcing rather than contradicting each other?

The development of conscious mind in a universe where at an early (hot, dense) stage there was no such phenomenon evident is an extraordinary feature: it represents the emergence of truly new, physically-based phenomena in the historical evolution of the world and the universe. A corresponding remarkable process occurs in the development of each individual: the initial few cells that grow into the embryo do not support conscious processes (the needed high-level structure simply is not there), but consciousness has developed by the time the infant can speak. Although there are many gaps and difficulties at present, a scientifically-based explanation is emerging in both cases. This essay (with others in this volume) claims that such an explanation does not exclude a simultaneous further level of explanation where this all happens as part of a process of revelation of an underlying transcendent principle of the universe ("The Word," John 1:1),[76] which is thereby seen to have a kenotic

---

[76] William Temple, *Readings in St John's Gospel*; my essay in *QCLN*.

nature.[77] It also supports the view that to make this convincing we should take a strong realist position on religion that includes an active process of revelation, both historically and today, which allows us to have intimations of transcendence both through specifically revelatory experiences and through a broader awareness of the nature of the everyday world around us. There is no contradiction between such a view and our present-day understanding of science and cosmology in general, and of neuroscience in particular.

There is no claim made here that this view is original; indeed the reverse is true: what is presented is not only similar to much suggested by others in this volume, it can also be regarded as a presentation of a very old traditional view (based in that strand in many religions that has a kenotic *telos* and ethic). What may be useful is a presentation of this position in a present-day context that takes seriously modern science (including neuroscience), as well as spiritual experience. That is what has been attempted here.

Acknowledgment. I thank all the participants of the meeting for their interesting comments. I have tried to incorporate their ideas, or at least give adequate responses to their criticisms, in this revised version of the original essay.

---

[77] *MNU*; Ellis, "God and the Universe."

CONTRIBUTORS AND CONFERENCE PARTICIPANTS[1]

Michael A. Arbib, Director of the USC Brain Project, Professor and Chairman of Computer Science, and Professor of Neuroscience, Biomedical Engineering, Electrical Engineering, and Psychology, University of Southern California, Los Angeles, California, USA.

Ian G. Barbour is retired from teaching at Carleton College, Minnesota, USA, where he has been Professor of Physics, Professor of Religion, and Bean Professor of Science, Technology and Society.

Leslie A. Brothers, Associate Research Professor of Psychiatry and Biobehavioral Sciences, UCLA Neuropsychiatric Institute, University of California, Los Angeles, California, USA.

Philip Clayton, Associate Professor and Chair of the Philosophy Department, California State University, Sonoma, California, USA.

Anne M. Clifford, Associate Professor, Department of Theology, Duquesne University, Pittsburgh, Pennsylvania, USA.

George V. Coyne, S.J., Director, Vatican Observatory, Vatican City State, Italy.

George F.R. Ellis, Professor of Applied Mathematics, University of Cape Town, Rondebosch, South Africa.

Joel B. Green, Dean of the School of Theology and Professor of New Testament Interpretation, Asbury Theological Seminary, Wilmore, Kentucky, USA.

Peter Hagoort, Max Planck Institute for Psycholinguistics and the University of Nijmegen, The Netherlands.

Stephen Happel, Associate Professor, The Catholic University of America, Washington, District of Columbia, USA.

Michael Heller, Professor of Philosophy, Pontifical Academy of Theology, Cracow, Poland.

Marc Jeannerod, Institut des Sciences Cognitives, Centre National de La Recherche Scientifique, Bron, France.

Fergus Kerr, Regent, Blackfriars, University of Oxford, and Honorary Senior Lecturer, Divinity, University of Edinburgh, Scotland.

Joseph E. LeDoux, Professor of Neural Science and Psychology, The Center for Neural Science, New York University, New York, New York, USA.

---

[1] Joel Green and Fergus Kerr, although not participants in the conference, were invited to contribute to this volume.

John C. Marshall, External Scientific Staff, Medical Research Council (UK), Neuropsychology Unit, University Department of Clinical Neurology, Radcliffe Infirmary, Oxford, England.

Theo C. Meyering, Associate Professor, Department of Philosophy, Leiden University, The Netherlands.

Mortimer Mishkin, Chief, Section on Cognitive Neuroscience, Laboratory of Neuropsychology, National Institute of Mental Health, Bethesda, Maryland, USA.

Nancey Murphy, Professor of Christian Philosophy, Fuller Theological Seminary, Pasadena, California, USA.

Arthur Peacocke, Director, Ian Ramsey Centre, Oxford, England, Warden Emeritus of the Society of Ordained Scientists, and formerly Dean of Clare College, Cambridge, England.

Ted Peters, Professor of Systematic Theology, Pacific Lutheran Theological Seminary and the Graduate Theological Union, and Program Director of the CTNS Science and Religion Course Program, Berkeley, California, USA.

John Polkinghorne, Past President and now Fellow, Queen's College, Cambridge, and Canon Theologian of Liverpool, England.

W. Mark Richardson, Associate Professor of Theology, General Theological Seminary, New York, New York, USA.

Robert John Russell, Professor of Theology and Science in Residence, Graduate Theological Union, and Founder, and Director of the Center for Theology and the Natural Sciences, Berkeley, California, USA.

William R. Stoeger, S.J., Staff Astrophysicist and Adjunct Associate Professor of Astronomy, Vatican Observatory, Vatican Observatory Research Group, Steward Observatory, University of Arizona, Tucson, USA.

Thomas F. Tracy, Professor of Religion and Chair, Department of Philosophy and Religion, Bates College, Lewiston, Maine, USA.

Fraser Watts, Starbridge Lecturer in Theology and Natural Science, University of Cambridge, England.

Wesley J. Wildman, Associate Professor of Theology and Ethics, Boston University, Boston, Massachusetts, USA.

Name Index